T0138117

Proceedings in Adaptation, Learning and Optimization

Volume 8

About this Series

The role of adaptation, learning and optimization are becoming increasingly essential and intertwined. The capability of a system to adapt either through modification of its physiological structure or via some revalidation process of internal mechanisms that directly dictate the response or behavior is crucial in many real world applications. Optimization lies at the heart of most machine learning approaches while learning and optimization are two primary means to effect adaptation in various forms. They usually involve computational processes incorporated within the system that trigger parametric updating and knowledge or model enhancement, giving rise to progressive improvement. This book series serves as a channel to consolidate work related to topics linked to adaptation, learning and optimization in systems and structures. Topics covered under this series include:

- complex adaptive systems including evolutionary computation, memetic computing, swarm intelligence, neural networks, fuzzy systems, tabu search, simulated annealing, etc.
- machine learning, data mining & mathematical programming
- hybridization of techniques that span across artificial intelligence and computational intelligence for synergistic alliance of strategies for problem-solving
- aspects of adaptation in robotics
- agent-based computing
- autonomic/pervasive computing
- dynamic optimization/learning in noisy and uncertain environment
- systemic alliance of stochastic and conventional search techniques
- all aspects of adaptations in man-machine systems.

This book series bridges the dichotomy of modern and conventional mathematical and heuristic/meta-heuristics approaches to bring about effective adaptation, learning and optimization. It propels the maxim that the old and the new can come together and be combined synergistically to scale new heights in problem-solving. To reach such a level, numerous research issues will emerge and researchers will find the book series a convenient medium to track the progresses made.

More information about this series at http://www.springer.com/series/13543

George Leu · Hemant Kumar Singh
Saber Elsayed
Editors

Intelligent and Evolutionary Systems

The 20th Asia Pacific Symposium, IES 2016, Canberra, Australia, November 2016, Proceedings

Editors
George Leu
School of Engineering and Information Technology, Australian Defence Force Academy
The University of New South Wales
Canberra, ACT
Australia

Saber Elsayed
School of Engineering and Information Technology, Australian Defence Force Academy
The University of New South Wales
Canberra, ACT
Australia

Hemant Kumar Singh
School of Engineering and Information Technology, Australian Defence Force Academy
The University of New South Wales
Canberra, ACT
Australia

ISSN 2363-6084 ISSN 2363-6092 (electronic)
Proceedings in Adaptation, Learning and Optimization
ISBN 978-3-319-84076-5 ISBN 978-3-319-49049-6 (eBook)
DOI 10.1007/978-3-319-49049-6

Printed on acid-free paper

This Springer imprint is published by Springer Nature
The registered company is Springer International Publishing AG
The registered company address is: Gewerbestrasse 11, 6330 Cham, Switzerland

Preface

This PALO volume constitutes the proceedings of the 20th Asia Pacific Symposium on Intelligent and Evolutionary Systems (IES 2016). The symposium first took place in 1997 in Canberra, and since then, the series has become a prestigious incubator of research ideas, as well as facilitator of research collaborations. The symposium aims to bring together researchers and practitioners from countries of the Asia-Pacific region in the fields of intelligent systems and evolutionary computation to present ongoing work, exchange ideas, and discuss future collaboration. In 2016, IES was held again in Canberra to celebrate its 20th anniversary.

IES 2016 was hosted by The University of New South Wales (UNSW), at its Canberra campus in the Australian Defence Force Academy (ADFA). It was collocated with two other events: the 24th National Conference of the Australian Society for Operations Research and the Defence Operations Research Symposium. The event included a number of plenary talks, special sessions, oral presentations, and industry workshops for a valuable interaction with researchers and practitioners in the field.

Out of the 51 submissions initially received, 36 were selected to be included in the final proceedings. Each submission was reviewed by 2–4 members from our international program committee.

We would like to thank the steering, organizing, and program committees for their efforts in supporting the symposium. The support and assistance from UNSW, Springer, and EasyChair are gratefully acknowledged.

Canberra, Australia
November 2016

George Leu
Hemant Kumar Singh
Saber Elsayed

Organizing Committee

Conference Chair

George Leu, The University of New South Wales, Australia

Proceedings Chairs

Saber Elsayed, The University of New South Wales, Australia
Hemant Kumar Singh, The University of New South Wales, Australia

Special Session Chairs

Saori Iwanaga, Japan Coast Guard Academy, Japan
Bing Xue, Victoria University of Wellington, New Zealand

Registration Chair

Kathryn Kasmarik, The University of New South Wales, Australia

Local Arrangement Chair

Erandi Lakshika, The University of New South Wales, Australia

Publicity Chair

Jiangjun Tang, The University of New South Wales, Australia

Sponsorship Chair

Naeem Janjua, The University of New South Wales, Australia

Program Committee

Hussein Abbass, The University of New South Wales, Australia
Sreenatha Anavatti, The University of New South Wales, Australia
Ahmed Arefin, CSIRO, Australia
Md Asafuddoula, The University of New South Wales, Australia
Mohamed Bader, University of Portsmouth, UK
Yukun Bao, Huazhong University of Science and Technology, China
Regina Berretta, The University of Newcastle, Australia
Kalyan Bhattacharjee, The University of New South Wales, Australia
Tom Cai, The University of Sydney, Australia
Ripon Kumar Chakrabortty, The University of New South Wales, Australia
Jonathan Chan, King Mongkut's University of Technology Thonburi, Thailand
Shelvin Chand, The University of New South Wales, Australia
Gang Chen, Victoria University of Wellington, New Zealand
Stephen Chen, York University, Canada
Winyu Chinthammit, University of Tasmania, Australia
Sung-Bae Cho, Yonsei University, Korea
Carlos Coello Coello, CINVESTAV-IPN, Mexico
Swagatam Das, Indian Statistical Institute, India
Suranjith De Silva, The University of New South Wales, Australia
Essam Debie, Zagazig University, Egypt
Kusum Deep, Indian Institute of Technology Roorkee, India
Jeremiah Deng, University of Otago, New Zealand
Grant Dick, University of Otago, New Zealand
Kathryn Kasmarik, The University of New South Wales, Australia
Junbin Gao, The University of Sydney, Australia
Amr Ghoneim, Helwan University, Egypt
Ayman Ghoneim, Cairo University, Egypt
Garry Greenwood, Portland State University, USA
Christian Guttmann, Institute of Value Based Reimbursement System, Sweden
Ahsanul Habib, The University of New South Wales, Australia
Noha Hamza, The University of New South Wales, Australia

David Howard, CSIRO, Australia
Quang Huynh, The University of New South Wales, Australia
Muhammad Iqbal, Victoria University of Wellington, New Zealand
Hisao Ishibuchi, Osaka Prefecture University, Japan
Monjurul Islam, The University of New South Wales, Australia
Saori Iwanaga, Japan Coast Guard Academy, Japan
Yasushi Kambayashi, Nippon Institute of Technology, Japan
Hiroshi Kawakami, Kyoto University, Japan
Masao Kubo, National Defense Academy, Japan
Paul Kwan, University of New England, Australia
Erandi Lakshika, The University of New South Wales, Australia
Kittichai Lavangnananda, King Mongkut's University of Technology Thonburi, Thailand
Ickjai Lee, James Cook University, Australia
George Leu, The University of New South Wales, Australia
C.P. Lim, Deakin University, Australia
Jing Liu, Xidian University, China
Michael Mayo, University of Waikato, New Zealand
Yi Mei, Victoria University of Wellington, New Zealand
Efrén Mezura-Montes, University of Veracruz, Mexico
Saber Mohammed Elsayed, The University of New South Wales, Australia
I. Moser, Swinburne University of Technology, Australia
Nasimul Noman, The University of Newcastle, Australia
Kazuhiro Ohkura, Hiroshima University, Japan
Mahamed Omran, GUST
Akira Oyama, Japan Aerospace Exploration Agency, Japan
Somnuk Phon-Amnuaisuk, Brunei Technological University, Brunei
Kai Qin, RMIT University, Australia
Ibrahim Radwan, Seeing Machines, Australia
Inaki Rano, University of Ulster, UK
Tapabrata Ray, The University of New South Wales, Australia
Paolo Remagnino, Kingston University, UK
Karam Sallam, The University of New South Wales, Australia
Hiroshi Sato, National Defense Academy, Japan
Friedhelm Schwenker, Ulm University, Germany
Karthik Sindhya, University of Jyväskylä, Finland
Hemant Kumar Singh, The University of New South Wales, Australia
Andrea Soltoggio, Loughborough University, UK
Andy Song, RMIT University, Australia
Kang Tai, Nanyang Technological University, Singapore
Jiangjun Tang, The University of New South Wales, Australia
Ben Vermeulen, Hohenheim University, Germany
Markus Wagner, The University of Adelaide, Australia
Peter Whigham, University of Otago, New Zealand
Bing Xue, Victoria University of Wellington, New Zealand

Jianhua Yang, Western Sydney University, Australia
Tomoko Yonezawa, Kansai University, Japan
Forhad Zaman, The University of New South Wales, Australia
Mengjie Zhang, Victoria University of Wellington, New Zealand

Contents

An Evolutionary Optimization Approach for Path Planning of Arrival Aircraft for Optimal Sequencing

Md Shohel Ahmed, Sameer Alam and Michael Barlow

Abstract In this paper, we present an evolutionary optimization based path planning algorithm at Terminal Airspace (TAS) that provides a near optimal aircraft arrival sequence at Final Approach Fix (FAF). The sequence obtained minimizes the inter-arrival time as well as provides conflict free path planning to an Air Traffic Controller (ATC). A classic Genetic Algorithm (GA) based optimization technique with conflict detection and resolution is developed. Conflict between any two aircraft is detected based on their future arrival time at the waypoint and resolved by stretching the gap between those two aircraft. The proposed algorithm is compared with the traditional GA. Results indicate that the proposed approach obtains a near optimal solution compared to the traditional GA based algorithm which does not consider TAS constraints.

Keywords Terminal airspace · Way-point manoeuvring · Optimal path planning · Optimal aircraft sequence · Conflict detection and resolution

1 Introduction

Terminal Area Airspace (TAS) is the airspace surrounding a controlled aerodrome where aircraft transition from the descend phase to the approach phase. TAS is also one of the most resource-constrained elements of an air transportation network as all air traffic converges in TAS and is sequenced for landing [1]. Aircraft sequencing in TAS is a highly challenging task due to complex manoeuvering constraints

---

Md.S. Ahmed (✉) · S. Alam · M. Barlow
School of Engineering and Information Technology, University of New South Wales Australian Defence Force Academy, Canberra, Australia
e-mail: md.ahmed@student.adfa.edu.au

S. Alam
e-mail: s.alam@adfa.edu.au

M. Barlow
e-mail: m.barlow@adfa.edu.au

G. Leu et al. (eds.), *Intelligent and Evolutionary Systems*, Proceedings in Adaptation, Learning and Optimization 8, DOI 10.1007/978-3-319-49049-6_1

(i.e., restricted speed, altitude and movement). During the busiest times, operation on this safety critical environment reduces ATC efficiency especially sequencing and manoeuvering the aircraft and eventually degrades the efficiency of the system [2]. One study shows that inefficient sequencing techniques in the TAS area resulted in, on average, 18 min of delay for 19 % of European flights [3].

The increasing demand of air traffic is stressing the capacity of the current Air Traffic Management System (ATMS). This is likely to cause both safety and performance degradation in the near future. It is believed that by increasing the level of automation, the efficiency of the ATMS can be enhanced. This may assist ATC to handle the increased traffic demand in a more reliable way.

Therefore, increasing the automation of ATM components, an automatic decision support techniques, is imperative to meet future needs and might increase the overall system performance.

In recent years, many optimization based algorithms and technique were proposed for aircraft sequencing as an automatic decision support tool [4, 5]. Most of the approaches available in the literature have goals that were simplistic to obtain the best sequence and hence provide high throughput. However, in practice, obtaining those optimal sequences might be a very challenging task due to the frequently changing environment, complex network structure and cost consideration (i.e., shifting position, vectoring a long way, holding a long time etc.). A survey of the literature has failed to discover any other works that use a path planning based approach for solving the aircraft sequencing problem. However, a path-planning approach may deliver near optimal results while addressing and dealing with real-world complexities such as limited capacity to shift sequence. Figure 1 presents the algorithm based optimization sequence that is prevalent in the literature and the real world scenario that is expected to be achieved.

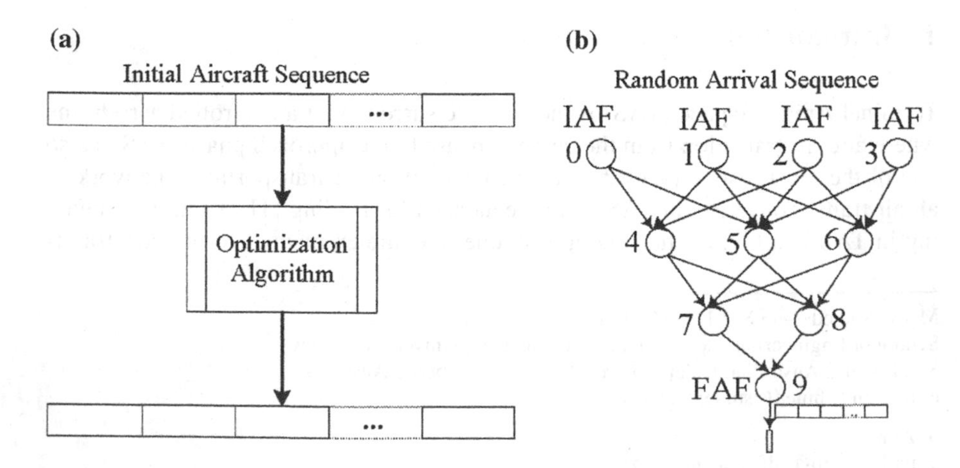

Fig. 1 **a** Optimal sequence (traditional optimization technique). **b** Expected optimal sequence

To achieve the expected arrival sequence, ATCs (the Approach controller) have to juggle the arrivals through the Standard Terminal Arrival Routes (STAR) along with the desired safe separation. A STAR is a flight route defined and published by the Air Navigation Service Provider (ANSP) that usually covers the phase of a flight that lies between the last point of the route filled in the flight plan and the first point of the approach to the airport, normally the Initial Approach Fix (IAF) [6]. Hence, a STAR connects the en route phase with the approach phase of the flight. Having an optimal sequence and implementing it in a TAS provides two different outcomes, given the complexity of the environment. First, increased probability of potential conflict at different waypoint. Second, it decreases the ATC efficiency. In this paper, we have been motivated to obtain the expected near optimal sequence at FAF by traversing through waypoint, given all the TAS constraints (i.e., speed limit, altitude limit and separation etc.). We develop an evolutionary optimization based path planning algorithm that provides an optimal aircraft sequence at the FAF. Our algorithm also predicts potential conflict at future waypoints and uses a path stretch technique to resolve the conflict. We call this algorithm Terminal-Airspace Traversing Algorithm (TATA). This algorithm finds the near optimal path for each aircraft, resolves potential conflict and maintains safe separation. The proposed algorithm also provides a detailed manoeuvring guidance to the ATC.

We compare and analyse our approach (i.e., TATA VS traditional GA) on several random arrival sequences in terms of the inter-arrival time of the sequences. The remainder of the paper is organized as follows: Sect. 2 explains the problem formulation. Section 3 outlines a methodology and proposed algorithm. Section 4 presents the experimental design. Results and analysis are discussed in Sect. 5. In Sect. 6, we present our conclusions and future work.

2 Problem Formulation

We subdivide the problem formulation into two stages:

- The optimization model for arrival sequencing using traditional GA
- The optimization model for arrival sequence using path planning (TATA)

2.1 Optimization Model for Arrival Sequencing Using Traditional GA

The aircraft-sequencing problem is to minimize the inter-arrival time between two consecutive arrivals and hence the total inter-arrival time. The optimization model for the arrival sequence is formulated as follows:
A: set of all aircraft in a sequence
S_{ij}: minimum safe separation between two aircraft i and j

L: length of the final approach path
O_i: runway occupancy time of aircraft i
V_i: approach speed of aircraft i
V_j: approach speed of aircraft j
T_{ij}: inter-arrival time between aircraft i,j

The inter-arrival time between aircraft i and j can be determined by Eqs. (1) and (2) [7]. The runway occupancy time of the aircraft are presented in Table 1b.

$$T_{ij} = Max[\frac{L + S_{ij}}{V_j} - \frac{L}{V_i}, O_i] \ when \ V_i > V_j \quad (1)$$

$$T_{ij} = Max[\frac{S_{ij}}{V_j}, O_i] \ when \ V_i \leq V_j \quad (2)$$

Objective Function: The objective is to minimize the inter-arrival time of each sequence. The objective function and the corresponding fitness function of the traditional GA based optimization problem is as follows:

$$\Psi = \sum_{i,j=1}^{k} \sum_{i,j=1}^{k} P_{ij}.T_{ij} \quad (3)$$

$$Min \sum_{i=1}^{n} \Psi \quad (4)$$

$$Fitness = Max(1/\Psi) \quad (5)$$

where k is the number of aircraft classes (i.e., 3 for Heavy, Medium and Light classification), P_{ij} is the probability of the arrival of aircraft i followed by j and n is the total aircraft in the sequence. Equations (3, 4 and 5) determines the optimal sequence based on traditional GA.

2.2 *Optimization Model for Arrival Sequencing Using Path Planning (TATA)*

Aircraft are sequenced by traversing through the terminal airspace and the final sequence is obtained at the FAF. Let the arrival time of aircraft i and j at waypoint p be ${}_pt_i$ and ${}_pt_j$ respectively. Each aircraft needs to maintain a safe separation at each waypoint.

Objective function: The objective function of the optimization problem is as follows:

$$\Psi = \sum_{i,j=1}^{k} \sum_{i,j=1}^{k} P_{ij}.T_{ij} \tag{6}$$

$$Min \sum_{i=1}^{n} \Psi \tag{7}$$

$$Fitness = Max(1/\Psi) \tag{8}$$

Subject to the constraints that,

$$|_{p}t_{i} -_{p} t_{j}| \geq S_{ij} \tag{9}$$

TATA achieve the near optimal sequence through Eqs. (6, 7, 8 and 9).

3 Methodology

The proposed methodology consists of two major phases. Figure 2 shows the conceptual diagram of the methodology.

- A method for generating the optimal arrival sequence using GA
- TATA for optimal arrival sequence

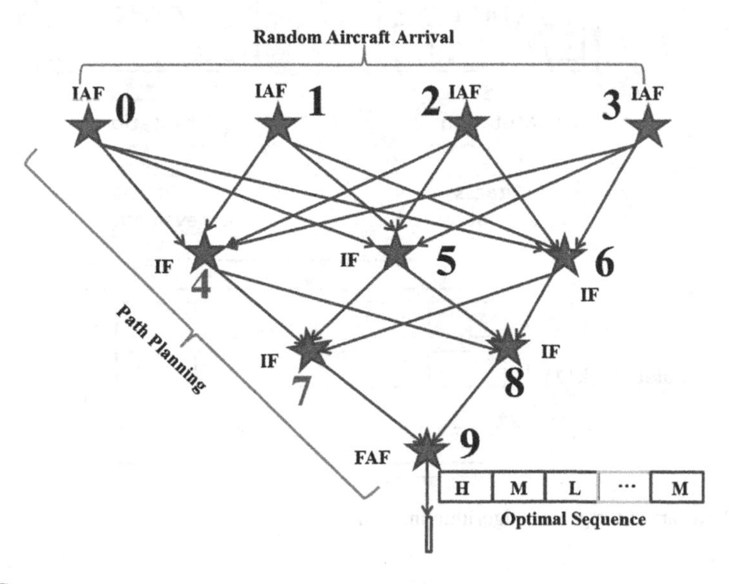

Fig. 2 Conceptual diagram of the methodology

3.1 A Method for Generating the Optimal Arrival Sequence Using GA

Figure 3 shows the flowchart of the traditional genetic algorithm implementation for optimal sequence derivation. The initial aircraft arrival sequence is generated using a mixed distribution of the aircraft as presented Table 1a. This distribution represents the aircraft mix for a typical spoke airport (in a Hub-Spoke network) [8]. The random

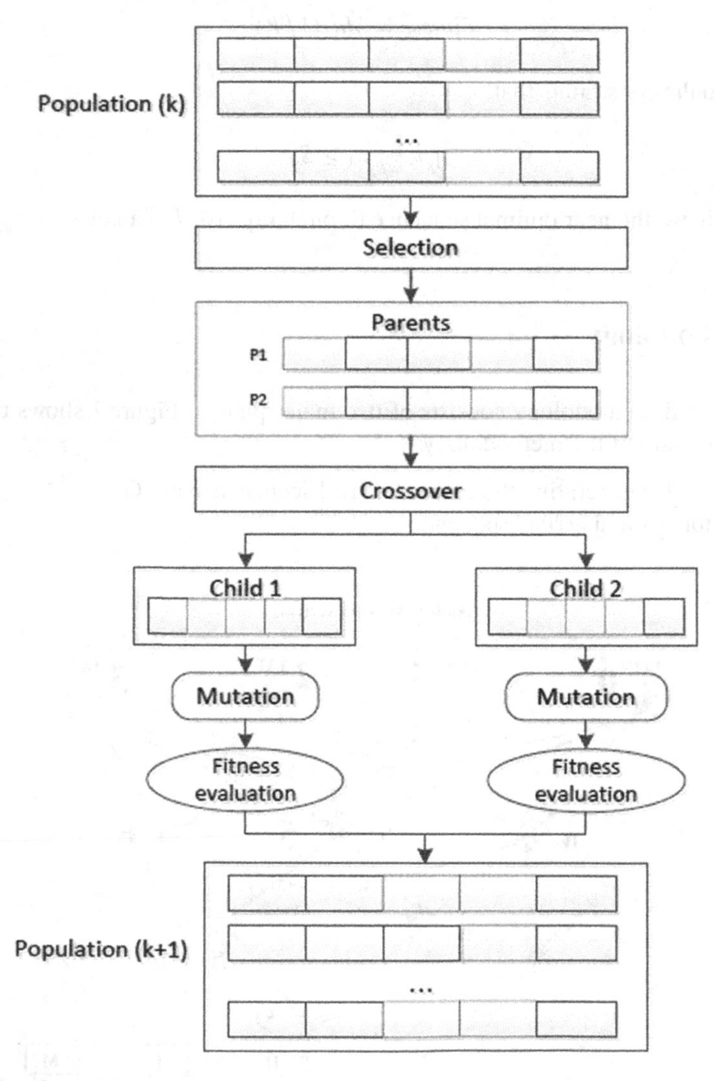

Fig. 3 Flowchart of the genetic algorithm implementation

Table 1 (a) Aircraft mix distribution (%) (b) Runway occupancy time (Sec)

	Heavy (H)	Medium (M)	Light (L)
(a) (%)	20	40	40
(b)	60	55	50

arrival sequence is modelled as a chromosome where each gene of the chromosome represents the aircraft wake category (heavy, medium or light).

GA based optimization considers inter-arrival time as the sole optimization criteria without involving the manoeuvring complexity of TAS. The evolutionary process attempts to maximize the fitness function through genetic operations. In this implementation, according to the problem characteristics and the chromosome size the initial population size is considered 50 [9].

Two parent chromosomes (out of a population of K parent chromosomes) are selected (using tournament selection) to undergo a genetic operation (i.e., crossover and mutation). The elitism algorithm is used to determine the survival of parent and offspring in the new generation [10]. The fitness of the offspring is evaluated using a fitness function. The fitness generation curve shows in Fig. 8a that the fitness value does not further improve after 200 generation. This is because, after 200 generations the produced sequence's total inter-arrival time remained unchanged. However, we continued our evolution up to 400 generations as an evidence of convergence of the solution.

- Selection
  Two individuals are chosen from the population using a selection operator. The preference is given to fitter individuals, allowing them to pass on their genes to the next generation. Fitness is determined by Eq. (5). The lower the inter-arrival time of a particular sequence the higher the fitness value.
- Crossover
  Fig. 4 shows the crossover procedure. In this crossover method a subset of the gene is selected from the first parent and then that subset is added to the offspring. The missing genes are then added to the offspring from the second parent by ensuring that the total number and types of genes (i.e., aircraft) remain equivalent to the parents'. To make this explanation a little clearer, consider the example in Fig. 4. Note here that a subset of the genes (i.e., M, L, H) of the offspring is taken from the parent 1 chromosome. Next, the remaining genes are taken from parent 2 sequentially.
- Mutation
  Mutation is used to maintain genetic diversity from one generation of the population to the next generation. In this implementation, swap mutation is used. With swap mutation two gene's positions in the chromosome are selected in a random fashion. Swap mutation is only swapping of pre-existing genes, it never creates a new gene. Eventually, once the population is not producing offspring that are noticeably different from the previous generation, it is assumed that the population converges to a set of solutions to the problem.

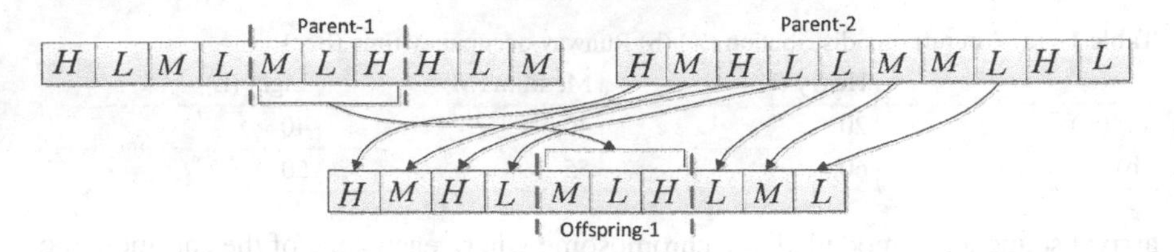

Fig. 4 Crossover procedure of GA based optimization

Determining the population size and mutation rate for GA is problem specific [11]. Too high a mutation rate increases the probability of searching more areas in the search space, however, it prevents the population from converging to an optimum solution. On the other hand, too small a mutation rate may results in premature convergence. To prevent both premature convergence and local optima, a small mutation rate of 0.015 is used [12].

3.2 *TATA for Optimal Arrival Sequence*

The flowchart of the TATA approach is shown in Fig. 5. The proposed method utilizes the evolutionary algorithm technique to obtain the near optimal sequence that ensures a conflict free path at TAS.

A. Initial Aircraft Sequence and Activation Time

The population of possible solution sequences are generated randomly according to a poisson arrival rate λ. Aircraft arrival events occur at IAF continuously and independently of one another. Each flight is activated at IAF by following a poisson arrival process. The initial aircraft sequence is randomly generated by a mix of three classes of aircraft (Heavy, Medium, and Light) using the distribution as shown in Table 1a. The probability distribution of the number of homogeneous poisson arrival events in a fixed interval gives the cumulative function of an exponential distribution as,

$$F(t) = \begin{cases} 1 - e^{-\lambda t}, & t \geq 0 \\ 0, & t < 0 \end{cases} \tag{10}$$

Given the inverse of the exponential equation $y = 1 - e^{-\lambda t}$, we can write for the next arrival time t in terms of $y \in (0, 1)$,

$$t = F^{-1}(y) = -\frac{1}{\lambda} ln(1 - y) \tag{11}$$

Equation (11) gives the continuous activation time at IAF of each random aircraft expressed in terms of the arrival rate.

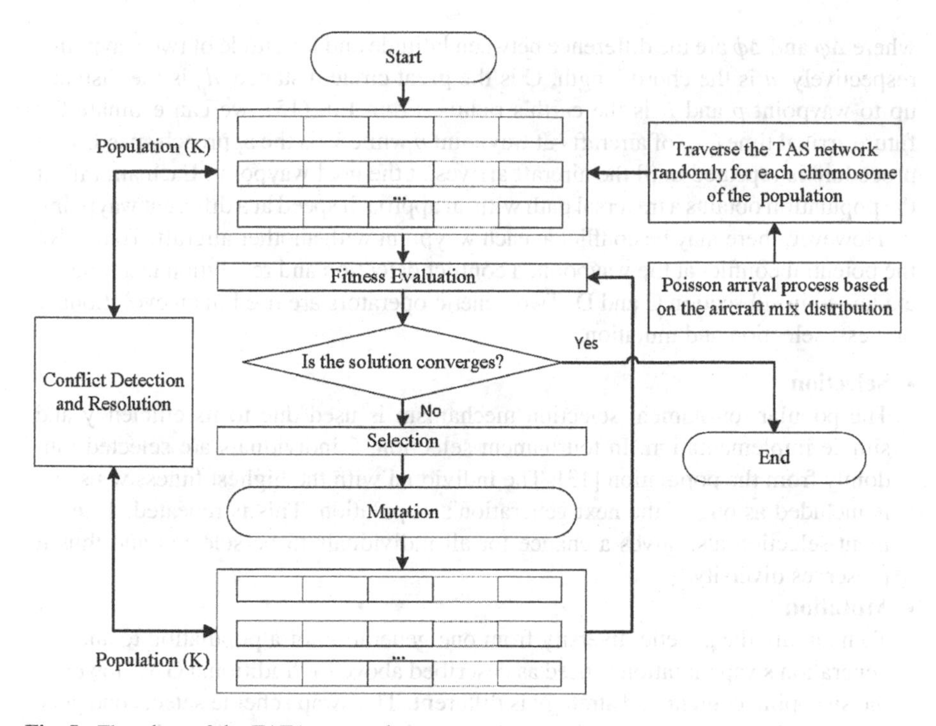

Fig. 5 Flowchart of the TATA approach

B. TATA Based Path Planning Algorithm

The TAS is considered as a network of waypoints. We assume that all the way-points are static and that the distance between all waypoint pairs is known. Equations of motion are used for calculating the position of the aircraft at any time t. After activating at the IAF, each flight finds the next possible waypoint towards the destination (FAF). The next waypoint is selected randomly from the next available connected waypoint. At each waypoint the traversing aircraft calculates the distance and the arrival time between the current and next waypoint. The path distance and the arrival time are estimated as follows,

$$a = sin^2(\Delta\varphi/2) + cos\varphi_1 cos\varphi_2 sin^2(\Delta\phi/2) \tag{12}$$

$$c = 2.atan2(\sqrt{a}, \sqrt{(1-a)}) \tag{13}$$

$$d_p = R.c \tag{14}$$

$$_pt_i = \frac{d_p}{V_i} \tag{15}$$

where $\Delta\varphi$ and $\Delta\phi$ are the difference between latitude and longitude of two waypoints respectively. a is the chord length, C is the great circle distance, d_p is the distance up to waypoint p and R is the earth's radius. From Eq. (15), we can estimate the future arrival time ($_pt_i$) of aircraft i at waypoint p while V_i is the approach speed. This procedure is repeated until the aircraft arrives at the goal waypoint. Each aircraft of the population obtains a traversal path with an approach speed at a different waypoint.

However, there may be conflict at each waypoint with another aircraft. To resolve the potential conflict at the waypoint, a conflict detection and resolution technique is explained in subsection C and D. Two genetic operators are used in an evolutionary process: selection and mutation.

- **Selection**
  The popular tournament selection mechanism is used due to its efficiency and simple implementation. In tournament selection, 5 individuals are selected randomly from the population [13]. The individual with the highest fitness wins and is included as one of the next generation's population. This is repeated. Tournament selection also gives a chance for all individuals to be selected and thus it preserves diversity.
- **Mutation**
  To maintain the genetic diversity from one generation of a population to another generation swap mutation is used as described above for traditional GA. However, the swapping content and strategy is different. The swap scheme selects one gene within a chromosome at random and then selects a waypoint from the traversed path of that gene (aircraft) randomly as well. If there is another available path instead of the selected waypoint, swap these contents. An example of the swap mutation procedure is shown in Fig. 6. Note here that from the parent chromosome, a gene is chosen randomly as indicated by the dotted lines. The selected gene (i.e., aircraft) has four waypoints in its chosen path ($2 \rightarrow 5 \rightarrow 7 \rightarrow 9$). The selected random waypoint is 5 which has two available next waypoints i.e., 7 and 8 as indicated in the dotted box. After swapping the waypoints, an offspring is produced. Note that the offspring's gene sequence has also changed as an outcome of the swap.

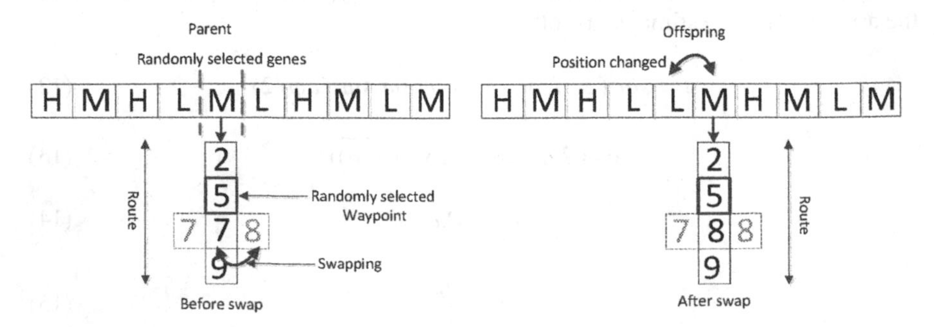

Fig. 6 Swap mutation operator

C. Conflict Detection Technique

In this paper, a prediction based conflict detection model is introduced. Conflict is a situation where two aircraft come closer than a certain prescribed distance to one another. The safety distance is determined by means of a minimum allowed horizontal separation and a minimum vertical separation. In this model, we consider the horizontal separation as conflict detection metrics and the potential conflicts are predicted at different waypoint based on an aircraft's future arrival time.

We assume that each aircraft follows its flight plan moving along the straight line joining successive waypoints p_{k-1} and p_k with the prescribed speed V_i. The nominal arrival time ${}_kt_i$ of an aircraft i at waypoint k is

$${}_kt_i = \frac{\|p_{k-1} - p_k\|}{V_i} \quad (16)$$

where $\|p_{k-1} - p_k\|$ is the distance between two waypoint. A potential conflict between aircraft i and j at waypoint k is predicted if,

$$|{}_kt_i - {}_kt_j| \leq \delta \quad (17)$$

where ${}_kt_j$ is the arrival time of aircraft j and δ is the minimum separation.

D. Conflict Resolution

A path stretching technique is used to resolve the potential conflict. The objective of path stretching is to maintain a smooth motion along the trajectories. Two approaches are used. i.e., speeding the aircraft that will arrive first or slowing the aircraft that will arrive second. Suppose the arrival time of two conflicting aircraft at waypoint p_k is ${}_it_k$ and ${}_jt_k$ respectively. The resolution advisories (i.e., required adjustment of speed) is estimated by following equation.

$$dt = |{}_it_k - {}_jt_k| \quad (18)$$

$$\Delta t = \delta_k - dt \quad (19)$$

$$v_{exp} = \frac{d_i}{{}_it_k + \Delta t} \quad (20)$$

$$\Delta v = |v_{exp} - {}_{ap}v_i| \quad (21)$$

$${}_{ap}v_i = \begin{cases} {}_{ap}v_i - \Delta v, & \text{if } {}_it_k \geq {}_jt_k \\ {}_{ap}v_i + \Delta v, & \text{if } {}_it_k < {}_jt_k \end{cases} \quad (22)$$

Here, dt is time gap, Δt is the required time adjustment, δ_k is the minimum separation. Expected speed v_{exp} can be estimated by Eq. (20) where d_i is the distance between aircraft i and next connecting waypoint. Therefore, we can estimate the required approach speed by Eq. (22) where Δv is the required speed adjustment.

4 Experimental Design

The optimization model is evaluated through simulation. The performance of the model is evaluated using a mixture of arrivals by taking into consideration that all the ATC separation rules are satisfied. Table 2 shows the summary of the experimental parameters used. The simulation is conducted 30 times to observe closeness of the obtained sequence with traditional GA.

Table 3a shows the pair-wise ATC separations from arrival to arrival in seconds. In practice arrival separation is measured as nautical miles (NM) and departure separation is measured in seconds. In this implementation, we convert the separation distance into time (seconds) for simplicity of computation. Table 3b presents the probability matrix of two consecutive aircraft based on their wake-type.

Table 2 Experimental set-up

Parameter	Values
Mutation rate	0.015
Chromosome size	20
Population size	50
Tournament size	5
Generation	400
No. of simulation	30

Table 3 (a) Arrival-arrival (sec) (b) Probability (P_{ij})

Trail				
(a) Separation				
		H	*M*	*L*
Lead	*H*	90	120	120
	M	60	60	60
	L	60	60	60
(b) Probability				
		H(0.2)	*M*(0.4)	*L*(0.4)
Lead	*H*(0.2)	0.04	0.08	0.08
	M(0.4)	0.08	0.16	0.16
	L(0.4)	0.08	0.16	0.16

Table 4 Near optimal path and obtained sequence at FAF

IAF seq	GA seq	FAF seq	IAF time	FAF time	Optimal path
L	L	H	6.83	23.03	$3 \to 6 \to 7 \to 9$
M	M	M	5.35	23.66	$0 \to 4 \to 7 \to 9$
L	H	L	7.15	30.69	$0 \to 5 \to 8 \to 9$
M	L	M	14.56	32.89	$3 \to 6 \to 8 \to 9$
L	M	M	9.61	34.92	$1 \to 4 \to 8 \to 9$
M	H	L	5.26	37.79	$1 \to 5 \to 8 \to 9$
L	M	H	17.28	40.04	$1 \to 6 \to 7 \to 9$
H	H	L	17.64	41.21	$3 \to 6 \to 7 \to 9$
L	L	M	24.16	42.47	$0 \to 4 \to 8 \to 9$
L	L	L	10.31	42.82	$2 \to 6 \to 8 \to 9$
M	M	M	15.74	43.65	$2 \to 6 \to 8 \to 9$
H	L	L	20.52	44.09	$3 \to 5 \to 8 \to 9$
L	L	M	19.89	45.17	$2 \to 5 \to 8 \to 9$
M	M	L	15.62	48.12	$1 \to 4 \to 8 \to 9$
L	L	L	25.14	48.71	$3 \to 5 \to 7 \to 9$
M	L	L	26.62	50.16	$0 \to 5 \to 8 \to 9$
L	M	H	28.44	51.2	$1 \to 5 \to 8 \to 9$
M	M	M	31.65	51.65	$2 \to 5 \to 8 \to 9$
H	L	M	26.4	53.07	$3 \to 6 \to 7 \to 9$
M	M	L	28.92	54.64	$2 \to 5 \to 8 \to 9$

5 Result Analysis and Discussion

In this section, we present an illustration of the TATA model and a demonstration of the aircraft simulation. Figure 8a presents the convergence curve of the optimization problem. The visualisation shows that the fittest individual had not improved further after 200 generations.

The simulation result presented in this section is the optimal sequence of GA based optimization and path planning based optimal sequence of the TATA algorithm. An optimal solution of TATA and GA based arrival sequence is shown in Table 4. Notice that the arrival sequence of the GA and arrival sequence of the path planning algorithm (TATA) are very close. The TATA algorithm also provides the estimated arrival time at FAF. Separation is maintained between all aircraft at all waypoints at all times. A significant contribution of the TATA algorithm is to provide detailed guidance to the ATC i.e., the path planning, maximization of runway capacity and the estimated arrival time at FAF as shown in Table 4.

To observe the mutual closeness of the GA based optimal sequence and the TATA sequence, the simulation is conducted 30 times. The average inter-arrival time is shown in Fig. 8b. Note that the path planning based optimal sequence took slightly greater time, however it is only 0.51 %. Finally, we analyze the time-space diagram of the GA based optimal sequence and the TATA based optimal sequence. Figure 9

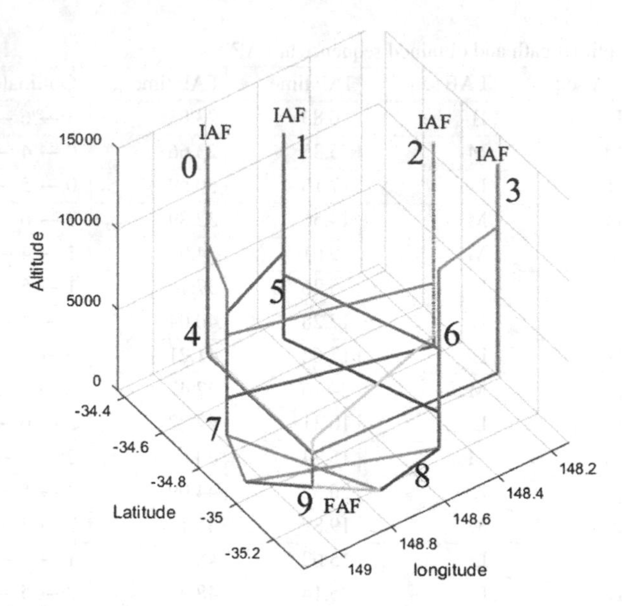

Fig. 7 3D traversed network

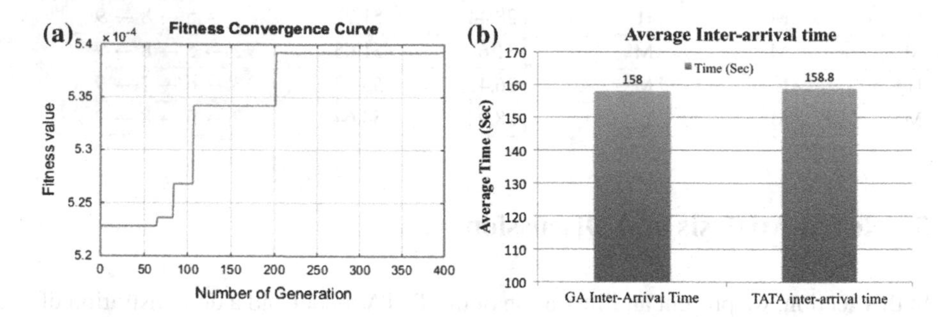

Fig. 8 Simulation result. **a** Fitness generation curve. **b** Average inter-arrival time/window (Seconds)

shows the comparison of the obtained optimal sequence from the traditional GA approach and the TATA based approach. Note here, an interesting result for the best possible sequences in both the cases is that similar wake category aircraft are positioned side by side to reduce the inter-arrival time. A 3D trajectory network is depicted based on the traversed path of the TATA based optimal sequence as presented in Fig. 7.

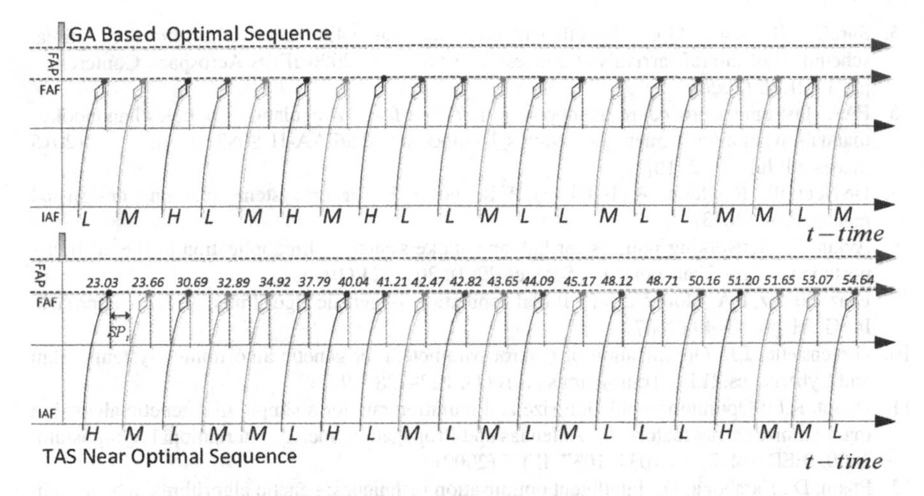

Fig. 9 Time-space diagram of GA optimal sequence VS TATA near optimal sequence

6 Conclusion

In this paper, we addressed one of the common challenges faced by ATC in TAS. How to plan the arrival path of aircraft in transition airspace such that they are conflict free and their inter-arrival time is minimized. State-of-the-art methods provide the optimal sequence which minimizes the inter-arrival time given an arrival sequence of aircraft, with population based search methods being highly effective. We proposed a GA based path planning technique which can not only achieve an optimal sequence but also address conflict between arriving aircraft and resolve them. The proposed algorithm fills an important gap in advising ATC on arrival aircraft path planing and sequencing to achieve a conflict free optimal sequence which reduces the inter-arrival time which in turn increases the runway capacity. However, this approach comes at the cost of some airborne delay which stems from aircraft speed manoeuvres for conflict resolution.

References

1. Khadilkar, H., Balakrishnan, H.: Integrated control of airport and terminal airspace operations. IEEE Transactions on Control Systems Technology 24(1), 216–225 (2016)
2. Erzberger, H., Paielli, R.A.: Concept for next generation air traffic control system. Air Traffic Control Quarterly 10(4), 355–378 (2002)
3. Fricke, H., Schultz, M.: Delay impacts onto turnaround performance. In: ATM Seminar (2009)
4. Ahmed, M.S., Alam, S.: An evolutionary optimization approach to maximize runway throughput capacity for hub and spoke airports. In: Australasian Conference on Artificial Life and Computational Intelligence. pp. 313–323. Springer (2016)

5. Saraf, A.P., Slater, G.L.: An efficient combinatorial optimization algorithm for optimal scheduling of aircraft arrivals at congested airports. In: 2006 IEEE Aerospace Conference. pp. 11. IEEE (2006)
6. FAA: Instrument procedure handbook. http://www.faa.gov/regulations_policies/handbooks_manuals/aviation/instrument_procedures_handbook/media/FAA-H-8083-16.pdf (2015 [accessed: July 21, 2016])
7. De Neufville, R., Odoni, A., Belobaba, P., Reynolds, T.: Airport systems: planning, design and management (2013)
8. Aykin, T.: Networking policies for hub-and-spoke systems with application to the air transportation system. Transportation Science 29(3), 201–221 (1995)
9. Diaz-Gomez, P.A., Hougen, D.F.: Initial population for genetic algorithms: A metric approach. In: GEM. pp. 43–49 (2007)
10. Grefenstette, J.J.: Optimization of control parameters for genetic algorithms. Systems, Man and Cybernetics, IEEE Transactions on 16(1), 122–128 (1986)
11. Haupt, R.L.: Optimum population size and mutation rate for a simple real genetic algorithm that optimizes array factors. In: Antennas and Propagation Society International Symposium, 2000. IEEE. vol. 2, pp. 1034–1037. IEEE (2000)
12. Pham, D., Karaboga, D.: Intelligent optimisation techniques: genetic algorithms, tabu search, simulated annealing and neural networks. Springer Science & Business Media (2012)
13. Goldberg, D.E., Deb, K.: A comparative analysis of selection schemes used in genetic algorithms. Foundations of genetic algorithms 1, 69–93 (1991)

A Game-Theoretic Approach to the Analysis of Traffic Assignment

Caixia Li, Sreenatha G. Anavatti, Tapabrata Ray and Hyungbo Shim

Abstract In order to improve the cooperation between traffic management and travellers, traffic assignment is the key component. In terms of the traffic assignment, it can be classified into two models based on the behavior assumption governing route choices: the User Equilibrium (UE) and System Optimum (SO) traffic assignment. By the definition of UE and SO traffic assignment, traffic users usually competitively choose the least cost routes to minimize their own travel cost, while system optimum traffic assignment requires traffic users work cooperatively to minimize overall cost in road network. Thus, the paradox of benefits between UE and SO makes both of them are not practical. Thus, a solution technique needs to be proposed to balance between UE and SO models, which can compromise both sides and give more feasible traffic assignments. In this paper, Stackelberg game theory is introduced to the traffic assignment, which can achieve the trade-off process between traffic management and travellers. Since the traditional traffic assignments have low convergence rates, the gradient projection algorithm is proposed to improve the efficiency of the traffic assignment.

Keywords Traffic management · Traffic assignment · Route choices · Stakelberg game theory

C. Li · S.G. Anavatti (✉) · T. Ray
UNSW Canberra, Campbell, Australia
e-mail: S.Anavatti@adfa.edu.au; agsrenat@adfa.edu.au

C. Li
e-mail: Cara.cxl@hotmail.com

T. Ray
e-mail: T.Ray@adfa.edu.au

H. Shim
Seoul National University, Seoul, Korea
e-mail: hshim@snu.ac.kr

G. Leu et al. (eds.), *Intelligent and Evolutionary Systems*,
Proceedings in Adaptation, Learning and Optimization 8,
DOI 10.1007/978-3-319-49049-6_2

1 Introduction

In the Wardrop's [1] first principle, the journey times in all routes actually used are equal or less than those which would be experienced by a single vehicle on any unused route. This first principle is often referred to as "user optimal" (UE) model. Under the "user optimal" model, each traveler act competitively and make their own optimal route for their own benefit. According to Wardrop's second principle, at equilibrium, the overall system journey time is minimum. This implies that each user behaves cooperatively in choosing his own route to ensure the most efficient use of the whole system, which is referred to as "System Optimal" (SO) model. Under the UE principle, traffic users competitively choose the least cost route to minimize their own travel cost, while the SO principle requires the traffic users work cooperatively to minimize their overall cost. Thus, the competitive and cooperative interaction between traffic information provider and traffic users can be interpreted as a game.

Game theory is related to several parties with different interests to decide the optimal choice. The benefit to each party not only depends on its own actions, but also on the choices of the other parties. For the UE, it consists of many travellers and each traveller is a game player to compete with each other to find the shortest path for his/her benefit. In terms of SO model, there is only the traffic control centre to control all the travellers on the roads. It is assumed all the travelers can cooperate with each other to get the minimized travel cost of the system. Therefore, users are fully competitive or fully cooperative in the Wardrop's theory.

In reality, both competition and cooperation among users exist in the traffic network. More general situation is proposed by Haurie and Marcotte [2], who present a relationship between non-cooperative Coumot-Nash (CN) and Wardrop's model. It denotes users belonging to a common player are fully cooperative, while different players are fully competitive. For the first situation it approaches to SO model and the latter one corresponds to the UE model. A mixed behavior situation is considered by Harker [3], where some distinct players are controlled by the CN players, while the other users follow the UE principle. Vuren et al. [4] studied the route guidance problem by combining the UE principle and SO principle in the traffic assignment model and different levels of information are incorporated into the model via a multiple user class Stochastic User Equilibrium (SUE). Wie [5] introduced a differential game model of Nash equilibrium on a congested traffic network and applied it to solve the dynamic mixed behaviour traffic network equilibrium problem [6]. The objective of the research is to establish the relationship between the Nash equilibrium and the dynamic user equilibrium. Friesz et al. [7] examined a certain class of dynamic games known as open loop differential Nash games. Kumar and Peeta [8] introduced the strategies to enhance path based static traffic assignment. A Stackelberg routing strategy is combined into the network optimum model by Korilis et al. [9]. A non-cooperative game framework combining the traffic control model with routing is proposed by Altman et al. [10], and more detailed study is done on uniqueness, efficiency and computational method of Nash equilibrium. In addition, multi-class equilibrium models are also studied by La and Anantharam [11], which is analogous to the multiple equilibrium behaviour models. The game theory algorithm is introduced in agent based cooperative decentralized airplane system by Sislak et al. [12].

The Stackelberg strategy safeguards each player against any attempts by the other players to deviate, but in a sequential manner. In other words, once the leader and the follower are specified; then the leader has no better choice than to select a Stackelberg strategy under his own leadership and the follower has no better choice than to react according to his reaction set. The leader has no incentive to cheat, since he knows that his control is continuously monitored by the follower. In this sense the Stackelberg strategy is also an equilibrium point. Thus, the relationship between traffic authority and traffic users can be modelled by the Stackelberg game model. The traffic manager is regarded as a leader and traffic users act as followers and the traffic manager makes this optimal strategy and let the users converge to their respective equilibrium.

In terms of the traffic assignment, the method of successive average (MSA) [13], the method of simple projection (SP) [4], the method of day-to-day swapping [14], and the method of modified alternative direction [15] are the most well documented. Although these algorithms are implemented in the transportation networks, their relative performance is still unclear.

Moreover, a simple optimization algorithm is required in view of the relationship of Stackelberg game between traffic management and drivers. Considering the implementation of the game based strategy between traffic management and traffic users, an efficient solution algorithm is required, since it needs to update route choice advices as often as possible. In recent years, Gradient projection algorithms have been found to outperform the Frank-Wolfe algorithm and the feasibility of applying the gradient projection algorithm to the traffic networks is demonstrated. Thus, a modified gradient projection algorithm is introduced in view of Stackelberg game between traffic management and drivers to mediate their relationship.

The remainder of this paper is organized as follows: Sect. 2 presents the game theory based traffic assignment for the road network. The gradient projection algorithm is introduced to improve the efficiency of traffic assignment in Sect. 3. Section 4 gives a description of the solution methodology for the integrated system. Numerical experiments are implemented in Sect. 5. Concluding remarks are given in Sect. 6.

2 Stackelberg Game Based Modelling

There are two decision variables in the game strategy, where one is a set of path flows caused by the drivers' route choice behaviour, and the other is a set of travel cost information as a result of traffic management strategy. The control variables are set by the information providers to gain their objectives, which can be combined in a mathematical form. Thus, the cooperation and competition relationship between traffic management and traffic users can be expressed as a mathematical minimization problem;

$$min_{\alpha}Z_{\alpha}(\alpha, x^{*}(\alpha)) \tag{1}$$

where α is a vector of traffic information influencing drivers' perception of the travel cost, and $x^{*}(\alpha)$ is a vector of traffic flows as a result of the traffic assignment which is a mathematical formulation of drivers' route choice problem. Traffic flow patterns can be achieved by solving the traffic assignment problem based on the information vector α determined by the above problem.

The traffic assignment representing drivers' route choice behaviour can also be expressed as a minimization problem;

$$\min_{x} Z_{x}(\alpha^{*}, x(\alpha^{*})) \tag{2}$$

The state of equilibrium of $(\alpha^{*}, x(\alpha^{*}))$ can be achieved by bi-level programming problem: For the upper level of Stackelberg game programming, find the optimal guidance indicator α for given link g that minimizes total system cost, for the given link g, drivers can make compromise with information providers on sub-paths to accept the route advices. Thus the total system cost for minimization is as follows:

$$\min_{\alpha_g} Z_U = \sum_{a \in A} C^a(x^a + \dot{x}^a) * (x^a + \dot{x}^a) \tag{3}$$

Subject to:

$$\dot{x}^a = x^a - \sum_k \sum_\tau \sum_i \sum_j r^{\tau}_{ijk} * \delta^{\tau ta}_{ijk} \tag{4}$$

where Eq. (4) is the constraint of the minimization process, representing the traffic volume change by previous traffic volume and the selected traffic volume. Equation (3) is the traffic volume $\dot{x}^a$ added to the current traffic flow x^a. r^{τ}_{ijk} is the number of vehicles on the link from previous node i to next node j at time τ and $\delta^{\tau ta}_{ijk}$ is the 0–1 time dependent link-path incidence variables corresponding to the number of vehicles assigned to each link of specific link a.

For the lower level of Stackelberg game programming, find the optimal traffic flow pattern to satisfy the UE condition;

$$minZ_L = \sum_{a \in A} \int_0^{x^a} C^a(w + \dot{x}^a) * \alpha_g dw \tag{5}$$

where C^a is the travel cost on link a with traffic volume x^a.

While the upper level problem is the system optimal traffic assignment, the lower level problem is a process of a user equilibrium traffic assignment problem. In order to improve the efficiency of the problem, the traffic assignment problem can be solved by the gradient projection algorithm, where the main target of the problem is to find the guidance indicator α for sub-paths. The sub-path represents the paths which can be perceived by drivers for a user-optimal redistribution under the system optimal traffic assignment to yield better system travel cost.

3 Gradient Projection Based Traffic Assignment

Gradient projection algorithm has been shown as an efficient algorithm for solving the traffic assignment. The GP algorithm is a path-based flow formulation, which cannot find auxiliary solutions in the link-flow space. The feasible space for the gradient projection algorithm is defined only by the non-negativity, since GP algorithm makes moves to the direction of the minimum of the Newton approximation. The update step can be expressed by the following interactive equation:

$$f_k^{rs}(n+1) = [f_k^{rs}(n) - \alpha(n)D(n)\nabla\dot{Z}(n)]^+ \tag{6}$$

where $\alpha(n)$ is the step size, $D(n)$ denotes a diagonal, positive definite scaling matrix, $\nabla\dot{Z}(n)$ is the gradient of the transformed objective function, and $[.]^+$ is the projection of the argument on the positive axis of the independent variables $f_k^{rs}(n)$. This operation of moving demand conservation constraints from the constraint to the objective function can make projection simpler. In the process of the operation, $f_k^{rs}(n)$ is partitioned into the least cost path flow $f_{\bar{k}_{rs}}^{rs}(n)$ and the non-least cost path flow $f_k^{rs}(n)$ in the path set K_{rs}.

$$f_{\bar{k}_{rs}}^{rs}(n+1) = q_{rs} - \sum_{k \in K_{rs}, k \neq \bar{k}_{rs}} f_k^{rs}(n+1) \tag{7}$$

where $\bar{k}_{rs}$ is the least cost path from origin to destination. The optimization problem can be transformed into the following form by substituting the least cost path flow $f_{\bar{K}_{rs}}^{rs}(n)$. Thus, the minimization process is related to find non-least traffic cost with non-least traffic cost flow;

$$min\, \dot{Z}(\dot{f}) \tag{8}$$

Subject to:

$$f_k^{rs} \geq 0, \forall k \in K_{rs}, k \neq \bar{k}_{rs}, r \in R, s \in S \tag{9}$$

where $\dot{f}$ is the set of non-least cost path flows for all origin to destination pairs. The objective value can be improved by moving in the negative gradient direction. The gradient of the transformed objective function is related to the set of non-least cost paths, and a diagonal scaling of the gradient direction can be achieved by the second derivatives of the independent variables.

$$\frac{\partial \dot{Z}}{\partial f_k^{rs}} = \frac{\partial Z}{\partial f_k^{rs}} - \frac{\partial Z}{\partial f_{\bar{k}_{rs}}^{rs}}, \forall k \in K_{rs}, k \neq \bar{k}_{rs}, r \in R, s \in S \tag{10}$$

where Z is the original objective function including both the least cost and non-least cost path. Each component of the gradient becomes the difference between the first derivative cost of a non-least cost path and the least cost path, where the first derivative of Z related to any path is link traversal cost based on the current traffic flow information.

$$\frac{\partial Z}{\partial f_k^{rs}} = \sum_{a \in A} c^a(x^a)\delta_k^{rsa} \tag{11}$$

$$\frac{\partial Z}{\partial f_k^{rs}} = \sum_{a \in A} c^a(x^a)\delta_{\bar{k}_{rs}}^{rsa} \tag{12}$$

Thus, the diagonals of the second derivatives of the transformed objective function are the differentiation of the gradients,

$$\frac{\partial^2 \dot{Z}}{\partial f_k^{rs2}} = \sum_{a \in A} c^{a'}(x^a)(\delta_k^{rsa} - \delta_{\bar{k}_{rs}}^{rsa})^2 \tag{13}$$

where $c^{a'}(x^a)$ denotes the first derivative of the link traversal time.

Let d_k^{rs} and $d_{\bar{k}_{rs}}^{rs}$ be the first derivative costs of path k and the least cost path $\bar{k}_{rs}$ of the origin to destination pair, the iterative flow update can be expressed as follows:

$$f_k^{rs}(n+1) = \max\{0, f_k^{rs}(n) - \frac{\alpha(n)}{s_k^{rs}(n)}[d_k^{rs}(n) - d_{\bar{k}_{rs}}^{rs}]\} \tag{14}$$

where $\alpha(n)$ denotes a scalar modifier. Once all the non-least cost paths are updated, the traffic flow on the least cost path is appropriately updated so that the demand is conserved.

4 Solution Methodology for the Stackelberg Game Based Traffic Assignment

In order to improve the efficiency of the game based traffic assignment, the Gradient Projection method is implemented. In the principle of the GP method, the flow f_k^{rs} is partitioned into the least cost path flow $f_{\bar{K}_{rs}}^{rs}(n)$ and the non-least cost path flow $f_k^{rs}(n)$ in the path set K_{rs}. Substituting the partition of the least cost path flow and non-least cost path, the path flow f_k^{rs} in the game based traffic assignment is partitioned into user equilibrium least cost path flows and system optimal least cost path flows.

For the different objectives of the UE model and SO model, the least cost path flows on the UE model and SO model corresponding to the same O-D (origin to destination) pair are different. Let the traffic flow on the least cost path between UE model and SO model be the least cost path flow $f_{\bar{K}_{rs}}^{rs}(n)$ and the other one be the non-least cost path flows $f_k^{rs}(n)$. Moreover, the paths with traffic cost between the two least cost of the UE model and SO model are defined as the sub-paths and stored as the path set K_{rs}. Substituting the least cost path flow $f^{rs}_{\bar{K}_{rs}}(n)$ given by Eq. (12) for each O/D pair into the objective function, the optimization problem can also be formalized by the form of Eq. (13). Then, the modified GP method can be implemented in the traditional implementation process.

Thus, the Stackelberg game based integrated system can be improved by the Gradient Projection method to achieve efficient traffic assignment. The solution algorithm, as shown in Fig. 1 gives the framework for the Stackelberg game based traffic assignment, can be described as follows:

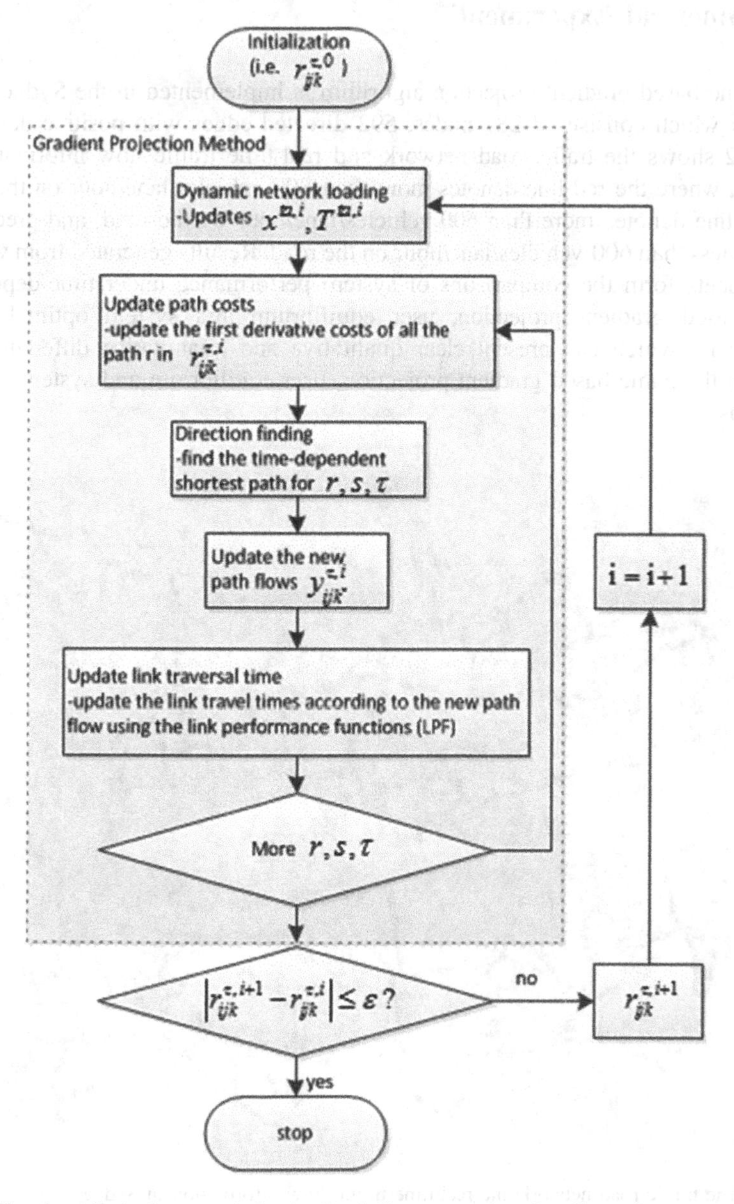

Fig. 1 The algorithmic framework for the Stackelberg game based traffic assignment

(1) Pre-assignment
(2) Find sub-paths associated with traffic management and traffic users
(3) Update optimal guidance indicator value to satisfy system optimal traffic assignment
(4) Update sub-paths, indicator, and total cost
(5) Stop, if no more sub-paths, otherwise go to step 2.

5 Numerical Experiment

The game based gradient projection algorithm is implemented in the Sydney road network which consists of 287 nodes, 592 directed edges with positive demands. Figure 2 shows the traffic road network and real-time traffic flow information in Sydney, where the red line denotes more than 800 vehicles/lane/hour on the road, orange line denotes more than 600 vehicles/lane/hour on the road, and green line denotes less than 600 vehicles/lane/hour on the road. Results generated from various experiments form the comparisons of system performance under time-dependent game based gradient projection, user equilibrium and system optimal traffic assignment, which can present clear qualitative and quantitative differentiations between the game based gradient projection, user equilibrium and system optimal solutions.

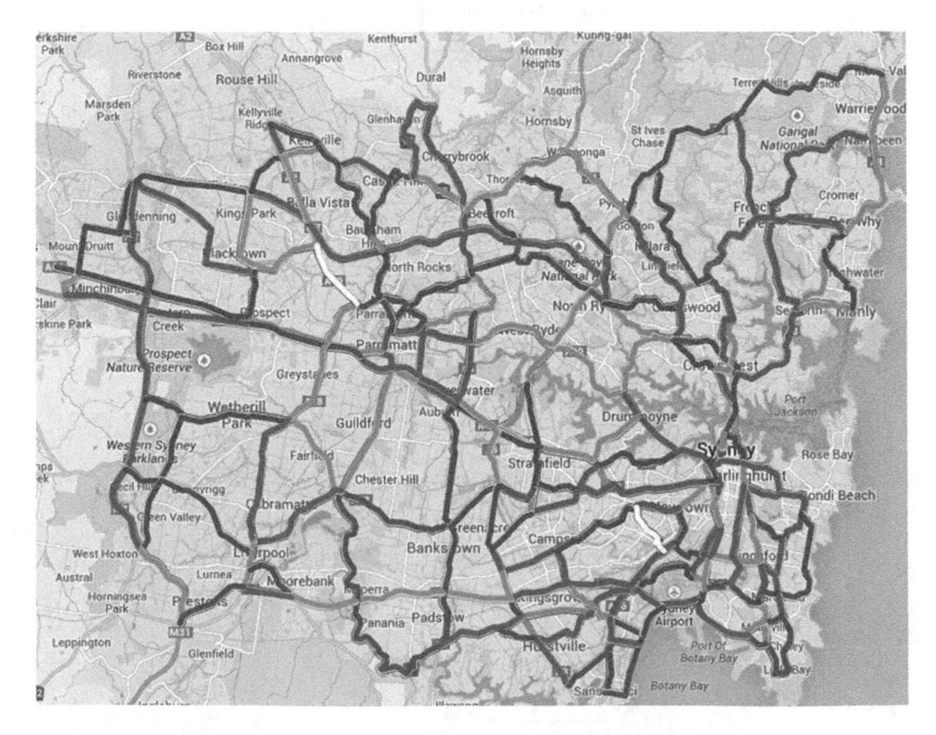

Fig. 2 The traffic road network and real-time traffic flow information in Sydney

5.1 Convergence and Computation Performance of the Gradient Projection Method

In order to manifest the effectiveness of gradient projection algorithm, the gradient projection algorithm is compared with Frank-Wolfe algorithm and the method of successive averages (MSA) algorithm regarding convergence and computation performance. In term of the Frank-Wolfe algorithm [13], it is one of the promising algorithms for traffic assignment, since it can make full use of the network structure of the road networks. The search directions of the Frank-Wolfe algorithm usually tend to be perpendicular to the steepest descent directions of objective function as the iteration proceeds. For the method of successive averages algorithm [14], it is one of the most widely used solution methods in simulation-based dynamic traffic assignment. This method relies on predetermined step sizes without requiring derivative information, so that it can obviate the need to solve one-dimensional line search problems for finding the optimal move size.

Figure 3 shows the Solution convergence of the Gradient Projection method Compared with the Frank-Wolfe algorithm. Both the Frank-Wolfe (FW) algorithm and the gradient projection (GP) algorithm are initialized with zero flows on all the links in the road network. In the FW model, it uses the all-at-once flow update, where the total link-flow pattern are adjusted after the traffic demands from all O/D pairs are assigned to the network, while the GP model updates the flow pattern one O/D at-a-time, that is, the total link-flow pattern is revised after the assignment of an O/D pair before continuing to the next O/D pair. Since different traffic flow patterns are updated, the convergence rates are also different. As shown in Fig. 3, the GP converges faster than FW. Typically, the 5th or 6th iteration in GP corresponds to the 10th iteration in FW. Actually, FW slowly approaches to the minimum solution, and the objective value of FW in the 100th iteration is exactly the same as the 10th iteration. However, the GP can quickly approach to the minimum solution.

Fig. 3 Solution convergence

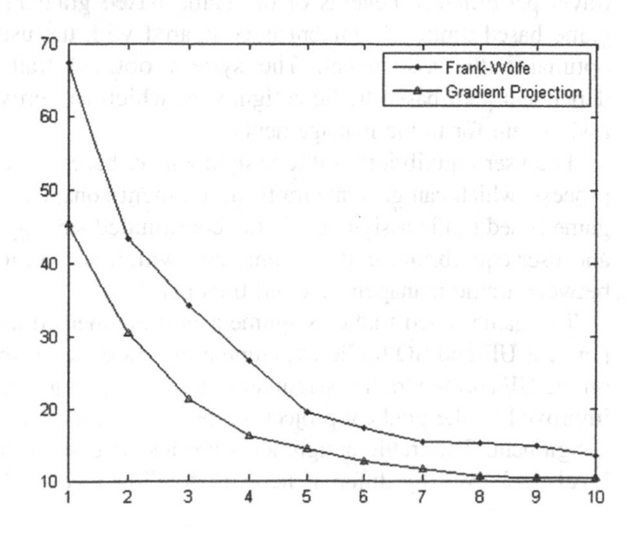

Table 1 Computation time and number of iterations to convergence for various size of road network

–	10 nodes, 42 links, 170 O/Ds	36 nodes, 92 links, 670 O/Ds	100 nodes, 375 links, 5670 O/Ds	287 nodes, 592 links, 51670 O/Ds
–	Iteration/time (s)	Iteration/time (s)	Iteration/time (s)	Iteration/time (s)
FW	43/17.63	56/25.46	161/58.76	268/95.56
MSA	31/19.92	59/23.12	139/57.6	336/113.21
GP	6/2.41	9/4.37	13/11.39	18/45.23

In order to test the computation of the Gradient Projection method, the GP method is tested on various sizes of grid networks. Table 1 shows the performance associated with computation time and number of iterations for the Frank Wolfe algorithm, the method of successive averages (MSA) and the Gradient Projection algorithm tested on various sizes of road network, ranging from 10 to 592 nodes. From the comparison results, it shows that GP takes much less iterations and less computation time than the other two algorithms to reach the same objective value in all the situations. As the size of the road network increases, the ratio of computation time comparing with FW and MSA decreases. However, it is still more efficient than the conventional FW and MSA methods. Moreover, it suggests that the GP method in the decomposed networks of smaller size can achieve significant benefit in computation time.

5.2 Travel Performance Comparison with SO and UE

Since the implementation of the game based traffic assignment is based on the interaction process between UE and SO traffic assignment, in order to manifest the travel performance benefits of the game based gradient projection algorithm, the game based traffic assignment is compared with the user equilibrium and system optimal traffic assignment. The system optimal traffic assignment is a time-dependent path based traffic assignment, which can provide system optimal traffic assignment for traffic managements.

The user equilibrium traffic assignment is based on the route choices selection process, which can generate traffic assignment considering divers' preferences. The game based traffic assignment is the coordinated strategy between system optimum and user equilibrium traffic assignment, which is used to compromise the benefits between traffic managements and travellers.

The game based traffic assignment is implemented as follows: Firstly, an independent UE and SO traffic assignment are modelled respectively; Secondly, based on the UE and SO traffic assignment, the Stackelberg game based traffic assignment improved by the gradient projection method is compared with the individual traffic assignment. The traffic assignments are tested under different network congestion levels, achieved by different network loading levels. The network loading factor

Table 2 Loading factors and the corresponding number of generated vehicles for the numerical experiments

Loading factor	Number of generated vehicles
1.0	19346
1.4	27168
1.8	34887
2.0	38762
2.2	42631

denotes the ratio of the total number of vehicles generated in the road network during the traffic assignment periods compared to a given reference number (19346 vehicles over 15 min period represent a loading factor of 1.0). In the numerical experiments, five loading factors are considered, namely, 1.0, 1.4, 1.8, 2.0 and 2.2. Table 2 shows the number of vehicles generated for each loading factor.

Table 3 shows the system performance under the time dependent SO, UE and Stackelberg game based traffic assignment associated with different loading factors. Based on the principle of the traffic assignment, when the road network is relatively uncongested (at low loading levels), the average travel time of vehicles in the road network is relatively close. As the loading factor is increased, congestion can be caused and the average travel time increases with the loading factor. The results show that the average travel time is significantly increased with the increase of the loading factors, while the average travel distance makes limited variation under the various loading levels. It indicates that the greater traffic congestion is the primary cause of the higher system travel time instead of the travel routes. Moreover, the

Table 3 Comparison results of SO, UE and GAME based traffic assignment associated with various loading factors

Loading factor	Av. travel time (min)	Total travel time (h)	Av. travel distance (km)	Total travel distance (km)	Av. speed (kmph)
1.0	21.46	6759.419	15.43	290508.8	44.44073
1.4	21.62	9599.536	15.58	414777.4	44.53774
1.8	27.83	15971.75	16.73	573659.5	37.36899
2.0	35.64	22744.63	17.46	663784.5	30.69394
2.2	41.24	28991.71	18.43	764689.3	28.11377
1.0	21.46	6759.079	15.43	290435.8	44.39473
1.4	21.65	9613.12	15.53	413774.4	44.31926
1.8	28.36	16279.92	16.49	565286.6	36.18717
2.0	37.94	24230.50	17.24	655256.9	28.56410
2.2	43.65	30704.05	18.37	762131.5	26.55086
1.0	21.46	6759.209	15.43	290474.8	44.43034
1.4	21.63	9603.854	15.56	414600.1	44.45197
1.8	28.04	16093.65	16.51	565984.4	36.61781
2.0	36.17	23086.82	17.33	658745.5	30.03728
2.2	41.31	29751.75	18.41	763836.7	27.39725

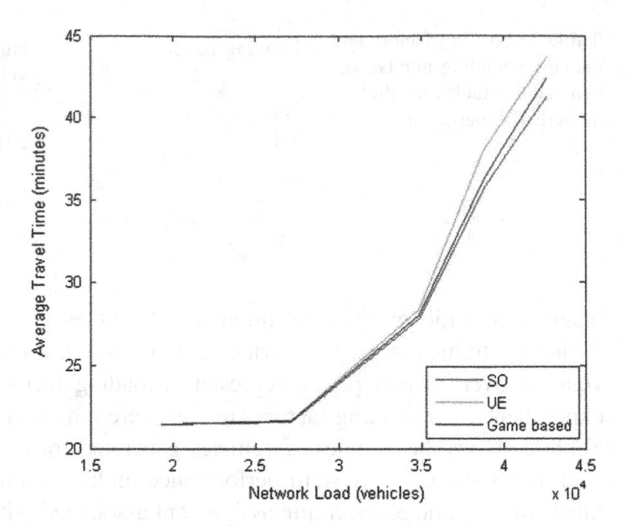

Fig. 4 Average travel time comparison between UE, SO and game based traffic assignment

average travel distance increases with the loading level increased, suggesting an increase number of drivers assigned to longer travel routes. In addition, the average travel distances under UE are less than average travel distances under SO for various loading factors, indicating some drivers are assigned to longer routes under SO in order to reduce traffic congestion to achieve system optimal benefit.

For different objectives of the UE and SO models, Stackelberg game based traffic assignment is proposed to balance the benefit between UE and SO in order to relieve traffic congestion and reduce travel cost. The results under the Stackelberg game based traffic assignment show that the game based traffic assignment makes compromise between UE and SO. Figures 4 and 5 compare the average travel time and average travel distance under UE, SO and Stackelberg game based traffic

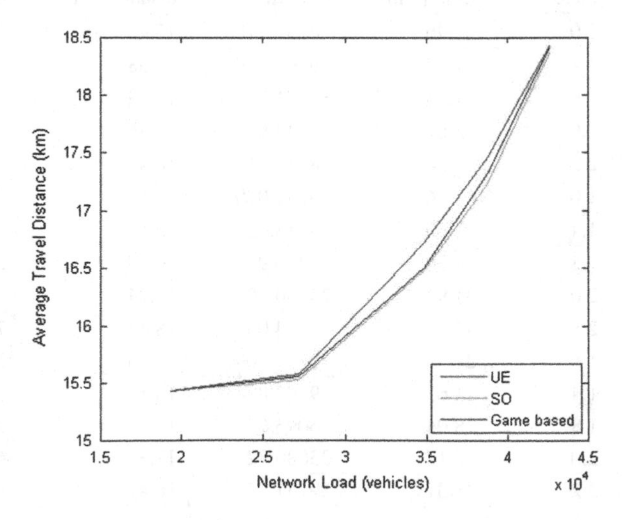

Fig. 5 Average travel distance comparison between UE, SO and game based traffic assignment

assignment. The results show that limited variation in both average travel time and average travel distance is caused by the game based traffic assignment at the lower loading levels.

The reason is that the traffic is not congested at lower loading levels, and travel costs as a result of the UE and SO traffic assignment are identical to each other, thus, limited variation is caused by the game based model. With the increase of the loading factors, the average travel time approaches the SO model, indicating a number of vehicles assigned to SO model to relieve traffic congestion on the roads at the expense of the increase of average travel distance.

6 Conclusion

In this paper, a Stackelberg game based traffic assignment is applied to deal with the cooperation and competition relationship between UE and SO. While the traffic management requires the traffic users cooperatively to achieve SO traffic assignment, traffic users competitively make route choices based on UE traffic assignment. Thus, the cooperation and competition relationship between traffic management and traffic users can be modelled by the game theory model, which can balance the benefit between traffic managements and travellers. Comparing with the Nash game theory, the Stackelberg game theory with the leader and follower player can better model the relationship between traffic authorities and traffic users. Moreover, the Gradient Projection algorithm is introduced to improve the efficiency of the game based traffic assignment.

The integrated system is implemented in the Sydney road network, and the introduced gradient projection algorithm can improve convergence rates and computation time comparing with Frank-Wolfe algorithm and MSA method. The system performance results show that route choices given by the game based redistribution can compromise between traffic management and traffic users to avoid congested routes and reduce travel time in the road network.

References

1. Wardrop, J. G.: Some theoretical aspects of road traffic research. In ICE Proceedings: Engineering Divisions, volume 1, pages 325–362. Thomas Telford. Road paper (1952).
2. Haurie, A., and Marcotte, P.: On the relationship between nashcournot and wardrop equilibria. Networks, vol. 15 (3), pp. 295–308 (1985).
3. Harker, T.P.,: Multiple equilibrium behaviours on networks, Transportation Science, 22(1), pp. 39–46, 1988.
4. Vuren, T. V., Vliet, D. V. and Smith, M. J.: Combined equilibrium in a network with partial route guidance. In Traffic control methods. Proceedings of the fifth NG foundation conference, California. (1990).

5. Wie, B.-W.: A differential game model of nash equilibrium on a congested traffic network. Networks, vol. 23(6), pp. 557–565 (1993).
6. Wie, B.-W.: A differential game approach to the dynamic mixed behavior traffic network equilibrium problem. European Journal of Operational Research, 83(1), pp. 117–136 (1995).
7. Friez, T.L., Mookherjee, R., and Yao, T.,: Securitizing congestion: the congestion call option, Transportation Research, Part B: Methodological, 42(5), pp. 407–437, 2008.
8. Kumar, A. and Peeta, S.: Strategies to Enhance the Performance of Path-Based Static Traffic Assignment Algorithms, Computer-Aided Civil and Infrastructure Engineering, 29:5, pp. 330–341 (2014).
9. Korilis, A.Y., Lazar, A.A., and Orda, A.: Achieving network optima using Stackelberg routing strategies, IEEE/ACM Transactions on Networking, 5(1), pp. 161–173, 1997.
10. Altman, E., Basar, T and Srikant, R.: Nash equilibria for combined flow control and routing in networks: Asymptotic behavior for a large number of users. Automatic Control, IEEE Transactions on, 47(6), pp. 917–930 (2002).
11. La, R. J. and Anantharam, V.: Optimal routing control: Repeated game approach. Automatic Control, IEEE Transactions on, 47(3), pp. 437–450 (2002).
12. Sislak, D., Volf, P. and Pechoucek, M.: Agent-based cooperative decentralized airplane-collision avoidance. Intelligent Transportation Systems, IEEE, 12(1), pp. 36–46 (2011).
13. Tong, C. O. and Wong, S. C.: A predictive dynamic traffic assignment model in congested capacity-constrained road networks. Transportation Research Part B: Methodological, 34(8), pp. 625–644 (2000).
14. Altman, E., Basar, T, Jimenez, T and Shimkin, N.: Competitive routing in networks with polynomial costs. Automatic Control, IEEE Transactions on, 47(1), pp. 92–96 (2002).
15. Huang, H. J. and Lam, W. H.: Modeling and solving the dynamic user equilibrium route and departure time choice problem in network with queues. Transportation Research Part B: Methodological, 36(3), pp. 253–273 (2002).

Impact of ALife Simulation of Darwinian and Lamarckian Evolutionary Theories

Yuliya Betkher, Nuno Nabais and Vitor Santos

Abstract Until nowadays, the scientific community firmly rejected the Theory of Inheritance of Acquired Characteristics, a theory mostly associated with the name of Jean-Baptiste Lamarck (1774–1829). Though largely dismissed when applied to biological organisms, this theory found its place in a young discipline called Artificial Life. Based on the two models of Darwinian and Lamarckian evolutionary theories built using neural networks and genetic algorithms, this research presents a notion of the potential impact of implementation of Lamarckian knowledge inheritance across disciplines, including biology, computer science and philosophy. There is an evidence that Lamarckian organisms can have wide practical application across several different domains, therefore this type of research should be allowed and encouraged. However, even though Lamarckian evolutionary algorithm already holds major benefits for various disciplines and promises even more, its implementation in Artificial Life needs regulation to avoid malevolent use.

Keywords Artificial life • Philosophy • Genetic algorithms • Darwinism • Lamarckism

1 Introduction

The behavior of natural organisms in the real world is not fixed across their lifespan. Through interactions with the environment, they gain experience and develop a tendency to repeat the actions that bring pleasure or benefit, and to avoid those that lead to danger or pain.

Y. Betkher (✉) · N. Nabais · V. Santos
NOVA IMS Information Management School, Campus de Campolide, 1099-085 Lisbon, Portugal
e-mail: m2014005@novaims.unl.pt; hypercaine92@gmail.com

N. Nabais
e-mail: nunocastronabais@gmail.com

V. Santos
e-mail: vsantos@novaims.unl.pt

G. Leu et al. (eds.), *Intelligent and Evolutionary Systems*,
Proceedings in Adaptation, Learning and Optimization 8,
DOI 10.1007/978-3-319-49049-6_3

At the same time, organisms are not born in a blank state—they develop according to the information in their genes, which are inherited from the ancestors and selected through the struggle for existence [1].

However, only what is inborn in their own heritage can be transferred along with their genes. The acquired characters will not be encoded in the genes, and therefore will not be directly passed to the offspring—according to Darwinism, all the knowledge that biological organisms have gained should be developed over again by each new generation.

Here a simple question arises: what if living beings could get past this limitation? What if children could pick up where parents left off developing their expertise, health, coordination and reflexes, each generation building on the last to reach out for higher and higher goals?

This attractive, although hypothetic process is called Lamarckian inheritance, a long-discredited mechanism of evolution. Through learning, individuals would experience certain adaptive changes and acquire new traits that would be directly transmitted to their offspring. Although it was largely dismissed as a valid theory for natural systems, Lamarckian evolution found its place and proven effective within computer applications [2].

But science moves ahead, and in the nearest future, the implementation of Lamarckian evolution may turn into something bigger that just a cybernetic adventure. Even though our commonly accepted definition of life does not yet recognize any current simulations or applications as alive, it may not always stay that way. The opinions regarding this matter vary, but according to the strong ALife position, first introduced by Neumann in [3], life can be abstracted away from any particular medium.

In this research, we are going to evaluate the potential impact of the implementation of Lamarckian evolution, in particular, of the inheritance of skills and knowledge, given the possibility of creating life within computational environment. Instead of being just an engineering problem, it becomes a cross-disciplinary topic that creates numerous philosophical questions and implications.

In order to demonstrate that it is possible to implement Lamarckian evolution in a computational environment, we refer to previously created models and present our own, where a neural network is regarded as a learnable individual [4], and genetic algorithms [5] are applied to the population of such individuals based on mechanisms of natural evolutionary processes and genetics. Using a focus group of researchers from computer science, biology and philosophy, we have validated the model and evaluated its potential impact on different matters in our lives, including technology, ethics, life and society.

2 Background and Problem Identification

Throughout the history, scientists have studied evolution for the same reasons that they have learned any other discipline—the thirst for knowledge, the desire to understand the past and predict the future, and the necessity to organize our world.

Evolution, especially the understanding of how organisms evolve through natural selection, has always been an area of science with various practical applications [6].

But nowadays, with the emergence of artificial intelligence, computational neuroscience and transhumanism, evolutionary studies have acquired fundamentally different ethical and social significance that extends beyond simple curiosity. Already existing roadmap on whole brain emulation [7] attempts to achieve software intelligence by copying the function of biological nervous systems into software. This approach produces numerous ethical issues that should affect responsible policy for developing the field. Animal emulations have controversial moral status, and a principle of analogy is suggested for judging treatment of virtual animals. Various considerations of developing and utilizing human brain emulations are discussed [8].

Among the latest published books, taking inspiration from self-awareness in humans, the new notion of computational self-awareness as a fundamental concept for designing and operating computing systems has been introduced [9]. The basic ability of such self-aware computing systems is to gather information about their state and progress, learning and maintaining models containing knowledge that enables them to reason about their behavior. Self-aware computing systems will have the ability to utilize this knowledge to effectively and autonomously adapt and explain their behavior in dynamic environments.

Although the accuracy of predictions of future developments in AI and ALife is difficult to evaluate, according to Ray Kurzweil himself, 89 out of 108 predictions he made so far were entirely correct by the end of 2009. An additional 13 were what he calls "essentially correct" (meaning that they were likely to be realized within a few years of 2009), for a total of 102 out of 108. Another 3 are partially correct, 2 look like they are about 10 years off, and 1, which was tongue in cheek anyway, was just wrong [10].

While at present, whole brain emulation seems an unfeasibly ambitious challenge, the necessary computing power and various scanning methods are rapidly developing. Large-scale computational brain models are a very active research area, at present reaching the size of mammalian nervous systems [11–14]. Whole brain emulation can be considered the logical endpoint of current trends in computational neuroscience and systems biology [8].

The implementation of Lamarckian evolution in future ALife systems such as virtual lab animals has wide practical application across disciplines. We have examined the trends and role of such evolution in engineering, computational biology and ethics, and based on this analysis, came to several assumptions that are to be evaluated by the focus group.

2.1 Engineering

In a dynamic and unpredictable environment such as real world, it is very difficult to construct intelligent machines or computer programs that would perfectly manage to produce desirable results from the very beginning. Therefore, an approach based on adaptive computation or evolutionary computation, where programs adapt themselves towards given situations through generating and testing, gained its popularity and significance [1].

Simulations of evolution using evolutionary algorithms originate from the work of Barricelli in the 1960s, continued by Fraser, who published a series of papers on simulation of artificial selection [15]. As a result of the work of Rechenberg, who used evolution strategies in the 1960s and early 1970s to solve complex engineering problems, artificial evolution became a widely recognized optimization method [16]. Genetic algorithms in particular became well-known through the writing of Holland [5]. As academic interest grew, dramatic increases in the power of computers allowed practical applications, including the automatic evolution of software [17]. Evolutionary algorithms are now applied in solving multi-dimensional problems more efficiently than computer programs developed by human designers, and also to optimize the design of systems [18].

A research area called artificial life [19] is a typical example that analyzes mathematical aspects of the dynamics residing in life in a synthetic way and tries to apply principles of natural systems (ranging from swarms of cells to human societies) as models for possible novel methods of adaptive computation. In software-based artificial life, neural networks are often applied in modeling the brain of an agent. Although traditionally more of an artificial intelligence technique, neural nets can be used for simulating population dynamics of organisms with an ability to learn [20]. Genetic algorithms are applied to such populations based on evolutionary and genetic mechanisms.

Due to the biological background, earlier attempts of artificial life modeling have always focused on a Darwinian evolution, based on competition of artificial beings in a computational environment, where new artificial organisms would appear only as the result of combining morphology of parents [21]. Until nowadays, Darwin's evolutionary models have been widely used in different scientific fields. Many of such implementations were motivated by the idea of constructing practical devices that have some of the useful features of living systems, such as robustness, flexibility, and autonomy [22].

At the same time, from the engineering point of view, it is not necessary to consider only Darwinian models. The possibility of heredity of acquired characteristics can be quite useful, and several studies have already shown the significant increase in performance of problem-solving systems using Lamarckian scheme [23, 24].

In evolutionary algorithms, the implementation of Lamarckian inheritance means that an individual can modify its genetic code during or after fitness evaluation, or lifetime. This idea has been used in several studies with particular success in problems where the application of a local search operator obtains a substantial improvement, e.g. traveling salesman problem [2]. The effectiveness and superiority of Lamarckian evolutionary algorithm has also been demonstrated for fixed tasks in stationary environments, even though Darwinian population adapts better to dynamic environments [1].

2.2 *Computational Biology*

Relatively recent research in cell biology has shown that the internal chemistry of living cells is a form of computation [25]. Such ideas are currently breaking boundaries between scientific disciplines and give rise to interdisciplinary sciences like computational biology, which involves the development and application of data-analytical and theoretical methods, mathematical modeling and computational simulation techniques to the study of biological, behavioral, and social systems [19].

A wetware computer is an organic computer (also known as an artificial organic brain or a neurocomputer) built from living neurons. Professor Ditto, at the Georgia Institute of Technology, is the primary researcher driving the creation of these artificially constructed, but still organic brains. One prototype is constructed from leech neurons, and is capable of performing simple arithmetic operations. The concepts are still being researched and prototyped, but in the near future, it is expected that artificially constructed organic brains, even though they are still considerably simpler in design than animal brains, should be capable of simple pattern recognition tasks such as handwriting recognition [26].

At the same time, while originally dismissed as non-feasible, Lamarckian evolution now appears more and more in biological systems ranging from microbes to mammals, and molecular mechanisms that might realize this mode of inheritance are being clarified. Epigenetics, a set of means to propagate a phenotypic change across generations, appears to provide a set of feasible molecular means that may realize Lamarckism. In addition, several mechanisms exist which may allow the phenotype to instruct the genotype at a given environment. Recent advances in molecular evolution have been surveyed and realistic means have been presented to engineer Lamarckian organisms in the lab which might possess improved evolvability [27].

2.3 *Philosophy and Ethics*

Computational biology gives one the sense that we are at the threshold of yet another of civilization's "Spinoza moments" where the entire framework for thinking about life is dramatically, and irrevocably restructured. The idea that cellular membranes and contents may be functional equivalents of computers does not appear strange and implausible any longer. And even if the implementation of strong ALife is a matter of future, considering potential risks and their ethical impacts is an important aspect of research ethics, even when dealing with merely possible future radical technologies [8].

Evolutionary studies have provided us better understanding of ourselves and helped us find our own place on Earth with 1.8 million identified species, and possibly 10 million total species. The context of evolution gives an insight on how to behave among members of our own and other species. Evolution helps us understand the purpose and reasons for our physiology and anatomy [28].

Since the Darwinian theory of evolution gained widespread acceptance in the late 1800 s, scientists and philosophers have been looking for ways to relate traditional evolutionary theory to the way we live, interact with society, and think about our place in existence. Now the Lamarckian evolution within artificial life has become a relatively recent object worthy of philosophical attention [29]. Therefore, the new questions of particular interest in evolutionary philosophy are how much of an influence Lamarckian evolution in ALife would have on human behavior, and what are the philosophical implications of this evolution on issues that relate to ethics and morality.

2.4 Research Objectives

The main goal of this research is to identify the potential impact of implementation of knowledge inheritance in artificial organisms using Lamarckian scheme. The objectives being pursued in order to achieve this goal are the following:

1. Build two artificial life models of Darwinian and Lamarckian knowledge inheritance processes using genetic algorithms and artificial neural networks. Using this example, prove that such implementation is possible in computational environment.
2. Evaluate possible impact of knowledge inheritance in artificial organisms on life and society, considering latest trends across disciplines.
3. Understand the philosophical, social and ethical implications of Lamarckian evolution in ALife.

3 Research

The inception on species capable of Lamarckian learning does have the potential to do great harm, as well as good. From the philosophical point of view, this research may be seen as playing God and even distorting the essence of life, instead of allowing life to emerge through natural processes and perhaps by nature's will. From a more practical point of view, some irreversible horrors may come creeping out of the flask on the laboratory bench, once a new Lamarckian specimen is introduced to the natural environment. These issues should give pause even to those who normally embrace advances in science with enthusiasm.

3.1 Fundamentals

In 2010, two American biologists Craig Venter and Hamilton Smith have made a bacterium that has an artificial genome—creating a living creature with no ancestor

[30]. According to Craig Venter himself, this cell has not yet found any practical applications, but it enables a change in philosophy, it is a proof of concept. But the proof of concept that we can potentially create and modify living creatures the way we want was key, otherwise it is just speculation and science fiction.

At the same time, there have been several successful attempts to model Lamarckian evolution in computer science and engineering [1, 31], e.g. in automated docking [31]. This model of evolution has been used for boosting search in particular kind of applications, however, the cost associated with the evaluation of the objective function with the use of Lamarckian evolution was an issue to consider. Such models have a broad range of applications over several different domains, e.g. optimization in engineering.

If the two mentioned approaches were combined, and there were indeed purely artificial organisms that would learn through interactions with the environment, numerous ethical and philosophical implications would arise. While some may regard the creation of purely artificial organisms as a defining moment in the history of biology, others may claim that the risks could outweigh the benefits.

3.2 *Proposal (Model)*

This model is not attempting to explain how a brain of a pure Lamarckian being would work. It aims to demonstrate a possibility of creating a robot, whose learning would have a Lamarckian component, since at least some of its knowledge would be passed to the offspring. Instead of presenting an automaton with single-layer neural networks like in AntFarm [32], we will train a multilayer neural network using backpropagation (of errors) or the generalized delta rule, since a multilayer net can lean any continuous mapping to an arbitrary accuracy [33]. Training a network will include the feedforward of the input training pattern, the backpropagation of the associated error, and the adjustment of the weights.

Consider thc following nelwork, in which we can formulate both feedforward propagation and backpropagation as a series of matrix multiplies. From now on, we are going to index matrices as A(i), where A refers to the type of matrix and (i) is an index of the position of the matrix in the network (we can also have (i $\rightarrow$ j) for a weight matrix connected layer i to layer j). The only exceptions are the input data matrix X and the output of the network Y. We denote the value of an element in row i and column j of some matrix A(k) with Aij(k) Dolhansky [34].

The defined automata can be made of several areas, part of them would corresponding to a neural net [35] and making up a neurological system: vision, hearing, touch, and internal sensing; the rest are physical characteristics. In AntFarm [32], agents of the same colony have identical genetic codes. This is not what we are looking for. The automata in this model have some, although minor differences in their connections between neurons, which better represents such in real biological systems.

The neurological system of automata consists of organs (vision, hearing, touch, internal sensing), through which it receives inputs (like synapses) in the neurons nm about the environment. Based on the weights wp, q, which stand for knowledge, a mathematical function would determine the activation of the neuron. Another function (which may be identical) computes the output of the artificial neuron and lead to an action of automata, like movement, eating, reproducing, or breathing.

Changing weights in the model represents learning. Automaton will learn throughout its life, which means all its weights wi, j will change in order to let it make better decisions (as a result of its learning experience). Cwi, j is a learning matrix with weights that change with time. The backpropagation algorithm will be used to compute the necessary corrections. The algorithm can be decomposed in the following four steps [35]: feed-forward computation, backpropagation to the output layer, backpropagation to the hidden layer, and weight updates.

During feedforward, each input unit Xi receives an input signal and broadcasts this signal to the each of the hidden units Z1, Zp. Each hidden unit then computes its activation and sends its signal zj to output units. Output units Yk compute their activation yk to form the response of the net for the given input pattern [33]. Note that the network will not be fully connected, just like our human brain.

The environment can be represented as a bi-dimensional plane 500×1000, where the automata can move right, left, up and down. The letter A stands for the automata, while O stands for food and Z stands for predators. In the Fig. 1, there is an example of what neural networks of parent A1 could be. The inputs from the environment reach input neurons on the left, get to a hidden layer, and based on knowledge hidden in weights, output neurons trigger actions on the right.

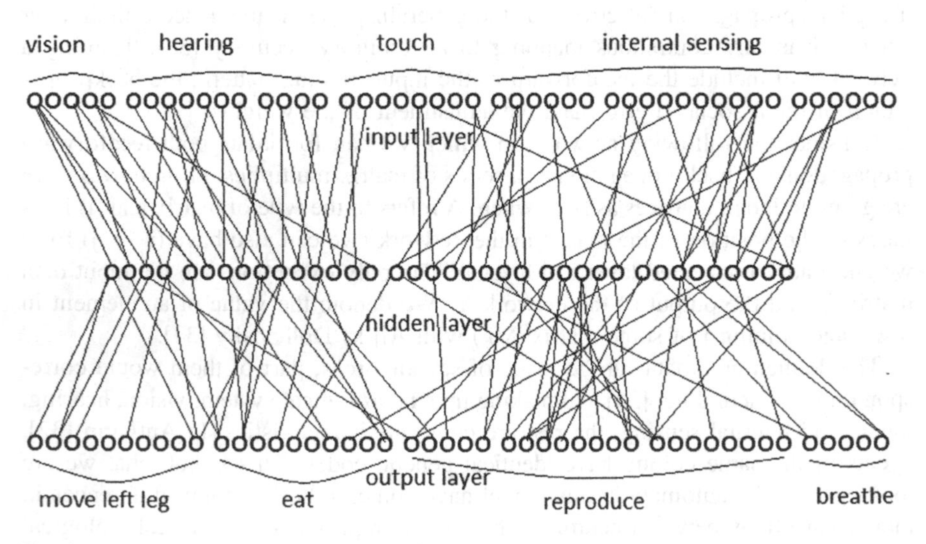

Fig. 1 Neural network of parent A1

Parents							
A1	n1 n14 ∅	n2 n18 ∅	n5 n14 ∅	n6 n15 ∅	n8 n18 ∅	n10 n23 ∅	n13 n21 ∅
A2	n2 n18 ∅	n4 n14 ∅	n5 n14 ∅	n6 n15 ∅	n7 n17 ∅	n9 n20 ∅	n13 n21 ∅
			point of crossover				
Children							
A3	n1 n14 ∅	n2 n18 ∅	n5 n14 ∅	n6 n15 ∅	n7 n17 ∅	n9 n20 ∅	n13 n21 ∅
A4	n2 n18 ∅	n4 n14 ∅	n5 n14 ∅	n6 n15 ∅	n8 n18 ∅	n10 n23 ∅	n13 n21 ∅

Fig. 2 Darwinian crossover (piece)

The neural network of the second parent would be very similar to such of the first parent. We assume that the amount of the neurons is roughly the same, the difference is in the location of the connections between them, and it should not be very big.

According to Darwinist approach, automaton is born with heuristic values that are inherited through a crossover, however, it has almost no knowledge at all (see Fig. 2). The weights of its neural network are chosen pseudo-randomly.

This means that instead of a crossover, the initial weights would be replaced by some heuristic (inborn) values appropriate just for automaton's survival, e.g. breathing or eating. In the Figs. 2 and 3, there will be only pieces of the whole network in such form (in reality, there are many more neurons, this is the demonstration of the concept).

In a Lamarckian neural network, however, weights will be inherited as well. They can be transferred to children directly (as in Fig. 3) or through any mathematical function, e.g. average of corresponding weights.

Parents							
A1	n1 n14 $w_{1,14}$	n2 n18 $w_{2,18}$	n5 n14 $w_{5,14}$	n6 n15 $w_{6,15}$	n8 n18 $w_{8,18}$	n10 n23 $w_{10,23}$	n13 n21 $w_{13,21}$
A2	n2 n18 $w_{2,18}$	n4 n14 $w_{4,14}$	n5 n14 $w_{5,14}$	n6 n15 $w_{6,15}$	n7 n17 $w_{7,17}$	n9 n20 $w_{9,20}$	n13 n21 $w_{13,21}$
			point of crossover				
Children							
A3	n1 n14 $w_{1,14}$	n2 n18 $w_{2,18}$	n5 n14 $w_{5,14}$	n6 n15 $w_{6,15}$	n7 n17 $w_{7,17}$	n9 n20 $w_{9,20}$	n13 n21 $w_{13,21}$
A4	n2 n18 $w_{2,18}$	n4 n14 $w_{4,14}$	n5 n14 $w_{5,14}$	n6 n15 $w_{6,15}$	n8 n18 $w_{8,18}$	n10 n23 $w_{10,23}$	n13 n21 $w_{13,21}$

Fig. 3 Lamarckian crossover (piece)

3.3 Argument

In our culture, there is a widespread image of a superior artificial mind conquering the planet. In reality, Darwinian or Lamarckian, an artificial species will more likely become another species' lunch. The risks about creating artificial life are exaggerated.

When the news about Craig Venter's achievement came out, people started to become worried about the dangers that ALife could bring. Such worries are of the same nature as popular beliefs that natural is good and artificial is bad. In fact, nothing could be further from the truth—malaria is very natural yet disastrous, while antibiotics are man-made, but very handy sometimes.

However, Lamarckian artificial life can be risky if released into natural environment, given the examples of already existing failed interventions in natural design, mostly because the dimensions of its danger are never known in advance. Therefore, even if the implementation of synthetic Lamarckian evolution is allowed, it should be regulated and licensed in order to avoid malevolent use.

For human beings, however, the possibility of inheriting knowledge would have its major drawbacks. Often our parents are not the people we want to inherit knowledge from, especially in a form of a random combination of their skills instead of a catalog. Moreover, it is nice to inherit knowledge from a person like Einstein, but all the pains, fears and mental issues of our parents would be inherited too, which would have a major negative impact on our lives. In addition, our brains do not work like hard drives that store information the way it is over the years. Our memories get distorted over time, we forget some details and come up with new ones. Passing such flawed knowledge through generations does not seem like a good idea.

While Lamarckian evolution has wide application in ALife and can be surely used in experiments in artificial environment, for human beings, Darwinian evolution seems to make more sense. At the end of the day, we are all living in dynamic environments where inheriting knowledge is often useless—we keep re-learning again and again, and the ability to unlearn old skills and study everything anew seems to be the new literacy.

4 Conclusions

There is an evidence that Lamarckian organisms can have wide practical application across several different domains, therefore this type of research should be allowed and encouraged. But even though Lamarckian evolutionary algorithm already holds major benefits for humanity and promises even more, this implementation needs regulation. For now, potential benefits seem to outweigh risks, however, the risks are unknown and the required investment might be an issue.

Firstly, it is absolutely necessary to prevent malevolent use of the research. It can be abused in numerous ways, e.g. applied in the creation of biological weapon and bioterrorism. Secondly, since the project will require a lot of investment, the research may become driven by profits for harmful purposes rather than benefits to humanity as a whole. Thirdly, safety measures should be taken before releasing the Lamarckian species into natural environment. This will be needed not only for the sake of safety of natural biodiversity, but also to help the Lamarckian organisms survive outside the lab.

Nowadays, even for a non-scientist it seems quite easy to distinguish a living organism from a non-living, except probably for viruses, whose status is still questionable. The agreed on definition of life, however, does not exist, and the inception of artificial species, in this case, artificial Lamarckian species makes it even murkier. It is still unclear whether this implementation will defeat the divinity of life or concept of the soul and prove that there is no magic spirit of vitality. Until now, man did not manage to create life from scratch, only to manipulate it, so the question remains open.

The inception of Lamarckian organisms may eliminate some existing philosophical concepts, such as free will, because inherited knowledge contradicts its definition, however, what matters is the ability to choose which stays put. In this sense, the definition of free will in philosophy might be changed or expanded, considering the possibilities that are arising in artificial life.

This type of research is still rather related to computer science and does not prove or disprove the theory of creationism. Therefore, by conducting this kind of studies, we may not embrace the concept of God or understand the origins of our species. Also we may not understand the relationship between us and the creator any better, as well as find out whether the creator has ever existed at all.

References

1. Sasaki, T. and Tokoro, M.: Comparison between Lamarckian and Darwinian Evolution on a Model Using Neural Networks and GAs. Knowledge and Information Systems (KAIS): An International Journal, Springer. (2000).
2. Ross, B. J.: A Lamarckian Evolution Strategy for Genetic Algorithms. In Lance D. Chambers, editor, Practical Handbook of Genetic Algorithms: Complex Coding Systems, Boca Raton, FL: CRC Press, volume III, pp. 1–16. (1999).
3. Neumann, J. v.: The General and Logical Theory of Automata. Collected Works Volume V: Design of Computers, Theory of Automata and Numerical Analysis. Ed. A.H. Taub. New York: Pergamon, p. 288–326. (1963).
4. Rumelhart, D. and McClelland, J.: Parallel Distributed Processing. MIT Press, Cambridge, Mass. (1986).
5. Holland, J. H.: Adaptation in Natural and Artificial Systems. Ann Arbor MI University of Michigan Press (Vol. Ann Arbor). (1975).
6. Bull, J. J. and Wichman, H. A.: Applied Evolution. Annual Review of Ecology, Evolution, and Systematics, 32, 183–217. (2001).

7. Sandberg, A. and Bostrom, N.: Whole Brain Emulation: A Roadmap Technical Report. Future of Humanity Institute, Oxford University. (2008).
8. Sandberg, A.: Ethics of brain emulations. Journal of Experimental & Theoretical Artificial Intelligence, (May), 1–19. (2014).
9. Lewis, P.R., Platzner, M., Rinner, B., Torresen, J., and Yao, X.: Self-aware Computing Systems. An Engineering Approach. Springer, Natural Computing Series. (2016).
10. Wang, B.: Ray Kurzweil Clarifies his Vision of Reverse Engineering the Brain and Developing Artificial Intelligence from Principles Gleaned from Brain Science, Next Big Future: Coverage of Disruptive Science and Technology (2010). http://www.nextbigfuture.com/2010/08/ray-kurzweil-clarifies-his-vision-of.html?m=1.
11. Djurfeldt, M., Lundqvist, M., Johansson, C., Rehn, M., Ekeberg, O., and Lansner, A.: Brain-scale simulation of the neocortex on the IBM Blue Gene/L supercomputer. IBM Journal of Research and Development, 52, 31–41. (2008).
12. Eliasmith, C., Stewart, T. C., Choo, X., Bekolay, T., DeWolf, T., Tang, Y., and Rasmussen, D.: A large-scale model of the functioning brain. Science, 338, 1202–1205. (2012).
13. Markram, H.: The blue brain project. Nature Reviews Neuroscience, 7, 153–160. (2006).
14. Preissl, R., Wong, T. M., Datta, P., Flickner, M. D., Singh, R., Esser, S. K., Modha, D. S.: Compass: A scalable simulator for an architecture for cognitive computing. Proceedings of Supercomputing 2012, Salt Lake City, November 10 –16, 2012. (2012).
15. Fraser, A. S.: Monte Carlo analyses of genetic models. Nature 181 (4603): 9–208. (1958).
16. Rechenberg, I.: Evolutionsstrategie – Optimierung technischer Systeme nach Prinzipien der biologischen Evolution (PhD thesis) (in German). Fromman-Holzboog. (1973).
17. Koza, J. R.: Genetic Programming. MIT Press. (1992).
18. Jamshidi, M.: Tools for intelligent control: fuzzy controllers, neural networks and genetic algorithms. Philosophical Transactions of the Royal Society A 361 (1809): 1781–808. (2003).
19. Huerta, M., Haseltine, F., Liu, Y., Downing, G., and Seto, B.: NIH Working Definition of Bioinformatics and Computational Biology. BISTIC Definition Committee. (2000).
20. Kumar, D. and Bhatnagar, R.: An approach implements artificial intelligence into human life with new technologies and application. International Journal on Emerging Technologies 1(2). (2010).
21. Hinton, G. E. and Nowlan, S. J.: How Learning Can Guide Evolution. Complex Systems, 1, 495–502. (1987).
22. Bedau, M. A.: Leonardo. The Scientific and Philosophical Scope of Artificial Life. MIT Press Vol. 35, No. 4, p. 395–400. (2002).
23. Grefenstette, J. J., Ramsey, C. L., and Schultz, A. C.: Learning sequential decision rules using simulation models and competition. Machine Learning, 5(4), 355–381. (1990).
24. Davidor, Y.: Genetic Algorithms and Robotics. World Scientific Series in Robotics and Intelligent Systems: Volume 1. (Weizmann Inst. Sci., Israel). (1991).
25. Bray, D.: Wetware: A Computer in Every Living Cell. Yale University Press: New Haven, Connecticut. (2011).
26. Borresen, J. and Lynch, S.: Neuronal computers. Nonlinear Analysis, Theory, Methods and Applications, 71(12). (2009).
27. Pilpel, Y.: Realizing Lamarckian Evolution. Center for Bits and Atoms. MIT Media Lab, E14–633. (2016).
28. Moritz, C.: Biology 1B—Evolution Lecture 1. Introduction to Evolution. UC Berkeley. (2010).
29. Stewart, R. C.: The Journal of Evolutionary Philosophy. The Academy of Evolutionary Metaphysics. (2005).
30. Gibson, D. G., Glass, J. I., Lartigue, C., Noskov, V. N., Chuang, R.-Y., Algire, M. a, … Venter, J. C.: Creation of a bacterial cell controlled by a chemically synthesized genome. Science (New York, N.Y.), 329(5987), 52–56. (2010).
31. Morris, G. M., Goodsell, D. S., Halliday, R. S., Huey, R., Hart, W. E., Belew, R. K., and Olson, A. J.: Automated Docking Using a Lamarckian Genetic Algorithm and an Empirical Binding Free Energy Function. Journal of Computational Chemistry, 19, 1639–1662. (1998).

32. Collins, R. J. and Jefferson, D.: Antfarm: Towards Simulated Evolution. In Artificial Life II (pp. 579–601). Retrieved from http://citeseer.ist.psu.edu/collins91antfarm.html. (1992).
33. Fausett, L.: Fundamentals of Neural Networks: Architectures, Algorithms, and Applications. Prentice-Hall. (1994).
34. Dolhansky, B.: Artificial Neural Networks: Matrix Form (Part 5). ML Primers, Neural Networks. Retrieved from http://briandolhansky.com/blog/2014/10/30/artificial-neural-networks-matrix-form-part-5. (2014).
35. Domeniconi, C.: Proposal of a Darwin-Neural Network for a Robot Implementation. Perspectives in Neural Computing, Taylor Ed., p. 186–193. (1996).

A Local Search Algorithm for Saving Energy Cost in Duty-Cycle Wireless Sensor Network

Huynh Thi Thanh Binh, Vo Khanh Trung, Ngo Hong Son, Eryk Dutkiewicz and Diep N. Nguyen

Abstract Wireless Sensor Networks (WSNs) have been recently used for various applications. Due to the distributed and (often) unattended nature of the nodes after deployment, the lack of energy and the interruptive process in each sensor are the two major problems of WSN systems. Hence, designing a protocol which not only improves system performance but also lowers sensors' energy consumption so as to maximize the network lifetime is very much desirable. The network lifetime maximization problem was known to be NP-Hard. This paper addresses the Minimum Energy-Multicasting (MEM) problem in Duty-Cycle Wireless Sensor Networks (DC-WSNs) in which sensors cyclically switch between on/off (wake/sleep) modes. To that end, we propose a local search algorithm and compare its performance with the best algorithm so far called GS-MEM over the four datasets designated for the MEM problem. The experimental results show that our proposed algorithm significantly outperforms GS-MEM in terms of energy cost.

H.T.T. Binh (✉) · V.K. Trung · N.H. Son
School of Information and Communication Technology,
Hanoi University of Science and Technology, Hanoi, Vietnam
e-mail: binhht@soict.hust.edu.vn

V.K. Trung
e-mail: trungvokhanh@gmail.com

N.H. Son
e-mail: sonnh@soict.hust.edu.vn

E. Dutkiewicz · D.N. Nguyen
School of Computing and Communications, University of Technology Sydney,
Ultimo, Australia
e-mail: eryk.dutkiewicz@uts.edu.au

D.N. Nguyen
e-mail: diep.nguyen@uts.edu.au

G. Leu et al. (eds.), *Intelligent and Evolutionary Systems*,
Proceedings in Adaptation, Learning and Optimization 8,
DOI 10.1007/978-3-319-49049-6_4

1 Introduction

Recent advances in micro-electro-mechanical (MEMs) technologies have made sensor motes affordable for most applications. That allows popular deployment of wireless sensor networks (WSNs) in a wide range of applications (both civil and military), e.g., environment monitoring, seismic analysis, surveillance, internet of things (IoT). So far, the ability to widen the communication range and increase the performance's flexibility allows Wireless Sensor Networks (WSNs) deployment in complicated terrain or hostile environment. For such applications where battery replacement/replenishing is impossible or prohibitively expensive, it is critical to design the network in an energy-efficient manner.

There are usually two major components in WSNs, a base station (or data gateway) and a set of sensor nodes that capture environmental parameters such as temperature, humidity and pressure. Each wireless sensor node consists of two separate parts: a data processor and a radio transceiver. After being deployed over a pre-defined area, sensor nodes continuously monitor the environment and send sensing data either directly or through relay nodes to the base station. Information collected from sensors helps detect and analyze environmental changing models. Therefore, WSNs widely appear in many aspects of life such as environment, health, military, industry, agriculture, etc. However, sensor nodes have some limitations. The sensors' energy source is too low to maintain network and the reparation for damaged sensors is difficult to execute. For example, in a battlefield, military WSN systems play an important role in transporting information to the command center or individual soldiers. If some sensor nodes run out of battery, they will be unable to collect or send data; thus, the enemy observing process will be interrupted. For this reason, saving sensor's energy is needed.

To save energy, a wireless sensor is designated not to operate continuously but to switch its transceiver between on/off (or working/idle) modes during an operating cycle. While being in the working state, the transceiver is activated such that collected data is transmitted to other nodes and the sensor could also receive data packets from others. Meanwhile, during the idle state the transceiver is turned off and that sensor can't communicate with others. By doing so, a great amount of energy is saved and the network lifetime can be prolonged. This type of WSN is often referred to as duty-cycle wireless sensor networks (DC-WSN). This paper focuses on saving energy cost in DC-WSNs.

The rest of this paper is organized as follows. Related works and the problem formulation are presented in Sects. 2 and 3, respectively. Our proposed algorithms are in Sect. 4. Section 5 presents experiments along with computational and comparative results. Conclusion and future works are in Sect. 6.

2 Related Works

In WSNs, multicasting energy refers to the total energy consumption including both transmitting and receiving energy from a source node to all terminal ones on a multicast tree. Reducing this energy cost is one of the main objectives when operating WSN systems and is often referred to as the Minimum Energy-Multicasting (MEM) problem. This problem has been intensively studied and tackled by using approximate algorithms [1–4]. However, these methods are only applicable to WSNs consisting of standardized static sensors whose sensing function is performed continuously throughout the system operation, which causes a significant loss of energy. Later, the MEM problem is investigated in DC-WSNs, where duty-cycle sensors are turned on/off only at a specific time slot [5, 6]. Su et al. [6] suggested two optimization algorithms named oCast and BD-oCast to resolve the MEM problem in a narrowed DC-WSNs model that contains nodes possessing consecutive working time slots. However, the two algorithms require exponential time w.r.t. the number of destination nodes. For this reason, oCast and BD-oCast can only solve the MEM problem with a small number of nodes. Han et al. [5] proposed four algorithms: Shortest Path Tree (SPT), Approximate Minimum Steiner Tree (AMST), Minimal data overhead Tree (MNT) and an approximation algorithm for MEMTCS denoted TCS in order to solve MEM problem on a general DC-WSN model. On one hand, the first three algorithms are all executed in two steps as follows. The first step is to construct a multicast tree relying on the structure of WSNs' graph. The second step establishes the transporting schedule for each sensor node in the multicast tree found in the first step. Because two steps are executed separately, the performance of the first three proposed algorithms is quite modest. On the other hand, TCS algorithm builds up an expanded graph based on the initial network graph and the working schedule. The minimum Steiner tree on the graph is then found and the model is mapped to the tree in order to obtain the final solution. The performance of the TCS algorithm is relatively good but heavily depends on the quality of the Steiner tree. Huynh et al. [7] introduced the Ratio-function-based Tree Construction and Scheduling (R-TCS) and another algorithm called (Group Steiner's Tree applied in MEM problem); GS-MEM algorithms based on the expanded graph model. R-TCS is a new version of TCS algorithm. Better quality multicast tree is produced when utilizing this method to solve the MEM problem. Furthermore, GS-MEM turns the MEM problem into a Group Steiner Tree problem, which can be effectively solved using heuristic algorithms [2–4, 8, 9]. Nonetheless, two algorithms R-TCS and GS-MEM may require significant computational time when a large number of sensor nodes involves.

This paper proposes a local search algorithm for the MEM problem in DC-WSN that aims to conserve the energy cost. Our experiments on four datasets [5] showed that the proposed algorithm is more energy efficient than R-TCS and GS-MEM with shorter execution time compared to that of GS-MEM.

3 Problem Formulation

3.1 Duty-Cycle Wireless Sensor Networks Model

A WSN is represented by a single un-weighted scalar graph $G = (V, E)$. Nodes in V position on a 2-dimensional space coordinates plane. If two random nodes stay in their communication area, a link between them is created. Each node owns the same initial energy. In DC-WSN, all nodes are operated cyclically and each cycle is divided into K equal periods. In order to save energy cost, nodes are only activated in a period belonging to a set $\Gamma(u) \subset \{1, 2, 3, \ldots, K\}$ where $\Gamma(u) \neq \emptyset, \forall u \in V$. In other words, every node u can be awakened to transfer the information at any period but they only receive data at periods covered by the set $\Gamma(u)$.

3.2 Multicast on Duty-Cycle Wireless Sensor Network

Considering a set of terminal nodes $M \subset V$; in a multicast implementation process, data packets are transported from a source node $s \in M$ to other terminal nodes. With T being a random child tree of the graph G, we denote:

- $V(T)$ and $E(T)$ as the sets of vertices and edges, respectively.
- $nl(T)$ as a set of internal nodes of the tree T.
- $child(u, T)$ as a set of nodes which are children of a node u on T.

The tree T is considered a multicast tree of G if it is a child tree rooted at s and at the same time, all terminal nodes in M belong to T. Figure 1 describes a DC-WSN graph that contains a set of terminal nodes {1, 5, 6, 7} including the root node $s = 1$, the time slices set $\Gamma(u)$ written next to each node and a multicast tree shown with bold edges in the picture.

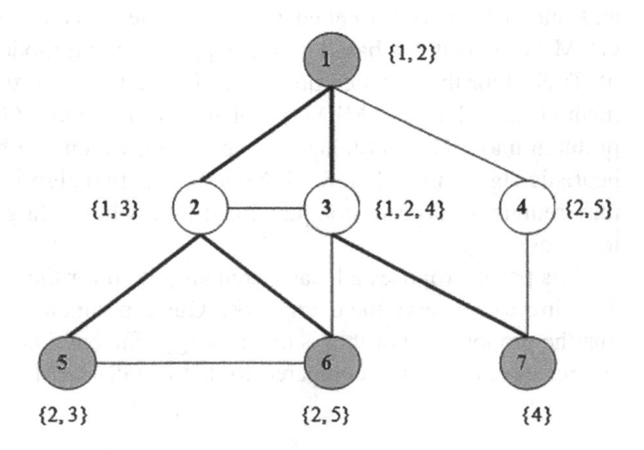

Fig. 1 A multicast tree represents a DC-WSN

Definition 1 (*Hitting set* [5, 10]) Let $C = \{A_1, A_2, \ldots, A_n\}$

A set *F* is called the hitting set of *C* if *F* holds at least one member in each subset of *C*. The hitting set having the smallest number of members is called the *minimum hitting set* of *C* and is denoted as MHS(*C*).

Definition 2 (*Possible transport schedule* [5]): A function *B*: *nl*(*T*) $\rightarrow 2^{\{1,2,\ldots K\}}$ (*T* is a multicast tree) is called a possible transport schedule of *T* if $\forall u \in nl(T)$. In this case, *B*(*u*) is the hitting set of $\{\Gamma(v)|v \in child(u,T)\}$.

Let e_s and e_r be the transmitting and receiving energy of each data packet in every node in *V* respectively (e_s, $e_r > 0$). The energy consumption in a multicast session includes two parts: transmitting part and receiving part. According to Definition 2, the total sending energy consumption in a multicast session following the transport schedule *B* on a multicast tree *T* equals $\sum_{u \in nl(T)} |B(u)| \cdot e_s$. Besides every T'_s nodes except *s* being able to receive data packets, their transmitting energy in a multicast session depends only on the number of nodes and equals $|V(T) - 1| \cdot e_r$.

Given a DC-WSN formulated as a graph $G = (V, E)$, a set $M \subset V$ consists of terminal nodes and a base station *s*. In each multicast session, the total energy consumed is calculated as follows.

$$\Pi(T_{opt}, B_{opt}) = \sum_{u \in nl(T_{opt})} |B_{opt}(u)| \cdot e_s + (|V(T_{opt})| - 1) \cdot e_r \quad (1)$$

The MEM problem in DC-WSN now boils down to finding the multicast tree T_{opt} of *G* and a schedule B_{opt} with respect to the tree T_{opt} such that $\Pi(T_{opt}, B_{opt})$ is minimized. For example, taking the multicast tree *T* shown in Fig. 1 in form of bold lines, a possible schedule B_1 are $B_1(1) = \{2, 3\}$, $B_1(2) = \{3, 5\}$, $B_1(3) = \{4\}$. In case this schedule is applied with $e_s = 100$ and $e_r = 15$, the total energy consumed within one multicast session would be $5 \times 100 + 5 \times 15 = 575$. However, if the schedule B_2: $B_2(1) = \{1\}$, $B_2(2) = \{2\}$, $B_2(3) = \{4\}$ is applied on the same multicast tree, it would take up a total energy of only $3 \times 100 + 5 \times 15 = 375$. Hence, the multicast tree *T* and the schedule B_2 lead to a more energy-efficient solution for transporting from one terminal node to others in one multicast session compared with that of the combination of *T* and B_1.

4 Proposed Algorithm

The main idea of the algorithm is to create a multicast tree and find the best transport schedule. In each step, the best transport schedule of the neighbor multicast tree is compared with that of the chosen tree. If the neighbor multicast tree holds the schedule that consumes less energy than that of the current tree, we will set the neighbor tree as the new chosen tree.

4.1 Searching for the Best Transport Schedule of Multicast Tree

For a $G = (V, E)$ graph, a multicast tree $T \subset G$ and a best transport schedule B: $nl(T) \rightarrow 2^{\{1,2,\ldots,K\}}$ on T, we aim to find B to minimize the total energy consumed in each multicast session.

To find B, one needs to find a minimum hitting set belonged to the timeslot set of all child nodes in each tree's node (MHS). The MHS problem is formulated as follows.

Considering a set C which includes $\{A_1, A_2, \ldots, A_n\}$ set, $F \subset A_1 \cup A_2 \cup \cdots \cup A_n$ is called the hitting set of C if F consists of at least one member of each set in C, meaning that $F \cap A_i \neq \varnothing \forall i = \overline{1, n}$. The *hitting set* having the smallest number of members is called the minimum hitting set of C, denoted by MSH(C).

The MHS problem is proven to be NP-Hard [1, 3, 11]. In this work, we propose a greedy algorithm to solve this problem.

- Initializing a set called $H = \varnothing$.
- Repeating the following steps:
  - Step 1: Chose $c*$ members so that the number of $A_i \in C$ satisfies that $c* \in A_i$ is the largest.
  - Step 2: Add $c*$ into the set H and remove all A_i set having $c*$ from the set C. While C is not empty, repeat this process continuously until no members is remained in C. After that, H will then be the minimum hitting set.

In the worst case, the proposed algorithm must visit all set in $|C|$. Therefore, the complexity is $O(K \cdot |C|)$, where K is the maximum number of elements in each set. Because there are V vertices in the multicast tree, the best transport schedule is found when the MHS problem is solved for $|V|$ times. Hence, the complexity is $O(K \cdot |V| \cdot |V|) = O(K \cdot |V|^2)$.

4.2 Finding the Base Solution

The solution can be depicted by the multicast tree T_0, a graph G and the best transport schedule B_0.

We set the weight of each edge in graph G to one then find the shortest path from the source node s to all the terminal nodes u ($u \in M$, M is the set of terminal nodes). After that, T_0 tree is constructed ($T_0 = (V_0, T_0)$) from all paths found. Each path is then separately evaluated. If a vertex v or an edge e appears on the path but they have not been put in V_0 or E_0, v and e are put into V_0 and E_0 respectively.

The breadth first search algorithm is applied for finding the shortest path from the source node s to others and its complexity is $O(|V| + |E|)$. Thus, the complexity of the algorithm equals $O(|V| + |E|)$.

4.3 Solution Initialization

One of the most crucial elements that strongly affects the quality of the local search algorithm is finding neighbor solutions of a given one. All neighbor solutions should diverse in order to maintain a large enough search space that potentially holds the optimal solution, but the quantity should be limited at a high enough amount so as not to affect the algorithm's speed.

Each solution of MEM DC-WSN problem is encoded as a multicast tree T and the best transport schedule B, denoted $S = (T, B)$. A solution S' being a neighbor of S is derived from the transition operation $f: S \longrightarrow S'$, satisfying $S' = (T', B')$ (T' is another multicast tree on G and B' is the best transport schedule of T').

Four transition operations $f: S \longrightarrow S'$ are proposed.

(1) *Swapping Two Nodes on the Tree*

Choose two random nodes u and v on the multicast tree T and swap them when the following conditions are satisfied:

- Either u or v is a source node.
- After swapping, the new multicast tree T' belongs to G.

After T' is found, we continue to find the best transport schedule B' on T' to obtain the solution $S' = (T', B')$. If $\Pi(T', B') < \Pi(T, B)$, S will be replace by a neighbor solution S'.

Therefore, this transition swaps two nodes and checks the existence of the edges. The number of edges is equal to the number of vertices near by the considered vertex. Each vertex is close to $|V| - 1$ vertices so the complexity of the first transition is $O(|V|)$.

Figure 2 illustrates a node swap transition between node 3 and node 5 on the tree. After the transition, node 3 could be omitted from the multicast tree despite being a leaf node because it is not a terminal node and the its rejection does not trigger additional time slot for data transmission. In other words, the new tree allows a full communication from all terminal nodes to the root regardless of node 3's presence. Deleting node 3 also leads to lower multicast energy. The original tree has the transport schedule $B(1) = \{1, 2\}$ and $B(3) = \{2, 4\}$ with the multicast energy equals $4E_s + 4E_r$. The new tree after transition has the transport schedule $B(1) = \{1, 3\}$ and $B(5) = \{2\}$ with a smaller multicast energy of $3E_s + 3E_r$.

(2) *Replacing a Node by Another Node that Has not Presented in the Current Tree.*

Consider a random node u on the multicast tree T and a node v which does not belong to T but to the graph G. The condition for u to replace v on T is:

- Node u is neither a source nor a terminal node.
- After u is replaced by v, the new tree T' satisfies $T' \subset G$

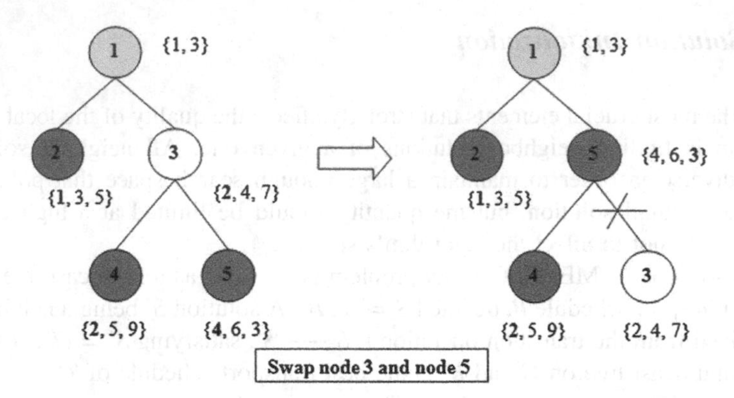

Fig. 2 Swapping two nodes

If two conditions above are satisfied, u could be replaced by v, constructing a new tree T'. Again, the mission is to find the best transport schedule B' and the neighbor solution $S' = (T', B')$. Then S' is set as the new solution if $\Pi(T', B') < \Pi(T, B)$.

The complexity of the second transition is the same as that of the first one. Considering the existence of $|V|$ vertices, the complexity of this transition is $O(|V|)$.

Figure 3 demonstrates a node replacement transition applied on node 3. To be specific, node 3 is replaced by node 6, which was not belonged to the multicast tree but existed in graph G. Similar to Fig. 2, the transport schedule of the original tree was $B(1) = \{1, 2\}$ and $B(3) = \{2, 4\}$ with the multicast energy equals $4E_s + 4E_r$. Meanwhile, the new tree after transition has the transport schedule $B(1) = \{1\}$ and $B(6) = \{2, 3\}$ with a smaller multicast energy of $3E_s + 4E_r$.

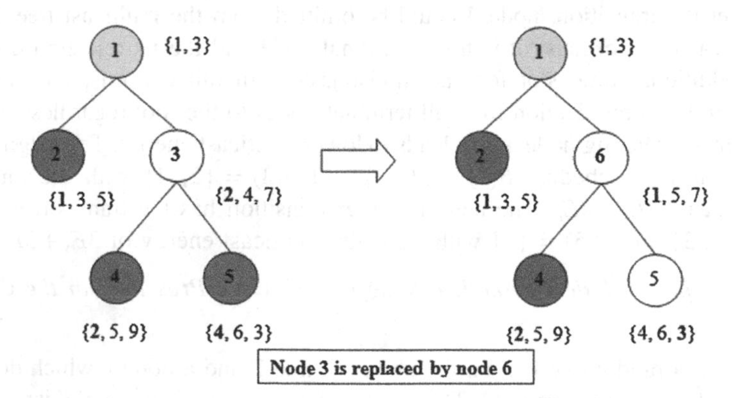

Fig. 3 Replacing a node

(3) *Removing Edges and Finding a New Connection*

This transition starts with visiting a terminal node u on the multicast tree T. The next step is finding the path P from u to an ancestor node which is the closest terminal node to u. On the path P there are two tasks to accomplish:

- Remove all vertices P (except its ancestor nodes)
- Remove all edges on P (except the edges coming from or to the ancestor u).

On graph G, the depth first search (DFS) algorithm is used to find a new path P' from u to a random node in T. T' is completed after adding vertices and edges into T. T' is regarded as a clue to find the new best transport schedule B' and neighbor solution $S' = (T', B')$. Similar to two transitions above, S' is set as the new solution if $\Pi(T', B') < \Pi(T, B)$.

The complexity of the third transition is equal to that of the DFS algorithm, which is $O(|V| + |E|)$.

Removing edges and finding a new connection, the third transition, is shown in Fig. 4. The edge connecting the source and node 6 is removed, thus a new connection between the terminal node number 2 and the source must be formed. In this case, node 2 is then linked directly to the source node. By this way, the multicast energy is significantly reduced. The original tree has the transport schedule $B(1) = \{1, 3\}$, $B(3) = \{3, 5\}$ and $B(6) = \{2\}$ with the multicast energy equals $5E_s + 5E_r$. After the transition, the transport schedule becomes $B(1) = \{3\}$ and $B(3) = \{2, 3\}$ with a much smaller multicast energy of $3E_s + 4E_r$.

(4) *Moving a sub-tree*

The fourth transition looks up nodes on a multicast tree T except the source node s. Let v be the parent node of u. The child node set of v is denoted *child* (T, v) $(u \in$ *child* $(T, v))$. The next step is to solve the *minimum hitting set* problem of $C = \{\Gamma(i) : i \in child(T, v)\}$ and $C' = C \backslash \{\Gamma(u)\}$. If $\text{MHS}(C) = \text{MHS}(C') + 1$ (meaning that, by removing child nodes of v, the minimum hitting set is reduced by

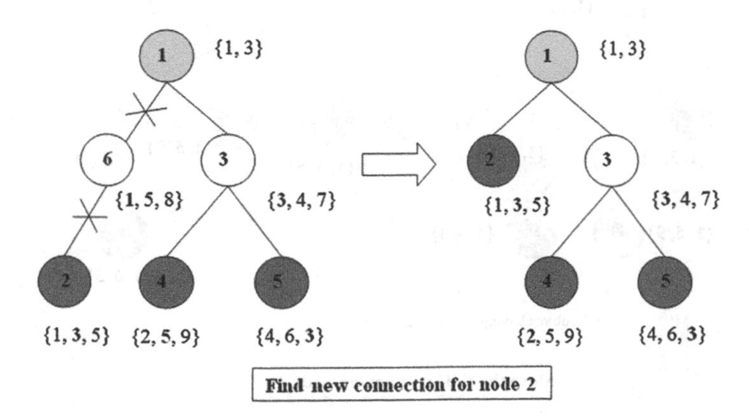

Fig. 4 Removing edges and finding a new connection

1, therefore the sending energy from node v to other nodes is reduced by e_s), the mission now is to attach the u is sub tree to v' on T so that the transmitting energy from v' to other nodes remains the same.

On the multicast tree T, implementing the information transfer process from source node s to leaf nodes which are not terminal ones is considered unnecessary. Consequently, after the transition, the multicast tree requires removing redundant nodes.

A new multicast tree T' is created when the transition finds u is sub tree and that sub tree is attached to a new position on the multicast tree T.

This fourth transition requires checking the maximum number of nodes u (up to $|V|$) and solving the *minimum hitting set* problem of each node. As a result, the complexity of the transition is $O(|V| \cdot K \cdot |V|) = O(K \cdot |V|^2)$, in which K is the working cycle of each node.

Figure 5 illustrates the fourth transition where the edge connecting node 4 and node 6 is removed and a new connection between the source node and node 4 is formed. It is worth noticing that the transition of node 4 also applies to its sub-tree as well, as shown in the figure. The transmission energy of node 6 is decreased by E_s because it can send data to node 5 in the fourth time slot. And since the multicast energy of node 1 remains unchanged, node 1 only need to send data in the fifth time slot and every terminal node receives it (including node 5 as being explained before).

Figures 2, 3, 4 and 5 illustrate four transitions proposed in this paper, in which red node is the source node and terminal nodes are colored blue.

Amongst four transition methods, the first and the second transitions can easily lead to unstable state where two nodes u and v keep swapping to each other continuously. In order to overcome the problem, TABU search is applied. Each pair of swapped nodes is added into a forbidden list, thus they are not allowed to be used within N repetition steps. After N steps, the swapped pair is removed from the list

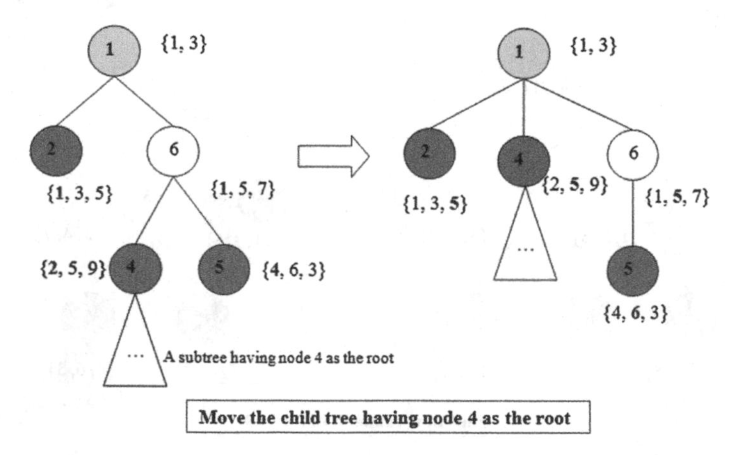

Fig. 5 Moving a sub-tree

and the transform process starts over. TABU search helps reduce a big amount of unnecessary transition steps and improves the algorithm's performance.

4.4 Complexity of Proposed Algorithm

The local search algorithm initializes the first solution, then the algorithm is repeated for L times, each execution operates all four transition functions to generate neighbor solutions and update the best one. The complexity of each step is given above, thus the total complexity of LSMEM algorithm is $O(|V| + |E| + L (|V| + |V| + |V| + |E| + K\ |V|^2)) = O(L\ \ K\, |V|^2)$.

5 Experimental Results

5.1 Problem Instances

The datasets used in this experiment are constructed in [5] and they have been applied in [7]. The sensor network graph includes N nodes randomly initialized in a 1000 m × 1000 m plane. Each node possesses a communication radius $R = 300$ m, transmitting energy $e_s = 100$, receiving energy $e_r = 15$ and the working circle $K = 20$ timeslots. Nodes will randomly choose some time slices in K. The ratio between the number of time slices and terminal nodes changes in different datasets.

Dataset number 1, 2 and 3 consist $|V| = 100, 200, 300$ vertices respectively (Table 1). The number of time slices in each working circle from 3 datasets is stable and equal to 5. Terminal node's number fluctuates between 5 and 100 % of the total nodes. The last dataset has 200 nodes. Terminal nodes are kept at a steady amount which takes up to 50 % of nodes in the network. The time slices ratio lies between 5 and 60 % of the total 20 time slices in a working circle.

Table 1 Parameters of the datasets

Dataset	1	2	3	4
#Nodes	100	200	300	200
#Terminal nodes (%)	5–100	5–100	5–100	50
#Time slices ratio (%)	25	25	25	5–60

5.2 System Configuration

The program demonstrating the proposed algorithm is written in Java and run on a computer having Intel Core i5 2.4 GHz chip and 4 GB RAM.

5.3 Computational Results

The proposed algorithm will be compared with R-TCS and GS-MEM algorithms [7]. Each algorithm executes 5 times on each dataset. The multicast energy consumption and average running time are reported in Tables 2 and 3 respectively, in which $|V|$ is the number of nodes and K is the number of time slices.

Figures 6 points out that the LSMEM algorithm performs better than R-TCS and GS-MEM algorithms in terms of saving energy cost during a multicast session.

To be specific, the energy cost of LSMEM algorithm used in dataset 1 is especially the lowest throughout 20 test times as shown in the Fig. 6a. From Fig. 6b and c with data taken from Table 2, it is seen that the proposed algorithm is better than the existing ones in term of multicast energy.

Moreover, Fig. 6b shows that LSMEM algorithm saves more energy than R-TCS and GS-MEM algorithms do (roughly 70 % better than R-TCS and GS-MEM on dataset 2). However, on dataset 3, the proposed algorithm and R-TCS have similar energy cost, while the energy cost of GS-MEM is not shown in [7] because GS-MEM consumed too much time. The energy cost of LSMEM in the final dataset is also lower than that of R-TCS and GS-MEM algorithms, which is 83.3 and 91.7 % better than R-TCS and GS-MEM respectively in dataset 4.

Figure 7 illustrates that LSMEM is faster than GS-MEM but slower than R-TCS. However, LSMEM only takes one minute to execute all datasets. In this case, the proposed algorithm takes significantly more time than the other algorithms at test

Table 2 Multicast energy optimization ratio when applying proposed algorithm compare with R-TCS and GS-MEM

Data-set	Test times	$\|V\|$	K	R-TCS (%)	GS-MEM (%)
1	20	100	20	100	100
2	20	200	20	70	75
3	20	300	20	50	100
4	12	200	20	83.3	91.7

Table 3 Running time optimization ratio when applying proposed algorithm compare with R-TCS and GS-MEM

Data-set	Test times	$\|V\|$	K	R-TCS (%)	GS-MEM (%)
1	20	100	20	35	100
2	20	200	20	40	100
3	20	300	20	35	100
4	12	200	20	0	100

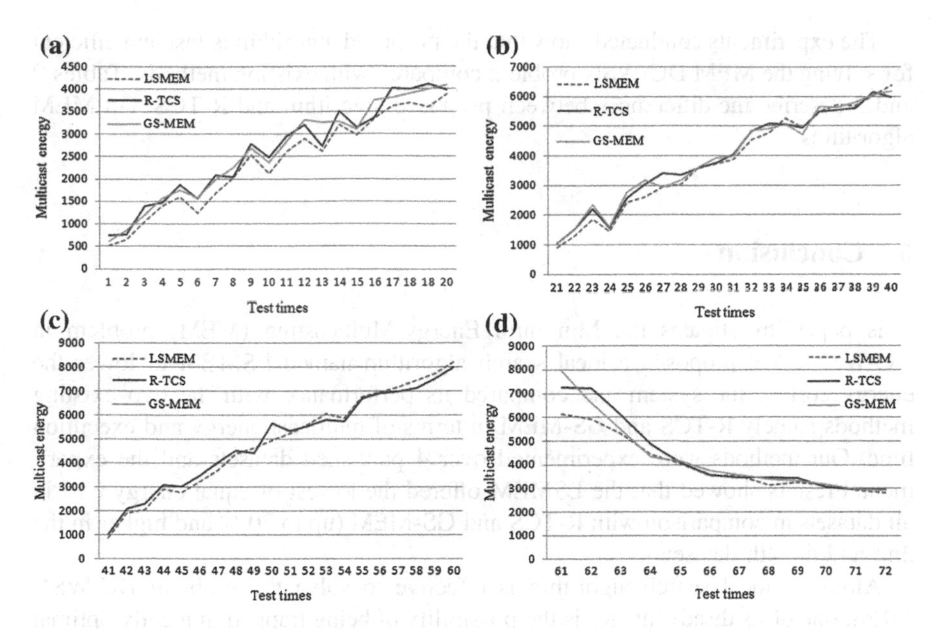

Fig. 6 Multicast energy of LSMEM, R-TCS, GS-MEM on dataset from 1 to 4 corresponding to **a**, **b**, **c** and **d** respectively

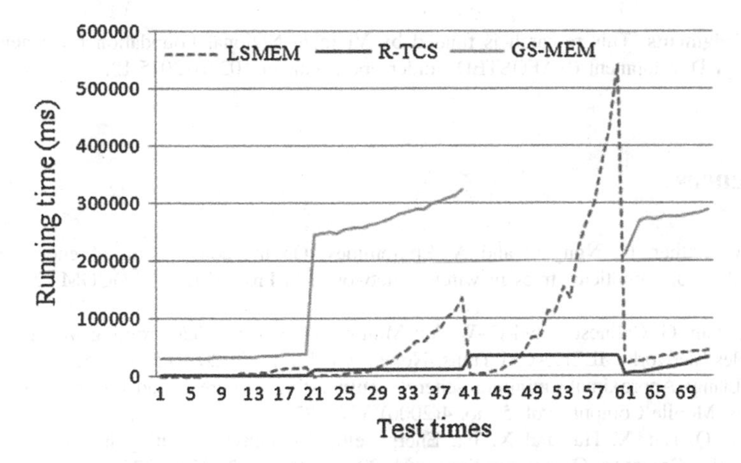

Fig. 7 Running time of LSMEM, R-TCS, GS-MEM in different tests

instances between 57 and 61 because those input sets have the quantity of terminal nodes on the initial graph G being nearly equal to the total number of nodes, leading to higher computational demand (higher number of calculations).

The experiments conducted show that the proposed algorithm is fast and efficient for solving the MEM DC-WSN problem compared with existing methods. Tables 2 and 3 describe the differences between proposed algorithm and R-TCS, GS-MEM algorithms.

6 Conclusion

This paper investigates the Minimum Energy-Multicasting (MEM) problem in DC-WSNs. We proposed a local search algorithm named LSMEM to lower the energy cost of the system and compared its performance with the two existing methods namely R-TCS and GS-MEM in terms of multicast energy and execution time. Our methods were experimented over 4 published datasets and the experimental results showed that the LSMEM offered the lowest or equal energy cost in all datasets in comparison with R-TCS and GS-MEM (up to 70 % and higher in the 2nd and the 4th datasets).

Although local search algorithm is effective to solve the problem DC-WSN MEM, one of its disadvantages is the possibility of being trapped in locally optimal points. In the future, the authors would develop a genetic algorithm with an effective heuristic initialization technique in order to increase the opportunities to find the global optimal.

Acknowledgments This research is funded by Vietnam National Foundation for Science and Technology Development (NAFOSTED) under grant number 102.01-2015.12.

References

1. J. Wieselthier, G. Nguyen, and A. Ephremides: On the construction of energy-efficient broadcast and multicast trees in wireless networks. in Proc. IEEE INFOCOM (2000) 585–594.
2. P.-J. Wan, G. Calinescu, and C.-W. Yi: Minimum-power multicast routing in static ad hoc wireless networks. IEEE/ACM Trans. Netw., vol. 12, no. 3 (2004) 507–514.
3. W. Liang: Approximate minimum-energy multicasting in wireless ad hoc networks. IEEE Trans. Mobile Comput., vol. 5, no. 4(2006) 377–387.
4. D. Li, Q. Liu, X. Hu, and X. Jia: Energy efficient multicast routing in ad hoc wireless networks. Computer Communications, vol. 30, no. 18 (2007) 3746–3756.
5. K. Han, Y. Liu, and J. Luo: Duty-Cycle-Aware Minimum-Energy Multicasting in Wireless Sensor Networks. IEEE/ACM Transactions on Networking. vol. 21, no. 3 (2013) 910–923.
6. L. Su, B. Ding, Y. Yang, T. F. Abdelzaher, G. Cao, and J. C. Hou: ocast: Optimal multicast routing protocol for wireless sensor networks, in Proc. IEEE ICNP (2009)151–160.
7. T. D. Nguyen, T. B. Huynh and H. S. Ngo: Group Steiner Tree Model for Energy Efficient Multicast in Duty-Cycle Wireless Sensor Networks, in The Fifth IEEE International Conference on Communications and Electronics (2014) 244–249.

8. G. Ausiello, A. D'Atri, and M. Protasi: Structure preserving reductions among convex optimization problems, Journal of Computer and System Sciences, vol. 21, no. 1 (1980) 136–153.
9. B. Awerbuch: Distributed Shortest Paths Algorithms (Extended Abstract), in Proc. ACM STOC (1989) 490–500.
10. M. R. Garey, and D. S. Johnson: Computers and Intractability: A Guide to the Theory of NP Completeness, W. H. Freeman, New York (1979).
11. D. S. Johnson: Approximation algorithms for combinatorial problems. Proc. ACM STOC (1973) 38-49.

Obstacle Avoidance for Multi-agent Path Planning Based on Vectorized Particle Swarm Optimization

Sumana Biswas, Sreenatha G. Anavatti and Matthew A. Garratt

Abstract This paper deals with an approach to path planning by obstacle avoidance for multi-agent systems. An effective framework is presented based on the Particle Swarm Optimization (PSO) method; an evolutionary computation (EC) technique that uses the dynamics of the swarm to search the solutions for the optimization problems. It describes the path replanning technique and obstacle avoidance for autonomous multi-agent systems. A simultaneous replanning concept is incorporated into the path planning to avoid both static and dynamic obstacles. This proposed algorithm reduces the computational time of the path planning. In the dynamic environment, the numerical results show that the Simultaneous Replanning Vectorized Particle Swarm Optimization (SRVPSO) algorithm is effective and also efficient for multi-agent systems.

Keywords Path planning • Particle swarm optimization • Obstacle avoidance • Multi-agent systems

1 Introduction

In recent years path planning is one of the research hotspots in the field of autonomous agent systems. Apart from searching for the optimum waypoints to reach the destination, advanced path planning is also responsible for ensuring safe and reliable navigation in a complex environment which may involve different types of static and dynamic obstacles.

S. Biswas (✉) · S.G. Anavatti · M.A. Garratt
Australian Defence Force Academy, The University of New South Wales, Canberra, Australia
e-mail: Sumana.Biswas@student.adfa.edu.au

S.G. Anavatti
e-mail: S.Anavatti@adfa.edu.au

M.A. Garratt
e-mail: M.Garratt@adfa.edu.au

G. Leu et al. (eds.), *Intelligent and Evolutionary Systems*, Proceedings in Adaptation, Learning and Optimization 8,
DOI 10.1007/978-3-319-49049-6_5

Compared to single agent systems, multi-agent systems can provide more flexibility, strong adaptation capacity and high reliability [1] in completing a successful mission. For this reason, multi-agent systems have been widely used in many complex and hazardous scenarios. Obstacle avoidance plays an important role in the context of managing multiple agents [2, 3]. Evolutionary path planning algorithms like PSO [4] are most prominent in the research area of the multi-agent system, where the path evolves based on the obstacles.

The existing approaches for solving the problem of motion planning for multiple agents can be divided into two namely centralized and decentralized [5]. In a centralized approach, a single unit gives instructions to other units and supervises the correct performance of a cooperative task. Centralized methods can be highly efficient for a small group of agents. However, control of agents becomes difficult or impossible as the group grows in size. In contrast, the decoupled approach inspired by biological societies, does not have an agent that controls the whole system and is a generalization of behavior-based control over multiple robots. No central planning unit is necessary for this system and every robot can calculate its own path [3, 6].

Many methods have been developed to tackle the problem of multi-agent path planning. Some of the known architectures are potential field method [7], genetic algorithms [8] etc. Yongqiang et al. [9] introduce a new control law based on stream function and hierarchical associations to avoid obstacles for multi-agent system. For effective motion planning of multiple mobile agents, Clark et al. [10] used dynamic networks that coordinate centralized planning. A dynamic rectangular roundabout collision avoidance scheme based on human behaviour can be used for collision detection and avoidance [11]. An A*-Dijkstra-Integrated algorithm is promoted by Zhang and Zhao [12] to make multiple agents moving parallel without any collision. An improved D* Lite algorithm with a fast replanning techniques is proposed by Peng et al. [13]. Roy et al. [14] use prediction principle for dynamic obstacle avoidance in multi-agent motion planning. However, most of the existing methods have some problems, such as potential field method cannot be able to find an optimum path in a closed barriers environment [15]. Some other conventional methods like A*, D* lite have the lack of flexibility and the popular genetic algorithm shows low solution efficiency [16].

Recently, swarm agent systems based on the concept of swarm intelligence have become one of the challenging research areas of multi-agent systems. Swarm intelligence is a part of evolutionary computation. Technically swarm intelligence is regarded as a collective behavior of decentralized, self-organized systems, natural or artificial. In a swarm agent system, population of simple agents is fully autonomous but a combination of these agents has the capacity to solve more complex problems. They are inspired from the nature, especially biological systems. Particle swarm optimization (PSO) has become an excellent optimization tool, using the concepts of swarm intelligence. PSO is a metaheuristic and random search algorithm inspired by the social behavior of bird flocking or fish schooling. It has been widely used in the field of multi-agent exploration. In PSO, the population is initialized with a random solution. Like other evolutionary algorithms, PSO searches for the optimum solution by updating generation. In this method, agents follow very simple rules, no centralized control structure to detect how individual agents should behave but the interactions between such agents lead to the emergence of "Artificial Intelligence" global behavior, which is not known to other agents [17]. The algorithm ensures a diversity of response and adheres to the principle of stability [18]. It generates a high quality solution with less computational time [19]. The advantages of PSO are that, it has few parameters to adjust and it can converge in a very fast manner [20]. In optimization problems, PSO gets better results in a faster and cheaper way compared with other methods, such as genetic algorithm [21]. By the quality of rapid searching and easier realization, PSO attempts to solve the drawbacks of conventional/existing methods [15]. Wang et al. [22] used PSO for obstacle avoidance in a soccer robot system. An intelligent PSO based real time multi-agent path planning is proposed by Mohamed et al. [23]. The authors used on board camera image for obstacle avoidance. For a two agent system, Lu and Gong [24] proposed a novel local path planning method based on PSO. In this method, the obstacle avoidance system is based on sensor system with limited detection range.

Generally, the main difficulties for multi-agent path planning problem are time efficiency, collision avoidance and adaptability. When a number of agents share a common workspace surrounded by different types of obstacles, high level path planning is required to avoid collisions [10]. Quick response to the random dynamic obstacles is the other major concern of path planning problem. Moreover, path planning by basic PSO algorithm faces a problem of trapping in local optima. Therefore, to find an optimum solution, a path planning system that efficiently avoid both static and dynamic obstacles (collisions) as well as rapidly replan the path to reach the destination is required.

In this paper, a path planning approach based on particle swarm optimization algorithm is presented where the replanning should be done to avoid both stationary and moving obstacles. In the proposed, simultaneous replanning vectorized PSO (SRVPSO) algorithm, one agent considers other agents as dynamic obstacles and applies collision avoidance strategy. At first, it assigns a very high cost to the fitness function of the particles within the collision zone. As a result, the velocities of those particles get increased rapidly with the augmentation of the parameters from PSO to leave the collision region. This algorithm also has the ability to overcome the drawback of basic PSO. The main focus of the paper is efficient collision avoidance.

The rest of the paper is organized as follows: At first, representation of the environment is provided in Sect. 2. An overview of basic PSO and SRVPSO algorithm, with the key technique in the implementation of the algorithm is presented in Sect. 3. In Sect. 4. the numerical results are described. Finally, conclusion and some possible future work are suggested in Sect. 5.

2 Problem Description

Multi-autonomous agents are deployed to carry out a search mission in a complex environment. The working environment consists of different types of obstacles. In the working space, the agents can start from any position and can go any of the targeted points. The agents should plan a path from a start location to destination and the path cannot pass through any obstacles. Agents have to avoid both static and dynamic obstacles and also should avoid collision among them. The agents have to efficiently replan their paths based on the change of the environment due to the presence of dynamic obstacles. This multi-agent system will not follow the central control flight formation type path planning. In the given environment, the agents can plan their collision-free paths according to their own judgment of the dynamic environment. Based on PSO algorithm, the agents will avoid collisions and successfully plan their optimum path.

3 Particle Swarm Optimization

The PSO method was first proposed by James Kennedy and R.C. Eberhart in 1995 [18]. It is a population based stochastic optimization method based on the social behavior of some species to locate the desired position in a group activity. Based on the social psychology, PSO represents socio-cognition of human and artificial agents. It combines local search with global search methods by balancing exploration and exploitation.

In PSO, the workspace is initialized with random particles in which each particle represents a potential solution to the problem. Each particle of the swarm randomly searches in a multidimensional search space. They update themselves with the best solutions of their own experience as well as the social information gathered from other particles. The best solutions of the particles are evaluated by the fitness function of their current location. The swarm changes its position dynamically throughout the optimization process until an optimum solution is reached. Each particle is randomized with a velocity. Particles change their positions by updating their velocities. The procedure of velocity update is influenced by three factors namely, particles current motion, particles previous experience and the influence of the whole swarm as shown in Eq. (1). Figure 1 describes the various factors that influence particles movement in PSO. Particles constantly update their velocities and positions, until they have reached the goal position.

Let, a PSO swarm consists of a set of particles (S). Each particle changes its position with time. Position and velocity of the ith particle at time step ‘t’ are respectively-

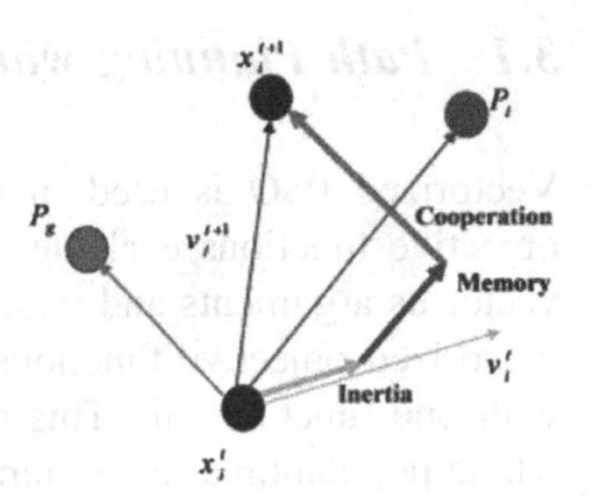

Fig. 1 Particle movement in PSO [25]

$$x_i(t) = (x_{i,1}(t), x_{i,2}(t), x_{i,3}(t), \ldots\ldots x_{i,n}(t)) \text{ and}$$
$$v_i(t) = (v_{i,1}(t), v_{i,2}(t), v_{i,3}(t), \ldots\ldots v_{i,n}(t)).$$

Each particle updates its velocity and position according to the following equations:

$$\mathbf{v}_i(t+1) = \underbrace{\omega \times \mathbf{v}_i(t)}_{\text{Inertia}} + \underbrace{c_1 \times rand_1() \times (PBest_i - \mathbf{x}_i(t))}_{\text{Cognitive}} + \underbrace{c_2 \times rand_2() \times (GBest - \mathbf{x}_i(t))}_{\text{Socialinfluence}} \tag{1}$$

$$x_i(t+1) = x_i(t) + v_i(t+1) \tag{2}$$

where, ω = inertia weight factor, c_1 = cognition parameter, c_2 = social parameter and $rand_1()$, $rand_2()$ = are independent random variables uniformly distributed in [0, 1]. *PBest* is the personal best position achieved by the particle and *GBest* is the globally best position achieved by the swarm.

There are three problem dependent parameters, the inertia weight (ω) and the acceleration coefficients c_1 and c_2. 'ω', controls the exploration and exploitation of particles and the acceleration coefficients c_1 and c_2 influence the personal and global leaders on the search process [26]. These parameters have a significant effect on the performance of PSO algorithm. Kennedy and Eberhart [18] proposed to use $c_1 = c_2 = 2$ and also Shi and Eberhart [27] suggested to use the value of 'ω' from 0.8 to 1.4 [28]. A large inertia weight (ω) facilitates a global search and a small inertia weight has greater local search ability. The *rand()* terms ensures good convergence by avoiding local optima.

Optimization is conducted using an objective function defined by the Euclidean distance between the starting point (x_1, y_1) and the goal point (x_2, y_2).

$$\text{Distance} = \sqrt{(x_1 - x_2)^2 + (y_1 - y_2)^2} \tag{3}$$

In PSO, the algorithm evaluates the fitness function of all the particles and the solution is optimized by iterative improvements of the positions of particles [29].

3.1 Path Planning with SRVPSO

Vectorized PSO is used in this research work. In the case of basic PSO, the objective functions evaluate one point at a time. They consider a single numeric vector as arguments and return a single number. On the other hand, in VPSO the vectorized objective functions evaluate all the points in a search pattern at once, with one function call. This means, the vectorized objective function expects the whole population as an argument and returns a vector of objective function values. The input of the vectorized objective function is a matrix and it generates a vector of functional values. In the case of searching, the programming code of VPSO is shorter as it avoids loops and runs faster than basic PSO with loops.

To simplify the representation, some assumptions are considered: Due to heterogeneous shape of the autonomous vehicle and obstacles, a circle is used to enclose the vehicles and obstacles. To ensure a collision free path, a safety gap is considered by extending the enclose radius of both the vehicles and obstacles. The vehicles are assumed as points and the size of the vehicle is added to the radius of the obstacle.

3.2 Collision Avoidance Strategy

During the searching for optimum path, the agents need to maintain a safe distance from the obstacles.

The distance between the obstacles and the current position of the agents must be ≥ 0.

The SRVPSO algorithm follows two steps to avoid collision with both static and dynamic obstacles. The first step is to increase the fitness value (cost) of those particles within the collision zone, with a high positive amount of functional value and the second step is to increase the velocity of those particles to come out from the zone.

3.3 Update the Fitness Value

The obstacle avoidance is achieved by PSO in a reactive fashion, unlike any other conventional method which computes the whole path at every instance. The path can be changed on the fly when any risk of collision is detected. The velocity of the static obstacles is zero. The particles don't need to consider the velocity of the dynamic obstacles to compute the path. That is the advantage of the reactive planning. At every instance agents look at where the obstacles are and compute a path based on that. Hence, the computation time of PSO is not affected by the presence of dynamic obstacles. In this paper, the path replanning occurs to avoid any types of obstacles. At first, to maintain a collision free path all the particles of the swarm have to detect the positions of the obstacles, then they calculate the distance between the obstacles and the particles. A collision occurs if

$$D^{a}_{obs} \leq 0 \tag{4}$$

where, D^{a}_{obs} represents the distance between agents and obstacles, then it considers a risk of collision. Here, the negative value of D^{a}_{obs} means that the positions of the particles are inside the collision zone. The fitness value of the swarm particles within the collision zone is assigned by a higher positive functional value. As the optimization is considered as a minimization problem, the particles with higher fitness vales are not considered as the best fitness function at all. So the path within the collision zone is avoided.

3.4 Update the Velocity of the Collision Zone Particles

In order to escape the particles from the collision zone, the values of the parameters (ω, c_1, c_2) in Eq. (1) are increased. Therefore, the velocities of those particles are also increased and they leave the collision zone. The pseudocode of the proposed algorithm can be expressed as follows:

Step 1: Initialize the parameters: ω, c_1, c_2, the maximum number of iteration, swarm size etc.
Step 2: Particles, initial velocity = 0, initial position = 0
For each particle
Step 3: Evaluate the fitness value from the vectorized objective function
Step 4: For each particle set
Local best fitness = current fitness and
Local best position = current position
Set global best fitness = min{ local best fitness}
If (current fitness> PBest fitness)
Assign current fitness = new PBest
Else
Keep previous PBest
End
Assign best PBest = GBest
Step 5: Update the particles velocity and position according to equation (1) and (2).
Step 6: If the distance between the obstacles and the current position of the agents' ≤ 0 ; replan the path.
- Assign a high positive functional value for the particles within the collision zone.
- Increase the velocity of those particles to move on from the collision zone
Step 7: If all the particles converge at the global optimum solution (e.g. target position) within maximum iteration go to step 8; otherwise turn to step 3.
Step 8: End

4 Results and Discussions

In this section, several tests are performed on different cases to validate the feasibility of the proposed method. The SRVPSO algorithm is coded in MATLAB R2015a and tested on a Windows computer with Intel(R) Core(TM) i7-4770 CPU @3.40 GHz and 16.0 GB of RAM.

The parameters of PSO used in tests are as follows:
$\omega = 1.0$, $c_1 = c_2 = 2.0$, Population size of the swarm = 64 and
Maximum iterations = 100
Parameters considered for obstacle avoidance strategy are as follows:
$\omega = 2.0$ and $c_1 = c_2 = 3.0$

In obstacle avoidance strategy, the velocities of the particles are increased by increasing the inertia parameter and the acceleration coefficients. Therefore, the particles can move away from the region of collision.

In all the experiment the agents attempt to find the shortest path by avoiding any types of obstacles.

4.1 Path Planning by Obstacle Avoidance

A 2D environment is considered as a workspace. For obstacle avoidance, we considered different types of working environment. The map size of the environment is considered as a 120 × 120 square unit with different types of obstacles. The solid circles represent the static obstacles and the hollow circles represent the dynamic obstacles (Figs. 2, 3, 4, 6, 7, 8, and 9). The hollow circles sequence represents the trajectory of dynamic obstacles. The dynamic obstacles can have constant speed or they can be randomly moving objects. The dots/points on the path denote the agents. In the SRVPSO algorithm, a 3–5 % tolerance zone is considered. A rectangular goal region is plotted instead of a single goal location to show the tolerance zone in the destination.

Figures 2, 3 and 4 describe the path planning and obstacle avoidance strategy of a single agent system in three different environments. The dotted line represents the optimal path to reach the destination.

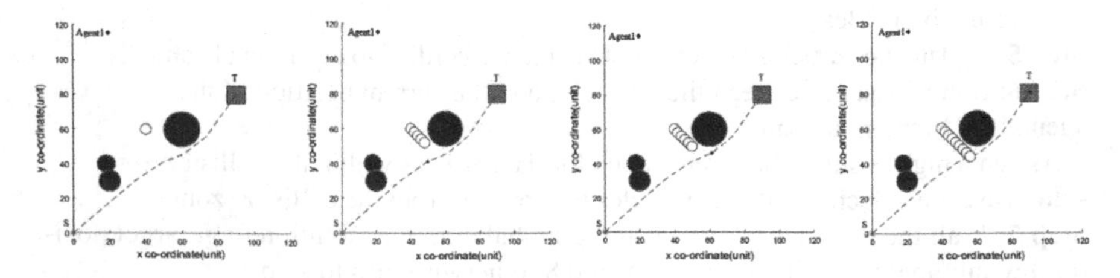

Fig. 2 Sequences of path planning for single agent system

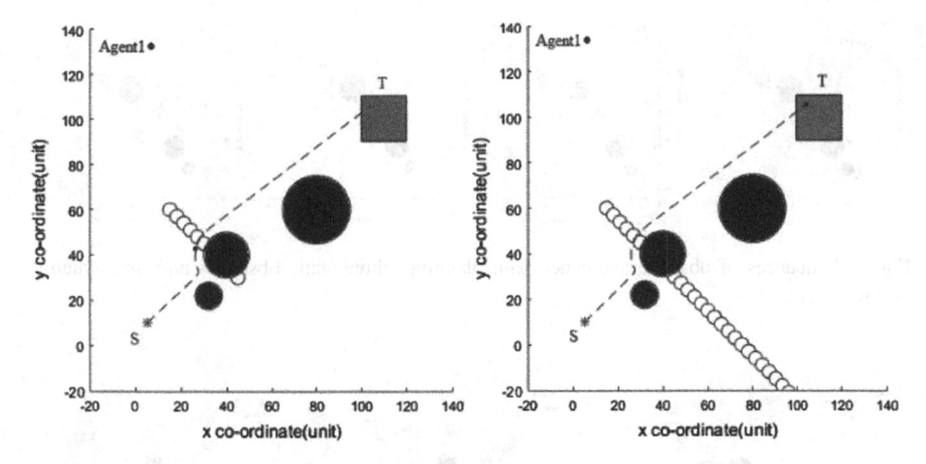

Fig. 3 Sequences of obstacle avoidance (3 static and 1 dynamic obstacles)

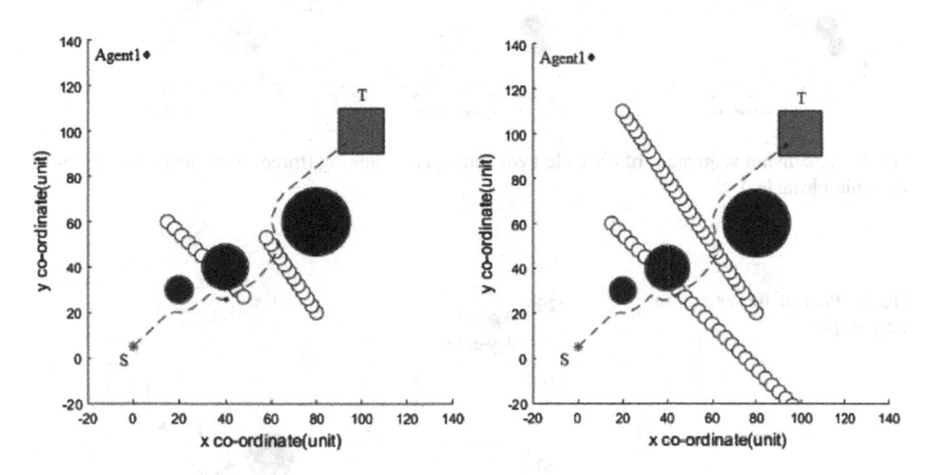

Fig. 4 Sequences of obstacle avoidance (3 static and 2 dynamic obstacles)

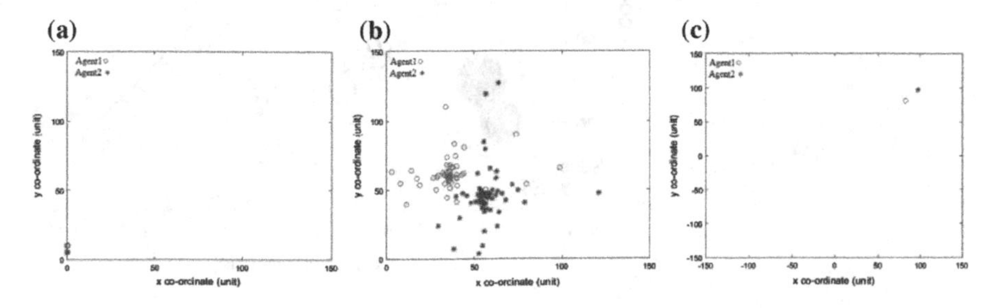

Fig. 5 Searching procedure of particles

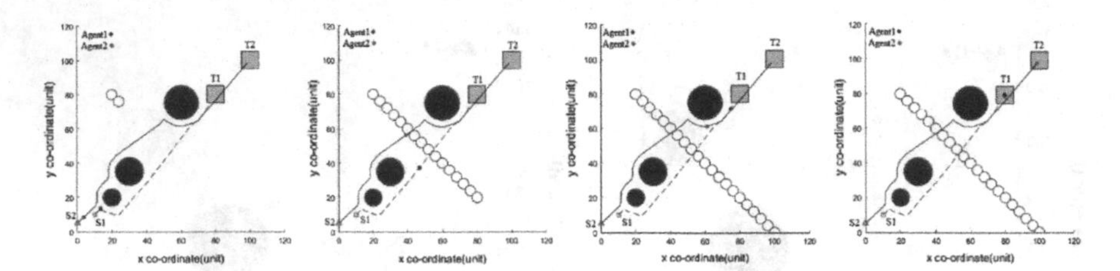

Fig. 6 Sequences of obstacle avoidance path planning (three static obstacles and one dynamic obstacle)

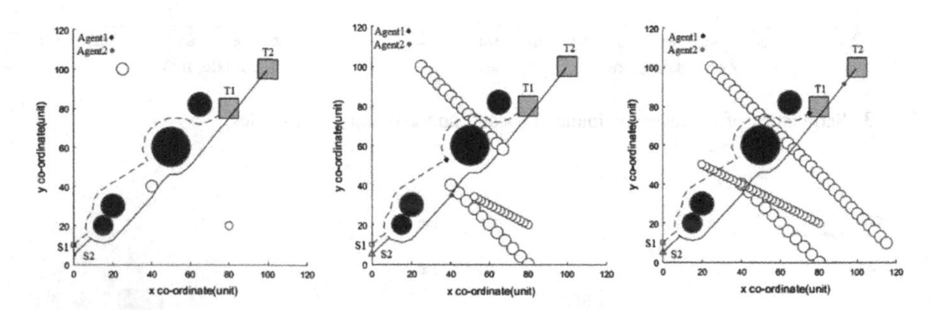

Fig. 7 Screenshot sequences of obstacle avoidance path planning (three static obstacles and three dynamic obstacles)

Fig. 8 Path of the agents in the complex

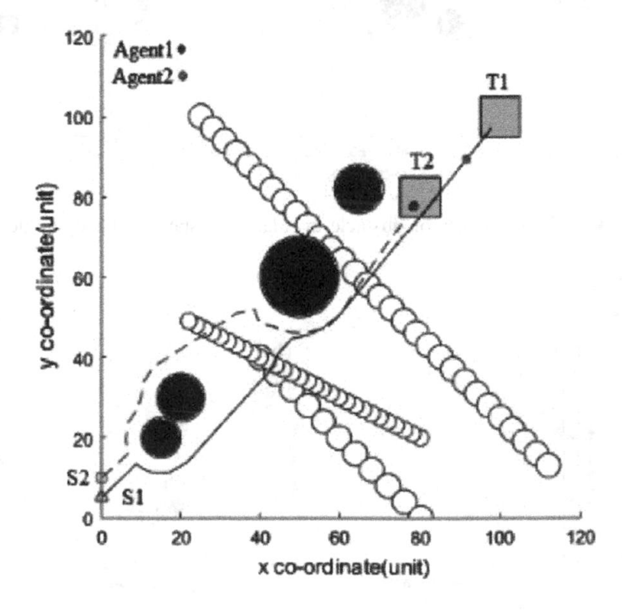

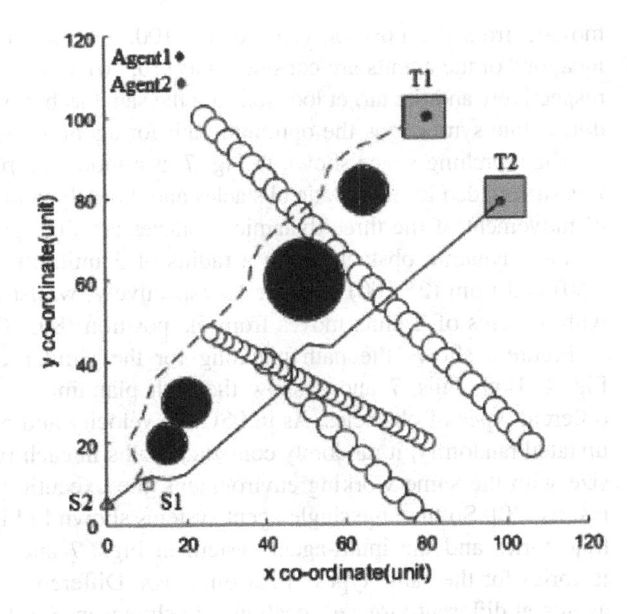

Fig. 9 Multi-agent path planning at dynamic environment different targeted location

In Fig. 2 the starting point and the target point of the agent are (0, 0) and (90, 80) respectively. The environment is surrounded by three static obstacles and one dynamic obstacle. The dynamic obstacle has a radius of 3 units and is moving from the point (50, 55) to the point (68, 37) at a fixed speed.

The working environment of Fig. 3 is as same as Fig. 2. Figure 4 shows the results of the algorithm in a complex environment, where there are three static and two dynamic obstacles. The starting and goal position of the agent are the same as before.

4.2 Path Planning with Multi-agent Systems

During experiments, we run the algorithm about 50 times, every time all the particles converge in the targeted location and we get the optimum solution. As an example, we consider a two agent systems path planning. Figure 5 shows the screen shots of the random path searching procedure of two agents. The agent's movement environment is considered as 150 × 150 square unit. Figure 5a shows the starting points (0, 10) for agent 1 and (0, 5) for agent 2. In Fig. 5b according to PSO, particles of both agents are randomly distributed in the searching space and in Fig. 2c, all the particles converge in the goal point (80, 80) and (100, 100) for agent 1 and agent 2 respectively.

The screenshots for the sequences of obstacle avoidance path planning are shown in Fig. 6. In this case, the working environment is surrounded by three static obstacles and one dynamic obstacle. The dynamic obstacle, with radius 3 units is moving from the position (20, 80) to (100, −1) at a constant speed. The initial locations of the agents are considered as (10, 10) and (0, 5) for agent 1 and agent 2 respectively and the target locations are the same as before. The continuous line and dotted line symbolizes the optimum path for agent 1 and agent 2 respectively.

The searching space shown in Fig. 7, is a more complex dynamic environment. It is surrounded by four static obstacles and three dynamic obstacles. The directions of movement of the three dynamic obstacles are different.

Two dynamic obstacles with a radius of 3 units move from (40, 40) to (110, −60) and from (25,100) to (121, 4) respectively, whilst the other moving obstacle with a radius of 2 units moves from the position (80, 20) to (16, 52).

Figure 8 shows the path planning for the similar environment as shown in Fig. 4. Both Figs. 7 and 8 show the path planning of multi-agents by avoiding different types of obstacles. As in PSO the velocity and position of the particles are updated randomly, it randomly constructs paths in each run. Even on the same map size with the same working environment, the execution might not give the same results [30]. So that, the single agent systems shown in Figs. 2 and 3 gives different trajectories and the multi-agent system in Figs. 7 and 8 also show different trajectories for the same types of surroundings. Different trajectories for autonomous agents at different targeted locations are shown in Fig. 9. In this case, the starting point and goal position for agent 1 are (10, 10) and (80,100) and for agent 2 the positions are (0, 5) and (100, 80).

In every test run, the SRVPSO algorithm easily handles the moving and static obstacles in a very wide range of environments. In addition, it deals with multi-agents. One robot is regarded as a dynamic obstacle of the other one. If agents find any obstacles on their way, they simultaneously replan their path and reach the destination. From the above results, it is clear that the agents safely and successfully have found their solution.

5 Conclusions

In this paper, we have proposed a novel method that we have named the Simultaneous Replanning Vectorized PSO (SRVPSO) algorithm for multi-agent path planning. This optimization algorithm has a safe navigation capability by ensuring a collision free path. The research shows that the proposed algorithm is easy to implement and it can easily find an optimum solution by avoiding any types of obstacles. Moreover, this algorithm is able to optimize the path very efficiently. The test results validate the feasibility of the proposed algorithm. In the near future, the SRVPSO algorithm will be simulated with real vehicle dynamics and implemented in a real world scenario.

References

1. Wang, D., Wang, H., Liu, L.: Unknown environment exploration of multirobot system with the FORDPSO. Swarm and Evolutionary Computation, vol. 26, pp. 157–174 (2016)
2. Chang, D. E., Shadden S. C., Marsden, J. E., Olfati-Saber, R.: Collision avoidance for multiple agent systems. Proceedings of the 42nd IEEE conference on Decision and Control, pp. 539–543 (2003)
3. Francis, S. L. X., Anavatti, S. G., Garratt, M.: Real time cooperative path planning for multi autonomous vehicles. In Advances Computing, Communications and Informatics (ICACCI), International Conference paper on, IEEE, pp. 1053–1057 (2013)
4. Guo, D. W., Zhou, C. G. and Liu, M.: A hero evolutionary algorithm hybridizing from PSO and GA. In Computational Methods, Springer Netherlands, pp. 1075–1080 (2006)
5. Latombe, J.C.: Robot Motion Planning. Kluwer Academic Publishers, Boston, MA., ISBN 0-7923-9206-X (1991)
6. Parker, L. E.: Path planning and motion coordination in multiple mobile robot teams. In Encyclopedia of Complexity and System Science, Meyers, R. A., Ed, pp. 5783–5800. Springer, (2009)
7. Keron,Y., Borenstein, J.: Potential field methods and their inherent limitations for mobile robot navigation. Conference on Robot Automation, California, pp. 1398–1404, (1991).
8. Hu, Y., Yang, S. X., A knowledge based genetic algorithm for path planning of a mobile robot. In Proceedings of IEEE Intl. Conference on Robotics and Automation, New Orleans, pp. 4350–4355 (2004)
9. Yongqiang, B., Wenzhi, X., Hao, F., Jixiang, L., Jing, C.: Obstacle avoidance for multi-agent systems based on stream function and hierarchical associations. In Control Conference (CCC), 31st Chinese, pp. 6363–6367, IEEE (2012).
10. Clark, C. M., Rock, S. M., Latombe, J-C.: Motion planning for multiple mobile robot systems using dynamic networks. In Robotics and Automation, Proceedings. ICRA'03, IEEE international Conference on, vol. 3, pp. 4222–4227, IEEE (2003)
11. Liu, F., Narayanan, A.: A human-inspired collision avoidance method for multi-robot and mobile autonomous robots. In International conference on Principles and Practice of Multi-Agent Systems. LNCS, vol. 8291, pp. 181–196. Springer, Heidelberg (2013)
12. Zhang, Z., Zhao, Z.: A multiple mobile robots path planning algorithm based on A-star and Dijkstra algorithm. International Journal of Smart Home, vol. 8, no. 3, pp. 75–86, (2014)
13. Peng, J. H., Li, I. H., Chien, Y. H., Hsu, C. C., Wang, W. Y.: Multi-robot path planning based on improved D* Lite Algorithm. In Networking, Sensing and Control (ICNSC), 2015 IEEE 12th International Conference on (pp. 350–353). IEEE. (2015)
14. Roy, S., Banerjee, D., Majumder, G., Konar, A., Janarthanan, R.: Dynamic obstacle avoidance in multi-robot motion planning using prediction principle. International journal of Advanced computer Engineering & Architecture, vol. 2 no. 2, pp. 269–283 (2012).
15. Ma, Q., Lei, X., Zhang, Q.: Mobile robot path planning with complex constraints based on the second order oscillating particle swarm optimization algorithm. In Computer Science and Information Engineering, WRI World Congress on, vol. 5 pp. 244–248, IEEE (2009)
16. Li, T. J., Yuan, G. W., Wang, F. J.: Behavior control of multiple robots exploring unknown environment. In 4th IEEE Conference on Industrial Electronics and Applications, pp. 1877–1882, IEEE (2009)
17. Zhang, Y., Wang, S., Ji, G.: A comprehensive survey on particle swarm optimization algorithm and its applications. Mathematical Problems in Engineering, (2015)
18. Kennedy, J., Eberhart, R.: Particle swarm optimization. Proceedings IEEE International Conference on Neural Networks, pp. 1942–1948. IEEE Computer Society Press, Los Alamitos (1995)

19. Wang, Y., Chen, P., Jin, Y.: Trajectory planning for an Unmanned ground vehicle group using augmented particle swarm optimization in a dynamic environment. Proceedings of the IEEE International Conference on Systems, Man and Cybernetics, san Antonio, TX,USA, pp. 4341–4346 (2009)
20. Qiaorong, Z., Guochang, G.: Path planning based on improved binary particle swarm optimization algorithm. In IEEE Conference on Robotics, Automation and Mechatronics pp. 462–466, IEEE (2008)
21. Eberhart, R. C., Shi, Y.: Comparison between genetic algorithm and particle swarm optimization. In Proceedings of the 7th International Conference on Evolutionary Programming VII, pp. 611–616, Springer Berlin Heidelberg (1998)
22. Wang, L., Liu, Y., Deng, H., & Xu, Y.: Obstacle-avoidance path planning for soccer robots using particle swarm optimization. In IEEE International Conference on Robotics and Biomimetics, pp. 1233–1238, IEEE. (2006)
23. Mohamed, A. Z., Lee, S. H., Aziz, M., Hsu, H. Y., Ferdous, W. Md.: A proposal on development of intelligent PSO based path planning and image based obstacle avoidance for real multi-agents robotics system application. In Electronic Computer Technology (ICECT), International Conference on, pp. 128–132, IEEE. (2010)
24. Lu, L., Gong, D.: Robot path planning in unknown environments using particle swarm optimization, Natural Computation, ICNC '08. Fourth International Conference on, Vol. 4, 18–20, pp. 422 – 426 (2008)
25. Saway, A. A. E., Hendawy, Z. M., Shorbagy, M. A. E.: Reference point based TR-PSO for multi-objective environmental/economic dispatch, Scientific Research: Applied Mathematics, vol. 4, no. 5, pp. 803–813 (2013)
26. Hao, Y., Zu, W., Zhao, Y.: Real-time obstacle avoidance method based on polar coordination particle swarm optimization in dynamic environment. Second IEEE Conference on Industrial Electronics and Applications, pp. 1612–1617 (2007)
27. Shi, Y., Ebertart, R.: A modified particles swarm optimizer. In Evolutionary Computation Proceedings, IEEE World Congress on Computational Intelligence., The IEEE International Conference on, pp. 69–73 (1998)
28. Venter, G., Sobieski, J. S.: Particle swarm optimization. American Institute of Aeronautics and Astronautics (AIAA) Journal, vol. 41, no. 8, pp. 1583–1589 (2003)
29. Parsopoulos, K. E., Vrahatis, M. N.: Particle swarm optimization and intelligence: advances and applications. Hershey, USA: Information Science Reference, (2010)
30. Oral, T., Polat, F.: MOD* lite: an incremental path planning algorithm taking care of multiple objectives, IEEE Transactions on Cybernetics, vol. 46, no. 1, pp. 245–257 (2016)

Resource Constrained Multi-project Scheduling: A Priority Rule Based Evolutionary Local Search Approach

Ripon K. Chakrabortty, Ruhul A. Sarker and Daryl L. Essam

Abstract This paper considers a static resource constrained multi-project scheduling problem (RCMPSP) with two lateness objectives: project lateness and portfolio lateness. To solve the RCMPSP, we have proposed an evolutionary local search heuristic that uses a variable neighborhood (ELSH-VN) approach. The heuristic is further analyzed by incorporating different priority-rules. To judge the performance of these priority rule based heuristics, an extensive simulation-based analysis has been conducted with different scenario-based schedules. For the experimental study, we have considered a standard set of 77 generated RCMPSP test instances of 20 activities. The experimental analysis indicates that the proposed heuristic is able to solve multiple projects with reasonable computational burden. The influence of the variation of resource distribution and resource contention on the algorithm's performance for different priority rules is also analyzed and discussed.

Keywords Multi-project scheduling • Heuristics • Priority rules • Resource constraints

1 Introduction

The resource constrained project scheduling problem (RCPSP) involves the scheduling of project activities so that given temporal constraints between activities are satisfied, that the prescribed resource capacities are not exceeded, and a given objective, e.g., the project duration (i.e., makespan) is minimized. However, in contemporary enterprises, single project settings are rare today. Hence, issues involving the simultaneous management of multiple projects (or portfolio of projects) have become more prevalent. According to Payne [1], up to 90 % of all

---

R.K. Chakrabortty (✉) · R.A. Sarker · D.L. Essam
School of Engineering and Information Technology,
University of New South Wales, Canberra 2600, Australia
e-mail: ripon.chakrabortty@student.adfa.edu.au

G. Leu et al. (eds.), *Intelligent and Evolutionary Systems*,
Proceedings in Adaptation, Learning and Optimization 8,
DOI 10.1007/978-3-319-49049-6_6

projects worldwide are executed in a multi-project context. Considering this demanding fact, this paper addresses the resource constrained multi-project scheduling problem (RCMPSP), which is an extension of RCPSP, and is considered as the simultaneous scheduling of two or more projects which demand the same scarce resources. Meanwhile, when dealing with multiple projects, two approaches have been used in earlier researches: (1) a single project approach, using dummy activities and precedence arcs to combine multiple projects into a single mega-project or (2) a multi-project approach, maintaining each RCMPSP and a separate critical path per project [2]. Here, this paper took the second approach.

In spite of having high relevance with modern project management practices, RCMPSP has not been studied as comprehensively as single-project scheduling. Moreover, as a generalization of the RCPSP, RCMPSP is also considered to be NP-hard [3], meaning that there are no known algorithms for finding optimal solutions in polynomial time. When using a multi-project approach, two general approaches are used to solve RCMPSPs, namely exact methods and heuristic procedures. Among the few, exact methods are the similar earlier works of Drexl [4]. But those exact methods are limited to solving small problem instances and are impractical for solving large RCMPSPs. On the other hand, most of the heuristic methods used for solving RCMPSPs belong to the class of priority rule based methods [5]. In summary, while various studies have identified potentially important characteristics of RCMPSP and proposed various priority rules, the variety of results and their disagreements, have left project managers lacking clear guidance on which priority rule to use in a particular situation.

Although using a large variable neighborhood search (LNS) approach for various problems, such as the vehicle routing problem [6] and travelling salesman problem [7] already exists, to the best of our knowledge no heuristic of this type has yet been proposed for RCMPSP. To fill this gap, we propose a new variable neighborhood search based heuristic, which we named evolutionary local search heuristic with variable neighborhood (ELSH-VN) for RCMPSP. As opposed to only considering activity sequences or generating sub-problems, this ELSH-VN approach considers overall partial schedules that are generated from a parallel schedule generation scheme.

The primary objective of this research was to develop an evolutionary algorithm for solving RCMPSPs. Besides of that, justifying feasibility of different priority rules for this type of NP-hard problems (i.e., RCMPSPs) was another secondary target. Hence, we address static RCMPSP with two tardiness objectives, defined as average project delay and average portfolio delay. We have employed a parallel schedule generation scheme to generate an initial baseline schedule of RCMPSP, which was then applied to our proposed evolutionary heuristic, ELSH-VN. The effectiveness of the solution scheme (i.e., ELSH-VN) was then demonstrated through extensive experimentation with 77 randomly generated problem instances. We then described ten different ELSH-VN alternatives, based on different priority rules, to demonstrate the superiority of any particular priority rule. Computational results have been summarized for all priority-rule based ELSH-VNs. Statistical analysis has also been carried out to further show the superiority of any particular priority rule over other ones. In brief, two key contributions have been made over the existing research on standard RCMPSPs. These are: firstly, we propose an approach for variable neighborhood diversification among the activities that represent an initial unbuffered schedule. We also show how to explore those schedules with evolutionary local searching scheme, particularly by enhanced swapping to find optimal or near-optimal results. Secondly, we have considered a set of priority rule based scheduling procedures with our proposed ELSH-VN for RCMPSP. We find significance differences in the performance of the priority rules which implies that several widely advocate priority rules generally do not perform as well as may be thought.

The structure of the paper is as follows: in Sect. 2 we define the basic RCMPSP with two lateness objectives. In Sect. 3, solution approaches and relevant algorithm designs with priority rules are discussed. The experimental studies, along with their results are in Sect. 4. Finally, we provide conclusions in the last section.

2 Problem Description

RCMPSP consists of several projects, where the individual projects have the characteristics of single project, and is similar to the mathematical formulation of Talbot [8]. Here we assumed that each project consists of i = 1 … I_v activities with deterministic and non-pre-emptible duration d_{vi}. We further assumed that (i) activities belonging to any particular project have unique characteristics and do not depend on other project's activities; (ii) there is a precedence relationship among the activities that belong to any particular project, and all predecessors must finish before an activity can start; (iii) resources considered here are only of renewable type; (iv) activities are non-pre-emptive (i.e., cannot be interrupted when in progress); (v) the parameters (durations, capacities and resource requests) are non-negative and integer valued; (vi) there is no dependency relationships among the projects, apart from resource sharing. The binary decision variable x_{vit} represents 1 if activity i of project v starts at time period t or 0 otherwise. RCMPSP involves finding a schedule for the activities (i.e., determining the start or finish times) that optimizes a performance measure, such as minimizing the average delay in all projects. For measuring delays, each project is associated with a due date. This due date is defined as the lower bound of project completion time without considering any resource constraints. Using the further considerations and notations, as given in the below nomenclature section, the conceptual mathematical formulation for RCMPSP is given below:

2.1 Nomenclature

V	*Set of projects, v = 1 ... V*
I_v	*Set of activities of project v, i = 1 ... I_v*
$Prec_v$	*Set of all precedence relationships of project v*
S_{vj}	*Set of all successors of activity j belonging to project v*
$\lvert S_{vj} \rvert$	*Number of successors for activity j in project v*
$k \in K$	*Renewable resources (e.g., Machines)*
NI_v	*Number of activities for any particular project v (last activity)*
R_k	*Capacity of renewable resource k*
r_{vik}	*Renewable resource k usage of activity i of project v*
d_{vi}	*Duration of activity i of project v*
ES_{vi}, LS_{vi}	*Earliest and latest finish time of activity i of project v*
T	*Total planning horizon/upper bound of all projects*
f_{vi}	*Finish time of activity i belonging to project v*
S_{vi}	*Start time of activity i belonging to project v*
CP_v	*Critical path duration of the v'th project without resource constraints*

$$\text{Min}\,\Phi(\text{F}) \tag{1}$$

$$\sum_{t=ES_{vj}}^{LS_{vi}} x_{vit} = 1\, for\, \forall i \in NI_v\, and\, \forall v \in V \tag{2}$$

$$\sum_{t=ES_{vb}}^{LS_{vb}} (t - d_{vb}) x_{vbt} \geq \sum_{t=ES_{va}}^{LS_{va}} t x_{vat}\, \forall (a, b) \in Prec_v\, and\, \forall v \in V \tag{3}$$

$$\sum_{v=1}^{V} \sum_{i=1}^{NI_v} \sum_{q=t}^{t+d_{vi}-1} r_{vik} x_{viq} \leq R_k\, \forall k \in K, \forall t \in T\, and\, \forall v \in V \tag{4}$$

$$x_{vit} \in \{0, 1\}\, \forall v \in V, \forall i \in I_v\, and\, \forall t \in T \tag{5}$$

The objective function (1) seeks to optimize a pre-specified performance measure set by the planner. Although a variety of objective functions have been used for RCMPSP, minimizing project duration is commonly used. However, in this study, we seek to minimize project or portfolio delays (tardiness). Hence we considered two different objective functions, which are defined in the following equations:

$$Average\, Project\, Delay\, (APD) = \frac{\sum_{v=1}^{V} a_{v/A_v}}{V} \times 100 \tag{6}$$

$$Average\, Portfolio\, Delay\, (APFD) = \frac{Max(A_1 + a_1, \ldots, A_V + a_V) - Max(A_1, \ldots, A_V)}{Max(A_1, \ldots, A_V)} \times 100 \tag{7}$$

Here in Eqs. (6) and (7), A_v represents the due date of project v, which equals the length of its resource unconstrained critical path (CP). Meanwhile, a_v represents the delay due to resource constraints. Constraint set (2) ensures that all activities of any particular project v are to be scheduled once and only once. Constraint set (3) implies the predecessor relationships for all activities within a project. Here no precedence relationship is considered within projects. Constraint set (4) limits the maximum level of renewable resources used by the activities belonging to any project v. For any particular time period t, the maximum amount of renewable resource usage should be within its availability limit. Finally, constraint set (5) forces the start times to be non-negative.

3 Algorithms

3.1 ELSH-VN

In this section, we present the procedural steps and basics of our proposed Evolutionary Local Search Heuristic with Variable Neighborhood (ELSH-VN). This ELSH-VN procedure is analogous to a descent search, or a hill-climbing approach, with a candidate list strategy. In contrast with a traditional searching strategy, the legal moves for active activities for this heuristic are only those which belong to the candidate list at each stage. The overall procedures of this heuristic are also comparable to Tabu search, in the sense that when a move or swapping is effected, the activity concerned is unlikely to be moved back to its old position, until it is encountered again while searching the sequence, or unless the resulting sequence leads to the best makespan so far. An evolutionary local search strategy was developed, where the primary focus was to increase the neighborhood search space with varied dimensions, and to allow activities to swap among both its direct and indirect successors and predecessors, even though it may take slightly more computational time than the traditional variable neighborhood approach.

For implementing that evolutionary local searching scheme among diversified neighborhood activities, at first, an initial feasible schedule matrix was generated by following some schedule generation scheme (in this paper we executed the parallel schedule generation scheme (PSGS) [9]). Later on, with that initial schedule, each activity was swapped or moved, in both forward and backward directions, while maintaining both their direct and indirect precedence relations. Following this iterative procedure, after meeting the stopping criteria, if possible, another updated schedule was generated with better makespan performance. It is worth noting however, that in either move strategy, not every move leads to a new schedule due to resource contradictions. Therefore, to find whether a move does lead to a change in schedule, at first, resource constraints should be satisfied with each of the newly generated schedules, by only calculating their start or finish times. If none changes, then the overall schedule will remain the same and it is worthless to re-compute it.

Moreover, extensive experimentation has shown that the better the initial solution with which ELSH-VN for RCMPSP is started, the better on average the final solution will be. For better clarification, the step-wise overall solution scheme of ELSH-VN is given below.

Algorithm: Step-wise procedure of ELSH-VN

```
Begin
  Define iter_max and run_max;   iter = 1; fit = ∞
    For run= 1 to run_max do
    Input V, I_v and all necessary data relating to RCMPSP
    Select the appropriate priority rule (default: random)
    Apply the PSGS heuristic and generate an initial feasible
    schedule matrix, f_sch
    Generate initial fitness matrix, fin_in for each project v
    Find the best schedules among f_sch by considering the least
    fin_in values for each v
    Generate the best schedule matrix (if multiple schedules
    show the same makespan), B_sch for each v
    Remove any duplicates from B_sch and that create the unique
    schedule matrix, B_schuniq for each v
    End
      Repeat until B_schuniq = {} do
      Input the sequence for the first schedule among B_schuniq
      (assume B_schuniq1)      for each v
      Select fit = makespan of B_schuniq1
        While iter ≤ iter_max or fit = LB_known do
        Execute enhanced move strategy
        Determine new schedule matrix, NB_sch for each v
        Remove duplicates and create unique schedule matrix,
        NB_schuniq
        Apply the same priority rule based parallel SGS again
        to calculate the resource feasible schedules matrix,
        NRB_sch
        Determine the least makespan value among all NRB_sch and
        call it MK_new for each v
            If MK_new,v < fit then fit = MK_new
            Elseif  MK_new,v = LB_known then stop
                    iter = iter + 1;
            End if
          End While
      End Repeat
  End
```

Table 1 Priority rules

Rule	Refs.	Formula
Random selection rule- RSR	[11]	Activities selected randomly
First come first served-FCFS	[12]	Min E_{vi}
Last come first served-LCFS	[2]	Max ES_{vi}
Shortest processing time-SPT	[13]	Min d_{vi}
Maximum processing time-MPT	[13]	Max d_{vi}
Shortest activity from shortest project—SASP	[14]	Min f_{vi}, where $f_{vi} = CP_v + d_{vi}$
Longest activity from longest project-LALP	[13]	Max f_{vi}
Minimum slack-MINSLK	[15]	Min SLK_{iv},
Maximum slack-MAXSLK	[15]	Max SLK_{iv}
Latest start time-LST	[10]	Min LS_{vi}, where $LS_{vi} = LF_{vi} - d_{vi}$

3.2 Priority Rules

In order to assess the performance of the proposed ELSH-VN, we have considered, for comparison, a set of scheduling procedures with different priority rules. Since different activities have different features (resource, duration combination), so selecting a set of (or single) active activities is always complicated. Therefore, proposing and switching some priority based rules for use among the active activities may resolve this issue and accelerate performance [10]. We compiled a set of 10 popular priority rules from the literature (shown in Table 1), some of which have already proved to be successful in a single project environment. The main motivation for selecting these different scheduling procedures was to see whether ELSH-VN with default activity selection rule (i.e., random selection rule) is competitive with other possible priority rules. As well as to see if any other combination of priority rules with ELSH-VN can give a better approach.

4 Experimental Results and Analysis

Extensive numerical experiments have been carried out with the aim of assessing the performance of our proposed ELSH-VN for RCMPSP. A comparative evaluation has also been made with the compiled set of 10 priority rules. The computational experiments have been performed on an Intel core i7 processor with 16.00 GB RAM and a 3.40 GHz CPU. All procedures were coded and solved in Matlab, R2015b. ELSH-VN was employed for generated RCMPSP instances, while the maximum iteration number, $iter_{max}$, was set as 1000 and the maximum number of algorithm runs, run_{max} was set as 30. The maximum solving time for all projects under each instance was set as 500 s. An in-depth analysis of the properties of RCMPSP, and the behavior of a scheduling method, is barely possible based on just a single RCMPS instance. Therefore, motivated by Browning and Yassine [16], we followed a pragmatic approach to generate additional test instances for RCMPSP, which will then be used to evaluate our proposed ELSH-VN in more detail.

4.1 Generation of Test Projects

Depending on different instance characteristics, we generated some test instances for RCMPSP. Among a few, Network complexity (C), Normalized average resource loading factor (NARLF) and Modified average utilization factor (MAUF) are the most important characteristics for generating multi-project instances. Details on these characteristics can be found in the earlier work of Browning and Yassine [2]. To maximize possible insights, we considered seven NARLF and 11 MAUF levels [17]. In addition to that, we considered a high project complexity measure for each project (i.e., C = 0.69) while all of the individual resources MAUFs were not equal (i.e., showing $\sigma^2_{MAUF} = 0.25$). Thus, we generated 7 × 11 = 77 test problems for this experiment. As standard problem generators and test sets such as ProGen/PSPLIB cannot create multi-project problems to these specifications, we used a test problem generator developed by Browning and Yassine [16]. For better understandings, a brief summary of this experimental set up and test project generation is outlined in Table 2.

4.2 Experimental Results for Priority Rule Based ELSH-VN

Table 3 summarizes the performance of our proposed ELSH-VN for our newly generated 77 RCMPSP instances. Here we considered four different posterior measures to justify the performance of ELSH-VN, namely: average project delay (APD) as percentage, average portfolio delay (APFD) as percentage, lower total

Table 2 Experimental design (C.F: Constant Factor; M.F: Main Factor)

C.F	Setting	M.F	Levels
V	3 projects per problem	*NARLF*	7 levels: −3, −2, −1, 0, 1, 2, 3
I	20 activities per project	*MAUF*	11 levels: 0–1.6 in increments of 0.1
R	4 types of resources per activity	*C*	1 level: HHH

Table 3 Performance of different priority rule based ELSH-VN

Parameters	RSR	FCFS	LCFS	SPT	MPT	SASP	LALP	MINSLK	MAXSLK	LST
APD	93.36	180.2	198.2	182.8	186.5	182.8	186.5	218.4	174.6	177.5
APFD	136.2	230.8	269.1	229.5	238.6	229.5	238.6	272.4	199.4	229.2
LTM	44.71	62.05	70.25	62.96	64.16	61.96	64.16	69.93	55.92	61.89
TCPU	0.125	0.173	0.176	0.16	0.195	0.160	0.19	0.207	0.189	0.182

makespan (LTM) after 30 runs and finally total computational time (TCPU) per run in seconds. As mentioned before, ten different priority rules were executed under ELSH-VN and the comparative results are outlined in Table 3. As of that table, the random selection rule based ELSH-VN showed best performance in comparison to other priority rules, while MAXSLK and LST are quite competitive. Irrespective of priority rules, the TCPUs for each heuristic are nearly consistent and per run time it is always 0.207 s or less.

Starting with APD, Fig. 1 shows the graphical representation on the impact of different MAUF values on APDs. As evident from this figure, the winning priority rule (i.e., the one with the smallest average percentage delay) is RSR. Surprisingly among LST, MPT, MAXSLK and SASP, there is no significance difference of their respective APD values and so they are apparently tied for second place. Obvious losers include MINSLK as the worst and LCFS as second worst. To investigate the margin of superiority for RSR over others, it is important to have a comprehensive statistical analysis on all proposed algorithms. Since our considered problems are complex in nature and there is a high chance of greater precedence constraints, priority rules that depend on precedence relationships might show lesser influence on minimizing APDs [2].

As evident from Fig. 2, a similar trend can also be found for APFDs. While APD affects the effects of delay on the projects individually, APFD only accounts for delays that lengthen the overall portfolio of projects. As a consequence, while individual project managers would care more about APDs, portfolio managers would have reason to focus on APFDs. Interestingly, here also RSR, LST and MAXSLK show the least (or lesser) APFD values than the other rules.

In addition to analyzing the influence of MAUF on minimizing project delays, we also carried out numerical experimentation on the project data to see the possible impact of NARLFs. The NARLF interaction plot shown in Fig. 3 exhibits the superiority of RSR at all levels, followed by MAXSLK and SASP. MINSLK and LCFS are placed as worst priority rules because of their inefficiency at minimizing APDs. Meanwhile, Fig. 4 plots the relative values of APFD for different levels of NARLF. Similar conclusion can be drawn for APFDs as well. However, as evident in

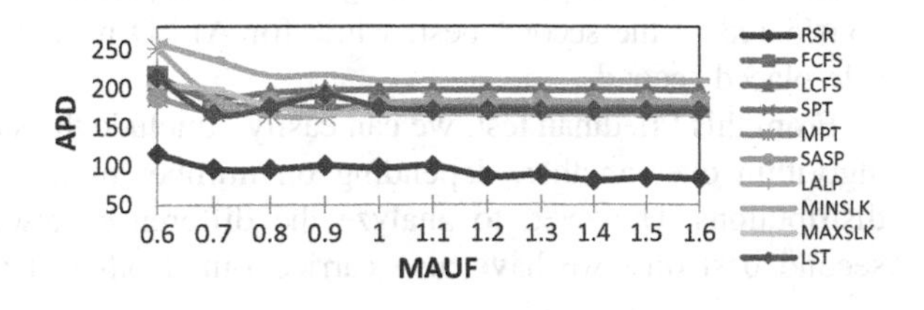

Fig. 1 APD versus MAUF level

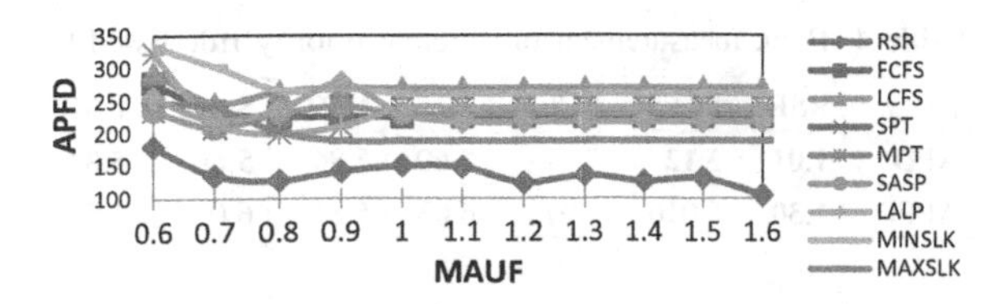

Fig. 2 APFD versus MAUF level

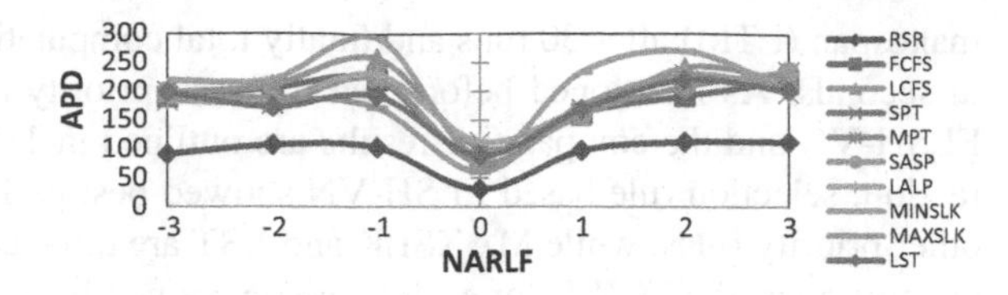

Fig. 3 APD versus NARLF level

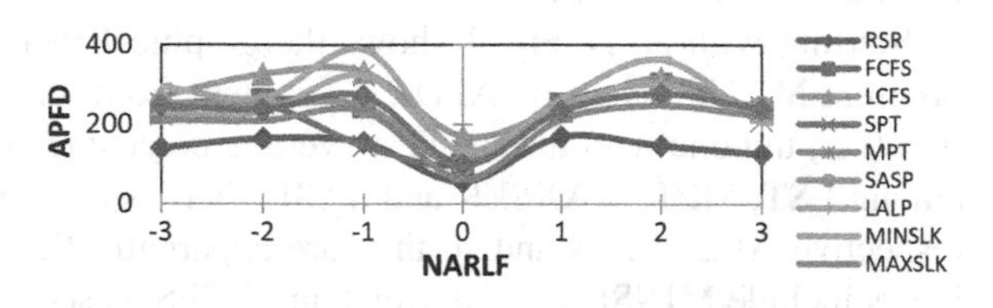

Fig. 4 APFD versus NARLF level

Fig. 4, RSR and MAXSLK show similar APFDs if the value of NARLF is zero. Again for NARLF equals −1, the MAXSLK priority rule shows even better results than the RSR priority rule, which can be considered an important finding of this research.

4.3 Statistical Comparison

As illustrated in the previous section, apart from the RSR priority rule, there is no firm evidence on the superiority of any particular priority rules over others. Hence to assist decision making and to evaluate the overall performance of all priority rule based ELSH-VNs, we used a ranking procedure based on a non-parametric test, called the Friedman test [18]. The Friedman test is often used for a randomized complete block design when the normality assumption is not satisfied or the data is ordinal. Under each instance, for each of the priority rule based ELSH-VN, relative ranks were calculated by considering the two major performance measures (APD and APFD) described in Sect. 4.2. Relevant rankings are summarized in Table 4. As evident from Table 4, RSR shows the least relative rank value, which shows its significant superiority. Considering APD as a performance measure, FCFS can be considered as the second best, while for APFD measures, the MAXSLK priority rule placed second.

From this Friedman test, we can easily conclude the superiority of any particular algorithm over another, depending on number of nodes and types of statistical distributions. However, to analyze the difference between the best one and the second best one, we have also carried out another statistical test, the Wilcoxon

Table 4 Rank measurement of different priority rule based ELSH-VN using friedman test

I.T	RSR	FCFS	LCFS	SPT	MPT	SASP	LALP	MINSLK	MAXSLK	LST
APD	**1.01**	**5.12**	7.17	5.60	5.84	5.60	5.84	8.42	**5.27**	5.14
APFD	**1.30**	6.010	7.97	6.03	5.81	6.03	5.81	7.91	**2.88**	5.27

Table 5 Wilcoxon signed ranks test results for comparing the best performing algorithms

Algorithms	Criterion	Better	Similar	Worst	p	Decision
RSR versus FCFS	APD	77	0	0	0.00	+
RSR versus MAXSLK	APD	77	0	0	0.00	+
RSR versus MAXSLK	APFD	67	0	10	0.00	+

signed ranks test [19]. Using a 5 % significance level, we assigned one of three signs (+, − and ≈), where "+" means that the first algorithm is significantly better than the second one, "−" means that it is significantly worse and '≈' represents that there is no significant difference between these algorithms. According to Table 5, it shows that the RSR priority rule convincingly outperformed its nearest competitive priority rules (i.e., FCFS or MAXSLK). We speculate that the superior performance of MAXSLK over others (except RSR) here may be due to the generally high constraining values of AUF which characterizes our generated instances. Moreover, while intelligence based priority rules work better for single mode RCPSPs, but for RCMPSPs, rules without intelligence (i.e., RSR) works better than others. The appropriate reasoning behind this superiority of RSR is still needs to be researched.

5 Conclusions

In this paper, we considered the RCMPSP with deterministic activity durations and temporal relations. To solve RCMPSP, we developed a variable neighborhood search based greedy algorithm, ELSH-VN. The suitability and appropriateness of that heuristic was then studied by incorporating ten different priority rules. In order to conduct computational experiments with RCMPSP, a large set of test instances has been generated. To judge the superiority of any particular priority-rule based ELSH-VN, we carried out a comprehensive statistical analysis based on both the Friedman and Wilcoxon tests. This statistical comparison, and extensive simulation-based experiment, revealed that the RSR outperformed others, while FCFS and MAXSLK algorithms are insignificantly superior to others, especially when the performance measures were like average project delay or average portfolio delay.

Because of its ongoing practical relevance, RCMPSP is quite common in modern industry, which makes efficient algorithms for them valuable. Decisions about which activities to do when (based on resource allocations) have a substantial effect on project completion time. Hence there is a necessity of choosing suitable priority rules to select appropriate activity. In this context, this paper dealt with ten different priority rules to compare and contrast. Importantly, our experimentation shows that previously published results do not always correspond with their performance on RCMPSP. While widely advocated rules such as MINSLK, SASP and LST did not perform well, our study revealed that the RSR outperformed others.

Future research is possible, for example, considering multiple modes to reflect alternative speeds of the production processes. Also, including additional priority rules, comparing with other schedule generation schemes, exploring other RCMPSP formulations (i.e., allowing activity pre-emption, stochastic activity durations, or uncertain resources) can also be important extensions of this paper.

References

1. Payne, J.H., *Management of multiple simultaneous projects: a state-of-the-art review.* International journal of project management, 1995. **13**(3): p. 163–168.
2. Browning, T.R. and A.A. Yassine, *Resource-constrained multi-project scheduling: Priority rule performance revisited.* International Journal of Production Economics, 2010. **126**(2): p. 212–228.
3. Blazewicz, J., J.K. Lenstra, and A. Kan, *Scheduling subject to resource constraints: classification and complexity.* Discrete Applied Mathematics, 1983. **5**(1): p. 11–24.
4. Drexl, A., *Scheduling of project networks by job assignment.* Management Science, 1991. **37** (12): p. 1590–1602.
5. Gonçalves, J.F., J.J. Mendes, and M.G. Resende, *A genetic algorithm for the resource constrained multi-project scheduling problem.* European Journal of Operational Research, 2008. **189**(3): p. 1171–1190.
6. Mouthuy, S., et al., *A Multistage Very Large-Scale Neighborhood Search for the Vehicle Routing Problem with Soft Time Windows.* Transportation Science, 2015. **49**(2): p. 223–238.
7. Mladenovic, N., R. Todosijevic, and D. Urosevic, *An efficient general variable neighborhood search for large travelling salesman problem with time windows.* Yugoslav Journal of Operations Research ISSN: 0354–0243 EISSN: 2334-6043, 2012. **23**(1).
8. Talbot, F.B., *Resource-constrained project scheduling with time-resource tradeoffs: The nonpreemptive case.* Management Science, 1982. **28**(10): p. 1197–1210.
9. Kolisch, R., *Serial and parallel resource-constrained project scheduling methods revisited: Theory and computation.* European Journal of Operational Research, 1996. **90**(2): p. 320–333.
10. Kolisch, R., *Efficient priority rules for the resource-constrained project scheduling problem.* Journal of Operations Management, 1996. **14**(3): p. 179–192.
11. Kolisch, R. and S. Hartmann, *Experimental investigation of heuristics for resource-constrained project scheduling: An update.* European journal of operational research, 2006. **174**(1): p. 23–37.
12. Patterson, J.H., *Alternate methods of project scheduling with limited resources.* Naval Research Logistics Quarterly, 1973. **20**(4): p. 767–784.
13. Alvarez-Valdes, R. and J.M. Tamarit, *Heuristic algorithms for resource-constrained project scheduling: A review and an empirical analysis*, in *Advances in project scheduling.* 1989, Elsevier Amsterdam. p. 113–134.
14. Kurtulus, I.S. and S.C. Narula, *Multi-project scheduling: Analysis of project performance.* IIE transactions, 1985. **17**(1): p. 58–66.
15. Davis, E.W. and J.H. Patterson, *A comparison of heuristic and optimum solutions in resource-constrained project scheduling.* Management science, 1975. **21**(8): p. 944–955.
16. Browning, T.R. and A.A. Yassine, *A random generator of resource-constrained multi-project network problems.* Journal of scheduling, 2010. **13**(2): p. 143–161.
17. Kurtulus, I. and E. Davis, *Multi-project scheduling: Categorization of heuristic rules performance.* Management Science, 1982. **28**(2): p. 161–172.
18. Zimmerman, D.W. and B.D. Zumbo, *Relative power of the Wilcoxon test, the Friedman test, and repeated-measures ANOVA on ranks.* The Journal of Experimental Education, 1993. **62**(1): p. 75–86.
19. Gehan, E.A., *A generalized Wilcoxon test for comparing arbitrarily singly-censored samples.* Biometrika, 1965. **52**(1–2): p. 203-223.

Genetic Programming with Embedded Feature Construction for High-Dimensional Symbolic Regression

Qi Chen, Mengjie Zhang and Bing Xue

Abstract Feature construction is an effective way to eliminate the limitation of poor data representation in many tasks such as high-dimensional symbolic regression. Genetic Programming (GP) is a good choice for feature construction for its natural ability to explore the feature space to detect and combine important features. However, there is very little contribution devoted to enhance the generalisation performance of GP for high-dimensional symbolic regression by feature construction. This work aims to develop a new feature construction method namely genetic programming with embedded feature construction (GPEFC) for high-dimensional symbolic regression. GPEFC keeps track of new small informative building blocks on best fitness gain individuals and constructs new features using these building blocks. The new constructed features augment the Terminal Set of GP dynamically. A series of experiments were conducted to investigate the learning ability and generalisation performance of GPEFC. The results show that GPEFC can evolve more compact models in an efficient way, has better learning ability and better generalisation performance than standard GP.

Keywords Genetic programming · Symbolic regression · Feature construction · Generalisation

Q. Chen (✉) · M. Zhang · B. Xue
School of Engineering and Computer Science, Victoria University of Wellington, Wellington, New Zealand
e-mail: Qi.Chen@ecs.vuw.ac.nz; chenqi1@ecs.vuw.ac.nz

M. Zhang
e-mail: Mengjie.Zhang@ecs.vuw.ac.nz

B. Xue
e-mail: Bing.Xue@ecs.vuw.ac.nz

G. Leu et al. (eds.), *Intelligent and Evolutionary Systems*, Proceedings in Adaptation, Learning and Optimization 8, DOI 10.1007/978-3-319-49049-6_7

1 Introduction

The importance of data representation has been recognized in machine learning for a long time. Poor input representation limits performance of learning algorithms. One approach to eliminating the limitation is to construct new high-level features. Feature construction is a process to discover relationships between features, and create novel features based on these relationships. These novel features will augment the feature space, and make the pattern to be more apparent, thus yielding better performance of learning algorithms [1, 2]. The natural ability of Genetic Programming (GP) [3] in detecting and combining important features has made it a good choice for feature construction. A large number of works have been devoted to this field [2, 4, 5].

Generalisation is a kind of ability by which learning algorithms produce solutions that can have a reasonable performance on the unseen data. It is one of the most important evaluation criteria for learning algorithms. For decades, generalisation is always a hot topic in many fields in machine learning [6, 7]. In GP for classification, many works have been devoted to the generalisation of GP. In GP for symbolic regression, however, before Kushchu published his work [8] in 2002, almost none of the works on GP based symbolic regression reported performance on unseen data. Until recent years, an increasing number of works have been devoted to enhance the generalisation of GP for symbolic regression [9–12]. However, there has not been any work on improving GP's generalisation using feature construction for symbolic regression.

In most GP-based symbolic regression tasks, there is typically one to ten input variables. We considered regression tasks having more than 100 features to be high-dimensional. For high-dimensional symbolic regression tasks, the underlying relationship between variables is more difficult to identify, and the risk of overfitting becomes high, thus it is difficult to generalise well [13, 14]. In this scenario, feature construction may help.

The overall goal of this work is to develop a new feature construction method using GP for high-dimensional symbolic regression tasks and investigate whether GP with the new feature construction method can improve its performance. Specific objectives are:

- whether GP with the new feature construction method can have better performance on training data over standard GP;
- whether GP with the new feature construction method can generalise well on unseen test data;
- whether GP with the feature construction method can induce more compact models; and
- how the feature construction method influences GP's computational cost.

2 Background

This section gives a brief background to GP for symbolic regression, GP-based feature construction, and building blocks in GP.

2.1 Genetic Programming for Symbolic Regression

Genetic Programming (GP) is a very popular approach for symbolic regression, which is a function identification process. The task of symbolic regression is to identify the relationship between the input variables and the outputs for a given dataset and express the relationship in a mathematical model. The ability of GP in automatically creating various programs and no requirement of predefined shapes and sizes of solutions makes it a desired approach for symbolic regression. A large number of improvements to GP-based symbolic regression have been developed on the literature every year [15–18].

2.2 Genetic Programming for Feature Construction

Many different GP-based feature construction methods have been presented in the literature [1, 2]. These methods can be classified into three categories: *filters*, *wrappers* and *embedded methods*. In *filter methods*, the fitness of the constructed feature is evaluated by a function, for example a function that acts as a surrogate classifier in classification tasks. In *wrappers*, instead of a specifically designed fitness function used in *filters*, the performance of another machine learning algorithm is used to evaluate the fitness of the constructed features. *Embedded methods* incorporate the feature construction process within the learning process and the constructed features take apart in the learning process of GP directly.

In previous research, approaches using GP for feature construction were mainly for classification problems. An early approach using GP for feature construction was introduced for the face recognition task by Vafaie and De Jong [19]. Since then, a number of contributions have been devoted to use GP for feature construction for achieving better performance with kinds of classifiers, such as C4.5 [20], Support Vector Machine [21] and K-Nearest Neighbour [22].

2.3 Building Blocks in Genetic Programming

GP with Automatically Defined Functions (GP-ADF) [23] is a very first method using the building blocks in GP. In GP-ADF, the program takes the form of a sequence of *m* program trees. The program trees in the sequence (except the first one) are referred to *Automatically Defined Functions* (ADFs). These ADFs are then used to extend the function set following the rule of using ADFs which occurs later in the sequence. GP with ADFs proposes an efficient way to evolve subtrees of programs. However, GP with ADFs selects building blocks by modifying ADFs at randomly chosen crossover points. In many cases, this might not be a good idea due to the low locality problem of ADFs.

Ballard and Rosca [24] proposed a method named *Adaptive Representation*, which aims to automatically extract common knowledge in the form of building blocks to extend the problem representation. In contrast to ADFs which manipulate building blocks randomly, Adaptive Representation takes a heuristic approach to selecting new blocks based on block fitness.

In summary, there are many works on GP-based feature construction to improve the classification performance. However, to date, there is no work devoted to discover useful building blocks and construct new features to GP for symbolic regression, thus helping improve its learning performance and generalisation ability, let alone high-dimensional symbolic regression problems.

3 Proposed Approach

In GP, a building block refers to a piece of code, a subtree or subprogram of the candidate programs [23, 25]. The quality of these simple pieces of code is an important factor for the success of GP. In this work, a new GP-based feature construction approach namely GP with embedded feature construction (GPEFC) is proposed. GPEFC uses building blocks as a way of representation transformation. The motivation of GPEFC is to detect good genetic materials(building blocks) and utilise them by constructing new features and introducing them into candidate programs. An overview of GPEFC is shown in Fig. 1. Compared with standard GP, GPEFC has two additional key components. While the first deals with how to detect and use

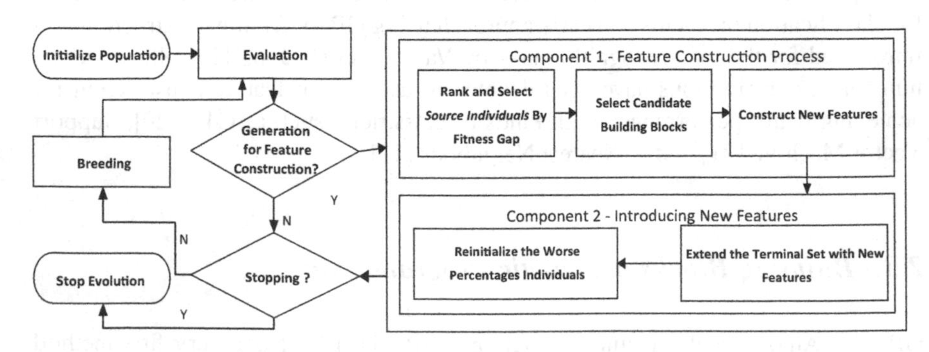

Fig. 1 Overview of GPEFC

good genetic materials to construct more informative new features, the second deals with how to use these new features to improve system performance.

3.1 Component 1— Feature Construction Process

The feature construction process consists of three sequential steps. First, discovering *source individuals* which are more likely to contain good building blocks. We assume that individuals that have higher fitness gain than their parent(s) are more likely to contain useful genetic materials, since higher fitness gain is presumably brought by combining more useful building blocks. Under the hypothesis, individuals are ranked according the fitness gain with its parent(s), which is defined by

$$FG = min(Fit(pi_1), Fit(pi_2)) - Fit(i) \quad (1)$$

where i is the individual, pi is/are the parent(s) of i, $Fit(i)$ is the fitness of i, $Fit(pi)$ is/are the fitness value(s) of pi, and $min(Fit(pi_1), Fit(pi_2))$ is the smaller fitness value of the parent(s) (smaller means better, as a minimisation function will be used). The larger FG, the better i. In addition, a parameter p is used to determine the percentage of individuals to be selected. A appropriate value of p is needed to maintain a trade-off between introducing useful building blocks and keeping the total number of them to be relatively small.

Second, detecting and collecting building blocks containing good genetic materials in the *source individuals*. Two key factors are taken into account, *depth* and *activeness*. While the *depth* is the number of levels in the building blocks, the *activeness* of a building block is defined to be the number of times it appears in all the *source individuals*. The threshold value for *depth* should be large enough to ensure the building blocks are useful while keeping the length of programs using these new features within manageable bounds. According to previous research [24, 26, 27], building blocks which have two to three levels are a good choice. In terms of *activeness*, each building block has an associate counter regarding its number of occurrences among all the *source individuals*. The importance of the building block is assumed to be related to this value. Based on this assumption, GPEFC splits all the building blocks within the limited *depth* into *important* ones and *unimportant* ones according to its *activeness* value and a threshold. The threshold is defined to be $\beta \times Num_p$ where Num_p is the number of *source individuals*, and β is the weight. β is simply set to be 0.5, so that the threshold defines a building block to be *importance* only when the number of individuals in which it appears is more than that of it does not appear.

The third step is using the output expression of these building blocks to construct new features directly. The values of the new feature should be transformed accordingly. Each distinguished building block can construct only one new feature. After the construction of all the available building blocks, the new features are ready to be used to extend the Terminal Set.

3.2 Component 2—Introducing New Features

The second key component is how to use the new features. For preparation, the new features are put into the feature pool and extend the Terminal Set. Then GPEFC uses these features in two ways. The first way is generating a number of new individuals randomly using the new extended Terminal Set. By doing this, GPEFC introduces new features into the population directly and increases the probability of using new constructed features. A parameter rp defines the percentage of worst individuals to be deleted in order to make room to the same number of newly generated individuals. These worst individuals should have highest error values compared to their counterparts in the current population. An appropriate value of parameter rp helps increasing the probability of introducing the new features into the population while does not destroy the good enough individuals evolved over generations. Secondly, by means of the mutation operator, subtrees containing new features will be introduced into the population of GP.

3.3 Evaluation Measure—Fitness Function

The performance of the GP individuals are evaluated by the *Normalised Root Mean Square Error* (NRMSE) on both the training set and the test set. It is calculated as shown in Eq. (2).

$$NRMSE = \sqrt{N/(N-1) \times MSE}/\delta_t \quad (2)$$

where N is the number of instances, MSE is the mean square error and δ_t is the standard deviation of the target outputs.

4 Experiment Design

This work used standard GP as a baseline for comparison. Both GP and GPEFC were tested on six real-world high-dimensional symbolic regression datasets.

4.1 Parameters

The experimental parameters are provided in Table 1. For GPEFC, there are three important parameters. They are the parameter α to determine how often GPEFC constructs new features, the parameter p which refers to the percentage of individuals having best fitness gain with their parent(s) and the percentage rp of the worst

Table 1 Parameters for GP and GPEFC

Parameter	Values	Parameter	Values
Population size	512	Generations	100
Crossover rate	0.9	Mutation rate	0.1
Elitism	0.01	Maximum depth	17
Initialisation	Ramped-half and half	Min-max initial depth	2–6
Function set	+, −, ∗,% protected	Terminal set	Features, constant ∈ [−1,1]
Generation interval α	5	Top (worst) individuals $p(rp)$	10 %

individuals in terms of their fitness to be deleted in order to give room to the new individuals. All the values of these important parameters are identified by empirical search. In this work, since the maximum number of generation is 100, α is set to be 5, 10, 20. Then, the trials pointed out that 5 is a reasonable value to α. At the same time, after a few trials the appropriate value to parameter p and rp are both set to be 10 %. GP and GPEFC have been run for 100 independent runs on each dataset.

4.2 Benchmark Problems

Since GP is still lacking benchmarks specifically designed for the research of generalisation, six real-world datasets are taken from previous research on generalisation of GP and UCI [28]. All the six datasets have a large dimensionality of the feature space. The numbers of instances and features of the six datasets are shown in Table 2. The first two datasets are problems in the field of pharmacokinetics. They consist of predicting the values of two different kinds of pharmacokinetics parameters: human oral bioavailability (represent as %F) and median lethal dose (LD50). For more detail of these two datasets, readers are referred to [29]. The third dataset, Diffuse Large-B-Cell Lymphoma (DLBCL) was collected from Rosenwald et al. [30], which is to predict the survival time of patients who have diffuse large-B-cell lymphoma after chemotherapy.

Table 2 Benchmark problems

Dataset	# Features	# Instance			Dataset	# Features	# Instance		
		Total	Training	Test			Total	Training	Test
%F	241	359	251	108	LD50	626	234	163	71
DLBCL	7399	240	180	60	CCUN	124	1994	1395	599
CCN	122	1994	1395	599	RLCT	384	53500	37450	16050

The other three datasets are all taken from UCI [28]. Two of the datasets are the Communities and Crime unnormalised dataset (CCUN) and the Communities and Crime normalised dataset (CCN). The task of them is to predict the per capita crimes. The final dataset is the Relative location of CT slices on axial axis (RLCT). The task of the dataset is to predict the relative location of the CT slice on the axial axis of the human body.

4.3 Training Sets and Test Sets

In order to investigate the generalisation performance of the two methods, we split each dataset into a training set and a test set (they are provided in DLBCL). All the experiments are conducted with 70 % of instances randomly selected from the dataset for training and the other 30 % instances forms the test set, which is a common way in many previous research [4, 29, 31].

5 Results and Discussions

This section presents experimental results of GP and GPEFC. The comparison will be presented in terms of NRMSE on the training sets and the test sets, the number of nodes and the number of distinguished features of the evolved solutions. The computational cost will also be compared. An non-parametric statistical significance test, the Wilcoxon test, is used in this work to compare the 100 runs NRMSE values of the best-of-run individuals in the training set and the test set of two methods. The significance level is 0.05.

Figure 2 displays the distribution of NRMSE of the 100 best-of-run individuals for the six datasets. The evolution plots of the training sets and the test sets over generations are shown in Figs. 3 and 4, respectively. All the evolution plots are based on the median value of the 100 runs of the best individual at every generation. Since the median value is more robust to outliers, it was preferred over the mean value [11].

5.1 Results of Program Size and Number of Features

The average program size and number of distinguished features of the 100 best-of-run individuals are shown in Table 3. On most of the problems, GPEFC has a much smaller program size than GP except for CCN and RLCT. This might due to the richness of the new constructed features. It is shown that GPEFC can evolve more compact solutions than GP in most cases. Compact models are generally easier to interpret and faster in execution.

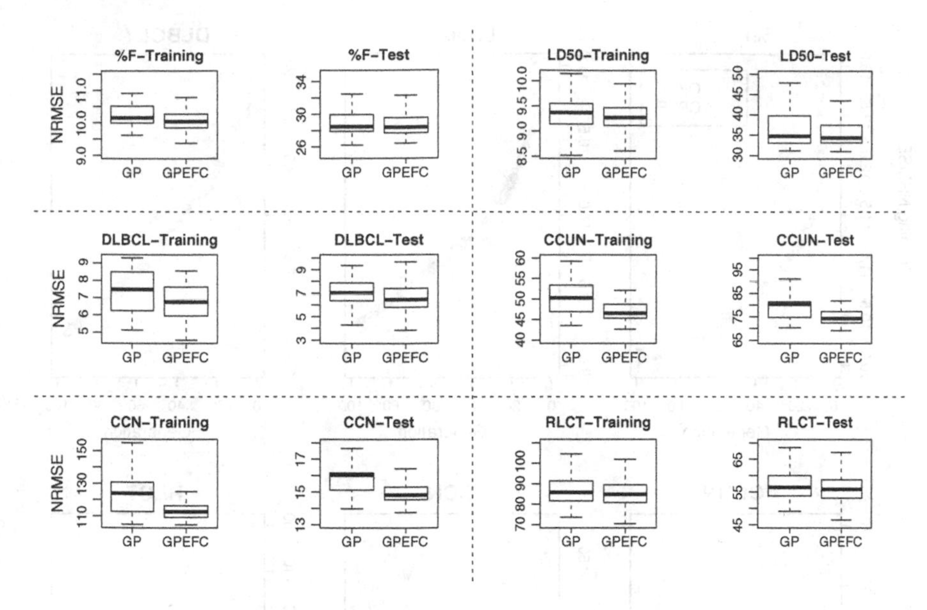

Fig. 2 Distribution of NRMSE of the 100 best of runs individuals

In terms of distinguished features, GPEFC has slightly larger average number of distinguished features in the evolved models on all the six problems. This is due to the dynamically extended Terminal Set, but the increase is very small and can almost be neglected.

The results show that GPEFC can utilise the feature construction process to compress different features together to discover hidden information among them, which allows it to evolve compact solutions to improve the performance.

5.2 *Results on the Training Sets*

As it shows in Fig. 3, it is unquestionable that GPEFC has smaller training errors than GP on all the test problems over generations. While GPEFC has slight performance gains on %F, LD50 and RLCT, it contributes to a quite dramatic performance improvement on the other three problems (DLBCL, CCUN and CCN). The error distances between GPEFC and GP increase along with the number of generations on most of the test problems except for RLCT. The major reason may lie in that the increasing number of new constructed features over generations may provide more useful information for the training process. Also this can explain why on RLCT which have over 35,000 instances does not have this trend, since the useful information can be provided by the large number of instances. In many cases, it is hard to obtain such a large number. The advantage of GPEFC on the training sets can

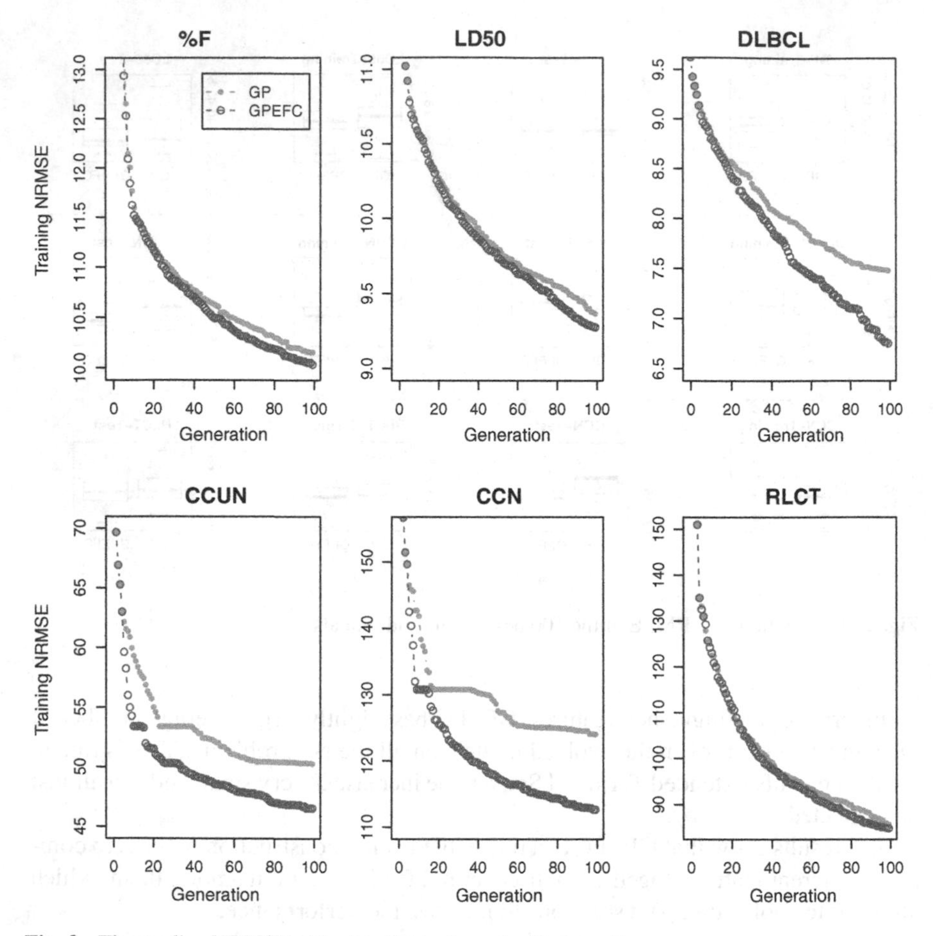

Fig. 3 The median NRMSE of best individuals on the *Training Sets*

also be confirmed by the training boxplots in Fig. 2, which display the distribution of NRMSE of the 100 best-of-run individuals on the training sets.

The results suggest that GPEFC which is equipped with introducing new constructed features into the candidate individuals and the Terminal Set is able to have a positive effect on accelerating the learning ability of GP.

5.3 *Results on the Test Sets*

The evolution plots of the test sets are presented in Fig. 4. It can be observed that the overall pattern is the same as the training sets. GPEFC is also superior to GP on all the six datasets. For the first two datasets (%F and LD50), overfitting occurs.

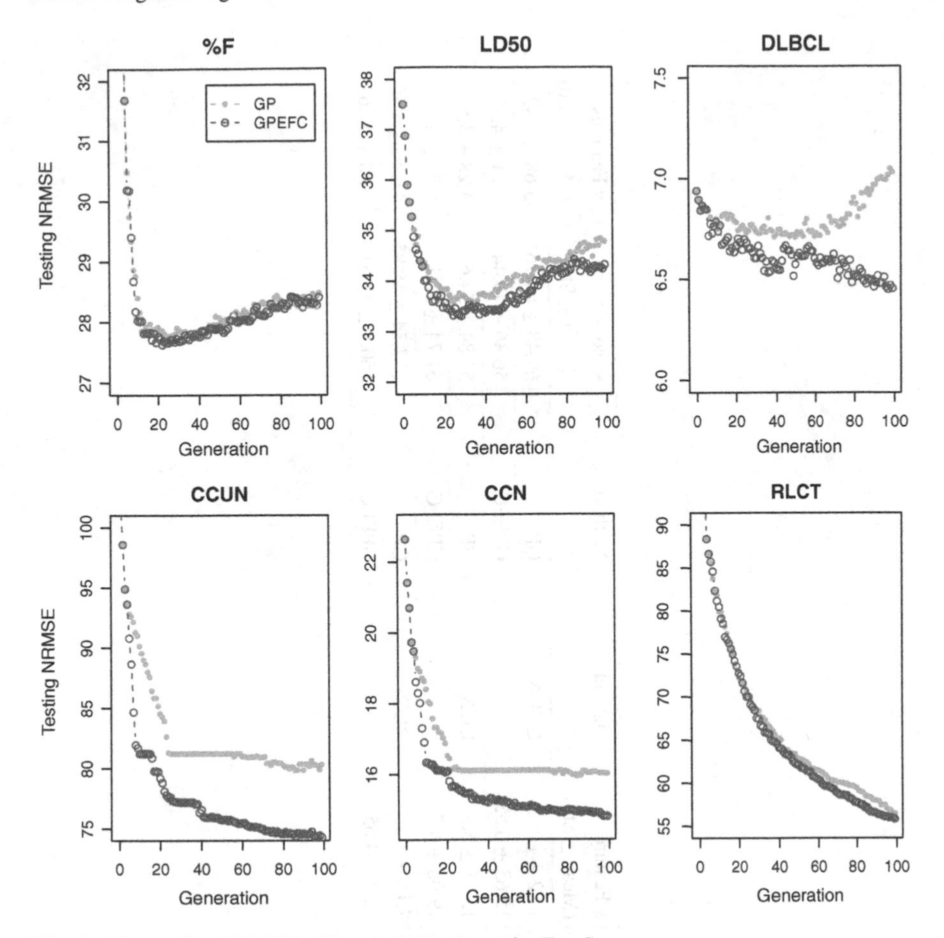

Fig. 4 The median NRMSE of best individuals on the *Test Sets*

This may be because the available features and instances of the two datasets do not provide enough useful information for the evolutionary process. This was confirmed by the extremely large average program size and number of distinguished features of the best individuals (Table 3), which are much higher than the other four datasets. While both GPEFC and standard GP have increasing test errors, GPEFC can still have lower NRMSE on both tasks, i.e. better generalisation ability.

In contrast to the first two tasks, GPEFC has good generalisation performance on the other four datasets. On these datasets, the pattern on the test sets is very similar to the training set. The distance of NRMSE between the two methods increases over generations. For DLBCL, CCUN and CCN, comparing to GP, GPEFC has a generalisation gain which is significant (p-values, DLBCL = 1.20E-8, CCUN = 1.45E-11, CCN = 2.28E-12). It needs to note that on DLBCL, while GP has overfitting problem, GPEFC can eliminate overfitting effectively and generalise dramatically well. On this task, the new constructed features may contain enough useful information

Table 3 Number of nodes and features

Dataset	Method	# Node (Mean ± Std)	# Features (Mean ± Std)	Dataset	Method	# Node (Mean ± Std)	# Features (Mean ± Std)
%F	GP	182.66 ± 65.29	13.2 2 ± 5.7	CCUN	GP	69.48 ± 64.76	5.68 ± 4.43
	GPEFC	157.44 ± 63.7	15.63 ± 6.53		GPEFC	56.36 ± 49.44	6.01 ± 4.72
LD50	GP	202.2 ± 76.63	15.98 ± 5.8	CCN	GP	51.84 ± 54.63	4.28 ± 3.52
	GPEFC	180.8 ± 68.17	19.08 ± 6.75		GPEFC	61.74 ± 52.1	6.21 ± 4.04
DLBCL	GP	96.02 ± 71.75	7.15 ± 5.29	RLCT	GP	157.34 ± 58.85	11.96 ± 4.55
	GPEFC	85.16 ± 58.78	7.54 ± 4.65		GPEFC	160.4 ± 57.51	14.0 ± 5.26

Table 4 Computational time

Dataset	Method	Time (Millisecond) (Mean ± Std)	Dataset	Method	Time (Millisecond) (Mean ± Std)
%F	GP	4401.01 ± 2006.13	CCUN	GP	8069.6± 7577.09
	GPEFC	3687.21 ± 1485.87		GPEFC	6803.75 ± 4411.12
LD50	GP	3553.41 ± 1703.16	CCN	GP	5639.76± 5480.00
	GPEFC	3211.18 ± 1364.41		GPEFC	6706.49 ± 5363.38
DLBCL	GP	1953.36 ± 1253.27	RLCT	GP	6.58E5 ± 2.07E5
	GPEFC	1859.13 ± 884.03		GPEFC	6.8E5 ± 2.03E5

that provided by candidate building blocks. This is quite different from the first two tasks. The contribution can be confirmed by the smaller program size and a bigger number of distinguished features. For RLCT, GPEFC can also have significantly better generalisation ability (p-value = 0.012), although the distance of NRMSE between the two methods is not as big as that on the other three datasets.

For all the tasks, GPEFC has a positive effect on improving the generalisation ability of GP. When overfitting happens, the detected building blocks are more likely to contain noisy information or information that is possibly unique to the training set, thus the test performance of GPEFC may decrease compared to its training performance. However, GPEFC is able to have much better generalisation performance than GP.

5.4 *Computational Cost*

The average computational cost of the 100 runs is shown in Table 4. It can be seen that for the first four problems, GPEFC has a lower computational cost than GP. This is not very intuitive, since compared to GP, GPEFC needs additional computational cost for feature construction, reinitializing a number of individuals and transferring the input values for the new features. However, the smaller average program size over the population of GPEFC decreases the evaluation cost. This is the major reason for the efficiency of GPEFC. This can also explain why on the last two problems, GPEFC is slightly computationally more expensive than GP.

5.5 *Results of Statistical Significance Tests*

The results of the statistical significance tests which compare the performance of GPEFC and GP on the training set and the test set are shown in Table 5. "–" ("+") shows GPEFC has significantly better(worse) performance than GP. "=" means they are similar.

Table 5 Result of statistical significance test

Dataset	GP versus GPEFC		Dataset	GP versus GPEFC	
	Training	Test		Training	Test
%F	–	=	CCUN	–	–
LD50	–	=	CCN	–	–
DLBCL	–	–	RLCT	–	–

6 Conclusions and Future Work

This work developed a new feature construction method for GP and investigates how it can influence the generalisation of GP for high-dimensional symbolic regression tasks. A set of experiments were conducted on six high-dimensional real-world symbolic regression datasets.

The experimental results show that, with appropriate design, GPEFC can evolve more compact models which have significantly better regression performance on the training sets and huge generalisation gains on the unseen data than GP. Furthermore, since GPEFC can reduce the size of models in the population, it is generally computationally less expensive than GP.

To the best of our knowledge, this is the first work on feature construction to enhance regression performance and generalisation ability of GP. Since there is still no systematic work on this field, we expect this work to attract the community's attention on exploring the topic of feature construction on GP for symbolic regression. In future, we will further investigate the key factor affecting the effectiveness of the detected building blocks and the relationship between the new features and generalisation ability. We also plan to incorporate an overfitting detecting mechanism, for example using a validation set, introducing early stop detection and model complexity measurement to GPEFC. It is expected to improve the generalisation of GP in a more effective way.

References

1. Ahmed, S., Zhang, M., Peng, L., Xue, B.: Multiple feature construction for effective biomarker identification and classification using genetic programming. In: Proceedings of the 2014 conference on Genetic and evolutionary computation. pp. 249–256. ACM (2014)
2. Neshatian, K., Zhang, M., Andreae, P.: A filter approach to multiple feature construction for symbolic learning classifiers using genetic programming. Evolutionary Computation, IEEE Transactions on 16(5), 645–661 (2012)
3. Koza, J.R.: Genetic programming: on the programming of computers by means of natural selection, vol. 1. MIT press (1992)
4. Krawiec, K.: Genetic programming-based construction of features for machine learning and knowledge discovery tasks. Genetic Programming and Evolvable Machines 3(4), 329–343 (2002)

5. Neshatian, K., Zhang, M., Johnston, M.: Feature construction and dimension reduction using genetic programming. In: AI 2007: Advances in Artificial Intelligence, pp. 160–170. Springer (2007)
6. Amari, S.i., Wu, S.: Improving support vector machine classifiers by modifying kernel functions. Neural Networks 12(6), 783–789 (1999)
7. Smola, A.J., Schölkopf, B.: A tutorial on support vector regression. Statistics and computing 14(3), 199–222 (2004)
8. Kushchu, I.: Genetic programming and evolutionary generalization. Evolutionary Computation, IEEE Transactions on 6(5), 431–442 (2002)
9. Castelli, M., Manzoni, L., Silva, S., Vanneschi, L.: A quantitative study of learning and generalization in genetic programming. In: Genetic Programming, pp. 25–36. Springer (2011)
10. Chen, Q., Xue, B., Shang, L., Zhang, M.: Improving generalisation of genetic programming for symbolic regression with structural risk minimisation. In: Proceedings of the 2016 on Genetic and Evolutionary Computation Conference. pp. 709–716. ACM (2016)
11. Gonçalves, I., Silva, S., Fonseca, C.M.: On the generalization ability of geometric semantic genetic programming. In: Genetic Programming, pp. 41–52. Springer (2015)
12. Uy, N.Q., Hien, N.T., Hoai, N.X., ONeill, M.: Improving the generalisation ability of genetic programming with semantic similarity based crossover. In: Genetic Programming, pp. 184–195. Springer (2010)
13. Glavan, M., Gradišar, D., Atanasijević-Kunc, M., Strmčnik, S., Mušič, G.: Input variable selection for model-based production control and optimisation. The international journal of advanced manufacturing technology 68(9-12), 2743–2759 (2013)
14. Smits, G., Kordon, A., Vladislavleva, K., Jordaan, E., Kotanchek, M.: Variable selection in industrial datasets using pareto genetic programming. GENETIC PROGRAMMING SERIES 9, 79 (2006)
15. Arnaldo, I., Krawiec, K., O'Reilly, U.M.: Multiple regression genetic programming. In: Proceedings of the 2014 conference on Genetic and evolutionary computation. pp. 879–886. ACM (2014)
16. Azad, R.M.A., Ryan, C.: A simple approach to lifetime learning in genetic programming-based symbolic regression. Evolutionary computation 22(2), 287–317 (2014)
17. Kommenda, M., Affenzeller, M., Kronberger, G., Burlacu, B., Winkler, S.: Multi-population genetic programming with data migration for symbolic regression. In: Computational Intelligence and Efficiency in Engineering Systems, pp. 75–87. Springer (2015)
18. Mousavi Astarabadi, S.S., Ebadzadeh, M.M.: Avoiding overfitting in symbolic regression using the first order derivative of gp trees. In: Proceedings of the Companion Publication of the 2015 on Genetic and Evolutionary Computation Conference. pp. 1441–1442. ACM (2015)
19. Vafaie, H., De Jong, K.: Genetic algorithms as a tool for restructuring feature space representations. In: Tools with Artificial Intelligence, 1995. Proceedings., Seventh International Conference on. pp. 8–11. IEEE (1995)
20. Otero, F.E., Silva, M.M., Freitas, A.A., Nievola, J.C.: Genetic programming for attribute construction in data mining. In: EuroGP. vol. 3, pp. 384–393. Springer (2003)
21. Li, D.C., Liu, C.W.: Extending attribute information for small data set classification. Knowledge and Data Engineering, IEEE Transactions on 24(3), 452–464 (2012)
22. Tran, B., Xue, B., Zhang, M.: Genetic programming for feature construction and selection in classification on high-dimensional data. Memetic Computing 8(1), 3–15 (2016)
23. Koza, J.R.: Genetic programming II: automatic discovery of reusable programs. MIT press (1994)
24. Ballard, D., Rosca, J.: Genetic programming with adaptive representations (1994)
25. Oppacher, U.M.O.F.: The troubling aspects of a building block hypothesis for genetic programming. Foundations of Genetic Algorithms 1995 (FOGA 3) 3, 73 (2014)
26. Kinzett, D., Johnston, M., Zhang, M.: Numerical simplification for bloat control and analysis of building blocks in genetic programming. Evolutionary Intelligence 2(4), 151–168 (2009)
27. Kinzett, D., Zhang, M., Johnston, M.: Analysis of building blocks with numerical simplification in genetic programming. In: Genetic Programming, pp. 289–300. Springer (2010)

28. Lichman, M.: UCI machine learning repository (2013), http://archive.ics.uci.edu/ml
29. Archetti, F., Lanzeni, S., Messina, E., Vanneschi, L.: Genetic programming for computational pharmacokinetics in drug discovery and development. Genetic Programming and Evolvable Machines 8(4), 413–432 (2007)
30. Rosenwald, A., Wright, G., Chan, W.C., Connors, J.M., Campo, E., Fisher, R.I., Gascoyne, R.D., Muller-Hermelink, H.K., Smeland, E.B., Giltnane, J.M., et al.: The use of molecular profiling to predict survival after chemotherapy for diffuse large-b-cell lymphoma. New England Journal of Medicine 346(25), 1937–1947 (2002)
31. Vanneschi, L., Silva, S., Castelli, M., Manzoni, L.: Geometric semantic genetic programming for real life applications. In: Genetic Programming Theory and Practice XI, pp. 191–209. Springer (2014)

The Convolutional Neural Network Model Based on an Evolutionary Approach For Interactive Picture Book

Saya Fujino, Taku Hasegawa, Miki Ueno, Naoki Mori and Keinosuke Matsumoto

Abstract Creating interactive picture books based on human "Kansei" is one of the most interesting and difficult issues in the artificial intelligence field. We have proposed a novel interactive picture book based on *Pictgent* (Picture Information Shared Conversation Agent) and *CASOOK* (Creative Animating Sketchbook). Since our system accepts human sketches instead of natural languages, a high degree of sketch recognition accuracy is required. Recently, convolutional neural networks (CNNs) have been applied to various image- recognition tasks successfully. We have also adopted a CNN model for the sketch recognition of the proposed interactive picture book. However, it takes a considerable effort to tune the hyperparameters of a CNN. In this paper, we propose a novel parameter tuning method for CNNs using an evolutionary approach. The effectiveness of the proposed method is confirmed by a computer simulation that uses, as an example, a scribble-object recognition problem for the interactive picture book.

Keywords Sketch recognition · Convolutional neural network · Evolutionary computation · Genetic algorithm · Scribble objects · Kansei engineering

S. Fujino (✉)
College of Engineering, Osaka Prefecture University,
1-1 Gakuencho, Nakaku, Sakai, Osaka 599-8531, Japan
e-mail: fujino@ss.cs.osakafu-u.ac.jp

T. Hasegawa · N. Mori · K. Matsumoto
Graduate School of Engineering, Osaka Prefecture University,
1-1 Gakuencho, Nakaku, Sakai, Osaka 599-8531, Japan

M. Ueno
Information and Media Center, Toyohashi University of Technology,
1-1 Hibarigaoka, Tempaku-cho, Toyohashi, Aichi 441-8580, Japan

G. Leu et al. (eds.), *Intelligent and Evolutionary Systems*,
Proceedings in Adaptation, Learning and Optimization 8,
DOI 10.1007/978-3-319-49049-6_8

1 Introduction

Understanding drawn picture information by a computer is one of the most important topics in the artificial intelligence field. We have developed an interactive picture book based on human "Kansei" by utilizing human sketches. Our proposed interactive picture book is based on two applications: *Pictgent* (Picture Information Shared Conversation Agent) and *CASOOK* (Creative Animating Sketchbook). Various studies have analyzed picture information by using computer algorithms. However, there exist three problems in those studies:

- It is not sufficient to collect freehand drawings from various users, especially children and people who are not adept at drawing.
- Several researchers have analyzed the features of existing pictures, such as color histograms and picture composition, but they ignore the mode of drawing [9].
- An important feature of pictures is the object class. One study focused on sketch recognition [3] and handwritten character recognition [10] to predict the object class. However, these approaches cannot be applied to freehand drawings because such drawings may be composed of various classes of objects, such as animals, characters, symbols, etc.

Since our system accepts human sketches instead of natural languages, a high degree of sketch recognition accuracy is required when considering the aforementioned problems. One of the most powerful approaches for solving those problems is the use of convolutional neural networks (CNNs) [6] with deep learning, which has shown great accuracy in various image-recognition tasks. We have introduced a CNN model as a sketch-recognition module into the proposed interactive picture book. However, the parameter tuning of CNNs requires a considerable effort.

In this paper, we propose a new parameter-tuning method for CNNs by using an evolutionary approach to solve a parameter tuning problem. To make a CNN useful, we have to consider both network structure and its numerical parameters, such as the filter size. A genetic algorithm (GA) [4, 5], which is a search-and-optimization algorithm based on the mechanism of natural evolution, is applied to the CNN parameter-tuning problem. We define AlexNet [6] as a basic structure of the CNN and apply GA to optimize the activation functions and tuning parameters. The effectiveness of the proposed method is confirmed by a computer simulation using a scribble-object recognition problem as an example. We also demonstrate an interactive picture book with a CNN model obtained by the proposed method.

2 Interactive Picture Book

In this section, we will show the outline of the proposed interactive picture book.

2.1 *Interactive Picture Book*

The main purpose of the proposed interactive picture book is to consider human Kansei to create an original story for different users. There are two main input data types from users in the interactive picture book: natural language and an image. Text data based on natural language are one of the most useful inputs. However, there exists an obvious limitation in the communication between users and systems with text-only information. For example, small children cannot use text languages, but they can communicate to the external world based on their emotions. On the other hand, drawings are one of the most intrinsic representations of individuals, regardless of age, national origin, or culture. Therefore, we have introduced a human-sketch-recognition module into the proposed interactive picture book.

Our interactive picture book is composed of two main parts: Pictgent, which represents the main frame of the system and controls the story transition, and CASOOK, which recognizes human sketches for the system (Fig. 1).

2.2 *Pictgent*

Pictgent is the main frame of our interactive picture book.

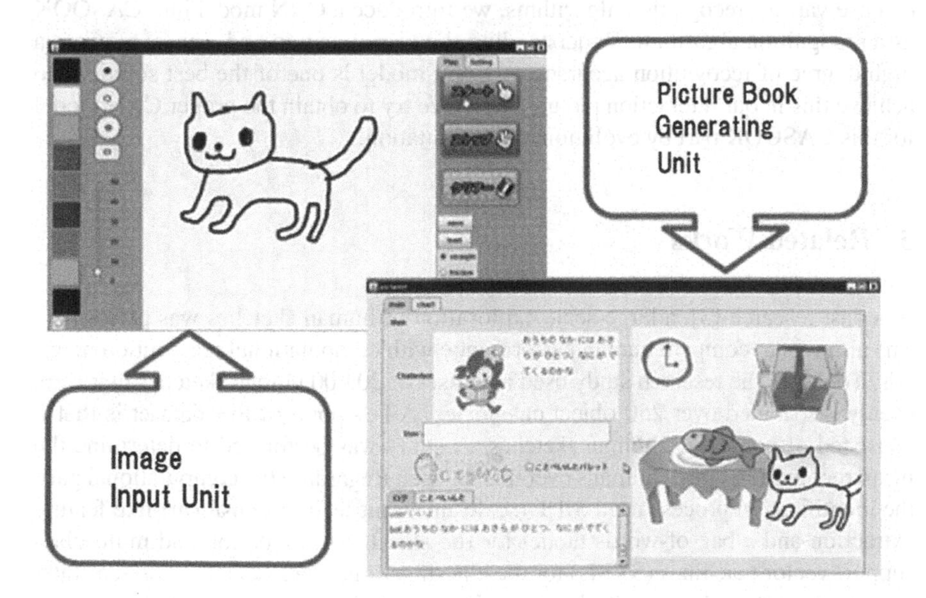

Fig. 1 Outline of interactive picture book

Pictgent consists of the following three modules:

Picture Module This module manages the picture information, which is an important item in this study. Each picture has its own model created by an object-oriented modeling technique. In this module, the picture model is written in XML format.

Scenario Module This module controls the transition of the scenarios in the picture book. To achieve an adequate transition, this module stores the user's various inputs such as natural-language text and/or a sketch.
Scenarios are written in XML format, which has a transition map and an answer example set.

Chat Module This module replies according to the user's input or asks the user some questions about the picture book story.

2.3 CASOOK

We used CASOOK for the user sketch inputs for the interactive picture book because Pictgent can only receive text information.

CASOOK consists of two modules: an interaction module and an analysis module. The interaction module includes functions that encourage the users to draw, and the analysis module contains methods that recognize user drawings. Since CASOOK can use various recognition algorithms, we introduce a CNN model into CASOOK as a recognition algorithm. Understanding the user intention and Kansei requires a high degree of recognition accuracy. A CNN model is one of the best solutions to achieve this in our interaction picture book. We try to obtain the proper CNN model for this CASOOK part by evolutionary computation.

3 Related Works

In a past research [3], a large-scale exploration of human sketches was performed. The aim was to compare human performance with computational recognition methods. To do so, the research study used a dataset of 20,000 unique sketches that were evenly distributed over 250 object categories. A key point of this dataset is that it consisted of non-expert human sketches. A study was performed to determine the mean recognition rate of humans over this dataset. Regarding the computational part, the identification process used SIFT (scale-invariant feature transform)-like feature extraction and a bag-of-words model for the sketch representation and multi-class support vector machines (SVMs) for the classification. The researchers used a state-of-the-art machine learning technique called parallel SVMs. Bertla et al. also proposed the recognition of sketches by a multi-class SVM classifier utilizing a binary decision tree [2]. However, there has been no research on sketch recognition for an interactive picture book using a CNN model.

Research on combining GA with CNN has been reported [12], but this study did not consider the hyperparameters of the CNN directly.

4 Convolutional Neural Network

In this section, we introduce a CNN model.

4.1 Outline of CNN

A CNN is one of the deep learning methods and is widely used in the image recognition field. In the ImageNet LSVRC contest, it was reported that approach of using CNN model is improved remarkable performance than those of conventional image recognition [6]. Now, a CNN is admitted as the one of powerful method of image recognition. The basic structure of CNN contains one or more convolutional layers, pooling layers, and fully connected layers. The fully connected layers are arranged after the convolutional and pooling layers for fine-tuning.

The CNN has the advantage of only using a small number of learnable parameters. The convolutional layers are the core building blocks with parameters of a set of learnable filters. An input image is passed to the convolutional layers, and dot products are computed between the entries of the filter and the image at any given position. The pooling layers are inserted into the architecture of the CNN as well. We used max-pooling which operates independently at every depth slice of the input and resizes using the MAX operation.

4.2 AlexNet

We now introduce the architecture of AlexNet [6], which is shown in Fig. 2.

AlexNet is one of the most popular architectures of a CNN. It contains eight learned layers: the first five are convolutional and the remaining three are fully connected. Using a softmax function, the last fully connected layer produces an output consisting of four class labels. The response-normalization layers follow the first and second convolutional layers. The pooling layers follow both the response-normalization layers and the fifth convolutional layer. To reduce overfitting of the data during the training stage, we use "dropout" [8, 11]. The idea is to drop out randomly hidden units and input features during the training of the neural network. This network uses dropout in the first two fully connected layers. Moreover, we rely on ReLU nonlinearity, which is applied to the output of every convolutional and fully connected layers.

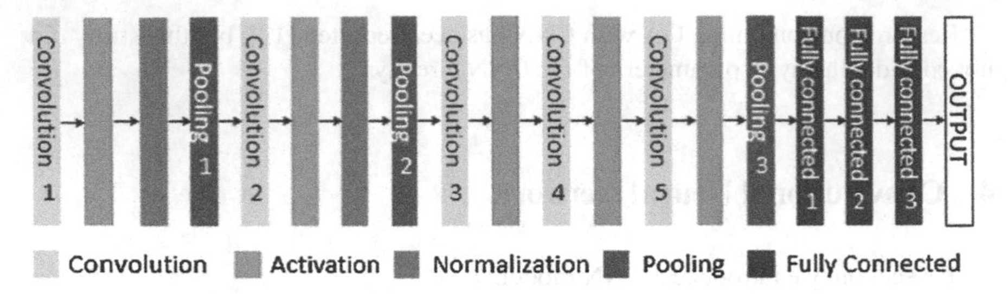

Fig. 2 The architecture of AlexNet

5 Evolutionary Approach for a CNN Model

The CNN is one of the best solutions for image recognition. However, setting the adequate structures and hyperparameters of the CNN requires a great effort. The performance of the CNN is strongly dependent on the network characteristics; therefore, finding better structures and hyperparameters is a very important issue. Although a grid search is one solution, a more efficient approach is required because of the huge hyperparameter space.

In this research, we have adopted evolutionary computations (ECs) as the optimization method for finding better CNN models. ECs are search-and-optimization algorithms based on the mechanism of natural evolution. We propose an evolutionary approach for finding a fine hyperparameter set for the CNN model.

5.1 Genetic Algorithm

To find a fine CNN model, we need to consider the following issues:

Structure This refers to the basic structure of the CNN such as the number and type of layers, type of activation function, and methods for fine-tuning.

Hyperparameters (integer) These are represented by an integer such as the number of filters, filter shape, max pooling shape, and convolutional layer shape.

Hyperparameters (real number) These are represented by a real number such as the weight value of the filter and that of the convolutional layers.

In this study, we only focus on two issues: *hyperparameters (integer)* and the type of activation function in the *structure*. We used a GA [4, 5], which is one of the popular ECs and has been applied to various types of problems. Although we strongly understand above setting is not enough to find novel deep architectures, this research stands the first step of combination of CNNs and ECs for creating more complex networks.

Table 1 The gene of individuals in GA

Design variables	Allele
The number of filter (NF)	16, 32, 64
Filter size (FS)	3, 5, 7, 9, 11
Pooling size (PS)	3, 5, 7
The number of node in fully connected layer 1 (NL1)	512, 1024, 2048, 4096
The number of node in fully connected layer 2 (NL2)	128, 256, 512, 1024
Batch size (BS)	10, 30, 100
Activate function with ReLU (Re)	1 (use), 0 (not use)

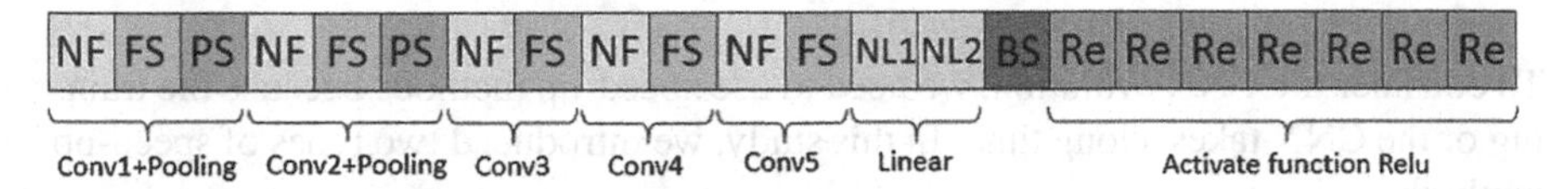

Fig. 3 An example of genotype

5.2 *Representation*

In a GA, solutions are represented as individuals and each individual has a chromosome. A symbol, which makes up the chromosome, is known as a gene, and the position of the gene in the chromosome is known as a locus, and possible genes in the loci are called alleles.

In this study, we set a unique number of alleles for each locus because the degree of contribution to the quality of a network can be dependent on each design variable. Table 1 shows the details of the representation.

Figure 3 shows an example of the chromosome of an individual. Each string in the locus is related to the strings in Table 1. The basic structure of the CNN is fixed according to AlexNet (see Sect. 4.2). In Fig. 3, the seven genes labeled as "Re" represent the flag of the ReLU functions, which are set just after five convolutional layers and two fully connected layers. If the gene Re is 1, the corresponding ReLU function is set after the target layer. Although AlexNet does not use a ReLU function in fully connected layers, our system can set a ReLU function in these layers.

5.3 *Fitness Function*

In this study, the fitness function $F(s)$ of the GA is the sum of the accuracy in a k-fold cross-validation using the training data, where s is an individual related to a certain CNN.

This optimization problem is represented as follows:

$$\arg\max_{s} F(s) = \sum_{i=1}^{k} f_{\mathrm{accuracy}_i}(s) \tag{1}$$

where s denotes an individual and $f_{\mathrm{accuracy}_i}(s)$ is the accuracy function of a k-fold cross-validation after i epochs.

5.4 Speed-Up Methods

To consider the CNN evolution, we need to use speed-up methods because the training of the CNN takes a long time. In this study, we introduced two types of speed-up methods.

The first method is called "genotype and fitness memory" in GA. Once the fitness of an individual is calculated, the relation of the genotype and the fitness value is recorded by the genotype and fitness memory method. If the same genotype appears in the GA, the fitness value is obtained from this memory method instead of recalculating the fitness. Although the cost of saving the genotype and fitness value is not low, this memory method reduces the total simulation time because of the heavy CNN training time.

The second method involves introducing a cutoff point into the CNN training stage. In this study, we give up the CNN training when the improved quantity of the accuracy rate of 10 epochs is less than 1 % compared to that of 1 epoch. We remove the useless training time of hopeless individuals by this method.

Algorithm 1 GENERATION-GA

```
Create random population
while |P'_i| < popsize do
    Select parents
    Apply crossover to them
    Add children to P'_i
for all individual ∈ P'_i do
    Apply mutation to it
while |P_{i+1}| < popsize − 1 do
    Apply survival selection to P'_i
    Add individual to P_{i+1}
Add elite to P_{i+1}
```

Fig. 4 One generation of GA. *popsize* is the number of individuals in a population, and P_i is a population in generation i

5.5 Algorithm

Figure 4 shows the algorithm for creating a generation in GA. GA has a crossover, a mutation, and a survival selection, respectively.

6 Experiments

Here, we show the experimental results of evolving a CNN model by using GA. The main purpose of this experiment was to confirm the validity of the proposed evolutionary method.

6.1 Target Problem

Since the main targets of our interactive picture book are children, we utilized a bitonal sketch image as the recognition problem for the CNN. Children start drawing before reading and writing; thus, sketches can be an appropriate teaching aid for early childhood development. Children can input their emotions as sketches into an interactive picture book, and a prepared CNN can recognize them. Owing to this recognition, certain transitions can be applied to the picture book story. For instance, if a child draws a frog, the system would recognize the animal and then transition the scene to an appropriate story about the frog, such as a rainy situation. In this experiment, the CNN model was evolved to recognize four classes (*face, animal, character, and symbol*) of the bitonal sketch image with a size of 28×28 because these four classes are the main categories of the children's drawing. Each class contained 250 items of data, so the total amount of data was $250 \times 4 = 1000$. We divided 1000 items of data into two as follows:

Training data in evolution Each class contained 225 items of data, so the total amount of training data was equal to 900. In parts of the GA, the fitness value in Sect. 5.3 was calculated by using the results of a threefold cross-validation after 100 epochs. Each individual was translated into a CNN model and trained by 600 items of data with 100 epochs. After this, we obtained the CNN accuracy and loss by utilizing 300 data items, which were the remainder of 900–600 data items. Because of the threefold cross-validation, we repeated this three times and obtained the fitness value using the equation above.

Test data In our test data, each class data contained 25 items. Since there were four classes ($25 \times 4 = 100$), the data were used to estimate the obtained CNN model represented by a unique individual in the final generation in GA.

6.2 Experimental Conditions

Table 2 shows the computer specifications used in the experiments. Table 3 shows the setting of the GA. In addition, we set the maximum number of generations of the GA to 20.

We use simple mutation that change gene to random allele because target problem is not so difficult. If the fitness landscape of problem become more complex, we have to rearrange the setting of mutation operator.

Results and Discussion Figure 5 shows the training and test accuracies. The training accuracy reached 1.0 in about 100 epochs and then fluctuated while keeping a high accuracy rate. The test accuracy reached 0.98 in about 30 epochs and then stayed at around 0.95. After 500 epochs, the test accuracy declined below 0.95 because of

Table 2 Computer spec of experiments

OS	Windows 7 Professional
CPU	Intel Core i7-4790K 4.00 GHz
Memory	32.0 GB
GPU	NVIDIA GeForce GTX 970

Table 3 Setting of GA

Generation size	20
Population size	20
Chromosome length	22
Crossover type	Uniform
Crossover rate	1.0
Mutation rate of each locus	$\frac{1}{L}$ (L is chromosome length)
Selection	Tournament selection
Tournament size	3

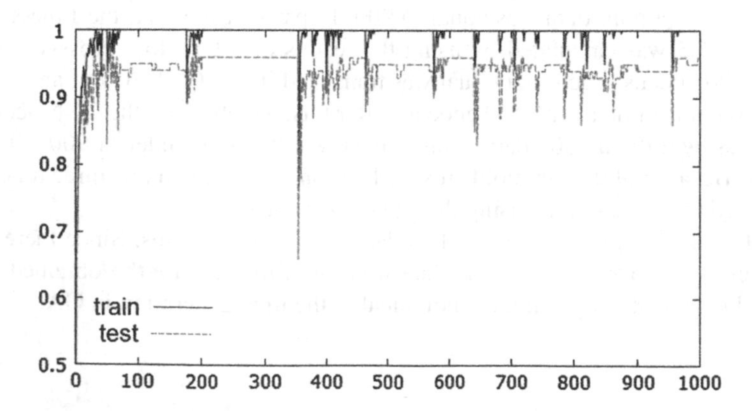

Fig. 5 Training and test accuracies

overfitting. Since the fitness in GA was evaluated after 100 epochs, only individuals with CNN hyperparameters that finished training before 100 epochs could survive. In this experiment, because the dataset was not so large, our system easily found fine individuals for CNN. We tried another fitness function using a negative value of the total losses instead of the total accuracy fitness in Sect. 5.3 and also obtained a high accuracy. The recognition task of the present dataset may be too easy for our proposed method. Applying our system to a more difficult dataset is one of the most important goals for further study. Figure 6 shows the training and test losses calculated by cross entropy. The training loss became almost 0.0 in about 30 epochs, but rose up occasionally. On the other hand, the test loss increased with oscillation. Actually, because of the small size of the test dataset, overfitting caused that oscillation.

Figure 7 shows an example of the training datasets. The character class contains only "Hiragana," which is the original Japanese character set. Figure 8 shows the two images in the test dataset that the CNN model with a 0.98 test accuracy failed to recognize. The left sketch represents a "dolphin" in the animal class, and the right

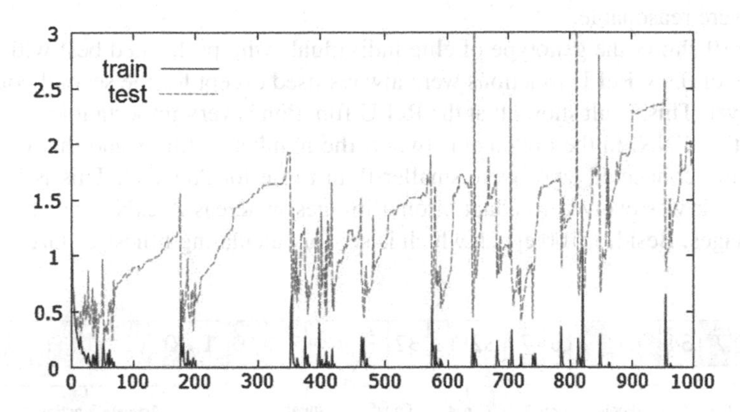

Fig. 6 Training and test losses

Fig. 7 Examples of training datasets

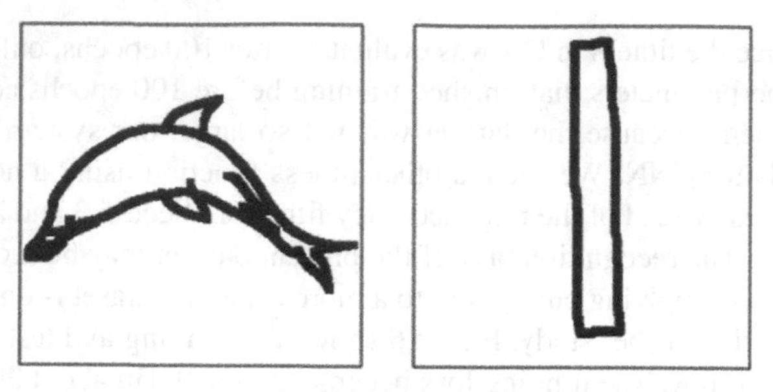

Fig. 8 Images leading to identification failure in test

sketch represents a "rectangle" in the symbol class. The trained CNN mistook the "dolphin" for a face class and the "rectangle" for a character class. Since there were few water creatures in our dataset, the dolphin seemed hard to recognize. The difference, however, between the rectangle and some characters was very small, so the results were reasonable.

Figure 9 shows the genotype of elite individuals who performed best with a test accuracy of 0.98. ReLU functions were always used except for the second convolutional layer. This result shows that the ReLU function is very important in obtaining an effective CNN. In the obtained network, the number of filters and that of nodes in the fully connected layer were smaller than those for AlexNet. This is because our network was evolved to adapt bitonal images, whereas AlexNet was tuned for RGB images. Besides, 100 epoch which is set for calculating fitness of GA is small

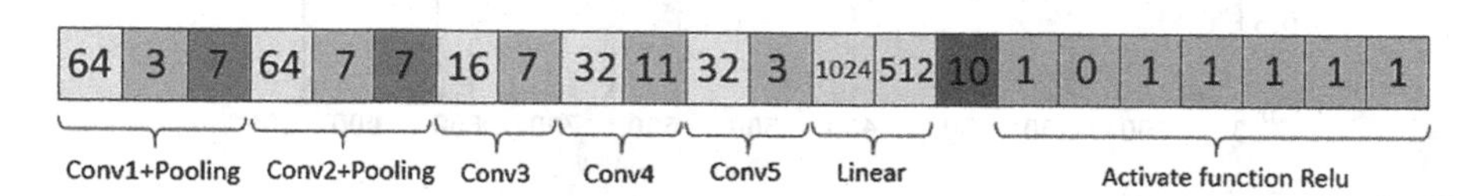

Fig. 9 The genotype of best individuals in test data

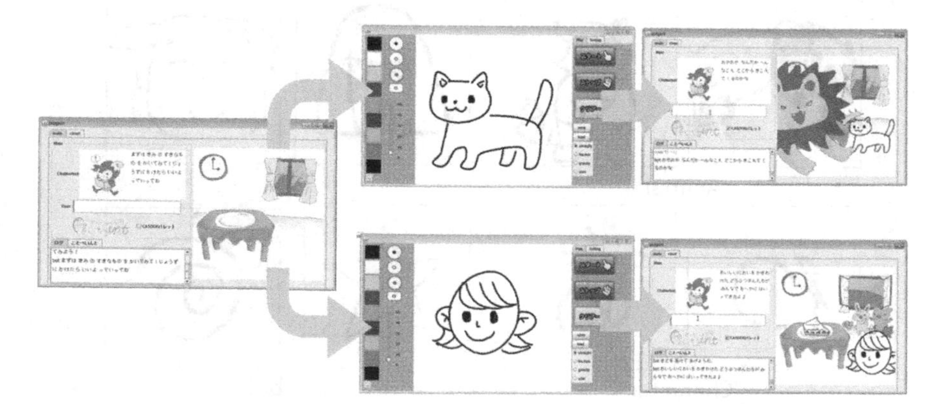

Fig. 10 Transition example of interactive picture book

to obtained large network. In this study, since fitness is calculated by Eq. (1), the landscape of objective function may be simple. We consider memory and/or computational cost of training network in fitness function, e.g. set some penalty for too large networks.

Figure 10 shows some examples of the output transition for our interactive picture book. If the user input is an animal, the scene transitions to a "lion appearing," whereas if the user input is a face, the scene transitions to "animals gathering." The results of the computer simulation show that the proposed method is effective in creating a CNN for our interactive picture book.

7 Conclusion

In this study, we proposed the novel method of making CNN models for the interactive picture book by using evolutionary approach. Using AlexNet as the basic CNN model, proposed method with GA can tune the hyperparameters of CNN models. Proposed method can also find more a simple network than AlexNet to adapt bitonal sketch images. Important future works are as follows:

- To extend the proposed method to various datasets such as photos or RGB images.
- Increasing the number of data in order to observe the search dynamics of ECs in detail.
- Comparison of using pure grid search technique versus the proposed method with considering both performance of the network and computational cost such as memory and CPU power.
- Tuning the real number hyperparameters by using CMA-ES.
- Extend ECs genotype to be able to represent complex network topology.
- The landscape analysis of search space based on various fitness functions.

Our final purpose is proposing the novel interactive picture book system. In our current system, the branches of story are only four because system can recognize only four types of image classes. Therefore, we have to increase the number of recognizable classes in order to show lots of story branches to users.

Acknowledgments A part of this work was supported by JSPS KAKENHI Grant, Grant-in-Aid for Scientific Research (C), 26330282. A part of this work was also supported by JSPS KAKENHI Grant, Grant-in-Aid for JSPS Fellows, 16J10941.

References

1. P. Bertola, N. Mori and K. Matsumoto: *Sketch Recognition for Interactive Multi-Agent System*, in *Institute of Systems Control and Information Engineers* 334–4 (2014)
2. P. Bertola, N. Mori, K. Matsumoto, Sketch Recognition for Interactive Multi-Agent System; Proc. of SCI'14, 334–4 (2014)

3. M. Eitz, J. Hays and M. Alexa: *How Do Humans Sketch Objects?, ACM Trans. Graph. (Proc. SIGGRAPH)*, Vol. 31, No. 4, pp. 44:1–44:10 (2012)
4. D. E. Goldberg: Genetic Algorithms in Search, Optimization, and Machine Learning, Addison-Wesley (1989)
5. J. H. Holland: Adaptation in Natural and Artificial Systems, The University of Michigan Press (1975)
6. A. Krizhevsky, S. Ilya and G. E. Hinton: ImageNet Classification with Deep Convolutional Neural Networks, Advances in Neural Information Processing Systems 25, pp. 1097–1105 (2012)
7. G. Madzarov, D. Gjorgjevikj, I. Chorbev, et al.: A Multi-class SVM Classifier Utilizing Binary Decision Tree, Informatica (Slovenia), Vol. 33, No. 2, pp. 225–233 (2009)
8. N. Srivastava, G. E. Hinton, A. Krizhevsky, I. Sutskever, and R. Salakhut-dinov: Dropout: A simple way to prevent neural networks from overfitting, Journal of Machine Learning Research, 15:1929–1958 (2014)
9. M. Suzuki and J. Gyoba: *Analyses of Factor Structure of Affective Impressions Produced by Line Drawings, Colors, Words and Combined Stimuli of Those Properties*, The Institute of Electronics, Information and Communication Engineering, Technical Report of IEICE. HIP, Vol. 103(166), pp. 57–62 (2003)
10. G. Vamvakas, B. Gatos, I. Pratikakis, N. Stamatopoulos, et al.: Hybrid Off-line OCR for Isolated Handwritten Greek Characters, in *Proceedings of the Fourth Conference on IASTED International Conference: Signal Processing, Pattern Recognition, and Applications*, SPPR'07, pp. 197–202 (2007)
11. S. I. Wang and C. D. Manning: Fast dropout training, JMLR W&CP 28(2): pp. 118–126 (2013)
12. Y. Zhining, P. Yunming: The Genetic Convolutional Neural Network Model Based on Random Sample International Journal of u- and e- Service, Science and Technology Vol. 8 No. 11 pp. 317–326 (2015)

Semi-automatic Picture Book Generation Based on Story Model and Agent-Based Simulation

Kiyohito Fukuda, Saya Fujino, Naoki Mori and Keinosuke Matsumoto

Abstract In the fields of artificial intelligence, automatic narrative generation has attracted considerable interest. Lots of studies on narrative generation have been reported. However, most of those do not consider stories and representation mediums separately. In this study, we focus on the picture book as the narrative because picture book is a mix of images and language. As a first step of automatic story generation without representation mediums, we propose a novel semi-automatic picture book generation method based on story model and agent-based simulation. The computational experiments are carried out to confirm the effectiveness of the proposed method.

Keywords Narrative engineering · Picture book generation · Story model · Agent-based simulation

1 Introduction

Recently, automatic narrative generation by the computer has attracted considerable interest as a challenging problem in the fields of artificial intelligence and natural language processing. Narrative [1] is a creation based on emotions of human and is comprised of a story and a representation medium. When we represent the story by language, narrative is called novels; and it is called comics when we represent the

K. Fukuda (✉) · S. Fujino · N. Mori · K. Matsumoto
Graduate School of Engineering and School of Engineering Osaka
Prefecture University, 1-1 Gakuencho, Sakai City, Osaka 599-8531, Japan
e-mail: fukuda@ss.cs.osakafu-u.ac.jp

S. Fujino
e-mail: fujino@ss.cs.osakafu-u.ac.jp

N. Mori
e-mail: mori@cs.osakafu-u.ac.jp

K. Matsumoto
e-mail: matsu@cs.osakafu-u.ac.jp

G. Leu et al. (eds.), *Intelligent and Evolutionary Systems*,
Proceedings in Adaptation, Learning and Optimization 8,
DOI 10.1007/978-3-319-49049-6_9

story in images mainly. In this study, we focus on the picture book which is one of the most simple comics, as narrative.

With a view toward realizing effective automatic narrative generation, lots of studies on automatic novel generation based on a case-based reasoning (CBR) [2–4] and problem-solving process [5], and automatic comic generation based on picture model to express pictures and images [6, 7] have been reported. However, most of those studies have a critical problem that they deal with stories and representation mediums together. If we can generate stories without dependency of representation mediums, we can generate various kinds of narratives which are same stories only using different representation mediums. Therefore, we have proposed the story model in order to model stories regardless of representation mediums.

In this study, we propose a novel semi-automatic picture book generation method based on story model and agent-based simulation (ABS). Here, ABS is a simulation to investigate actions and interactions between autonomous agents [8, 9]. A new picture book is generated by representing a story model which is generated by the simulation log of the ABS.

2 Story Model

We generate picture books based on our story model. The most important part of that is called a scene. There are lots of different scenes in the story model. That difference among scenes are represented as objects that appear in the scene. Each scene has actions and emotions of the object, time, and environment. The story is generated by selecting the next scenes from the all scenes set and transiting scenes from the start scene to the end scene. The ability of modeling branched stories is one of the important feature of our story model.

The scene has the following information that are necessary for story generation in hierarchical structure.

- The scene such as scene name, place and time.
- The objects that appear in the scene such as name, age and emotion.
- The relations between the objects such as action, identification and causal relation.

Figure 1 shows the scene structure of story model.

3 Proposed Method

In this study, we propose a semi-automatic picture book generation method based on story model and ABS. The proposed method proceeds as follows. Each step is explained in detail in Sects. 3.1–3.4.

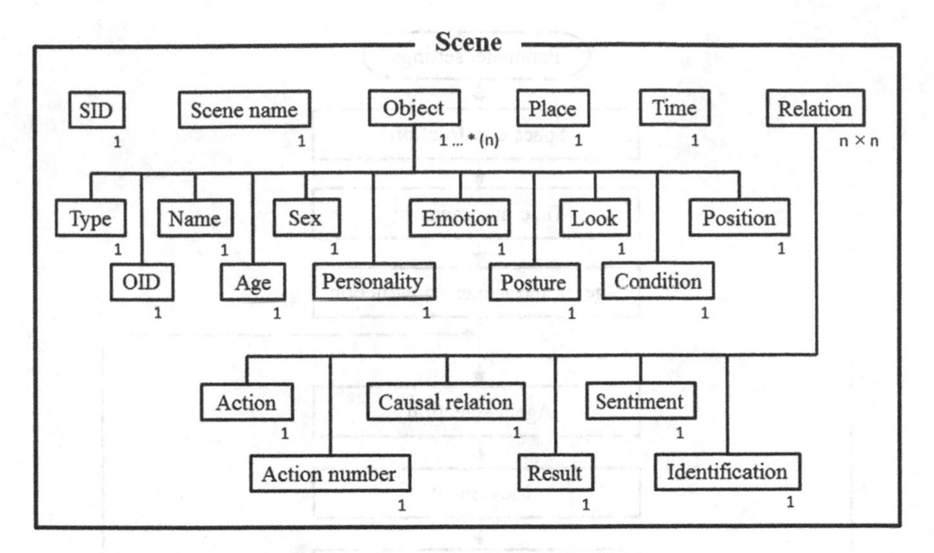

Fig. 1 Scene structure of story model

Step 1: Settings of the ABS
The environment and parameters of the ABS are set in order to generate log data. Several parameters of the ABS are determined by the user.

Step 2: Log data generation via ABS
Two types of log data, namely status logs and action logs are generated via the ABS.

Step 3: Story model generation
Story model is generated from action logs and status logs on the basis of a specific character's log data.

Step 4: Picture book generation
Picture book is generated based on the story model by using languages and images as representing mediums.

Steps 1–3 of the proposed method can be performed automatically. However, step 4 requires human's help in the current system. Thus, the former are referred to as the "computer part" and the latter is referred to as the "user part".

3.1 Setting of the ABS (Step 1)

Figure 2 shows the flowchart of the ABS algorithm. ABS is a simulation to investigate actions and interactions between autonomous agents and their influences on the environment. The ABS algorithm is implemented for the proposed method. Here we represent a real value according to the normal distribution $\mathcal{N}(\mu, \sigma^2)$ and the uniform distribution of $[a, b]$ as $w_1(\mu, \sigma)$ and $w_2(a, b)$.

The details of the ABS are explained as follows:

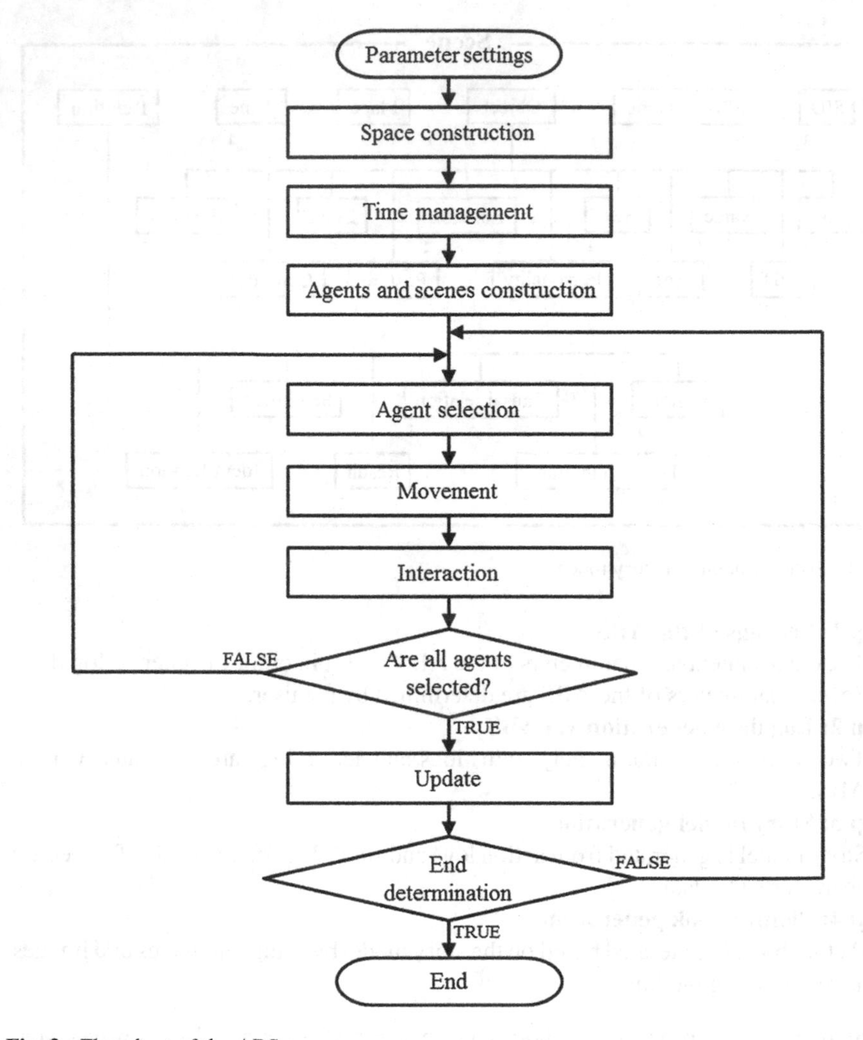

Fig. 2 Flowchart of the ABS

3.1.1 Parameter Settings

There are two types of parameters in the ABS, namely system parameters and user parameters.

System parameters:

x_{max} and y_{max}: The field size of the ABS.
t: The current number of turns.
N_p: The number of character agents.
N_i: The number of item agents.

N_s: The number of scenes.
N_a: Action number.

User parameters:

n: Each agent's name.
r: Each agent's role.
a: Each character agent's age.
s: Each character agent's sex.

3.1.2 Space Construction

In the ABS, the agent moving space is constructed by a field, which is two-dimensional torus-shaped spaces. All agents and scenes are set on random positions in a certain range of the field. Each of agents and scenes can not be set on the same position.

3.1.3 Time Management

In the ABS, time is managed by turn t and action number N_a. The turn t denotes a system parameter that manages the time course in the environment and is updated after all agents performed all actions. The action number N_a denotes a system parameter that manages the order of actions performed in the same turn and is updated each time some agent performs an action.

3.1.4 Scenes and Agents Construction

First, N_p character agents and N_i item agents are constructed.

A character agent is an agent that represents the main character in the stories generated by the proposed method and has the following specific internal states:

- attributes: age a, personality $P = \{p_1, p_2\}$ emotion e, health condition c, where a is a natural number, s is a binary of man or woman, $p \in P$ is a real number of $[1, 3]$, e is a real number of $[-1, 1]$, c is a real number of $[0, 100]$;
- friendship values toward other agent $F = \{f_1, f_2, \dots, f_{n_p+n_i}\}$ and familiarity values toward other agent $G = \{g_1, g_2, \dots, g_{n_p+n_i}\}$, where $f \in F$ and $g \in G$ are real numbers of $[-100, 100]$;
- the list of its own item agents.

An item agent is an agent that represents the item that is appeared in the stories and has the following specific internal state:

- information on character agent that is its owner.

Then, both character and item agents have the following internal state:

- attributes: names n, role in the story r, where r is "hero", "enemy", "ally", "food" or "weapon";
- influence value on other agent's movement v, where v is a real number of $[0, 200]$;
- position in the field x, y.

Second, N_s scenes are constructed. A scene is a special position in the field. That is important for the proposed method to generate story model and has the following parameters: role in the story r and influence value on other agent's moving v, where r is "event" or "goal" and v is a real number of $[0, 200]$.

3.1.5 Agent Selection

An agent is randomly selected from agents that do not perform an action in this turn. If there are character agents in them, agent is randomly selected from them. The selected agent is defined as i.

3.1.6 Movement

An agent i always performs an action of movement. It moves to the Neumann neighborhood in the direction of the value according to influence values of other agents and scenes, Euclidean distance between them, and its own personality. After i performed the action of movement, the action number is updated to $N_a = N_a + 1$.

3.1.7 Interaction

If an agent i is a character agent and it meets some conditions after movement, it performs an action of interaction with other agent. The algorithm of interaction is given below.

1. If a character agent i is set at the same position of other character agents in the field, a character agent is randomly selected form them and is defined as j. The friendship value and familiarity value of j toward i, represented by f^j_i and g^j_i and the emotion of j, represented by e^j are updated as follows.

$$f^j_{i,t} = f^j_{i,t} + w_1(0,3)p^j_{1,t} \tag{1}$$

$$g^j_{i,t} = g^j_{i,t} + |w_1(0,3)|p^j_{1,t} \tag{2}$$

$$e^j_t = e^j_t + \frac{w_1(0,3)p^j_{1,t}}{100}. \tag{3}$$

Furthermore, if the role of i is "hero" and it has an item agent that has the role corresponding to the j's role, its owner is changed from i to j. In addition, f_i^j and g_i^j are updated as follows.

$$f_{i,t}^j = f_{i,t}^j + w_2(10, 30)p_{1,t}^j \tag{4}$$

$$g_{i,t}^j = g_{i,t}^j + |w_1(0, 10)|p_{1,t}^j. \tag{5}$$

2. If i is set at the same position of item agents that is not belonged to any character agents in the field, an item agent is randomly selected from them and is defined as k. k's owner is changed to i and the emotion of i, e^i is updated as follows.

$$e_t^i = e_t^i + w_1(0, 3)p_{1,t}^i. \tag{6}$$

3. If the health condition of i, represented by c^i is below threshold and i have item agents whose role is "food", an item agent is randomly selected from them and is defined as ℓ. i loses ℓ and the health condition of i, c^i is updated as follows.

$$c_t^i = c_t^i + \frac{c_t^\ell}{2}. \tag{7}$$

After that, the health condition of ℓ, c^ℓ and position in the field are reset.
4. After i performed the action of interaction, the action number is updated to $N_a = N_a + 1$.

3.1.8 Update and End Determination of the ABS

A set of character agents is defined as J. The friendship value of a character agent i toward other character agent j, represented by f_j^i, the health condition of an agent k, represented by c^k and influence values of all agents and scenes, represented by v are updated as follows.

$$f_{j,t}^i = f_{j,t}^i + w_1(0, 1) \tag{8}$$

$$e_t^i = e_t^i + \frac{1}{100} \sum_{j \in J, i \neq j} w_1(0, 1) \tag{9}$$

$$c_t^k = \frac{499}{500} c_t^k \tag{10}$$

$$v_t = \begin{cases} 1.001 v_t & (r = \text{"goal"}) \\ v_t + w_1(0, 0.5) & (otherwise). \end{cases} \tag{11}$$

The objective of this update is representing that the importance of character's emotions, items and scenes in the story is changed by the time course. In addition, turn t is updated to $t = t + 1$.

Finally, if a character agent whose role is "hero" is set at the same position of a scene whose role is "goal", the ABS is halted. If not, it returns to the agent selection.

3.2 Log Data Generation (Step 2)

Two types of log data, namely action logs and status logs are generated via the ABS implemented by Sect. 3.1. The details of these log types are explained as follows:

Action log Action log information is saved if each agent performs a movement or an interaction in the ABS. The data of action log is {*turn, main agent, target agent, position of main agent, action type, action number, the change amount of each parameter caused by the action, item that is exchanged between agents by the action*}.

Status log The status log are saved at the start of each turn and at the end of the last turn in the ABS. These comprise the turn and all parameters of all agents and scenes.

3.3 Story Model Generation(Step 3)

The story model is generated from action logs and status logs. The algorithm of story model generation is as follows:

1. A character agent whose role is "hero" is defined as i. In addition, a set of scenes whose role is "event" is defined as J.
2. A set of maximum turn when i is set at the same position of scene $j \in J$ in the status logs is defined as T^{s}_{max}.
3. Sets of maximum and minimal turn when each different action are performed in the action logs are defined as T^{a}_{min} and T^{a}_{max}.
4. Start scene is generated using i's status log whose turn is 0.
5. The minimal turn in T^{s}_{max} is defined as t. After that, if there are action types that occurred at least once t' $(t' \leq t, t' \in T^{a}_{min})$. For each selected action types, investigate the latest action and action time $t'' \in T^{a}_{max}$, and one scene is generated from one action with using status log of turn 0 and t. Obtained scenes are sorted by in ascending order of each t'' of selected action.
6. T^{s}_{max} and T^{a}_{min} are updated to $T^{s}_{max} = T^{s}_{max} \backslash t$ and $T^{a}_{min} = T^{a}_{min} \backslash t'$. If $T^{s}_{max} \neq \Phi$, return to 5.
7. The end scene is generated using each agent's status logs whose turn is the last turn. Finally, the set of scenes generated in 4, 5 and 7 is defined as the story model.

Table 1 Correspondence of scene elements to the log data

Scene elements	Log data
SID	No data
Scene name	Scene and action type
Object	Agent
Place	Scene
Time	No data
Relation	Agent and scene
Type	Agent's type
OID	Agent's ID
Name	Agent's name
Age	Agent's age
Sex	Agent's sex
Personality	Agent's personality
Emotion	Agent's emotion
Posture	Scene
Look	Agent's emotion and health condition
Condition	Agent's health condition
Position	Scene and agent's role
Action	Action type
Action number	Action number
Causal Relation	No data
Result	Agent's parameters
Sentiment	Friendship value and familiarity value
Identification	Agent's role

The start scene is only one type. All story models have the same start scene. However, there are four types of end scene, namely simple-end, happy-end, bad-end, and unexpected-end. The end scene is selected from these prepared scenes in accordance with character agent's parameters such as friendship value, familiarity value, and position in the field.

Each scene is generated by selecting their elements from choices. The selection of them is in accordance with data of action logs and status logs corresponding to them. However, there are some elements that do not correspond to the log data. Table 1 shows the correspondence of scene elements to the log data.

3.4 Picture Book Generation (Step 4)

In this study, a page of the picture book is generated from a scene of the story model. It is constructed a picture and sentences. The algorithm of picture book generation is as follows:

1. The number of scenes of the story model generated by Sect. 3.3 is defined as N_{page} and the i-th scene in the story model is defined as $S_i (i = 1, 2, \ldots, N_{\mathrm{page}})$. Here, S_1 denotes the start scene and $S_{N_{\mathrm{page}}}$ denotes the end scene.
2. Set $i = 1$.
3. Suitable picture parts and sentence templates are selected by referring to S_i's elements from the prepared image and sentence template sets.
4. A picture is generated by arranging selected picture parts by human according to S_i. Sentences are generated by combining selected sentence templates and specific elements such as name or emotion of S_i. i-th page of picture book is generated by combining the picture and sentences.
5. $i = i + 1$. If $i \leq N_{\mathrm{page}}$, it returns to 3.
6. The picture book that is N_{page} pages is generated by the proposed method.

4 Computer Experiments

4.1 Experimental Method

To confirm the effectiveness of the proposed method, we generated several picture books by using the proposed method. Table 2 shows the experimental conditions. Table 3 shows the correspondence of object index called OID to object attributes. The outline of the story generated by the proposed method was prepared in advance. It was defined as follows:

Table 2 Experimental conditions

Field size: $x_{\mathrm{max}}, y_{\mathrm{max}}$	100, 100
Number of character agents: N_{p}	3
Number of item agents: N_{i}	2
Number of scenes: N_{s}	4

Table 3 Correspondence of OID to object attributes

OID	Name	Role	Sex	Age
1	Little red riding hood (Red)	"Hero"	Woman	10
2	Wolf	"Enemy"	Man	8
3	Hunter	"Ally"	Man	30
4	Meat	"Food"	Null	Null
5	Gun	"Weapon"	Null	Null

- One day, the hero was asked to go visit her grandmother who lives in a forest by her mother. Then some events happen on the way to her grandmother's house.

In this experiments, we prepared four types of the end of story (Fig. 3). They were defined as follows:

Simple-end: The hero arrives her grandmother's house without encountering the enemy.
Happy-end: The hero and the ally exterminate the enemy, and they arrives hero's grandmother's house and eat dinner with her.
Bad-end: The hero is eaten by the enemy in the way to her grandmother's house.
Unexpected-end: The enemy apologizes to the hero and becomes good friend with her, and they arrives her grandmother's house and eat dinner with her.

4.2 Experimental Results

First, we generated a picture book by using the proposed method in order to discuss the detail of the story model and picture book generated by the proposed method. We call the generated picture book "S_{sample}". Figure 4 shows picture book representation of S_{sample}. Tables 4, 5 and 6 show examples of the story model of S_{sample}. Figure 4 and Tables 4, 5 and 6 are generated in Japanese and translated into English. Tables 4, 5 and 6 correspond to Fig. 4. For example, the columns whose objects is O-2 or R-2

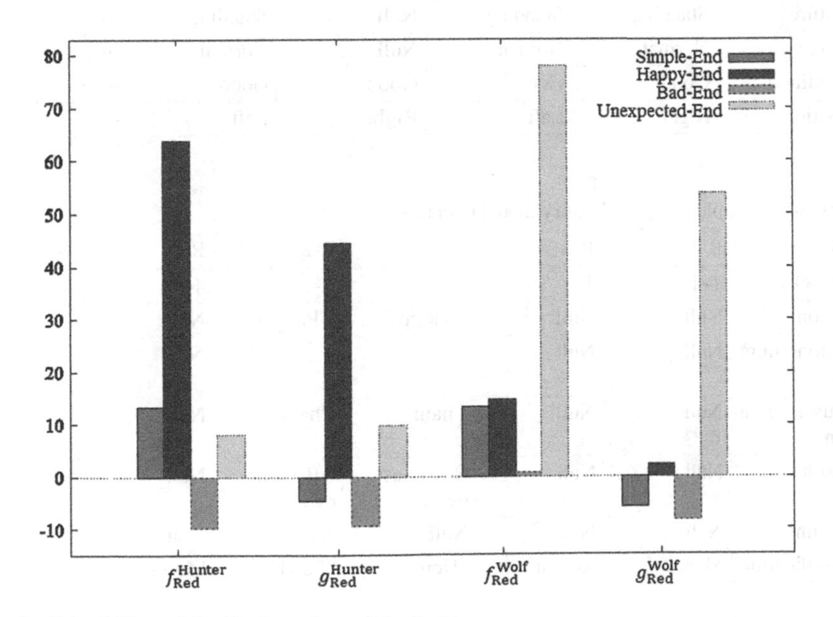

Fig. 3 Friendship and familiarity values of the last turn

Table 4 Example of S_{sample}'s story model (Scene)

SID	Scene name	Objects	Place	Time	Relations
1	Start	O-1	Start	Morning	R-1
2	Acquisition of meat	O-2	In the forest	Noon	R-2
3	Encounter with Wolf	O-3	In the forest	Noon	R-3
4	Encounter with Hunter	O-4	In the forest	Noon	R-4
5	Attack on Wolf	O-5	In the forest	Moon	R-5
6	Bad-end	O-6	In the forest	Night	R-6

Table 5 Example of S_{sample}'s story model (Object)

Objects	O-1	O-2	O-2	O-3	...
OID	1	1	4	1	...
Type	Character	Character	Item	Character	...
Name	Little red riding hood	Little red riding hood	Meat	Little red riding hood	...
Age	10	10	Null	10	...
Sex	Woman	Woman	Null	Woman	...
Personality	Curious	Curious	Null	Curious	...
Emotion	Normal	Normal	Null	Normal	...
Posture	Standing	Standing	Null	Standing	...
Look	Normal	Normal	Null	Normal	...
Condition	Good	Good	Good	Good	...
Position	Right	Left	Right	Left	...

Table 6 Example of S_{sample}'s story model (Relation)

RID	R-1	R-2	R-2	R-2	R-2	...
OID-OID	1–1	1–1	1–4	4–1	4–4	...
Action	Null	Null	Pick up	−Pick up	Null	...
Action number	Null	Null	1	1	Null	...
Causal relation	Null	Null	Chain	Chain	Null	...
Result	Null	Null	Becomes owner	−Becomes owner	Null	...
Sentiment	Null	Null	Null	Null	Null	...
Identification	Myself	Myself	"Hero"	"Food"	Myself	...

Table 7 Analysis of 10,000 stories

Type	Frequency	Mean number of scenes	Variance
Simple-end	3059	5.656	0.867
Happy-end	274	6.412	0.716
Bad-end	6637	5.49	0.854
Unexpected-end	30	5.800	0.945

in Tables 5 and 6 correspond to the second page of picture book in Fig. 4. We can understand that the story of Fig. 4 is that Little Red Riding Hood is eaten by the wolf in the way to her grandmother's house. The backgrounds, characters or items are defined by the story model in Fig. 4. The look, personality, emotion and position of the objects are also defined by the story model. Therefore, this shows that the proposed method can generate stories regardless of representation mediums because picture book is a mix of images and language. However, the stories generated by the proposed method have a low degree of freedom because the proposed method needs the outline of the story in advance. Therefore, we need to generate the outline of the story automatically.

Next, Table 7 shows the results of the analysis of 10,000 picture books generated by the proposed method. We classified them at the end of story and analyzed them by type of the end. There are four types of the end of story: simple-end, happy-end, bad-end and unexpected-end. We can understand that the difficulty in generating the picture book by the proposed method is different according to the end of story in Table 7. This shows that the proposed method can generate a wide variety of picture books. This is attributed to the difference by the change of result from the ABS such as the events order, whether they happen in the story, and character agent's parameters of the last.

Finally, Fig. 3 shows the friendship and familiarity values of the last turn for each end type. Figure 3 is mean of 30 picture books for each end type. It was found that some parameters of characters in the story of some end types are significantly different from ones of the other end types in Fig. 3. This suggests that stories of the picture books generated by the proposed method are controlled by some agent's parameters and actions in the ABS. Therefore, we consider that the proposed method can generate the only picture book for user by optimizing initial parameters and the amounts of change for each action in the ABS.

Especially, it was found that friendship value of the Hunter toward Little Red Riding Hood for a happy-end story and friendship value of the wolf toward Little Red Riding Hood for an unexpected-end story are higher than that ones for the other end in Fig. 3. The ally exterminates the enemy in order to help the hero in the prepared happy-end story and The enemy apologizes to the hero in the prepared unexpected-end story. These show that the proposed method can generate appropriate picture books because it is natural that the hunter helps Little Red Riding Hood who has

①

One day, curious Little Red Riding Hood was asked to visit her grandmother who lives in a forest. "Be careful because a bad wolf lives in the forest," her mother said. "OK. I'll be careful. So I'm going."

②

Immediately Little Red Riding Hood started walking in the forest, she became calm. At that time, she found a meat that looks like delicious in the middle of a road. "I got an idea! If grandmother eats this meat, she will get better." Little Red Riding Hood decided to bring and pick up it.

③

When Little Red Riding Hood walk in the forest for a while, she met Wolf. He looked like enjoyable. She didn't notice that he was the bad wolf who she was said to be careful. "Oh, hello, Little Red Riding Hood." "Will you visit your grandmother? you are great. She will be better soon," Wolf said. "Do you think so? I'm glad if that's right," she said and became friendly with him.

④

When the way to the house of Little Red Riding Hood's grandmother has passed the half, she became enjoyable. At that time, she met Hunter. He looked like downhearted. "Hello, Hunter," she said. "Oh, Little Red Riding Hood, do you go to your grandmother's house?" "Yes, I visit her." "I think that you go there with your mother because there is a bad wolf in this forest. I worry if you are only." "No problem! I'm not afraid of bad wolf," she said and became hostile towards him.

⑤

When Little Red Riding Hood arrived her grandmother's house in a little bit, Wolf suddenly appeared in front of her. "You always look like very delicious. I put up with some time ago but I can't put up with. I'll eat you!" "eeeeek!!" She screamed but there was no hunter who exterminates bad wolf in the neighborhood.

⑥

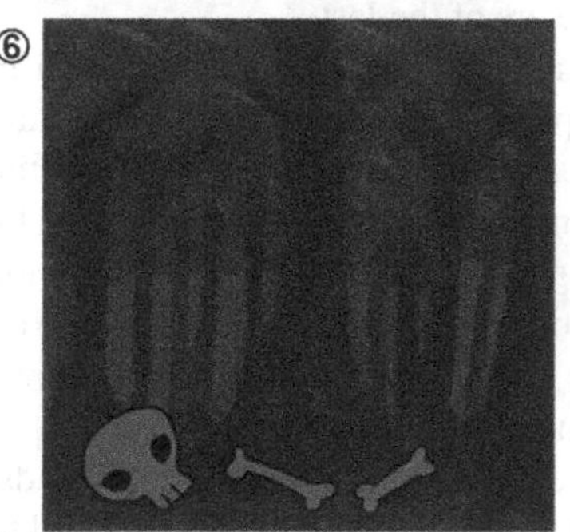

Night fell. A wolf howling is heard from the forest sometimes. There was no Little Red Riding Hood and girl's bones were put on the ground in the forest.

Fig. 4 Generated picture book $\mathcal{S}_{\text{sample}}$

good relationship with him and that the wolf apologizes to Little Red Riding Hood in order to make peace with her.

5 Conclusion

In this study, we proposed a novel semi-automatic picture book generation method based on story model and ABS as a first step of automatic narrative generation that does not depend on representation mediums. The computational experiments are carried out to confirm the effectiveness of the proposed method. The results of the computational experiments are listed below.

- The proposed method can generate appropriate picture books.
- The proposed method can generate a wide variety of stories.
- The proposed method can control stories in the settings of the ABS.

In future work, we plan to optimize the settings of the ABS by using evolutionary computation [10, 11] or machine learning [12, 13]. This will allow us to generate picture books based on human preferences and to quantitatively evaluate the story. In addition, current ABS proposes a solution where all agents select and perform actions according to their emotions, their personalities and so on. However, these elements have a weak influence while a strong influence is given by random numbers. Therefore, the implementation of a new ABS, where actions, emotions, personalities and roles would have a strong influence for each agent's actions is also an important future work.

Acknowledgments This work was supported by JSPS KAKENHI Grant, Grant-in-Aid for Scientific Research(C), 26330282.

References

1. Prince, G.: A Dictionary of Narratology. University of Nebraska Press (2003), https://books.google.co.jp/books?id=WvRNGan8YZcC
2. Gervǎs, P., Dǐaz-Agudo, B., Peinado, F., Hervǎs, R.: Story plot generation based on cbr. Knowledge-Based Systems 18(45), 235–242 (2005), http://www.sciencedirect.com/science/article/pii/S0950705105000407, aI-2004, Cambridge, England, 13th–15th December 2004
3. y Pěrez, R.P., Sharples, M.: Mexica: A computer model of a cognitive account of creative writing. Journal of Experimental & Theoretical Artificial Intelligence 13(2), 119–139 (2001), http://dx.doi.org/10.1080/09528130010029820
4. R., T.S.: The Creative Process: A Computer Model of Storytelling and Creativity. Psychology Press (1994)
5. Meehan, J.R.: The Metanovel: Writing Stories by Computer. Garland Publishing (1980)
6. Thawonmas, R., Shuda, T.: Comic Layout for Automatic Comic Generation from Game Log, pp. 105–115. Springer US, Boston, MA (2008), http://dx.doi.org/10.1007/978-0-387-09701-5_10

7. Ueno, M., Mori, N., Matsumoto, K.: 2-Scene Comic Creating System Based on the Distribution of Picture State Transition, pp. 459–467. Springer International Publishing, Cham (2014), http://dx.doi.org/10.1007/978-3-319-07593-8_53
8. Davidsson, P.: Multi Agent Based Simulation: Beyond Social Simulation, pp. 97–107. Springer Berlin Heidelberg, Berlin, Heidelberg (2001), http://dx.doi.org/10.1007/3-540-44561-7_7
9. Macal, C.M., North, M.J.: Tutorial on agent-based modeling and simulation. In: Proceedings of the 37th Conference on Winter Simulation. pp. 2–15. WSC '05, Winter Simulation Conference (2005), http://dl.acm.org/citation.cfm?id=1162708.1162712
10. Goldman, B.W., Punch, W.F.: Parameter-less population pyramid. In: Proceedings of the 2014 Conference on Genetic and Evolutionary Computation. pp. 785–792. GECCO '14, ACM, New York, NY, USA (2014), http://doi.acm.org/10.1145/2576768.2598350
11. Hansen, N.: The cma evolution strategy: a comparing review. In: Towards a new evolutionary computation. Advances on estimation of distribution algorithms. pp. 75–102. Springer (2006)
12. Mnih, V., Kavukcuoglu, K., Silver, D., Graves, A., Antonoglou, I., Wierstra, D., Riedmiller, M.: Playing Atari with Deep Reinforcement Learning. ArXiv e-prints (Dec 2013)
13. Sutton, R.S., Barto, A.G.: Introduction to Reinforcement Learning. MIT Press, Cambridge, MA, USA, 1st edn. (1998)

Where Does My Brand End? An Overlapping Community Approach

Ademir C. Gabardo, Regina Berretta, Natalie J. de Vries and Pablo Moscato

Abstract In this paper, we present a new Memetic Algorithm for overlapping community detection. We use a link-based clustering approach to detect the communities of edges in complex networks. To assess the quality of our method, we present experimental results for benchmark networks in comparison to other state-of-the-art algorithms. In addition, we present a case study of a co-purchasing product network from a brand-centric point of view to show the real-life utility of this new Memetic Algorithm.

Keywords Evolutionary computation · Memetic algorithms · Metaheuristics · Complex networks · Overlapping community detection

1 Introduction

'Where does my brand end?' This seemingly bizarre question may lie in the mind of marketing professionals but it may be so ill-posed that many would probably be too shy to ask it to their in-house analytic professionals. After all, to have a notion of a 'boundary' we need to have a notion of a *domain* where the brand 'resides'. In this sense, an oversimplification would be that a 'brand ends where another brand starts'. This statement is clearly wrong as today many brands are composed of a large number of highly dissimilar product lines which could significantly *overlap* with the

A.C. Gabardo (✉) · R. Berretta · N.J. de Vries · P. Moscato
School of Electrical Engineering and Computer Science, The University of Newcastle, Newcastle, Australia
e-mail: ademir.gabardo@uon.edu.au

R. Berretta
e-mail: regina.berretta@newcastle.edu.au

N.J. de Vries
e-mail: natalie.devries@newcastle.edu.au

P. Moscato
e-mail: pablo.moscato@newcastle.edu.au

G. Leu et al. (eds.), *Intelligent and Evolutionary Systems*, Proceedings in Adaptation, Learning and Optimization 8,
DOI 10.1007/978-3-319-49049-6_10

offerings of their competitors. Two brands then would be, collections of products of different types. There is then little hope to try to establish a metric over a myriad of different products and specify boundaries across these highly varied and putatively very dissimilar set of metrics.

Another possibility would be to try to identify what the 'brand image' is that consumers have in their mind and what the consumer perceives the 'brand personality' of that brand to be like. 'Brand personality' is a heavily studied topic in business and marketing domains and it is said to have a large impact on consumer perceptions of a brand [1]. Therefore, it becomes a key aspect of finding the "boundary" of a brand in comparison to other (similar) brands. For instance, companies like Apple are characterized by, and known for, the simplicity, but highly functional user interfaces designs. Another brand could be highly similar to Apple in terms of their brand image and personality, yet this other brand is in no way a competitor to Apple. They could be for instance in the clothing and fashion industry. This example shows that the boundaries we could find in these brand's "domains" may extend well into other product or industry domains making the problem at hand more complex.

In this paper we propose a new Memetic Algorithm (MA) for overlapping community detection and use it to inspect a *co-purchasing* network from Amazon.com. This interesting application shows the usefulness of our method in real-life domains and we will present the details of this domain in the following section. Following an approach which is quite intuitive, we can associate products with nodes of a graph. A pair of nodes that represent two products for which *at least one consumer* has purchased them during the same transaction on the website constitute an example of co-purchasing.

1.1 Complex Co-purchasing Networks

It is appropriate to think of co-purchasing as creating a weighted network, a mathematical abstraction in which the elements are represented by nodes and the links between elements are represented by edges [29]. The weight would then correspond to the number of instances in which a co-purchasing has occurred. For the purpose of finding a sharp boundary, however, we will work with an unweighted network. We will constrain our attention to the cases in which at least one co-purchasing event has occurred. The area of mining information from graphs and networks is wide including social networks [11], business networks [27], biological systems [15], and diverse other physical phenomena [11, 30]. One common approach in the area of network analytics is the identification of motifs with subgraphs of a small cardinality of nodes which has previously been applied to the study of co-purchasing networks by Srivastava in [36].

A good strategy would be to display products that lead to the core set of directed paths which lead to a highly valuable product. This said, the identification of these boundaries are becoming, far from being a bizarre question, a mathematical possibility thanks to the available online information about customers' co-purchasing patterns. The problem often is that these sets of information form large complex networks for which advanced analytical methods are necessary.

One important feature of complex networks is the presence of communities, groups of nodes which are densely connected among its members and sparsely linked to the rest of the network [29, 30]. Community detection could be used to investigate: a group of friends among a social network [5, 21], groups of connected electronic devices in communication networks [7, 32], customer engagement behaviour models [8], websites sharing common links [2], biological networks [15, 35], economic networks [12] and many others. In many real-life cases, communities in complex networks can be overlapping. This means that the definition of a community is extended and now it is possible that nodes can belong to more than one community [3, 19, 31, 33].

There are few studies of overlapping communities in co-purchasing networks. In 2015, Jebabli et al. [18] present a study using the Amazon co-purchasing network. Their study was not brand-specific and it aimed at comparing the results of several community detection algorithms with a ground-truth structure. More recently, Yamazaki et al. [38] have used co-purchasing graph data from YAHUOKU!, which is the largest online auction site in Japan. This method was also not brand-specific, is based on using node attributes and on the identification of maximal cliques. To our knowledge, ours is the first approach that aims at using the overlapping community detection problem for a brand-centric request of identifying the boundaries in the induced subgraph generated by the brand in the co-purchasing network. Detecting overlapping communities in large networks is not a trivial computational task [5, 19]. Although there is a wide list of community detection algorithms, the vast majority of them are designed to detect only disjoint partitions [5, 31, 33].

1.2 *Using Line Graphs to Detect Overlapping Communities in Complex Networks*

In this contribution we use link clustering with line graphs to detect the overlapping communities in complex networks by optimizing *modularity* to detect the community structure in a line graph. Introduced by Girvan and Newman [30], modularity is a measure of quality for network communities, which means that higher modularity implies good quality partitions. For a graph $G = (V, E)$, where V is a set of nodes, and E is a set of edges, the modularity can be expressed as:

$$Q = \sum_{c=1}^{k} \left[\frac{l_c}{m} - \left(\frac{d_c}{2m} \right)^2 \right], \tag{1}$$

where c represents a community, k is the number of communities, m is the number of edges in G, l_c is the number of edges inside a community and d_c is the sum of the degrees of the nodes inside a community.

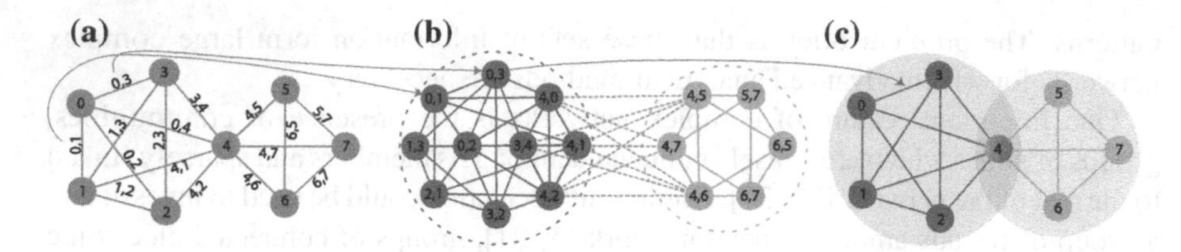

Fig. 1 **a** A graph $G(V,E)$, **b** the line graph $L(G)$ corresponding to G with community structure highlighted by *dotted circles*, and **c** the overlapping communities in G

Our algorithm performs the modularity optimization in a line graph $L(G)$ that corresponds to the adjacency between the edges of G, in such way that, each edge in G corresponds to a node in $L(G)$, and two nodes in $L(G)$ are connected if they share a node. Figure 1 shows a graph and its disjoint partitions, the corresponding line graph and the overlapping partitions recovered from the line graph.

2 MADOC, A Memetic Algorithm to Detect Overlapping Communities in Complex Networks

Community detection is known to be an NP-Hard computational problem [5, 11]. We chose Memetic Algorithms to tackle this problem as they are proven to be efficient metaheuristics capable of solving complex combinatorial problems such as the overlapping community detection problem [25, 26, 28].

Our link based Memetic Algorithm for Detecting Overlapping Communities (MADOC), employs the evolutionary mechanisms of selection, recombination, mutation, and replacement to evolve a population of individuals. Our aim is to maximize *Modularity* [30], our fitness function. The main steps performed by our proposed MA are listed in Algorithm 1.

2.1 Representation and Initialization

The representation used by our algorithm is *string-coding*, a simple and straightforward representation, in which, an individual is encoded using an array where each position represents the community that a node belongs [13, 26]. Given a graph $G = (V,E)$, where V is a set of nodes and E is a set of edges, the individual can be encoded with a string of size $n = |V|$ of values in the range $[0,n]$. For instance, a network with 8 nodes can be represented as the string $[1,1,1,1,2,2,2,2]$, where nodes 1 to 4 are in the community 1 and nodes 5 to 8 are in the community 2.

Initially, a population of 40 individuals is formed (see Sect. 3.1 for more details). In this initial population, 39 are *valid* uniform random solutions, that is, all com-

```
   Input: a graph G = (V, E);
   Output: the overlapping communities of G.
 1 InitializePopulation(pop, G);
 2 localSearch(pop);
 3 updateFitness(pop);
 4 while stopping criteria is not reached do
 5 |   parent_1 = selectFromElitePop(pop);
 6 |   parent_2 = tournamentSelection(pop);
 7 |   offspring = modularityCrossover(parent_1, parent_2);
 8 |   mutate(offspring);
 9 |   localSearch(offspring);
10 |   if getFitness(offspring) > getWorstFitness(pop) then
11 |   |   getWorst(pop) = offspring;
12 |   end
13 end
```

Algorithm 1: The schematic framework of MADOC.

munities present a single connected component. Our initialization scheme follows the approach of *egonets*, which starts from a centric node connected to its neighborhood to create an egocentric network [19, 34]. Algorithm 2 shows the initialization scheme.

```
 1 c = 1; //communities start from 1.
 2 counter = 0;
 3 while counter < |V| do
 4 |   seed = rand(1, |V|);
 5 |   counter = counter + 1;
 6 |   placeNodeInCommunity(seed, c);
 7 |   for each neighbor of seed do
 8 |   |   placeNodeInCommunity(neighbor, c);
 9 |   |   counter = counter + 1;
10 |   end
11 |   c = c + 1;
12 end
```

Algorithm 2: The steps performed to generate a new individual.

To complete the population, one of the individuals is created in a similar manner, however, the *seeds* are chosen according to the degree of the nodes, the first *seed* is the node with the highest degree in the network, the second *seed* is the highest degree node in the network not yet attributed to a community and so on. This scheme produces an individual with fitness above the average in the initial population, not yet with the best modularity, but with much higher quality than the individuals from random initialization.

2.2 Selection and Replacement

In each generation, MADOC creates a population *pop* of N_p individuals, where N_p is the size of the population. To create an *offspring*, two parents are recombined using the modularity based crossover. The first parent is selected with equal probability from the *elitePop*, a subpopulation corresponding to 10 % of the entire population with the best fitness individuals. The second parent is selected using tournament selection, in which a subpopulation *pop′* is selected with equal probability from the main population, next, the best individual of *pop′* is selected as the second parent.

The replacement strategy adopted by MADOC is simply replacing the worst individual with the *offspring*, but only if the fitness of the *offspring* is higher. This combination of selection and replacement consists of an elitist strategy for evolving the population.

2.3 Modularity Based Crossover

The recombination mechanism used by MADOC is a variation of the modularity based crossover [13, 26]. The modularity based crossover uses a priory list with the fitness of each community in both parents, the best communities are selected first, therefore, increasing the quality of the offspring. Our crossover operator differs in the sense that, instead of using a single global priority list for both parents, we use two priority lists, one for each parent, selecting the best community from each parent until it covers the graph. If a node is already set, a community is partially copied from one parent, as per the example in Fig. 2 (step 3). This strategy ensures that the offspring will inherit information from both parents, despite if the quality of the parents is highly uneven. Figure 2 shows the modularity recombination crossover used by MADOC.

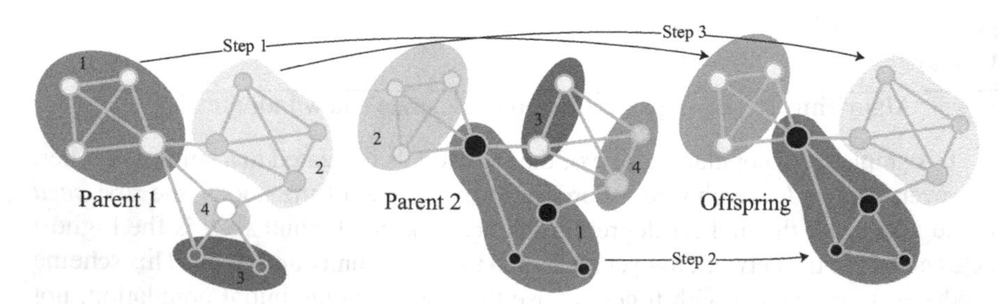

Fig. 2 The modularity based crossover, two parents are recombined into an offspring. Each parent has a priority list according to the quality of its communities. In step 1, the community with the highest modularity in parent 1 is copied to the offspring. In step 2, the community with the highest modularity in parent 2 is copied to the offspring. The process repeats until it covers the whole graph. In step 3, some nodes of the community in parent 1 are already assigned to another community in the offspring, therefore, only the remaining nodes are copied

2.4 Neighborhood Mutation

Mutation is the genetic operator used to preserve the diversity of the population [14, 17]. The mutation operator modifies a solution inserting a level of perturbation to the solution. A common approach for mutation is to change the value of the string in the individual with a random value, however, this can produce communities with disconnected components and therefore, invalid solutions. To overcome this issue, we introduce a new neighboring mutation strategy. Our mutation strategy is to choose a node with uniformly random probability, change its label to another community, next, select all its neighbors and place them in the same community, therefore ensuring that the individual is still a valid solution. The parameter mutation rate P_m controls the amount of perturbation inserted in the solutions, starting from 0.05, which represents a probability of 5 %, increasing in steps of 0.01 to a maximum of 0.15. The mutation is increased if there is no average fitness gain after 15 generations.

2.5 Local Search

The local search mechanism is important in order to improve the quality of the solutions with Memetic Algorithms. In our algorithm, the local search is applied to all individuals in every generation.

Similarly to label propagation methods, we update nodes communities according to the highest frequency label of neighbor nodes, if, the community is different from the current one. The local search algorithm traverses the graph node by node testing permutations which increases the fitness. If changing a node to another community results in better fitness this solution is adopted and the algorithm moves to the next node.

Calculate the total fitness at each permutation is computationally expensive, rather than, we use *delta fitness* to measure the gain of permutations. To achieve that, we store the fitness of each community in an auxiliary data structure, therefore, when moving a node v from community a to community b, we only need to compute the change in fitness of a and b to assess if the movement results in fitness gain. Algorithm 3 shows the steps performed by the local search.

Once a *node-community* permutation results in fitness 'gain', the permutation is *accepted* and the algorithm moves to the next node, regardless of the possibility of moving the node to other communities. This approach was adopted to reduce the number of permutations during the local search. Moreover, our local search is inspired by label propagation, in which the nodes tend to form communities with the highest frequency community label amongst its neighbors.

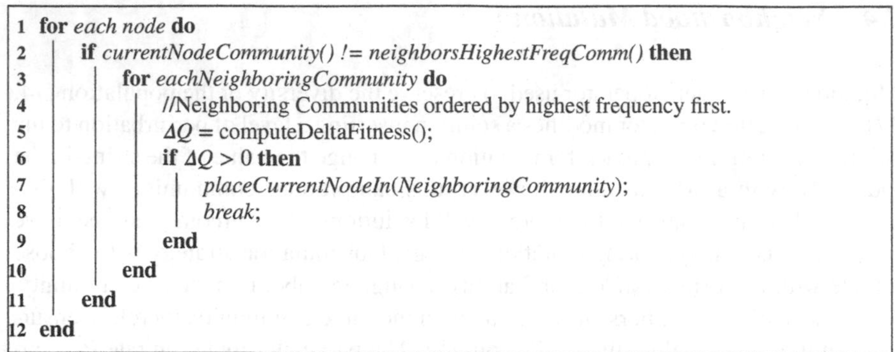

Algorithm 3: The steps performed during the local search.

3 Experimental Results

3.1 Parameters Used by MADOC

MADOC receives only two parameters, the population size N_p and the maximum number of generations without fitness improvement T_{max}. The amount of mutation P_m is dynamically set according to the population's average fitness. Table 1 shows two sets of parameters tested with MADOC. Set A follows the algorithms MLCD, MOGA-Net and MODPSO [22] and set B follows the algorithm MA-Net [26].

Three well-known benchmark networks were used to set the parameters used by MADOC, Zachary Karate Club, American College Football, and a LFR Synthetically generated network with 128 nodes and 1024 edges. Tests were performed over 50 independent runs for each benchmark network with the two sets of parameters summarizing 300 test runs.

Results showed no statistically significant difference between the two sets of parameters for the Zachary Karate Club network and the LFR benchmark. However, the **Set B** performed better for the American College Football benchmark network with statistically significant difference, with *p-value* = *0.04837*, also, producing higher average fitness and smaller standard deviation. The results presented were obtained with parameters following the values of **Set B**.

Table 1 Parameters used by community detection MAs. [a]The mutation rate is adjusted according to the fitness, [b]generations without fitness improvement

	Set A	Set B
Parameter meaning	MLCD, MOGA-Net, and MODPSO	MA-Net and MADOC
N_p: Population size	300	40
P_m: Mutation rate	0.15	0.05 to 0.15[a]
T_{max}: Maximum generations	200	30[b]

3.2 Results in Benchmark Networks

Since modularity optimization plays a key role in our algorithm, results are divided into modularity optimization for non-overlapping communities in benchmark networks and overlapping communities in benchmark networks. Table 2 shows a comparison between our algorithm and another state-of-the-art modularity optimization algorithms, we are able to find the best modularity found in the literature [4, 26] for each benchmark network.

In addition, we used the *Normalized Mutual Information* to measure the similarity between our results and the known community structure of benchmark networks. Table 3 shows the characteristics of the benchmark networks used to assess MADOC capacity of detecting overlapping communities.

Table 2 Modularity optimization results comparison table

Network		MOGA-Net	MODPSO	MLCD	MA-Net	MADOC
Zachary Karate Club	Max Q	0.4159	**0.4198**	**0.4198**	**0.4198**	**0.4198**
	AVG Q	0.3945	0.4182	**0.4198**	0.4190	**0.4198**
	Stdv	0.0089	0.0079	**0**	0.002	**0**
Lusseau's Dolphins	Max Q	0.5034	0.5265	**0.5285**	**0.5285**	**0.5285**
	AVG Q	0.4584	0.5255	**0.5285**	0.5230	0.5284
	Stdv	0.0163	0.0061	**0**	0.0040	0.0003
American College Football	Max Q	0.4325	**0.6046**	**0.6046**	**0.6046**	**0.6046**
	AVG Q	0.3906	0.6038	**0.6046**	0.6010	0.6045
	Stdv	0.0179	0.0011	**0**	0.030	0.0001
Jazz Musicians	Max Q	0.2952	0.4421	**0.4451**	**0.4451**	**0.4451**
	AVG Q	0.2929	0.4419	**0.4451**	**0.4451**	0.4446
	Stdv	0.0084	0.0001	**0**	**0**	0.0021

Table 3 Overlapping benchmark networks used to test MADOC

Network	Nodes	Edges	AVG degree	$L(G)$ nodes	$L(G)$ edges	$L(G)$ AVG degree
LFR benchmark-a	8	16	4.00	16	54	6.75
LFR benchmark-b	32	192	12.01	192	2112	22.01
Les Miserables	77	254	6.597	254	2808	22.11

Table 4 Results for overlapping communities with benchmark networks

Network	$L(G)$ Max Q	$L(G)$ Avg Q	Standard deviation	I_{norm}	Communities
LFR benchmark-a	0.2220	0.2220	0	1	2
LFR benchmark-b	0.4033	0.3749	0.0197	0.948	2
Les Miserables	0.5918	0.5815	0.0057	0.797	7

Table 4 shows the results obtained by MADOC in terms of similarity with the known benchmark network communities.

The maximum modularity, average modularity, and standard deviation are relative to fifty independent runs. The case study performed on the *Les Miserables* benchmark, a network of coappearances of characters in Victor Hugo's novel 'Les Miserables', in which, nodes represent characters and edges connect any pair of characters that appear in the same chapter of the book. The *Les Miserables* results are compared with the results[1] presented by Evans and Lambiotte [9, 10], with high similarity (I_{norm}: 0.797) between MADOC results and Evans and Lambiotte results. The overlapping communities also depict the central role of the character *Valjean*, which appears as a central character and participates in several communities. This result is also consistent with the results presented by He et al. [16], where the same number of communities is found, and the main character *Valjean* is a central node participating in diverse communities.

3.3 *Network of Co-purchases of Photographic Material at Amazon.com*

We analyse the overlapping communities of co-purchased products and the relationship between products and brands on the Amazon.com website. Through this example, we will get closer to answering the peculiar question *"Where does my brand end?"* and provide some insights that the MADOC algorithm can bring to its users. The dataset used in this experiment is a subset of a larger dataset comprised of 191,000 products crawled from the Amazon.com website [23, 24]. The metadata includes the product name and brand, price, sales rank, description and the co-purchasing links. The possible co-purchasing links include the *'also viewed'*, *'also bought'*, *'bought together'* and *'buy after viewing'*. We have concentrated our analysis on the network that can be formed by only using the *'also bought'* information to build a graph that links products and their corresponding brands.

[1]The data is available at: http://dx.doi.org/10.6084/m9.figshare.1573032.

We have chosen one particular brand to study due to its rich history; Kodak. Our brand-centric sub-network is the subgraph induced by the Kodak brand in the *'also bought'* co-purchasing network.

Kodak introduced the automatic snapshot camera over 110 years ago and holds several patents including the digital camera in 1976 [6]. Kodak had sales of $13.3 billion in 2006 and approximately 41,000 employees worldwide. Several authors point out that a sequence of bad corporate judgements caused the brand's devaluation. Kodak was ranked the fourth most valuable brand in the United States in 1996 but in 2007 it was no longer even on the top 100 most valued brands, ending up with bankruptcy in 2012 [20, 37]. This makes Kodak a useful case study to generate some user insights of MADOC.

In our network each node represents a product, the node ID is the Amazon Standard Identification Number (ASIN), and the 'label' is the product name and brand. The graph has **565 nodes** and **4,573 edges**, the line graph has **4,573 nodes** and **144,500 edges**, our algorithm partitioned the line graph into **11** communities with modularity $Q = \mathbf{0.645}$.

We start our discussion with community C2 as it provides an interesting analysis. Figure 3a shows the overlap of community C2 with C4 and the corresponding 'Wordle' images. These show the dominating products and brands in the overlapping section between these two communities and were created by inputting the product descriptions and metadata in the online 'Wordle' tool.[2]

It is clear that these communities (and their overlap) are largely dominated by Kodak products and its main competitors at the time of these purchases in the film area (e.g. Fujifilm, Ilford). Therefore, we are clearly still within Kodak's brand "boundaries" and close to their direct close competitors.

When we continue exploring the graph and take community C4 again with another overlapping community (C11) we get a different view (Fig. 3b). Looking at community C4 by itself we can see that it "bridges" most communities in the graph. It is mainly composed of Epson products (Kodak is for the first time a minority in the world cloud of brands) which is also apparent in the overlap shown in Fig. 3b.

Kodak is still one of the main brands featured in the overlapping section of these communities, however, we can see that Epson (one of the world's largest manufacturers of computer printers)[3] has become the most prominent brand. Alongside this, we can see a lot of products related to printing appear in the overlap, for instance; paper, photo sheets, inkjet and so forth. Many business analysts that have studied Kodak consider that Kodak tried to pursue the printing business for "too long", rather than diversifying, and this is what they claim largely led to their failure.[4] The information in this image shows that Epson is leading in this product space showing a somewhat blurry initial 'boundary' of Kodak. A sign that perhaps this could be *"where their brand ends"*.

---

[2]http://www.wordle.net.

[3]http://www.epson.com.

[4]http://www.forbes.com/sites/daviddisalvo/2011/10/02/what-i-saw-as-kodak-crumbled/#48f55b2820f5.

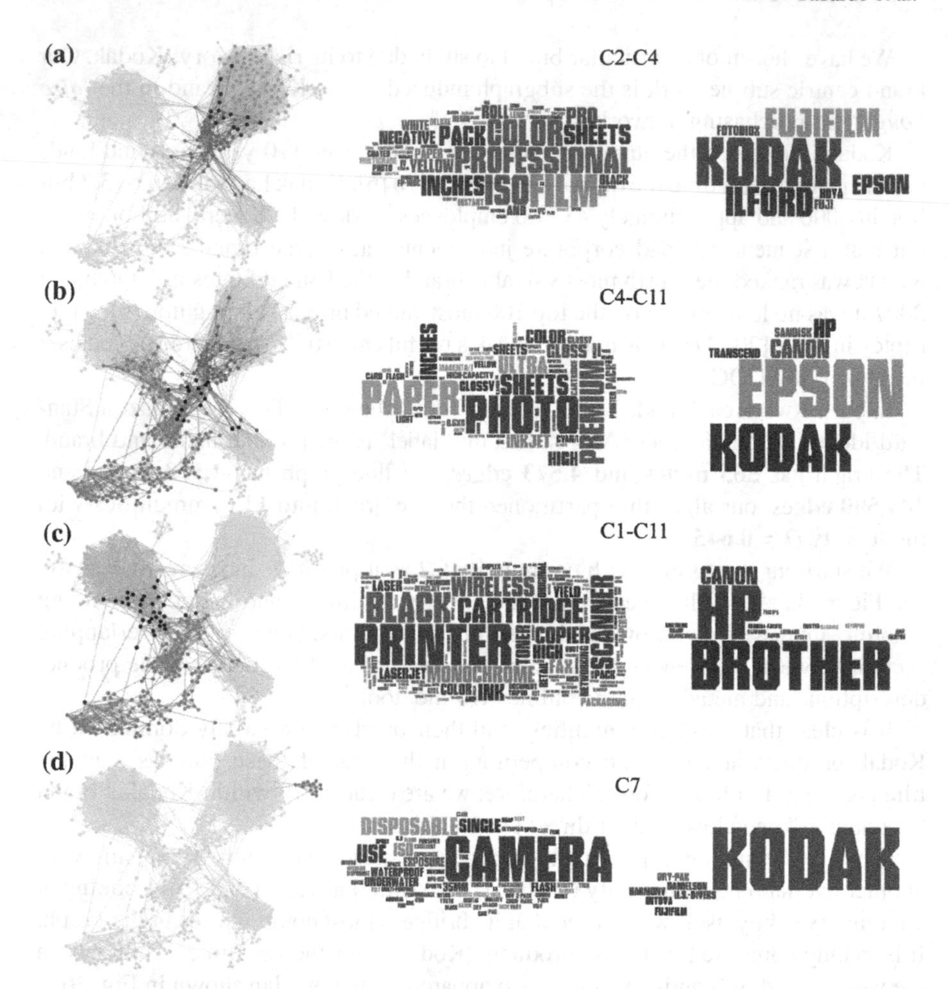

Fig. 3 Figure **a** shows the overlapping communities C2 and C4 on the *left*, in the *center* is the 'Wordle' of products for the intersection between C2-C4, and on the *right* is the 'Wordle' of the brands of this interestion. Figures **b** and **c** follow the same arrangement for communities C4-C11 and C1-C11 respectively. In Fig. **d**, the community C7 (highlighted in *green*) and its 'Wordle's' are also presented

Finally, if we move to yet another set of communities, C1 and C11 we see Kodak almost completely disappear from the overlapping product space as shown in Fig. 3c. Community C1 is largely dominated by printing products and brands. In Fig. 3c we can see that HP, Brother and Canon in particular are highly prominent. In this overlap it is even more clear that printing products significantly dominate this product space.

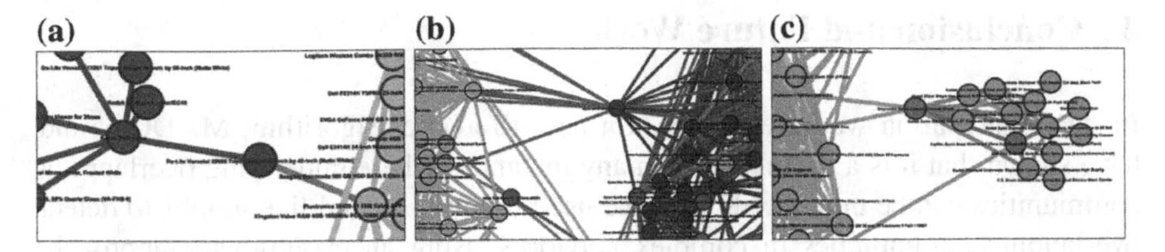

Fig. 4 A detailed look at some nodes that are in a specific overlapping location. In **a** the central node clearly belongs to both communities, the algorithm was able to depict this structure while preserving the community on the *left*. In **b**, the central node is overlapping different communities, where it shares several edges. And, in **c** the nodes sharing connections with other communities are overlapping, while the small community on the *right* is also depicted

The investigation of these three sets of communities shows that more information is hidden in the overlaps that helps identify brand "boundaries" and close competitors of Kodak. Interestingly, we have been able to find those areas of Kodak's brand and product offerings that led them to their devaluation as a result of strong competition. Furthermore, in this analysis we highlight one last interesting community; C7 which is shown in Fig. 3d.

This community seems to be a small segment of the market clearly dominated by Kodak with little competition. It seems to be that there are many 'Underwater', 'Disposable' and 'Waterproof' camera products which did not have many co-purchases associated to them in this community. We must remind the reader that we are analysing the graph induced by the Kodak brand and thus it is natural to have an unequal representation of other brands versus Kodak in the overall network. However, the lack of competition in this area is interesting with the exception of Intova (a current Hawaiian digital compact camara[5] which is a healthy competitor of the GoPro brand[6]). The reason we highlight this community is because it is clearly less densely connected to the other communities in the larger sections of the graph which in turn means that Kodak had less competition in this area. Together with some other small communities (e.g. C10 where the brand 'Led-Lenser' dominates) it provides some insights to the Kodak brand as these areas could indicate the missed opportunities of expansion markets possible for Kodak.

Finally, we would like to show that MADOC was able to accurately identify the community structure, preserving small communities and non-overlapping areas. Figure 4 shows some regions of the graph in more detail.

[5]http://www.intova.net.

[6]http://www.calypsoproductions.com.au/2014/01/gopro-vs-intova-video-comparison.

4 Conclusion and Future Work

In this contribution we have proposed a new Memetic Algorithm; MADOC, and have shown that it is a useful tool in many instances where underlying overlapping communities can be uncovered. We have used modularity and line graphs to detect overlapping communities in complex networks using an evolutionary approach. Throughout this contribution we have shown that boundaries are overlapping, not just in biological or natural networks, but also in complex marketing-domain co-purchasing networks. Boundaries of communities cannot be reduced to single disjoint network partitions, but rather a "blurry" overlap of interconnected and overlapping objects. Our future work includes omitting the line graph step by employing a metric capable of detecting overlapping communities as the fitness function for our algorithm. This will help to reduce the size of the instances used to compute the network communities and in turn improve running time. Finally, future work could also explore the many more applications of this method in a much wider variety of domains bringing more real-life solutions and decision support mechanisms for organizations and researchers.

Acknowledgments A.C. Gabardo is supported by CNPQ Brazil, scholarship number 204978/2014-9.

References

1. Aaker, J.L.: Dimensions of brand personality. Journal of Marketing Research 34(3), 347–356 (1997), http://www.jstor.org/stable/3151897
2. Adamic, L.A., Glance, N.: The political blogosphere and the 2004 us election: divided they blog. In: Proceedings of the 3rd international workshop on Link discovery. pp. 36–43. ACM (2005)
3. Ahn, Y.Y., Bagrow, J.P., Lehmann, S.: Link communities reveal multiscale complexity in networks. Nature 466(7307), 761–764 (2010)
4. Aloise, D., Cafieri, S., Caporossi, G., Hansen, P., Perron, S., Liberti, L.: Column generation algorithms for exact modularity maximization in networks. Physical Review E 82(4), 046112 (2010)
5. Barabási, A.L., Frangos, J.: Linked: the new science of networks science of networks. Basic Books (2014)
6. Bayer, B.E.: Color imaging array (Jul 20 1976), uS Patent 3,971,065
7. Boccaletti, S., Latora, V., Moreno, Y., Chavez, M., Hwang, D.U.: Complex networks: Structure and dynamics. Physics reports 424(4), 175–308 (2006)
8. de Vries, N.J., Carlson, J., Moscato, P.: A data-driven approach to reverse engineering customer engagement models: Towards functional constructs. PloS one 9(7), e102768 (2014)
9. Evans, T.: Information on Les Miserables network used in Evans and Lambiotte 2010. FigShare (10 2015), http://dx.doi.org/10.6084/m9.figshare.1573032
10. Evans, T., Lambiotte, R.: Line graphs, link partitions, and overlapping communities. Physical Review E 80(1), 016105 (2009)
11. Fortunato, S.: Community detection in graphs. Physics reports 486(3), 75–174 (2010)
12. Gabardo, A.C., Lopes, H.S.: Using social network analysis to unveil cartels in public bids. In: Network Intelligence Conference (ENIC), 2014 European. pp. 17–21. IEEE (2014)

13. Gach, O., Hao, J.K.: A memetic algorithm for community detection in complex networks. In: Parallel Problem Solving from Nature-PPSN XII, pp. 327–336. Springer (2012)
14. Goldberg, D.E., Holland, J.H.: Genetic algorithms and machine learning. Machine learning 3(2), 95–99 (1988)
15. Gomez Ravetti, M., Rosso, O.A., Berretta, R., Moscato, P.: Uncovering molecular biomarkers that correlate cognitive decline with the changes of hippocampus? gene expression profiles in alzheimer's disease. PloS one 5(4), e10153 (2010)
16. He, D., Jin, D., Chen, Z., Zhang, W.: Identification of hybrid node and link communities in complex networks. Scientific reports 5 (2015)
17. Holland, J.H.: Adaptation in natural and artificial systems: an introductory analysis with applications to biology, control, and artificial intelligence. MIT press (1992)
18. Jebabli, M., Cherifi, H., Cherifi, C., Hamouda, A.: Overlapping community detection versus ground-truth in AMAZON co-purchasing network. In: Yétongnon, K., Dipanda, A. (eds.) 11th International Conference on Signal-Image Technology & Internet-Based Systems, SITIS 2015, Bangkok, Thailand, November 23-27, 2015. pp. 328–336. IEEE Computer Society (2015), http://dx.doi.org/10.1109/SITIS.2015.47
19. Lancichinetti, A., Fortunato, S., Kertész, J.: Detecting the overlapping and hierarchical community structure in complex networks. New Journal of Physics 11(3), 033015 (2009)
20. Larish, J.J.: Out of Focus: The story of how Kodak lost its direction. CreateSpace Independent Publishing Platform, Luxembourg (2012)
21. Leskovec, J., Lang, K.J., Dasgupta, A., Mahoney, M.W.: Community structure in large networks: Natural cluster sizes and the absence of large well-defined clusters. Internet Mathematics 6(1), 29–123 (2009)
22. Ma, L., Gong, M., Liu, J., Cai, Q., Jiao, L.: Multi-level learning based memetic algorithm for community detection. Applied Soft Computing 19, 121–133 (2014)
23. McAuley, J., Pandey, R., Leskovec, J.: Inferring networks of substitutable and complementary products. In: Proceedings of the 21th ACM SIGKDD International Conference on Knowledge Discovery and Data Mining. pp. 785–794. ACM (2015)
24. McAuley, J., Yang, A.: Addressing complex and subjective product-related queries with customer reviews. arXiv preprint arXiv:1512.06863 (2015)
25. Moscato, P., Cotta, C.: A modern introduction to memetic algorithms. In: Handbook of Metaheuristics, pp. 141–183. Springer (2010)
26. Naeni, L.M., Berretta, R., Moscato, P.: Ma-net: A reliable memetic algorithm for community detection by modularity optimization. In: Proceedings of the 18th Asia Pacific Symposium on Intelligent and Evolutionary Systems, Volume 1. pp. 311–323. Springer (2015)
27. Naeni, L.M., De Vries, N.J., Reis, R., Arefin, A.S., Berretta, R., Moscato, P.: Identifying communities of trust and confidence in the charity and not-for-profit sector: A memetic algorithm approach. In: Big Data and Cloud Computing (BdCloud), 2014 IEEE Fourth International Conference on. pp. 500–507. IEEE (2014)
28. Neri, F., Cotta, C., Moscato, P.: Handbook of memetic algorithms, vol. 379. Springer (2012)
29. Newman, M., Barabasi, A.L., Watts, D.J.: The structure and dynamics of networks. Princeton University Press (2006)
30. Newman, M.E.: The structure and function of complex networks. SIAM review 45(2), 167–256 (2003)
31. Palla, G., Derényi, I., Farkas, I., Vicsek, T.: Uncovering the overlapping community structure of complex networks in nature and society. Nature 435(7043), 814–818 (2005)
32. Pastor-Satorras, R., Vespignani, A.: Evolution and structure of the Internet: A statistical physics approach. Cambridge University Press (2007)
33. Pizzuti, C.: Overlapped community detection in complex networks. In: Proceedings of the 11th Annual conference on Genetic and evolutionary computation. pp. 859–866. ACM (2009)
34. Rees, B.S., Gallagher, K.B.: Overlapping community detection by collective friendship group inference. In: Advances in Social Networks Analysis and Mining (ASONAM), 2010 International Conference on. pp. 375–379. IEEE (2010)

35. Solé, R.V., Pastor-Satorras, R.: 7 complex networks in genomics and proteomics. Handbook of graphs and networks: From the genome to the internet (2006)
36. Srivastava, A.: Motif analysis in the amazon product co-purchasing network. CoRR abs/1012.4050 (2010), http://arxiv.org/abs/1012.4050
37. Swasy, A.: Changing Focus: Kodak and the Battle to Save a Great American Company. Times Business, New York, NY (1997)
38. Yamazaki, T., Shimizu, N., Kobayashi, H., Yamauchi, S.: Weighted micro-clustering: Application to community detection in large-scale co-purchasing networks with user attributes. In: Bourdeau, J., Hendler, J., Nkambou, R., Horrocks, I., Zhao, B.Y. (eds.) Proceedings of the 25th International Conference on World Wide Web, WWW 2016, Montreal, Canada, April 11-15, 2016, Companion Volume. pp. 131–132. ACM (2016), http://doi.acm.org/10.1145/2872518.2889406

Analysis of Parameter-Less Population Pyramid on the Local Distribution of Inferior Individuals

Taku Hasegawa, Yuta Araki, Naoki Mori and Keinosuke Matsumoto

Abstract Unlike many GAs, the Parameter-less Population Pyramid (P3) is an optimization model that avoids premature convergence due to the pyramid-like structure of populations, and thus P3 can be applied to a wide range of problems without parameter tuning. However, in some problems, P3 cannot control the number of fitness evaluations in local search and in crossover, while adapting problem structures. Meanwhile, we have proposed a novel technique, called DII analysis. The computational complexity of applied problems can be estimated based on the number of local optima according to the results obtained using DII. In order to solve the problem of P3, we also have proposed combining P3 with DII analysis (P3-DII). In this study, we investigated the effect of DII analysis on balance between genetic operators. The performance of P3-DII was confirmed according to the computational experiments which were carried out taking several combinational problems as examples.

Keywords Genetic algorithm · Local search · Fitness landscape · Adaptation · Parameter-less

1 Introduction

Many evolutionary techniques such as genetic algorithms (GAs) employ parameters that provide user control of search dynamics. Therefore, reducing the number of these parameters or building a theory for tuning parameters adaptively are fun-

T. Hasegawa (✉) · Y. Araki · N. Mori · K. Matsumoto
Graduate School of Engineering, Osaka Prefecture University, 1-1 Gakuencho,
Sakai City, Osaka 599-8531, Japan
e-mail: hasegawa@ss.cs.osakafu-u.ac.jp

Y. Araki
e-mail: araki@ss.cs.osakafu-u.ac.jp

N. Mori
e-mail: mori@cs.osakafu-u.ac.jp

K. Matsumoto
e-mail: matsu@cs.osakafu-u.ac.jp

G. Leu et al. (eds.), *Intelligent and Evolutionary Systems*,
Proceedings in Adaptation, Learning and Optimization 8,
DOI 10.1007/978-3-319-49049-6_11

damental goals in the area of GAs. One approach is to learn the fitness landscape to develop a surrogate model [1]. GAs utilize the surrogate model to guide genetic operators during the irreversible search process [2], and then sometimes cause several problems including premature convergence due to a failure of prediction at one step.

The Parameter-less Population Pyramid (P3) [3] can avoid premature convergence due to the pyramid-like structure of the populations employed. P3 performs additional search according to P3 iterators, and thus P3 is a robust optimization model to the accuracy of the prediction obtained at each step. On the other hand, in several problems, P3 can not control the balance between the number of fitness evaluations in local search and in crossover, while adapting each problem structure. Moreover, we can not give the efficient distribution of the fitness evaluations between GAs and LS techniques because there is no parameter to do so.

In the present study, we have proposed a technique "analysis based on the Distribution of Inferior Individuals in the local neighborhood" (DII analysis), which facilitates adaptive search using P3 [4]. While no optimizer can solve all problems equally well [5], we focus on solving the problems which are structured as partial fitness functions with optimizers based on DII analysis. DII analysis predicts the computational complexity of the problem based on the features of a simple model by comparing the quality of the solutions in the local neighborhood. Furthermore, we also have proposed combining P3 with DII analysis (P3-DII) [4].

In order to show the effectiveness of P3 and P3-DII, in this study, we carried out taking several combinatorial problems as examples and showed that DII analysis can control the search dynamics of P3 while providing better results.

2 Parameter-Less Population Pyramid

P3 is an optimization algorithm that requires no user parameters. P3 employs a pyramid-like population structure rather than a single population of solutions. Each level of the pyramid comprises a population of solutions, which behaves as different generations during evolution. P3 utilizes these populations to improve the solutions by hill climbing and crossover operations.

2.1 *Algorithm*

As mentioned above, P3 maintains a pyramid-like population structure where each level of the pyramid comprises a population of solutions. More optimized solutions are found at the higher levels. The different levels comprise unique solutions such that $\forall_{i,j \in P, i \neq j} P_i \cap P_j = \phi$ where i and j are the levels of the pyramid. If new solutions do not exist in any levels of the pyramid, they can only be added to a level. For the details, please refer to [3].

3 DII Analysis

We propose a novel approach to estimate complexity of problems based on DII which is the distribution of the number of inferior individuals in the local neighborhood. In this paper, we focus on DII with a Hamming distance of 1, named DII-1. For simplification, we define the complexity analysis based on DII(DII-1) as just "DII(DII-1) analysis".

In DII-1 analysis, the procedure is divided into two parts, where the first estimates the distribution of inferior individuals with local search techniques and the second uses the features derived from the distribution based on a simple model, *Nt* landscape problem.

DII-1 analysis can be applied to any problem with a discrete representation, but we focus on problems with a binary domain for simplicity. In this section, we explain DII analysis with examples of the *Nk* landscape problem and the deceptive trap problem, which are described in Sect. 5, and the *Nt* landscape problem described in the next subsection.

The main symbols used in this section are as follows.
F: fitness function
f_i: partial fitness function
t: random functional size
N: chromosome length
s: individual
s_i: gene of individual s at locus i

3.1 Nt Landscape Problem

We define the *Nt* landscape problem as follows.

1. The fitness function F is given by sum of the partial functions f.
2. All of the partial functions are given by genes in t loci. The number of partial functions is N/t, where N is the chromosome length. N/t does not need to be an integer in our theory.
3. All of the partial functions are given by random variables based on the positions of the genes. The function values in partial functions are different from each other.

Nt landscape problem is a type of simplified *Nk* landscape with no overlapping genes of partial functions. In this study, we focus on the computational complexity derived from the value of t in *Nt* landscape problem. We refer to the value t as the "random functional size". Let f_i be the partial function given by the set of loci $L = \{(i-1)t+1, (i-1)t+2, \ldots, it\}$. We define the genes $S = \{s_{(i-1)t+1}, s_{(i-1)t+2}, \ldots, s_{it}\}$ in loci I with higher values of f_i than any combinations of genes at a Hamming distance of 1 in loci I as the partial local optima in the *Nt* landscape problem. DII-1 analysis predicts the computational complexity based on the partial local optima in the *Nt* landscape problem.

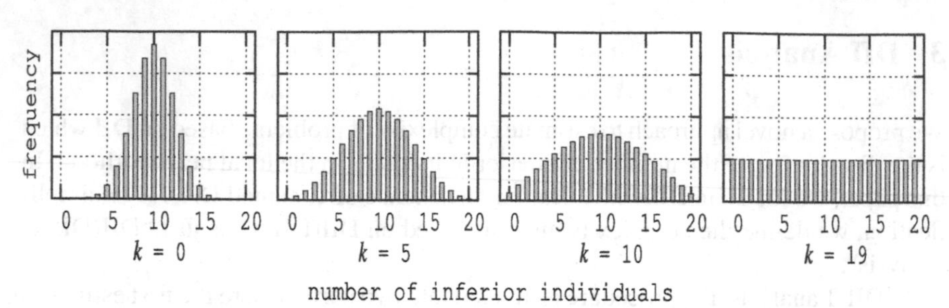

Fig. 1 Frequency with respect to the number of inferior individuals in the neighborhood with Hamming distance of 1

3.2 *Inferior Individuals*

The individuals inferior to individual s are defined as the individuals with not greater fitness than the fitness value of s. In this study, we focus on the number of individuals inferior to individual s in the neighborhood, which number is represented by I_s, and we define a local optimum as the genotype of individual s with $I_s = N$.

The distribution of inferior individuals represents many features related to the problem's complexity in GAs based on the rank of individuals, but we mainly consider an increase in the local optima according to an increase in the size of a partial function given by random variables. In the first example, we specify problems with a linear function and no fitness plateaus, i.e., the One Max problem and *Nk* landscape problem with $k = 0$. In these problems, the distribution of inferior individuals is a binomial distribution $\mathrm{B}(N, \frac{1}{2})$ because a partial function on a locus is independent of another locus. As another example, we specify a problem with a completely random fitness function such as the *Nk* landscape problem with $k = N - 1$. In this case, the probability that the number of individuals inferior to s equals $I_s (I_s = 0, 1, \ldots, N)$ is $\frac{1}{N+1}$, and the number of inferior individuals is given by the discrete uniform distribution $\mathrm{U_d}(0, N)$. Figure 1 shows the frequency distribution of the number of inferior individuals in the *Nk* landscape problem with $N = 20$. In this figure, the distribution of the number of inferior individuals in the neighborhood is affected by problem features, such as the value of k.

3.3 *DII-1 Analysis*

DII-1 analysis is divided into two steps. In the first step, DII-1 analysis estimates the distribution of inferior individuals with a beta-binomial distribution [6]. In the second step, DII-1 analysis estimates a random functional value t for applied problem, and estimates the maximum number of fitness evaluations in local search and crossover based on t.

3.3.1 Beta-Binomial Distribution

A random variable X is given by the beta-binomial distribution $\mathrm{BB}(n, \alpha, \beta)$, where a random variable X is given by the binomial distribution $\mathrm{Bi}(n, Y)$ and a random variable Y is given by the beta distribution $\mathrm{Be}(\alpha, \beta)$. The beta distribution $\mathrm{Be}(j, n - j + 1)$ gives the jth order statistic with a uniform distribution $\mathrm{U}(0, 1)$ in n trials, and the beta distribution is related to the order statistic of many distributions, which can be converted into a uniform distribution. If we select numbers from a set of integers $\{1, 2, \ldots, N'\}$ in n' trials, $X_j - j$ is given by the beta-binomial distribution $\mathrm{BB}(N' - n', j, n' - j + 1)$, where X_j denotes the jth order statistics.

The probability function $P(X = x)$ for the beta-binomial distribution $\mathrm{BB}(n, \alpha, \beta)$ is:

$$P(X = x) = {}_n C_k \frac{B(x + \alpha, n - x + \beta)}{B(\alpha, \beta)} \tag{1}$$

where $B(\alpha, \beta)$ is the beta function.

In this study, we simply estimate the parameters α, β for the beta-binomial distribution with an average μ and variance σ^2.

$$\alpha = \frac{-\mu^3 - \mu\sigma^2 + \mu^2 n}{\mu^2 - \mu n + \sigma^2 n} \tag{2}$$

$$\beta = \frac{\mu^3 + \mu\sigma^2 - 2\mu^2 n - \sigma^2 n + \mu n^2}{\mu^2 - \mu n + \sigma^2 n} \tag{3}$$

where the beta-binomial distribution is used for the estimation of the distribution of inferior individuals, the probability $P(X = N)$ in the beta-binomial distribution represents the estimated ratio of the local optima relative to the solutions in the search space. Figure 2 shows the ratio of the local optima relative to the variance σ^2 in the inferior individuals in the Nk landscape problem with $N = 20$ and $k = 0, 1, \ldots, 19$. The line in this figure shows the probability $P(X = N)$ in the beta-binomial distribu-

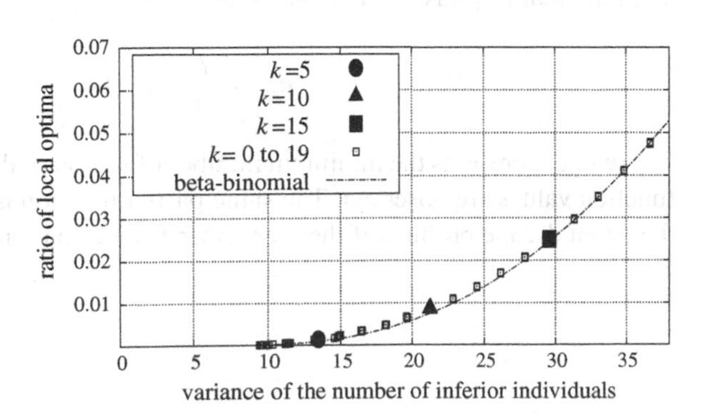

Fig. 2 Ratio of local optima with respect to the variance in the number of inferior individuals

tion, where the average $\mu = \frac{N}{2}$. Based on this figure, we can estimate the proportion of the local optima in the problem that have a similar structure to the *Nk* landscape problem.

3.3.2 Random Functional Size

Next, we estimate the random functional size t. The probability that a set of genes S_i is a partial local optimum is $1/(t+1)$. Then, the expected value of the number of local optima is $\left(2^t/(t+1)\right)^{\frac{N}{t}}$. The random functional size t is obtained from the solutions to the following equations if we have found the number of local optima N_{local}.

$$N_{\text{local}} = \left(\frac{2^t}{t+1}\right)^{\frac{N}{t}} \tag{4}$$

We estimate N_{local} at $2^t P(X = N)$, where $P(X = x)$ is the probability function of the beta-binomial distribution obtained in the previous section. The right-hand side member in Eq. 4 is monotone increasing where $t > 0$, and thus we use Newton's method to obtain the solution to this equation.

3.3.3 Maximum Number of Evaluations in Local Search Techniques

FIHC determines the local optima in $N + 1$ evaluations of problems with linear functions, but we cannot find the solutions to local optima until the number of fitness evaluation reaches almost $2N$. However, if $t = 1$ is given in the *Nt* landscape problem, FIHC can reduce the number of wasteful fitness evaluations. In particular, the distribution of fitness evaluations among operators should be efficient to obtain good performance for Genetic Local Search (GLS) [7]. Therefore, we use the following criteria to determine the number of fitness evaluations in local search techniques for an individual E_{LS} based on the value of t:

$$E_{\text{LS}} = \frac{N}{t}(2^t - 1) \tag{5}$$

where E_{LS} represents the minimum number of fitness evaluations where all of partial function values are observed. The same partial function is observed several times in the *Nt* landscape problem if there are over E_{LS} evaluations.

3.3.4 Number of Partial Local Optima for Crossover

$2^t/(t+1)$ represents the expected number of partial local optima for a partial function in the *Nt* landscape problem. Local search techniques can obtain a partial local optimum in $2^t/(t+1)$ types. Therefore, the crossover performed after local search is expected to recombine these partial local optima.

4 P3-DII

The P3-DII algorithm employs different algorithms compared with P3, as follows:

- Control the number of fitness evaluations employed in the local search technique.
- Control the number of fitness evaluations and the order of clusters during crossover.

4.1 *Local Search in P3-DII*

The number of inferior individuals I_s is observed based on N evaluations before applying FIHC. DII-1 analysis estimates the distribution of the number of inferior individuals based on the data during this procedure. For many problems, the search performance will be better if we apply the best improvement to individuals in this procedure, especially for deceptive trap problem, but we did not apply the best improvements in this procedure before applying FIHC to find the potential of DII analysis here.

In this study, we use the information in this sampling for the first step in FIHC although it was not used in previous study [4].

In P3, FIHC continues the fitness evaluations until the local optima are obtained, whereas P3-DII continues to apply FIHC until the local optima are obtained, or until E_{LC} fitness evaluations, obtained in Eq. 5, are performed in FIHC. Figure 3 shows the local search technique employed in P3-DII.

At least two data points are necessary for DII-1 analysis to estimate α and β in Eqs. (2) and (3). Therefore, we add two data points $I_s = 0.5N \pm \sqrt{0.25N}$ to the sampling data for DII-1 analysis in advance.

4.2 *Crossover in P3-DII*

In P3-DII, the crossover process differs at two points compared with P3: the order of the clusters and the number of fitness evaluations. During crossover, the clusters are tested in order with the nearest random functional size t first. The smallest first order employed by P3 helps preserve diversity, but the diversity in the population is

Algorithm 1 Local search technique in P3-DII

```
Create random solution s
number n_count ← 0
for all s' ∈ neighborhood of s do
    calculate fitness of s'
    if fitness of s' > fitness of s then
        n_count ← n_count + 1
add n_count to data set G_data
estimate random functional size t from G_data
calculate E_LS
number n_eval ← 0
while n_eval < E_LS and not finding local optima do
    apply FIHC to s with an evaluation
    n_eval ← n_eval + 1
```

Fig. 3 Algorithm describing the local search technique in P3-DII. G_{data} represents the data set of the number of inferior individuals

Algorithm 2 CLUSTER USAGE in P3-DII

```
Order useful based on cluster size, with the nearest t first.
for all C_i ∈ useful do
    for all d ∈ shuffled(P_i) do
        Copy d's gene values for C_i into the solution
        estimate random functional size t from G_data
        calculate E_C
        n_eval ← 0
        if solution changes then
            n_eval ← n_eval + 1
            if solution's fitness decreases then
                Revert changes
            else
                break
            if n_eval ≥ E_C then
                break
```

Fig. 4 Algorithm describing how clusters are used to perform crossover in P3-DII

on a rising trend with an increase in t. The order that employs the nearest random functional size t first helps primarily to build several blocks in a random function, which are usually the most difficult structures to optimize with local search techniques (Fig. 4).

The maximum number of fitness evaluations for crossover E_C is obtained as follows:

$$t' = \min(t, t_{\text{cluster}})$$
$$E_C = \max([\frac{2^{t'}}{t' + 1} - 1 + \delta_U], 1) \quad (6)$$

where $t_{cluster}$ denotes the cluster size in cluster usage and δ_U is a random variable given by the uniform distribution U(0, 1). P3-DII continues to perform crossover in a cluster until it reaches E_C, no candidate can be recombined in the cluster or the solutions are improved. $2^{t'}/(t'+1)$ represents the estimated number of partial local optima in a partial function, described in Sect. 3.3.4, where this procedure helps to avoid building blocks being generated rarely when the population size is larger.

5 Test Problems

In this section, we explain the test benchmark problems used to demonstrate the effectiveness of our proposed method.

5.1 *Deceptive Trap*

In the deceptive trap problem, the fitness function combines the sub-functions into separate traps with lengths of k bits where each trap is given by Eq. 7:

$$trap(n_{trap}) = \begin{cases} k - 1 - n_{trap} & (n_{trap} < k) \\ k & (n_{trap} = k) \end{cases} \tag{7}$$

where n_{trap} represents the number of genes with a value of 1 in the trap. Each partial function has one local optimum (all bits are 0) and one global optimum (all bits are 1). In this study, we set a trap size of $k = 7$.

It is easy for P3 to obtain optimal solutions to this problem because FIHC can optimize the sub-functions and crossover can optimize the combination of sub-functions based on few evaluations.

5.2 *Discretized Rosenbrock and Discretized Rastrigin*

The discretized Rastrigin problem and the discretized Rosenbrock problem convert the standard Rastrigin and Rosenbrock problems into gray-coded problems.

The standard Rastrigin evaluation function is as follows.

$$10n + \sum_{i=1}^{n}(x_i^2 - 10\cos 2\pi x_i) \qquad \forall x \in [-5.12, 5.12] \tag{8}$$

The Rastrigin problem has a highly multimodal landscape.

The standard Rosenbrock evaluation function is as follows.

$$\sum_{i=1}^{n-1}(100(x_{i+1} - x_i^2)^2 + (1 - x_i)^2) \qquad \forall x \in [-2.048, 2.048] \tag{9}$$

The Rosenbrock has a unimodal landscape and the global optimum lies in a narrow parabolic valley.

In these two problems, depending on the gray code size, the number of fitness evaluations increases until the local optimum is reached by local search [8]. We set the gray code size $g = 10$ or 20.

5.3 *Nk Landscape*

The function of the *Nk* landscape problem comprises evaluation tables. Each locus has an evaluation table and the partial function value depends on the linked k of other loci. The *Nk* landscape problem has a multimodal landscape and we can control the amount of epistasis in the landscape by changing k. A dynamic programming algorithm is polynomial in N and exponential in k to optimize *Nk* landscape problem [9]. We chose to set $k = 5$ or 7.

5.4 *Knapsack Problem*

This problem requires the maximization of knapsack items where the item weight $<$ weight limit of the knapsack. In this study, the fitness of solutions that exceed the weight limit is set as zero as a penalty. The knapsack problem has a complex landscape around optimal solutions because many solutions that exceed the weight limit are present around the optimal solutions.

6 Experiment 1

We analyzed the performance of P3 and P3-DII for 100 trials, where each run was limited to 30 million evaluations for seven kinds of benchmark problems. The experimental results comprise the average number of successful evaluations required for each algorithm and the speedup factor represented by $E_{\mathrm{P3}}/E_{\mathrm{P3-DII}}$ where $E_{\mathrm{P3-DII}}$ and E_{P3} are the average number evaluations in P3-DII and P3, respectively. We performed a pairwise Mann-Whitney U-test for deceptive trap problem with the maximum problem size.

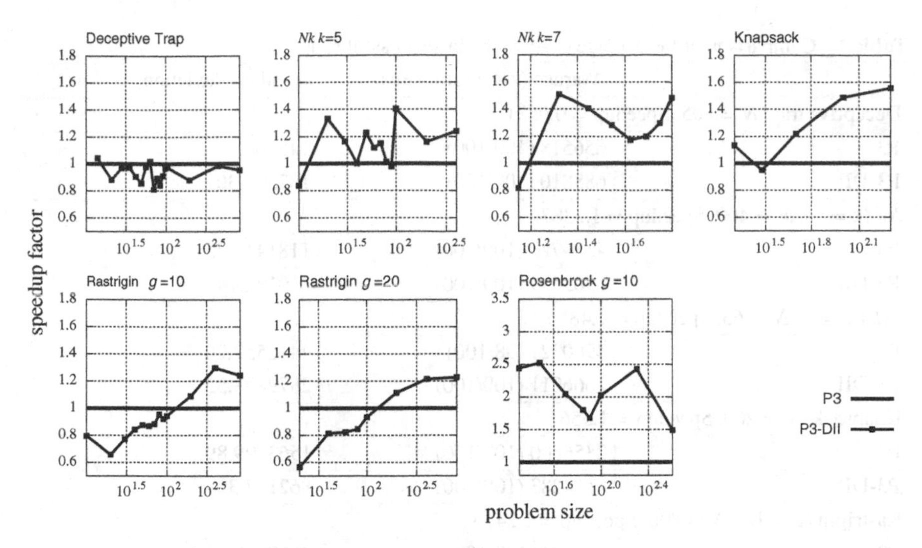

Fig. 5 Speedup factor for P3-DII compared with P3. Abscissa given on a log scale

6.1 Results and Discussion

Figure 5 shows the speedup factor for P3-DII compared with P3. Table 1 shows the average evaluation results and standard deviations in successful runs. Table 2 shows the p-values obtained using the Mann-Whitney U-test for deceptive trap problem.

P3 performed almost no wasteful evaluations in the deceptive trap problem, as specified in [3], and the performance of P3-DII was similar to that of P3. For the problem, P3-DII performed more wasteful evaluations than P3 due to the error when estimating DII-1 with a specific probability. However, in our experiments, the Mann-Whitney U-test did not detect any significant difference between the performance of P3 and P3-DII (p-value = 0.05) in the deceptive trap problem ($N = 805$).

The speedup factor increased with the problem size in the discretized Rastrigin problem. By contrast, the speedup factor appeared to decrease with the problem size in the Rosenbrock problem with $g = 10$, according to the results shown in Fig. 5. These results represent that P3 has several possibilities to outperform P3-DII in the discretized Rosenbrock problem with a larger problem size. However, P3 failed to obtain optimal solutions in all trials for the discretized Rosenbrock function with $g = 20$, even when the problem size was $N = 40$ which was the minimum size in our experiments. By contrast, P3-DII successfully obtained optimal solutions in all of our experiments. Many evaluations tended to be required by FIHC with the discretized Rosenbrock problem.

In the *Nk* landscape, knapsack, and Rastrigin problems, the speedup factor increased with the problem size. P3 outperformed P3-DII in most of the problems with a small size, which may have been caused by differences in the number of evaluations between crossover operations. In P3, the fitness evaluations were performed in

Table 1 Comparison of the average evaluations in successful runs

	Average (successful runs)	Standard deviation
Deceptive trap, $N = 805$, Speedup = 0.9531		
P3	656518 (100/100)	145708.87
P3-DII	688810 (100/100)	182392.30
Nk ($k = 5$), $N = 400$, Speedup = 1.2387		
P3	4339723 (100/100)	1181417.75
P3-DII	3503399 (100/100)	710192.14
Nk ($k = 7$), $N = 63$, Speedup = 1.4855		
P3	3960321 (98/100)	4418552.20
P3-DII	2666011 (100/100)	2879898.55
Knapsack, $N = 200$, Speedup = 1.5562		
P3	2356600 (100/100)	1563299.89
P3-DII	1514283 (100/100)	862157.36
Rastrigin ($g = 10$), $N = 800$, Speedup = 1.2416		
P3	134984 (100/100)	31370.79
P3-DII	108715 (100/100)	25754.56
Rastrigin ($g = 20$), $N = 800$, Speedup = 1.2274		
P3	245104 (100/100)	49367.82
P3-DII	199699 (100/100)	39771.45
Rosenbrock ($g = 10$), $N = 400$, Speedup = 1.4924		
P3	689007 (100/100)	499553.85
P3-DII	461664 (100/100)	429152.79
Rosenbrock ($g = 20$), $N = 200$, Speedup = NaN		
P3	NaN (0/100)	NaN
P3-DII	469896 (100/100)	291755.16

Table 2 p-value obtained using the Mann-Whitney U-test

Problem name	p-value
Deceptive Trap, $N = 805$	0.141714602
Nk ($k = 5$), $N = 400$	7.03379×10^{-13}
Nk ($k = 7$), $N = 63$	1.591985×10^{-3}
Knapsack, $N = 200$	1.12119×10^{-5}
Rastrigin ($g = 10$), $N = 800$	5.66293×10^{-11}
Rastrigin ($g = 20$), $N = 800$	8.11665×10^{-12}
Rosenbrock ($g = 10$), $N = 400$	1.13424×10^{-4}

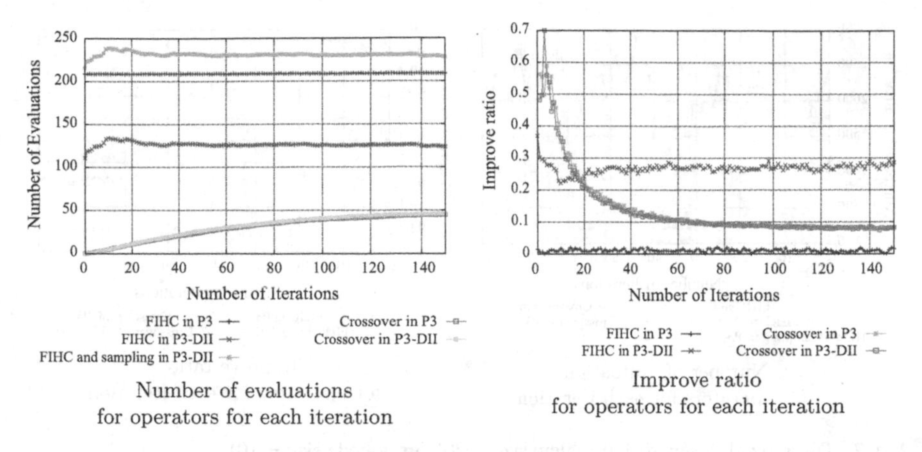

Fig. 6 Deceptive Trap (problem size = 105, trap size = 7)

a cluster during crossover. By contrast, P3-DII performed several fitness evaluations during crossover according to Eq. 6. Equation 6 is based on the estimated partial local optima, but it does not yield the optimal distribution of the fitness evaluations. In P3-DII, crossover recombines x times at most, where x is the estimated number of the partial local optima based on DII analysis, although some wasteful evaluations were performed when the problem size was small. However, P3 could obtain high quality solutions in an efficient manner when the problem size was small, and thus P3 might pass over the building blocks that are generated rarely, depending on the increase in the problem size and the population size in the pyramid. While the probability that a crossover operation in P3 will recombine a unique alternative is $O(E_i)$ where E_i is the number of P3 iterations, the probability of that in P3-DII is $O(N_p)$, where N_p is the number of partial local optima, if DII-1 analysis completely estimates partial local optima. This agreed with the results of our experiments, except the discretized Rosenbrock problem.

7 Experiment 2

We analyzed the number of fitness evaluations in local search and in crossover of P3 and P3-DII. The basic experimental conditions are as same as Experiment 1 and we focused on two problems: deceptive trap problem and discretized rosenbrock problem.

In this experiment, we investigated the average number of evaluations for each operator and the average number of improve ratio represented by $N_{improve*}/E_*$ where $N_{improve*}$ is the number of fitness improvements for an operator $*$ and E_* is the number of evaluations for an operator $*$.

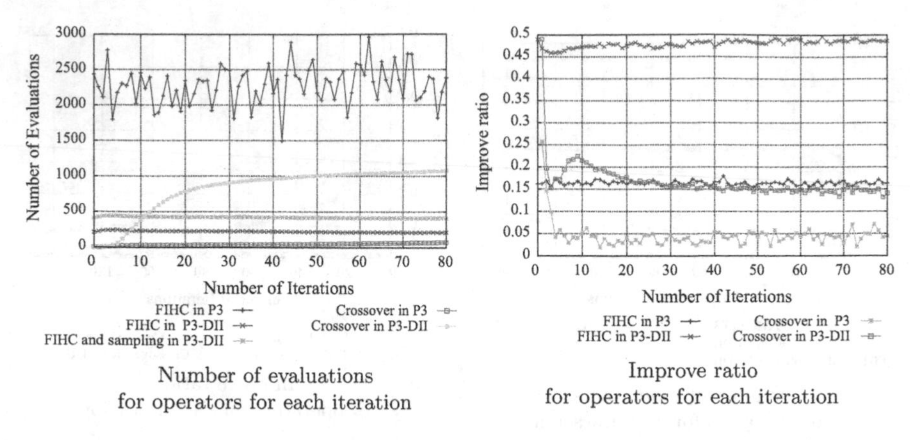

Fig. 7 Discretized Rosenbrock (problem size = 200, gray-code size = 10)

7.1 Results and Discussion

Figures 6 and 7 show the results for deceptive trap problem and for discretized rosenbrock problem.

P3-DII reduces the number of evaluations in FIHC and increases the improve ratio in FIHC for the deceptive trap problem. In addition, P3-DII keeps the number of evaluations in crossover, and improves ratio of P3-DII is almost the same as it is for P3 in crossover.

These results represent that FIHC from P3-DII provides a efficient building block for optimization as same as FIHC from P3 does because the number of evaluations in crossover depends on solutions found by FIHC. P3-DII can reduce the wasteful evaluations in FIHC and P3-DII can control the search dynamics.

On the other hand, proportion of sampling cost to the number of all the evaluations in FIHC is high for this problem. Therefore, P3-DII cannot outperform P3 for almost every problem size of deceptive trap problem although P3-DII can control the search dynamics as mentioned above.

In discretized rosenbrock, P3-DII also reduces the number of evaluations in FIHC and increases the improve ratio in FIHC. Moreover, P3-DII gains the number of evaluations in crossover and increases the improve ratio in crossover.

These results represent that P3-DII can control the balance between the local search and the crossover. In particular, P3 wastes the large number of fitness evaluations in FIHC and P3-DII can reduce it. Therefore, P3-DII can obtain a lot of efficient building block and can recombine them many times in crossover.

The previous results represent that P3-DII can control the balance between the local search and crossover, while adapting problem structures.

8 Conclusion

In this study, we showed the DII-1 analysis to estimate partial local optima based on the distribution of inferior individuals. DII-1 analysis gives a new point of view about the problem difficulty and search efficiency. We also showed P3-DII, which combines DII-1 analysis with P3 in order to control the fitness evaluations required for operators in P3.

To summarize this research:

- We show the DII analysis and P3-DII algorithm.
- P3-DII obtained optimal solutions that P3 failed to find in discretized rosenbrock problem. P3-DII's performance was almost the same as P3's one for the deceptive trap problem, where P3 performed almost no wasteful evaluations.
- P3-DII outperformed P3 in large size problems without the deceptive trap problem.
- We analyzed the search dynamics of P3-DII and confirmed that P3-DII control search dynamics while adapting problem structure.

In our future research, we plan to extend the control of fitness evaluations according to the population size. If the partial local optima are given, the necessary number of individuals required in a population can be estimated. The estimated number allows us to control the fitness evaluations according to the search state. Obtaining estimates for other models and building the theory for DII within an arbitrary Hamming distance are also fundamental issues.

Acknowledgments A part of this work was supported by JSPS KAKENHI Grant, Grant-in-Aid for Scientific Research(C), 26330282, by JSPS KAKENHI Grant, Grant-in-Aid for JSPS Fellows, 16J10941, and by Program for Leading Graduate Schools of Ministry of Education, Culture, Sports, Science and Technology in Japan.

References

1. Jin, Y.: A comprehensive survey of fitness approximation in evolutionary computation. Soft Comput. 9(1), 3–12 (Jan 2005), http://dx.doi.org/10.1007/s00500-003-0328-5
2. Rasheed, K., Hirsh, H.: Informed operators: Speeding up genetic-algorithm-based design optimization using reduced models. In: In Proceedings of the Genetic and Evolutionary Computation Conference. pp. 628–635. Morgan Kaufmann (2000)
3. Goldman, B.W., Punch, W.F.: Parameter-less population pyramid. In:Proceedings of the 2014 Conference on Genetic and Evolutionary Computation. pp. 785–792. GECCO '14, ACM, New York, NY, USA (2014), http://doi.acm.org/10.1145/2576768.2598350
4. Inoue, K., Hasegawa, T., Araki, Y., Mori, N., Matsumoto, K.: Adaptive control of parameter-less population pyramid on the local distribution of inferior individuals. In: Proceedings of the 2015 Annual Conference on Genetic and Evolutionary Computation. pp. 863–870. GECCO '15, ACM, New York, NY, USA (2015), http://doi.acm.org/10.1145/2739480.2754818
5. Wolpert, D., Macready, W.: No free lunch theorems for optimization. Evolutionary Computation, IEEE Transactions on 1(1), 67–82 (Apr 1997)

6. Shakil, M.: Using Beta-binomial Distribution in Analyzing Some Multiple-Choice Questions of the Final Exam of a Math Course, and its Application in Predicting the Performance of Future Students (2009)
7. Bosman, P.A.N., Thierens, D.: The roles of local search, model building and optimal mixing in evolutionary algorithms from a bbo perspective. In: Proceedings of the 13th Annual Conference Companion on Genetic and Evolutionary Computation. pp. 663–670. GECCO '11, ACM, New York, NY, USA (2011), http://doi.acm.org/10.1145/2001858.2002065
8. Rowe, J.E., Hidovic, D.: An evolution strategy using a continuous version of the gray-code neighbourhood distribution. In: Deb, K., Poli, R., Banzhaf, W., Beyer, H., Burke, E.K., Darwen, P.J., Dasgupta, D., Floreano, D., Foster, J.A., Harman, M., Holland, O., Lanzi, P.L., Spector, L., Tettamanzi, A., Thierens, D., Tyrrell, A.M. (eds.) Proceedings of the annual conference on Genetic and Evolutionary Computation - GECCO 2004. Lecture Notes in Computer Science, vol. 3102, pp. 725–736. Springer (2004), http://dblp.uni-trier.de/db/conf/gecco/gecco2004-1.html#RoweH04
9. Wright, A.H., Thompson, R.K., Zhang, J.: The computational complexity of n-k fitness functions. IEEE Trans. on Evolutionary Computation 4, 373–379 (1999)

Integrating Class Information and Features in Cluster Analysis Based on Evolutionary Distance Metric Learning

Wasin Kalintha, Satoshi Ono, Masayuki Numao and Ken-ichi Fukui

Abstract Most current applications of clustering only focus on a technological domain, e.g., numerical similarity, while overlooking human domain yield unnatural and incomprehensible results in a human point of view. Unsupervised clustering constructs based on the similarities of numerical features. This study decreases the gap between multiple disciplines that are concerned both computational artifact and the human understanding in order to construct a more understandable cluster structure by considering available class information as well as data features in the clustering. Hence, we applied Evolutionary Distance Metric Learning (EDML) in cluster analysis in order to simultaneously analyze both class label and features. This method is applied to the real-world problem of facial images and food recipes data. The analysis provided promising insights about the relation between class information and features of the data, overall cluster structure distribution, neighbor cluster relations, and the viewpoint of the cluster analysis. Finally, cluster analysis using EDML method can obtain a more intelligible cluster structure with neighbor relations, discover interesting insights, and particular cluster structure can be obtained according to the purpose of analysis. Precisely, these results cannot be achieved by unsupervised clustering.

W. Kalintha (✉)
Graduate School of Information Science and Technology, Osaka University,
8-1 Mihogaoka Ibaraki, Osaka 567-0047, Japan
e-mail: wasin@ai.sanken.osaka-u.ac.jp

S. Ono
Graduate School of Science and Engineering, Kagoshima University,
1-21-40 Kohrimoto, Kagoshima 890-0065, Japan
e-mail: ono@ibe.kagoshima-u.ac.jp

M. Numao · K. Fukui
The Institute of Scientific and Industrial Research (ISIR), Osaka University,
8-1 Mihogaoka Ibaraki, Osaka 567-0047, Japan
e-mail: numao@ai.sanken.osaka-u.ac.jp

K. Fukui
e-mail: fukui@ai.sanken.osaka-u.ac.jp

G. Leu et al. (eds.), *Intelligent and Evolutionary Systems*,
Proceedings in Adaptation, Learning and Optimization 8,
DOI 10.1007/978-3-319-49049-6_12

Keywords Distance metric learning · Semi-supervised clustering · Face recognition analysis · Recipe data analysis

1 Introduction

Cluster analysis [15] has played an important role in data mining and machine learning recently. Cluster analysis divides data into clusters according to their similarities between each other. Depending on the purpose of clustering, cluster analysis could obtain meaningful groups (e.g., capture the natural structure of the data), beneficial group, which is a useful initialization for another approach, or sometimes both. It has been extended to a wide variety of practical problems, such as, in biology [10, 14], information retrieval [3, 19], and so on.

Clustering algorithms are generally introduced only in an unsupervised learning. In real world application domains, the data usually contain some background information, e.g., pairwise constraint or class information (label). However, the traditional clustering algorithms have no way to make use of this information even it exists. Hence, the cluster structure is constructed only in the learning perspective, which often considers incomprehensible to human. One way to take advantage of this information is semi-supervised clustering [1, 11, 20] by integrating background information in the clustering.

Despite many approaches of semi-supervised clustering have been proposed, their aspiration is to divide the data into homogeneous subgroups (clusters) based on their similarities. Similarities, which normally estimated from the proximity between two data points (e.g., Euclidean distance), crucially affect the clustering result. Hence, distance metric learning (DML) has received much attention as the alternative distance metric recently, varieties of DML have been proposed in order to improve the accuracy of clustering and classification by learning an appropriate distance metric from the dataset [21].

Although, semi-supervised clustering with DML have a rich performance to improve the clustering accuracy; however, most of the clustering techniques are usually isolated from the visualization process which is beneficial in the cluster analysis. Meanwhile, visualization can help humanity to investigate the clusters and the relation between clusters in large data, and obtain some discovery or insights on the data. Thus, it cannot preserve the neighborhood relation of clusters in low-dimensional space as well as preserve class information as a human-point-of-view in each cluster.

This study applied EDML for cluster analysis by visualizing the cluster structure of EDML. It is constructed by considering the class label, which are available from human intervention, to guide clustering to be constructed in the human point of view, while preserving the features simultaneously. By taking advantage of evolutionary distance metric learning (EDML) [8], it allows us to propose this novel cluster analysis because of its capability, precisely, not only preserve the neighbor cluster relations, but also directly improve cluster accuracy in term of cluster validity index score.

The purpose of this study is to analyze the cluster structure by integrating class information and features in the cluster analysis based on evolutionary distance metric learning. In the first part, we review the literature about semi-supervised clustering, and DML. Next, the overview of EDML is introduced. Then, the EDML cluster analysis is applied to the real-world application: facial images dataset and food recipes dataset. Then, the cluster structure is visualized by a K-means clustering with K-nearest neighbor centroids graph (KMN-KNN) in order to explain how the proposed method aid the cluster analysis. Finally, we presented the benefits of the novel method which cannot be achieved by unsupervised clustering as follows:

- Comprehensiveness of cluster structure
- The relation between class information and features of the data
- Neighbor cluster relations
- Direction of the cluster analysis

2 Related Work

2.1 Semi-supervised Clustering

Unlike conventional clustering techniques, e.g., K-means clustering (KMN), which is unsupervised. It's possible that external knowledge, i.e., class label and pairwise constraints, are known for a subset of observation. Semi-supervised clustering [1, 5, 11, 16, 20] attempt to improve clustering quality by utilizing these provided knowledge, for example, COP-Kmeans attempts to introduce pairwise constraints to KMN. The cluster assignments of data points are forcibly modified to satisfy the constraints, and the centroids are updated based on the modified assignments. While MPC-Kmeans integrates distance metric learning (DML) with COP-Kmeans using a penalty function for violating the constraints.

2.2 Distance Metric Learning

Distance metric learning (DML) [21] attempts to optimize a metric to improve classification or clustering. Example approaches include nearest neighbor classification [18], and clustering [11, 20].

Figure 1a shows data points with three classes, i.e., circles, squares, and stars, and three initial partitions (or clusters) in Euclidean space. Note that one of the clusters has data points in all three classes. To cluster all data points correctly, the data space transformation stretches the partitions as shown in Fig. 1b.

DML attempts to learn a distance metric transform function based on auxiliary information, including class labels and pairwise constraints of must-links and cannot-links, in order to keep data within the same class close together, and the dif-

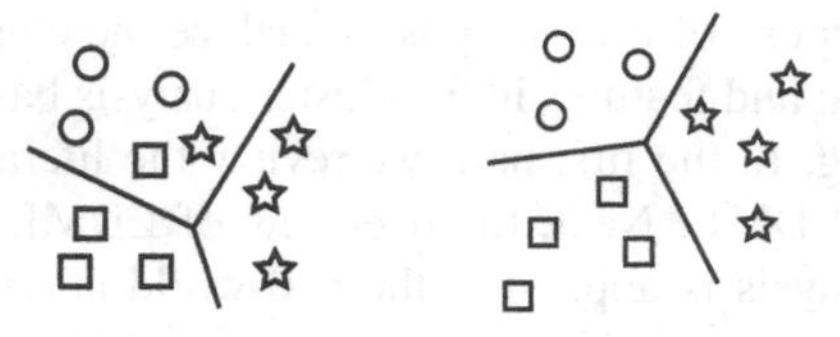

Fig. 1 Conceptual diagram of distance metric transformation [12]

(a) Euclidean space (b) After transformation

ferent class far apart. Our study is categorized as global distance metric learning [2, 4, 9, 20, 22], which attempts to learn an optimal transformation metric matrix for the whole dataset.

3 Evolutionary Distance Metric Learning

Evolutionary Distance Metric Learning (EDML) is originally proposed by Fukui et al. in 2013 [8]. EDML is an efficient framework that applies an evolutionary algorithm (EA) to firmly search for a sufficiently optimal distance metric transformation matrix, also known as, a distance metric learning method (DML). EDML is based on a clustering index with neighbor relation that simultaneously evaluates inter- and intra-clusters to improve clustering quality. In contrast to other semi-supervised clustering, which formulate a penalty function for constraints into an objective function, EDML, however, directly improves the cluster validity index, such as purity, F-measure, or entropy, depending on the clustering purpose, as an objective function, when class information is available. Moreover, the cluster validity index is smoothed by neighbor relations which refines neighboring cluster for better visualization and the data points of the same class tend to locate in neighboring clusters.

3.1 Global Distance Metric Learning

Similar to many other global distance metric learning methods, a Mahalanobis-based distance is used in EDML. Given a dataset $\mathcal{D} = \{\mathbf{x}_i = (x_{i,1}, \dots, x_{i,v})^t \in \mathbb{R}^v\}_{i=1}^N$, the Mahalanobis-based distance can be defined as follows:

$$d_{i,j}^2 = (\mathbf{x}_i - \mathbf{x}_j)^t \mathbf{M} (\mathbf{x}_i - \mathbf{x}_j), \tag{1}$$

where $\mathbf{M} = (m_{k,l})$ is a $v \times v$ matrix. In DML, the elements of $\mathbf{M}$ are variables to be learned that represent a transformation of the input data. In this case, $\mathbf{M}$ must be a symmetric positive semi-definite matrix to satisfy the distance propositions. Unlike original Mahalanobis distance, $\mathbf{M}$ is given by the inverse of the variance-covariance matrix of the input data, i.e., $\mathbf{M} = \Sigma^{-1}$.

The EDML optimizes a clustering index $Eval()$ (introduce in the next few sections) as follows:

$$\begin{aligned} &\text{Maximize } Eval(Clustering(d^2_{i,j})), \\ &\text{s.t. } |m_{k,k}| \geq \sum_{l(k \neq l)} |m_{k,l}|, \\ &0 < m_{k,k} \leq 1, \ -1 \leq m_{k,l} \leq 1 \ (k \neq l), \end{aligned} \tag{2}$$

where $Clustering(d^2_{i,j})$ denotes a clustering result obtained using a distance metric $d^2_{i,j}$—i.e., $Clustering() : \mathbf{x} \mapsto c \in \mathbf{C}$, where $\mathbf{C}$ is a set of cluster identifiers. Besides, matrix $\mathbf{M}$ must be a *diagonally dominant matrix*—i.e., $|m_{i,i}| \geq \sum_{j(i \neq j)} |m_{i,j}|$—and the diagonal elements must be positive to ensure that $\mathbf{M}$ is a positive semi-definite matrix and weak inequality holds.

3.2 Evolutionary Distance Metric Learning Framework

The EDML framework is summarized in Fig. 2. First, candidates of metric transform matrix $\mathbf{M}$ are generated using differential evolution (DE). Next, each obtained candidate $\mathbf{M}$ is individually used in Eq. (1) to acquire clusters with neighbor structures by any partition-based clustering algorithm. Then, class labels are utilized in order to evaluate the quality of the cluster structure through the neighborhood smoothing in the clustering index. This is followed by feeding the evaluated values back into DE as the fitness for each candidate $\mathbf{M}$. DE selects candidates based on the fitness to evolve and generate the next candidates by mutation and crossover with certain probabilities. These steps are repeated until the termination condition (e.g., the iteration limit) is satisfied. Finally, the optimal metric transform matrix $\mathbf{M}^*$ is obtained in terms of the most smoothed clustering index among the overall generations of candidates.

3.3 Cluster Validity Index with Neighbor Relations

This study focuses on using external criteria, that is provided by human interpretation of data. It is more beneficial to use external criteria because class labels are available in this study. This cluster validity index is used as *Eval*() in Eq. (2). In order to evaluate the overall cluster structure, we used neighborhood smoothing in the cluster validity index by adding a weighting function $h_{i,j}$, unlike conventional clustering validity that can only evaluate individual cluster quality, which is proposed by Fukui and Numao [7]. Weighted F-measure (*wFME*) is mainly used in this study.

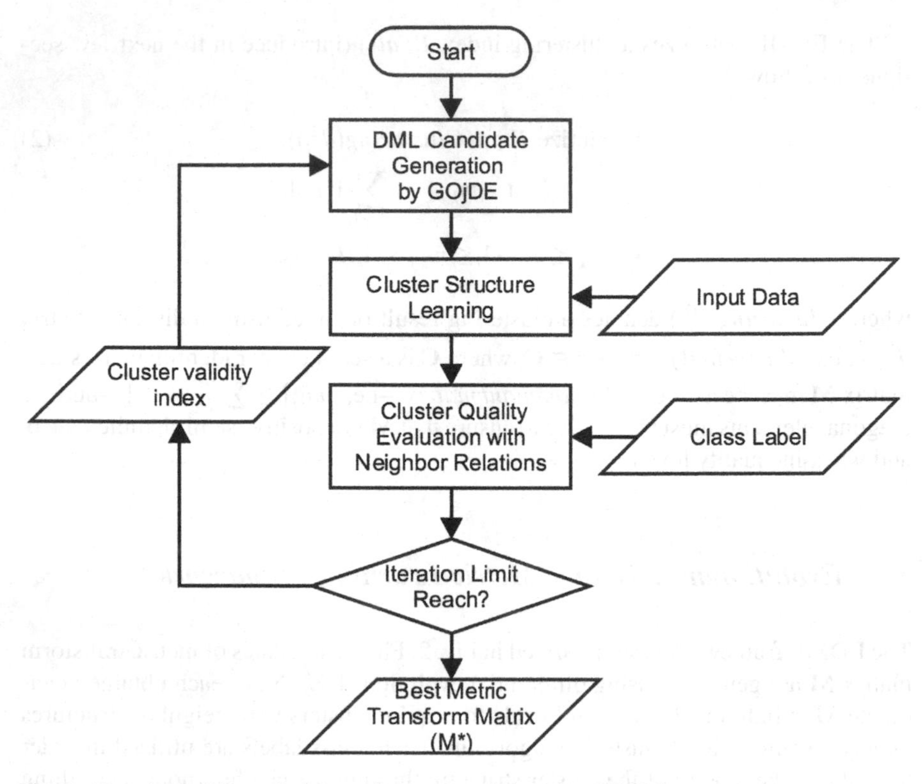

Fig. 2 Flowchart of the evolutionary distance metric learning (EDML)

3.3.1 Weighted F-Measure

Weighted F-measure (*wFME*) is an extension of set-based cluster validity indices [8]. Given dataset $\mathcal{D}$ with cluster set $\mathbf{C}$ and class set $\mathbf{T}$, let $N_{s,i}$ is the number of data points with class $s \in \mathbf{T}$ and the ith cluster $C_i \in \mathbf{C}$, N_s denotes the number of data points in class s, N_i denotes the number of data points in ith cluster, and N is the total numbers of data points in $\mathcal{D}$. These values are smoothed by $h_{i,j}$, as follows:

$$N'_{s,i} = \sum_{C_j \in \mathbf{C}} h_{i,j} N_{s,j}, \tag{3}$$

$$N'_i = \sum_{s \in \mathbf{T}} N'_{s,i} = \sum_{s \in \mathbf{T}} \sum_{C_j \in \mathbf{C}} h_{i,j} N_{s,j}, \tag{4}$$

By using Eqs. (3) and (4) Precision ($Prec(s, C_i)$) and Recall ($Rec(s, C_i)$) can be extended. Then, *wFME* can be obtained from the original formula as follows.

$$\mathrm{wFME}(\mathbf{C}) = \sum_{s \in \mathrm{T}} \frac{N_s}{N} \max_{C_i \in C} F(s, C_i), \tag{5}$$

where $F(s, C_i)$ is a harmonic average of $\boldsymbol{Prec}(s, C_i)$ and $\boldsymbol{Rec}(s, C_i)$.
And $\boldsymbol{Prec}(s, C_i) = N'_{s,i}/N'_i$, $\boldsymbol{Rec}(s, C_i) = N'_{s,i}/N_s$.

In this study, $h_{i,j}$ is a Gaussian function. Therefore, $h_{i,j} = \exp(-\boldsymbol{r}_{i,j}/\sigma)$, where $\boldsymbol{r}_{i,j}$ denotes the inter-cluster distance between C_i and C_j, and $\sigma(> 0)$ is a smoothing (neighborhood) radius.

3.4 Differential Evolution

Many real-world application typically involve a gigantic high-dimensional data. High-dimensional global optimization is one such high-complexity problem. Therefore, differential evolution, the state-of-the-art evolutionary algorithm for real-value problem has been utilized for several problems. In this study, self-adapting control parameters and generalized opposition-based differential evolution (GOjDE) [17], an effective concept in computational intelligence that can enhance differential evolution approaches by self-adapting control parameters and simultaneously computing its opposite solution while evaluating the current solution to provide another chance for finding a candidate solution closer to the global optimum, is used in order to manage the quality of candidate solutions. This study makes use of GOjDE due to faster computational and the confirmation of a better performance especially in a high dimensional problem.

3.5 K-Means Clustering with K-Nearest Neighbor Centroids Graph

K-means clustering with K-nearest neighbor centroids graph (KMN-KNN) [7] has been used in order to easily analyze obtained cluster structures, especially in the case of micro-clusters or cluster number is larger than class number. With this, we can study entire data structures via the inter-cluster connectivity and also visualize neighbor relations among cluster centroids.

3.6 EDML for Cluster Analysis

We addressed the limitation of state-of-the-art cluster analysis, which cannot preserve class information as a neighborhood relation by considering both features and class labels, which provide from a human point of view to guide clustering to be constructed correspondingly with human intuition based on evolutionary distance metric learning. By applying EDML for cluster analysis, it helps us to investigate the relationship between class and features, comprehensiveness of cluster analysis, neighborhood relation of cluster boundary and specify the direction of cluster analysis by constructing a particular structure according to the class category (e.g., age or gender), unlike any other semi-supervised clustering algorithm which is impossible to investigate and analyze these properties.

4 Cluster Analysis on Facial Images Data

4.1 Experimental Setup

This experiment performed on facial images from MIT face recognition project.[1] The original facial images are monochrome facial images with 99 features, calculated by coefficients measuring. We applied principal component analysis (PCA) for dimension reduction. Each face image contains a descriptor of each face, e.g., age, gender, skin color, facial expression, and property on their face. In our experiment, age, namely Child, Teen, Adult and Senior, was used as a class label for DML. Gender was also used as a class label, but only in the last experiment of this section. 200 images were selected in total, 50 images from each age category. Note that the number of dimension D depends on the number of variable in the distance transform matrix $\mathbf{M}$ in Eq. (1). In this dataset, we use the diagonal representation with 55 dimensions. Consequently, PCA is applied to reduce the features to 55. In addition, population size was set to 5 times the number of dimensions. We conducted five trials with 10,000 iterations (generations) per trial. K-means clustering with K-nearest neighbor centroids graph is utilized with EDML and without EDML for further analysis on the neighborhood relation of clusters. 20 clusters and 5 mutual neighborhoods were employed to obtain cluster structure. For better visualization, all labeled data are provided to EDML in this paper. The cluster structure was evaluated using weighted F-measure (*wFME*).

The KMN-KNN structures are individually visualized by Cytoscape[2] using edge-weighted spring embedded layout. We did cluster analysis and summarized interesting things in the following subsections.

4.2 Comprehensiveness of Cluster Structure

Before the visualization, we examine the numerical result first. EDML archived about 61 % accuracy in terms of *wFME*, obviously yielding an improvement of 10 %

---

[1]http://courses.media.mit.edu/2004fall/mas622j/04.projects/faces/.

[2]http://www.cytoscape.org/.

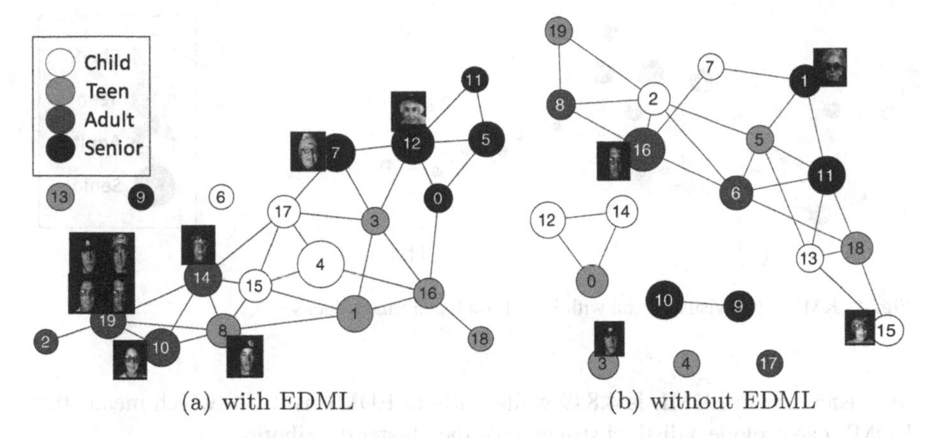

Fig. 3 Example of facial images in cluster

over the one without EDML. The clustering result by EDML shows that the images in the same age category are either in the same or neighboring clusters more than without EDML, and different age category are in the distant clusters.

We visualized KMN-KNN on the dataset results in Fig. 3 in order to investigate the micro-cluster. Let C_i denotes the *i*th cluster. C_i is represented by a node, where the edge between two clusters indicates a mutual neighborhood relation. The class label of each cluster is determined by the majority class of the samples in the cluster. The class label of each cluster is indicated by the brightness of the node, the brighter the younger, and vice versa. As well as the node's size indicates the number of majority class's sample.

Then we investigated the overall cluster distribution. In order to confirm the generality of visualization, we did 4 more trials (5 trials in total) in each dataset with EDML. 2 graphs are shown in Fig. 4. Similar trend can be obtained in every trial.

In Fig. 3a which is the visualization of EDML, overall distribution can be viewed as a clique, most of the cluster with the same class are intensely connected and appears to be neighbors because of the benefit of preserving age category while cluster is constructed. For example, in Fig. 3a Child (C_4, C_{15} and C_{17}), Teen (C_1, C_3, C_8, C_{16} and C_{18}), Adult (C_2, C_{10}, C_{14} and C_{19}) and Senior (C_0, C_5, C_7, C_{11} and C_{12}) class individually group together and has strong relation between the cluster with the same class than one with the different class. Also only a few individual nodes, no neighbor relation, appear here. Contrary, the without EDML results in Fig. 3b, the structures are complicated, the cluster with the same class divided into many parts, for instance, Child class is separated. Unlike the EDML results, many individual nodes appear in Fig. 3b, in which EDML is not applied.

Moreover, the number of majority class's sample in each micro-cluster, which is indicated by node's size is also investigated. In without EDML case, each cluster size is similar to each other, while they are varieties of cluster size in EDML. Numerically, we calculated the standard deviation of number of majority class sample in each figure to represent the distribution of the cluster size. The average standard deviation

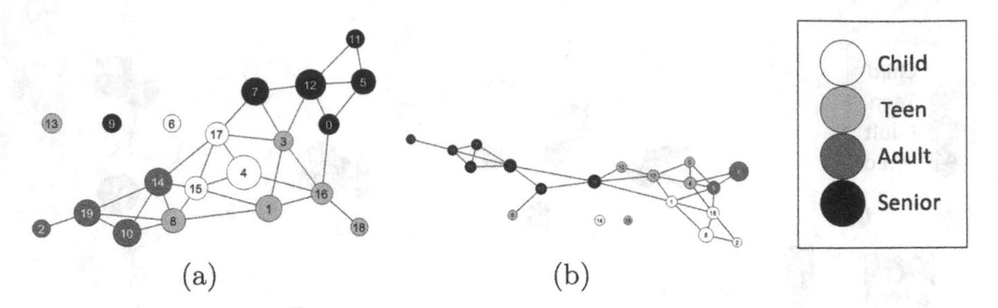

Fig. 4 KMN-KNN visualization with EDML on facial image data set

of cluster size in EDML is 4.849 while without EDML is 3.128 which means that EDML can provide a distinct structure of the cluster distribution.

Therefore, utilizing EDML to KMN-KNN provides a better view of cluster structure, for example, identical class clusters are intensely connected and appear to be neighbors, reduce the number of individual clusters, and affect distribution of samples. Because the cluster analysis using EDML could preserve class information in each cluster.

4.3 Investigating Relation Between Class and Features

By comparing Fig. 3a, in which EDML was applied, with Fig. 3b, in which did without EDML, and analyze the cluster structure. As a result, we recognized a pattern that images which contain a unique pattern such as persons wearing glasses are in neighbors and wearing a hat are in neighbors in Fig. 3a. In contrast to Fig. 3b that these pictures are located in the different region of the cluster structure. This is because of the benefit of EDML which can utilize class information (age category) in the clustering. Moreover, it preserves the neighbor relation between the same age category, as a result, it groups the clusters with the same age category close together while it preserves the similar image features, e.g. people wear glasses or hat. Therefore, clusters with the same age category are grouped together, and the similar feature images are gathered together in these groups as well, e.g., in Fig. 3a people who wearing glasses are in C_{10}, C_{14} and C_{19} or people who wearing hat in C_7 and C_{12}. These patterns cannot be occurred when without EDML (Fig. 3b).

4.4 Investigating Neighborhood Relation of Cluster Boundary

In order to investigate another benefit of this work that can preserve the neighborhood relations and provide the cluster boundary. Let assume that the cluster with same

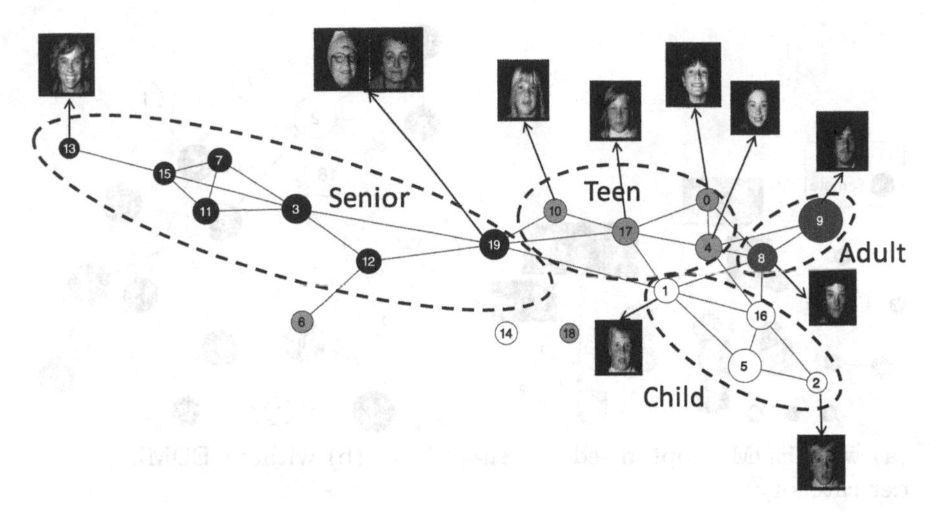

Fig. 5 Cluster structure boundary

label (nodes) which has a neighbor relation merge into a big cluster. We selected Fig. 4b and defined big clusters indicated by a dash line circle as in Fig. 5.

When we examined these big clusters, the features in the border node should be similar to the neighboring clusters of the other class. Result in the distribution of the sample in the big cluster that can divide a relatively younger or older in the class which can be seen from the sample of facial images in Fig. 5. If it connects to the older class, that cluster tends to be a relatively older face in that class, for example, we considered C_{10} (Teen) and C_{19} (Senior), the average age in C_{10} is relatively higher than the other cluster with the class label, and C_{19} are younger in the senior class. Moreover, C_4 and C_{17} has high probabilities to be a younger teen because it is connected to Child class. On the other hand, C_1 clustered the elder child, as it has neighbor relations with Teen, Adult and Senior class. Furthermore, when we considered Senior class, we could analyze each individual sub cluster, and show that C_{13} is relatively oldest senior because it has only relation with the same class (Senior) and has no relation to another class. Similarly to C_2 in Child class and C_9 in Adult class. Note that image features do not reflect only age, other possibilities are such as similar hair style, face shape, or they are families. Importantly, these boundaries cannot be found without EDML (Fig. 3b).

4.5 Changing the Viewpoint of Cluster Analysis

When the target of cluster analysis is changed, it is better to construct a new cluster structure according to the target class category. Since EDML has an ability to optimize class information, this method makes use of it to select a different class category

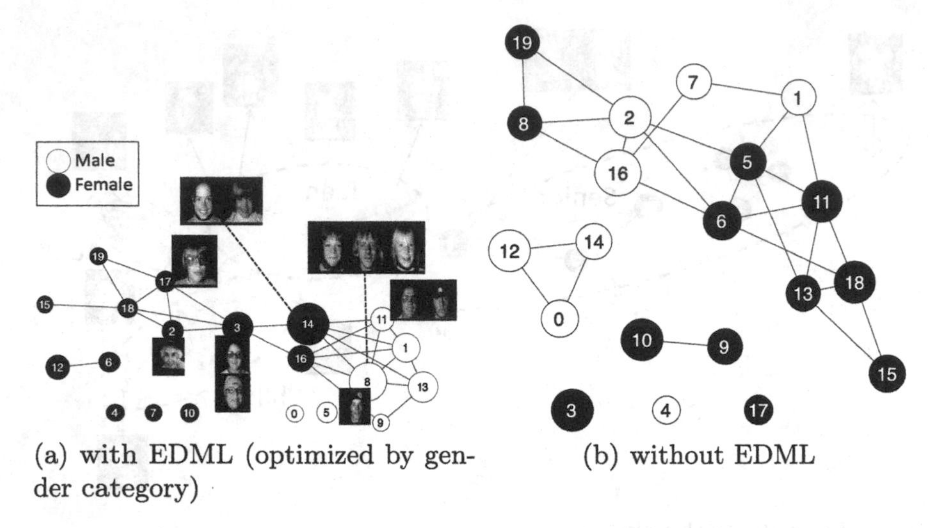

(a) with EDML (optimized by gender category) (b) without EDML

Fig. 6 KMN-KNN visualization on gender category

(e.g., age or gender) to optimize and change the viewpoint of cluster analysis. For example, using age category aids cluster analysis together with image features, and similarly in gender category.

In this experiment, we visualized the KMN-KNN according to the gender of sample: male and female, denoted by node's color (i.e., white and black respectively). The proposed method can provide a particular optimization for the gender analysis, which can be visualized as shown in Fig. 6a. Conversely, without EDML, as seen in Fig. 6b we notice changes only in the labels of the micro-clusters; however, we observe the same cluster structure as the age analysis in Fig. 3b.

Figure 6a, KMN-KNN preserves the neighbor relations and forms the clique in between the cluster with samples of identical majority class. Also, the border properties of clusters representing males with long hair are shown in close proximity with females who have short hair, shown in the figure by the dash line, due to their similar visual features. Lastly, we can recognize the unique patterns in each class: male wearing glasses C_{11}, male wearing hat (C_8 and C_{11}), women wearing glasses (C_3 and C_{17}) and women wearing hat (C_2 and C_3) which are located close together.

5 Cluster Analysis on Recipes Data

5.1 Experimental Setup

This experiment performed on food recipes from Cookpad,[3] the Japan's largest site for sharing their original recipes. The recipes are all in Japanese, each recipe con-

[3]http://cookpad.com/.

tains ingredients and how to cook. First, we select recipes from 10 preselected food categories (Seaweed, Vegetable, Fish, Noodle, Pasta, Egg, Salad, Soup, Sauce & Dressing, and Meat), which come from today's recipe category total 19236 recipes and 8398 ingredients. We applied a Japanese cooking ontology [13] to merge various ingredient names since the users freely write their ingredients. Then, we used random sampling to select about 220–400 recipes from each category to resolve the imbalance of the data. Consequently, total 3732 recipes and 129 ingredients were selected to use in this experiment. The features are extracted from ingredients of each recipe using term frequency-inverse document frequency (TF-IDF) and PCA is applied to reduce dimension to 10. The food categories is corresponded to data class label. As previous experiment *wFME* was used as an objective function. Unlike in facial images data set, we set cluster size to 50, and full matrix representation is applied instead, due to the limitation of computational time.

5.2 Experimental Results

The goodness of the cluster structure is preliminary evaluated by *wFME*. Although, EDML archived 0.261 of *wFME* in this data, it yielded an improvement of 30 % over the clustering without EDML. The improvement of *wFME* clearly illustrated the distribution of the same class recipes which are either in the same cluster or neighbor clusters more than the one without EDML. Conversely, the different recipe category seems to be located farther away.

Then, the KMN-KNN visualization is shown in Fig. 7. From the visualization, EDML still archived similar results like in facial image dataset. First, comprehen-

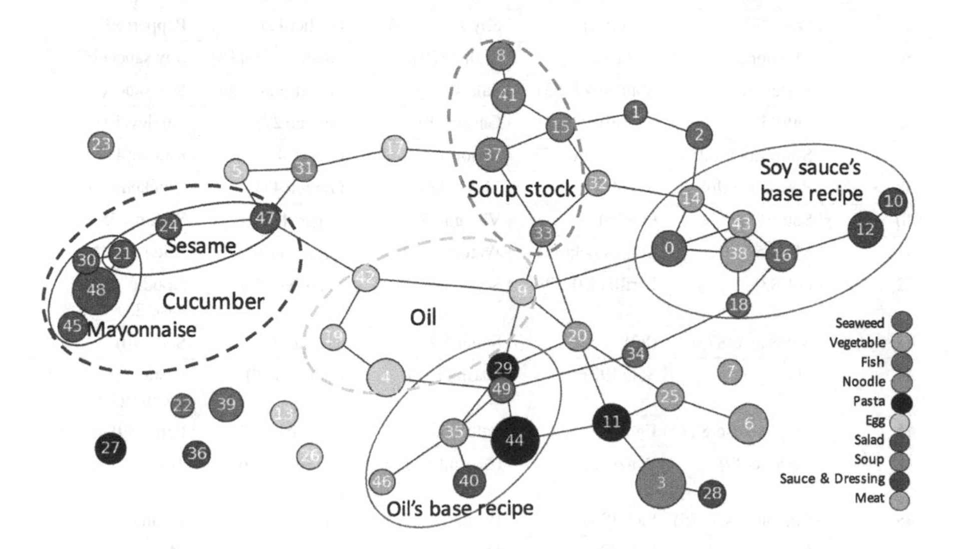

Fig. 7 KMN-KNN visualization on cookpad recipe's data set

siveness of cluster structure is preserved. It is evident from the visualization that preserving class category, while construct the cluster, benefit most of the clusters with the same class to be intensely connected and appears to be neighbors. These results are indicated by dash line ellipse in Fig. 7, e.g., Salad (C_{21}, C_{24}, C_{30}, C_{45}, C_{47} and C_{48}), Soup (C_8, C_{15}, C_{33}, C_{37} and C_{41}), and so on.

In order to investigate relationship between classes and features, the most occurrence ingredients in each cluster are investigated as shown in Table 1 (unnecessary

Table 1 Top 5 most occurrence ingredients in each cluster with number of occurrences (Translated from Japanese)

Cluster	Rank1	Rank2	Rank3	Rank4	Rank5
0	Soy sauce(97)	Soup stock(74)	Sugar(56)	Mirin(55)	Oil(53)
4	Egg(105)	Salt(74)	Oil(67)	Sugar(28)	Cheese(27)
8	Miso(42)	Soup stock(23)	Eggplant(7)	Fried tofu(5)	Cabbage(4)
9	Green onion(52)	Oil(39)	Soy sauce(33)	Vinegar(24)	Soup stock(20)
10	Soy sauce(66)	Sugar(58)	Vinegar(51)	Mirin(47)	Ginger(11)
12	Soy sauce(87)	Vinegar(80)	Sugar(77)	Oil(69)	Sesame(33)
14	Soy sauce(37)	Sugar(36)	Sake(33)	Mirin(27)	(25)
15	Miso(57)	Soup stock(28)	Oil(27)	Green onion(14)	Egg(13)
16	Soy sauce(97)	Oil(79)	Sugar(62)	Mirin(41)	Vinegar(38)
18	Soy sauce(52)	Sliced dried bonito(15)	Kombu(14)	Mirin(14)	Radish(10)
19	Egg(58)	Oil(26)	Noodle Soup base(19)	Wheat flour(19)	Cheese(12)
21	Mayonnaise(37)	Sesame(35)	Cucumber(18)	Sugar(11)	Soy sauce(10)
24	Sesame(50)	Cucumber(20)	Oil(20)	Ponzu(18)	Sugar(16)
29	Oil(127)	Salt(108)	Soy sauce(50)	Garlic(42)	Pepper(39)
30	Mayonnaise(65)	Tuna(22)	Radish(20)	Green onion(18)	Soy sauce(16)
33	Water(61)	Soup stock(53)	Salt(40)	Green onion(34)	Soy sauce(33)
35	Oil(95)	Salt(80)	Garlic(28)	Pepper(27)	Parsley(13)
37	Soup stock(52)	Green onion(51)	Miso(46)	Carrot(41)	Radish(41)
38	Soy sauce(96)	Sake(85)	Mirin(48)	Ginger(47)	Chicken(41)
40	Salt(88)	Oil(75)	Vinegar(58)	Sugar(42)	Sesame(35)
41	Miso(54)	Soup stock(45)	Water(26)	Green onion(25)	Fried tofu(16)
42	Oil(83)	Perilla(29)	Soy sauce(24)	Seaweed(23)	Noodle Soup base(23)
43	Soy sauce(87)	(82)	Sugar(72)	Mirin(52)	Sake(40)
44	Oil(231)	Salt(193)	Garlic(141)	Pepper(128)	Pasta noodles(106)
45	Mayonnaise(82)	Egg(45)	Salt(28)	Soy sauce(22)	Pepper(19)
47	Sesame(29)	Ponzu(27)	Cucumber(21)	Noodle Soup base(19)	Tomato(16)
48	Mayonnaise(175)	Salt(124)	Pepper(62)	Onion(53)	Potato(47)
49	Soy sauce(73)	Salt(72)	Oil(60)	Pepper(30)	Mayonnaise(25)

clusters are omitted due to space limitation). These results indicated that not only the same category recipes tend to locate close together, but recipes that contain similar ingredients (feature) are also located nearby either in the same or neighbor clusters. These clusters are denoted inside the red ellipse in Fig. 7. These could help the user to select the recipe according to the remaining ingredients, or what kind of taste they want, for example, soy sauce's base recipe: **Basil Chicken** (C_{38}), oil's base recipe: **Napolitan spaghetti** (C_{44}), and so on.

Lastly, the big cluster is indicated as in a dash line ellipse in order to explore thc cluster boundary. We again discovered as in the previous experiments that the features in the border node should be similar to the neighboring big clusters, as we can see from the relationship between C_9 (Egg) and C_{33} (Soup) that the intersection of these two clusters is **Green onion**. While in their own big clusters are **Oil** for Egg (C_4, C_9, C_{19} and C_{42}) and **Soup stock** for Soup (C_8, C_{15}, C_{33}, C_{37} and C_{41}) big clusters respectively. Moreover, we surprisingly discovered that the border is possible to occur in an individual big cluster. The Salad big cluster (C_{21}, C_{24}, C_{30}, C_{45}, C_{47} and C_{48}), indicated inside the blue dash line ellipse, was carefully investigated and found that C_{21} is the inner border of their own big cluster. By dividing the big cluster into two groups using C_{21}, indicated by blue ellipse, the similar features in C_{45}, C_{48}, C_{30} change to mayonnaise, and another group is sesame instead of cucumber. Thus, border between big clusters can be occasionally extended to the neighbor cluster of the border of the big cluster.

6 Conclusion

In this paper, we applied EDML for cluster analysis in order to construct comprehensive cluster structure integrating class label and features. Apparently, EDML can help investigating the relationship between class and features, for example, clusters with samples of the same class are grouped together in order to preserve the class and neighbor relations while features are also preserved. Moreover, overall cluster structure is well-organized, identical majority class groups together and reduce the number of individual clusters. Lastly, it can specify the direction of cluster analysis by constructing a particular structure according to the class category (e.g., age or gender). For the larger dataset, the data-parallel and distributed environment e.g., MapReduce [6] can be a candidate way.

Acknowledgments This work was partially supported by the cooperative research program of Network Joint Research Center for Materials and Devices. Moreover, the recipe data is provided from Cookpad and the National Institute of Informatics.

References

1. Bair, E.: Semi-supervised clustering methods. Wiley Interdisciplinary Reviews: Computational Statistics 5(5), 349–361 (2013)
2. Bar-Hillel, A., Hertz, T., Shental, N., Weinshall, D.: Learning distance functions using equivalence relations. In: Proc. the 20th International Conference on Machine Learning (ICML-03). pp. 11–18 (2003)
3. Beeferman, D., Berger, A.: Agglomerative clustering of a search engine query log. In: Proceedings of the Sixth ACM SIGKDD International Conference on Knowledge Discovery and Data Mining. pp. 407–416 (2000)
4. Bian, W., Tao, D.: Learning a distance metric by empirical loss minimization. In: Proc. International Joint Conference on Artificial Intelligence (IJCAI-11). pp. 1186–1191 (2011)
5. Bilenko, M., Basu, S., Mooney, R.J.: Integrating constraints and metric learning in semi-supervised clustering. In: Proc. of the 21st International Conference on Machine Learning. pp. 81–88. ACM (2004)
6. Dean, J., Ghemawat, S.: Mapreduce: Simplified data processing on large clusters. In: Proceedings of the 6th conference on Symposium on Operating Systems Design & Implementation (OSDI'04). pp. 137–150 (2004)
7. Fukui, K., Numao, M.: Neighborhood-based smoothing of external cluster validity measures. In: Proc. the 16th Pacific-Asia Conference on Knowledge Discovery and Data Mining (PAKDD-12). pp. 354–365 (2012)
8. Fukui, K., Ono, S., Megano, T., Numao, M.: Evolutionary distance metric learning approach to semi-supervised clustering with neighbor relations. In: Proc. of 2013 IEEE 25th International Conference on Tools with Artificial Intelligence (ICTAI). pp. 398–403 (2013)
9. Goldberger, J., Roweis, S., Hinton, G., Salakhutdinov, R.: Neighbourhood components analysis. In: Advances in Neural Information Processing Systems. pp. 513–520 (2004)
10. Guzzi, P.H., Masciari, E., Mazzeo, G.M., Zaniolo, C.: Information Technology in Bio- and Medical Informatics, chap. A Discussion on the Biological Relevance of Clustering Results, pp. 30–44. Springer International Publishing (2014)
11. Hertz, T., Bar-Hillel, A., Weinshall, D.: Boosting margin based distance functions for clustering. In: Proc. the 21st International Conference on Machine Learning (ICML-04). pp. 393–400 (2004)
12. Kalintha, W., Megano, T., Ono, S., Fukui, K., Numao, M.: Cluster analysis of face images and literature data by evolutionary distance metric learning. In: Proc. of the 35th SGAI International Conference on Innovative Techniques and Applications of Artificial Intelligence (AI2015). pp. 301–315. Springer (2015)
13. Nanba, H., Doi, Y., Tsujita, M., Takezawa, T., Sumiya, K.: Construction of a cooking ontology from cooking recipes and patents. In: Proceedings of the 2014 ACM International Joint Conference on Pervasive and Ubiquitous Computing: Adjunct Publication. pp. 507–516. UbiComp '14 Adjunct, ACM (2014)
14. Nugent, R., Meila, M.: Statistical Methods in Molecular Biology, chap. An Overview of Clustering Applied to Molecular Biology, pp. 369–404 (2010)
15. Tan, P.-N., Steinbach, M., Kumar, V.: Cluster Analysis: Basic Concepts and Algorithms. Addison-Wesley (2006)
16. Wagstaff, K., Cardie, C., Rogers, S., Schrdl, S.: Constrained k-means clustering with background knowledge. In: Proc. of the International Conference on Machine Learning (ICML-01). pp. 577–584 (2001)
17. Wang, H., Rahnamayan, S., Wu, Z.: Parallel differential evolution with self-adapting control parameters and generalized opposition-based learning for solving high-dimensional optimization problems. Journal of Parallel and Distributed Computing 73, 62–73 (2013)
18. Weinberger, K.Q., Blitzer, J., Saul, L.K.: Distance metric learning for large margin nearest neighbor classification. Journal of Machine Learning Research (JMLR) 10, 207–244 (2009)
19. Wen, J.R., Nie, J.Y., Zhang, H.J.: Clustering user queries of a search engine. In: Proceedings of the 10th International Conference on World Wide Web. pp. 162–168 (2001)

20. Xing, E.P., Ng, A.Y., Jordan, M.I., Russell, S.J.: Distance metric learning with application to clustering with side-information. In: Advances in Neural Information Processing Systems (NIPS). pp. 505–512 (2002)
21. Yang, L.: Distance metric learning: A comprehensive survey. Tech. Rep. 16, Michigan State Universiy (2006)
22. Zha, Z.J., Mei, T., Wang, M., Wang, Z., Hua, X.S.: Robust distance metric learning with auxiliary knowledge. In: Proc. International Joint Conference on Artificial Intelligence (IJCAI-09). pp. 1327–1332 (2009)

Multiple Additional Sampling by Expected Improvement Maximization in Efficient Global Optimization for Real-World Design Problems

Masahiro Kanazaki, Taro Imamura, Takashi Matsuno and Kazuhisa Chiba

Abstract Efficient global optimization (EGO), based on the expected improvement maximization of the Kriging model, is a suitable optimization method for designs based on time-consuming evaluations such as computational fluid dynamics. However, the original formulation of EGO can find only one additional sample point after an initial sample point is acquired for the initial Kriging model construction. Therefore, the efficiency of EGO is decreased even if the designer is able to evaluate several designs using a large parallel computer because additional samples can only be obtained sequentially through the additional sampling process. In this study, a multiple additional sampling method is proposed to improve the efficiency of the additional sampling process in EGO while maintaining the performance of exploration based on EI maximization. The design performance of EGO with MAS is investigated by solving a test problem first, and then an airfoil design problem, which is a single objective problem. The results are compared with the original EGO in terms of the convergence of the objective function and the diversity of the design variables. According to these problems, the proposed method acquires optimum solutions as effectively as the original EGO, while the diversity of the solutions is improved. In addition, the total design time can be reduced compared with the original EGO in a parallel computational environment.

M. Kanazaki (✉)
Graduate School of System Design, Tokyo Metropolitan University,
10-10 Asahigaoka, Hino, Tokyo, Japan
e-mail: kana@tmu.ac.jp
URL: http://www.sd.tmu.ac.jp/aerodesign/

T. Imamura
Department of Aeronautics and Astronautics School of Engineering,
The University of Tokyo, Tokyo, Japan

T. Matsuno
Graduate School of Engineering, The University of Tottori, Tottori, Japan

K. Chiba
Graduate School of Informatics and Engineering,
The University of Electro-Communications, Chofu, Japan

G. Leu et al. (eds.), *Intelligent and Evolutionary Systems*,
Proceedings in Adaptation, Learning and Optimization 8,
DOI 10.1007/978-3-319-49049-6_13

Keywords Efficient global optimization · Expected improvement · Multiple additional sampling · Real-world design problem

1 Introduction

In recent times, engineering optimization has been carried out using a combination of global optimization techniques, including evolutionary algorithms and high-fidelity/expensive evaluations such as computational fluid dynamics (CFD) [5–7]. On the other hand, such expensive functions require a time of between several hours and several days. Therefore, in view of the time cost it is not realistic to solve engineering problems using global optimization methods such as genetic algorithms (GAs) based on CFD results.

With this in mind, surrogate model-based global exploration represents one way to reduce the computational cost. Efficient global optimization (EGO) [1], which is based on the Kriging model has been proposed as one such solution. This approach defines the expected improvement (EI) based on the uncertainty of the Kriging model. The point of Maximum EI indicates that the possible design is near to an optimum point that can improve the model. Thus, optimum solutions can be determined using the maximum EI point as an additional sample. This method has been successfully utilized in aerospace design [4].

However, in this process only one additional sample can be determined by EI maximization. Therefore, EGO is not time effective in cases where a designer can carry out parallel evaluations using high performance computing. In this study, assuming that a designer has access to an environment capable of parallel evaluation, a multi-additional sampling method is proposed to improve the efficiency of EGO. The proposed method is demonstrated by solving test functions and an airfoil design problem using time consuming CFD.

2 Efficient Global Optimization (EGO)

Meta-heuristic based global optimization is useful in engineering design problems, such as in the multi-disciplinary design of aircraft and spacecraft, because the optimization can be carried out without knowing the gradient of the unknown function. Thus, several studies on engineering design have focused on meta-heuristic approaches, such as genetic algorithms (GAs). Meta-heuristic based optimization can acquire a global solution and design knowledge for arbitral design problems. However, because such methods require a number of evaluations, a long time may be required for the design process to terminate. This is a significant factor in aerodynamic design using CFD.

In response, EGO, a surrogate model based optimization, has been proposed [1], and its applicability has been demonstrated [4]. In EGO, a Kriging model is employed that can predict the multi-modal function with a high accuracy. The predicted solution $\hat{y}(\boldsymbol{x})$ for a design variable $\boldsymbol{x}$ can be expressed as

$$\hat{y}(\boldsymbol{x}) = \mu + \epsilon(\boldsymbol{x}), \tag{1}$$

where μ is a global model among sample points and $\epsilon((\boldsymbol{x}))$ is a deviation at $\boldsymbol{x}$. Optimization in EGO proceeds by solving the maximization problem of EI [1, 4], which can simultaneously consider optimality and uncertainty. In a minimization problem, the improvement $I(y((\boldsymbol{x})))$ in the minimum value y_{min} at $(\boldsymbol{x})$ can be expressed as

$$\begin{aligned} I(y((x))) &= \max\left(|y_{\text{min}} - \hat{y}|, 0\right) \quad (\hat{y} < y_{\text{min}}) \\ I(y((x))) &= 0.0 \quad (\hat{y} > y_{\text{min}}). \end{aligned} \tag{2}$$

EI at $\boldsymbol{x}$ can be defined as the expectation value of the function improvement, given by

$$EI \equiv E[I(\boldsymbol{x})] = \int_{-\infty}^{y_{\text{min}}} (|y_{\text{min}} - \hat{y}|)\phi(f)df, \tag{3}$$

where f is a Gaussian type stochastic variable obeying $N[(\hat{y}(\boldsymbol{x}), s^2(\boldsymbol{x}))]$, where s is the root mean square error (RMSE) and $\phi(f)$ is the probability density function (PDF). Because EI determines the expectation of the function improvement, possible optimum points can be acquired by solving EI maximization. EI often exhibits a multi-modal function, and the evolutionary algorithm is widely employed to solve this maximization problem.

3 Proposed Method: Multiple Additional Sampling

The original EGO can acquire one additional sample in each iteration, as shown in Fig. 1. The additional sample is obtained through maximization of expected improvement, as shown in Fig. 2. Thus, designers can only use computational resources for one evaluation in this case, while they can use a parallel environment in the initial sampling. In this case, available computational resources are not used efficiently during the additional sampling process, as shown in Fig. 3a. In addition, additional samples obtained by EI maximization improve the surrogate model around a local optimum when this point exhibits an optimum value. However, it often performs a local search that does not maintain the diversity. In such cases, many iterations are required to obtain additional samples in order to explore the global optimum.

In this study, multi-additional sampling (MAS) is proposed. In EGO with MAS, a sub-iterations is additionally included, as shown in Fig. 3b. In this sub-iteration, an additional sample obtained by an iteration is included, to create an updated Kriging model using the predicted point $(\boldsymbol{x}, \hat{y}(\boldsymbol{x}))$ as a temporal function value. Then, another additional sample is acquired by EI maximization. The EI value around a sample

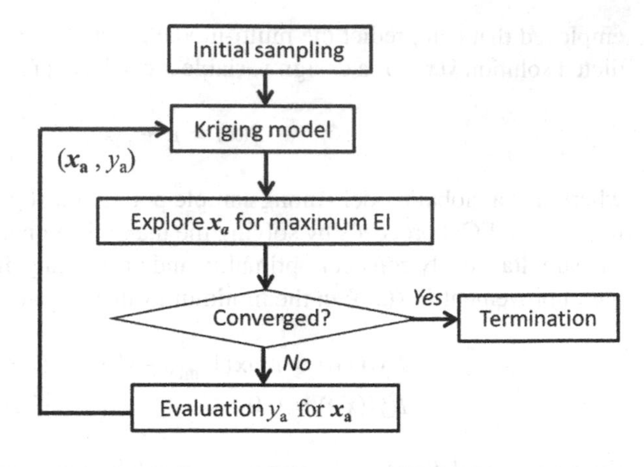

Fig. 1 Procedure of efficient global optimization

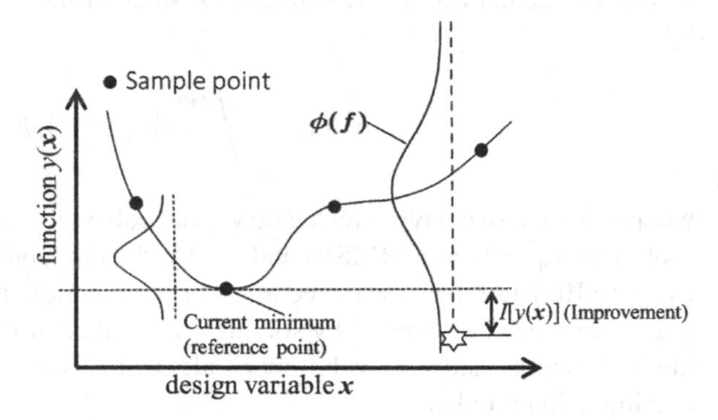

Fig. 2 Schematic illustration of expected improvement (EI)

point should be close to zero, because $\phi(\boldsymbol{x})$ should be close to zero. Therefore, global exploration is promoted, especially for a multi-modal function.

A schematic illustration of additional sampling in the proposed MAS method is presented in Fig. 4b. This process begins by obtaining $\boldsymbol{x}_{eimaax1}$ using EI maximization on the initial Kriging model. The predicted value $\hat{y}_{a1}$ is also calculated, and then the model is updated temporally using $(\boldsymbol{x}_{eimaax1}, \hat{y}_{a1})$. Because the EI value around $\boldsymbol{x}_{eimaax1}$ should not be larger, the next additional point $\boldsymbol{x}_{eimaax2}$ can be obtained. This process is iterated until an arbitrary number of additional samples are obtained. Then, the exact values $y_{a1}, y_{a2} \ldots$ are evaluated using a parallel evaluation environment for $\boldsymbol{x}_{eimaax1}, \boldsymbol{x}_{eimaax2} \ldots$ using the expensive function. The set of these additional samples is finally added to the dataset for the improvement of the model.

In this sub-iteration, only values $\hat{y}(\boldsymbol{x})$ predicted by the Kriging model are required. Thus, multiple additional samples can be acquired in a short time. In the main iteration, the exact values for these additional samples can be evaluated using a parallel evaluation. Thus, design using this EGO method can be completed in a shorter time than with the original EGO. Figure 3b illustrates the application of the parallel evaluation using the proposed EGO with MAS. The number of sub-iterations can be determined by the number of parallel evaluations.

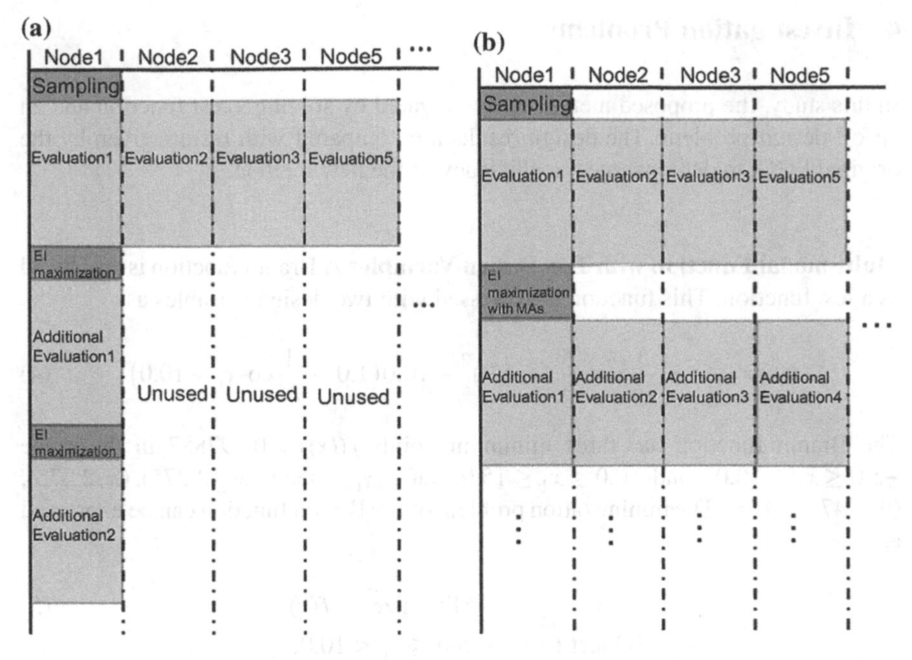

Fig. 3 Schematic illustration of the use of parallel computer. **a** Evaluation for original EGO, **b** evaluation for EGO with MAS

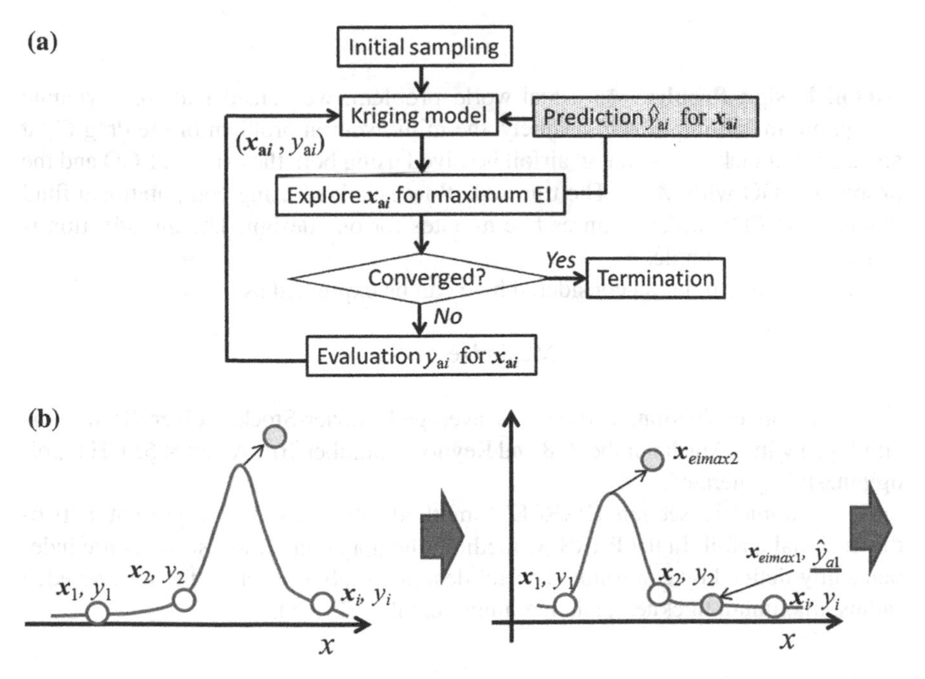

Fig. 4 Schematic illustration of the proposed multi-additional sampling in EGO. **a** Procedure of EGO with MAS. **b** Illustration of the sub-iteration

4 Investigation Problems

In this study, the proposed method is investigated by solving a test function and an airfoil design problem. The design results are compared with results given by the original EGO, to demonstrate the efficiency of the new method.

Multi-modal Function with Two Design Variables A Branin function is employed as a test function. This function is expressed with two design variables as

$$f(x) = \left(x_2 - \frac{5}{36}x_1^2 + \frac{5}{3}x - 6.0\right)^2 + 10.0\left(1.0 - \frac{1}{24}\cos x_1 + 10.0\right). \tag{4}$$

The Branin function has three minimum points $f((x)) = 0.397887$ in the range $-5.0 \leq x_1 \leq 10.0$ and $0.0 \leq x_2 \leq 15.0$, at $(x_1, x_2) = (-\pi, 12.275), (\pi, 2.275), (9.42478, 2.475)$. The minimization problem of the Branin function can be expressed as

$$\begin{aligned} \text{Minimize} :\ & f(x) \\ \text{Subject to} :\ & -5.0 \leq x_1 \leq 10.0, \\ & 0.0 \leq x_2 \leq 15.0 \end{aligned} \tag{5}$$

Airfoil Design Problem As a real-world problem, we consider an aerodynamic design optimization problem. Namely, the minimization problem of the drag C_d at an angle of attack of 2.0° for an airfoil is solved using both the original EGO and the proposed EGO with MAS. The transonic flow is solved using computational fluid dynamics (CFD), which requires five minutes for one design. The investigation is started using 89 samples.

The objective function considered here can be expressed as

$$\text{Maximize} :\ C_\mathrm{d}. \tag{6}$$

For airfoil evaluation, a Reynolds averaged Navier-Stocks solver (RANS) is employed with a Mach number 0.8 and Reynolds number 10^7. A 128×61 CH topology mesh is generated.

The parametric section (PARSEC) method [8] is used to represent a two-dimensional airfoil. In the PARSEC method, the upper and lower surfaces are independently defined by controlling several design variables, such as the leading edge radius, maximum thickness, and maximum camber (Fig. 5).

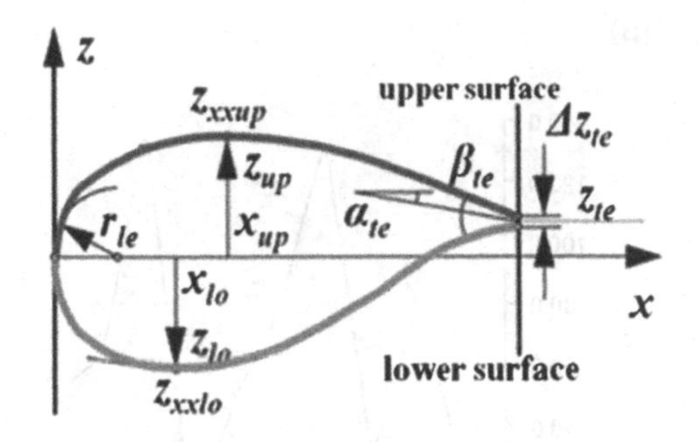

Fig. 5 Parameters for PARSEC method

5 Results and Discussion

5.1 Comparison of Global Exploration Performance

Minimization Problem of Branin Function This investigation is started using ten initial samples, which are obtained by Latin hypercube sampling (LHS). Then, the original EGO process (Fig. 1) and proposed EGO with MAS (Fig. 4) are performed independently, and the results of additional samples are compared. For the proposed EGO with MAS, the number of sub-iterations is set three; that is, three additional samples can be acquired for evaluations of exact solutions in each main iteration. The main iteration is performed four times, so that 12 additional samples are obtained in total. For the comparison, the original EGO process, which employs sequential sampling, is iterated 12 times.

Figure 6 presents a comparison of the convergence histories of the original EGO and the EGO with MAS. Each method exhibits a good convergence through the additional sampling process. The results suggest that the exploration performance of the proposed method does not deteriorate with temporal additional samples obtained by sub-iterations.

Figure 7 illustrates the variation resulting from additional sampling for each method. By comparing these figures, the differences in the function improvement process can observed. According to Fig. 7a, two minimum points can be determined quickly, but an additional minimum point can only be found after nine additional sampling processes. On the other hand, the proposed EGO with MAS found three minimum points after two additional samplings (after obtaining the additional sample #6). These results suggest that a higher diversity can be achieved with the proposed method using multiple-additional samples than with the original EGO.

Figure 8 shows the initial samples (blue diamond symbols) and additional samples (red square symbols) with sampling orders. As shown in Fig. 7, the original EGO explored intensively around two minimum points before the ninth sampling, while the third sample point was not explored. On the other hand, the proposed method explored three minimum points after the second additional sampling (additional sample #6.) These results suggest that the proposed method can achieve exploration with a higher diversity than that of the original EGO.

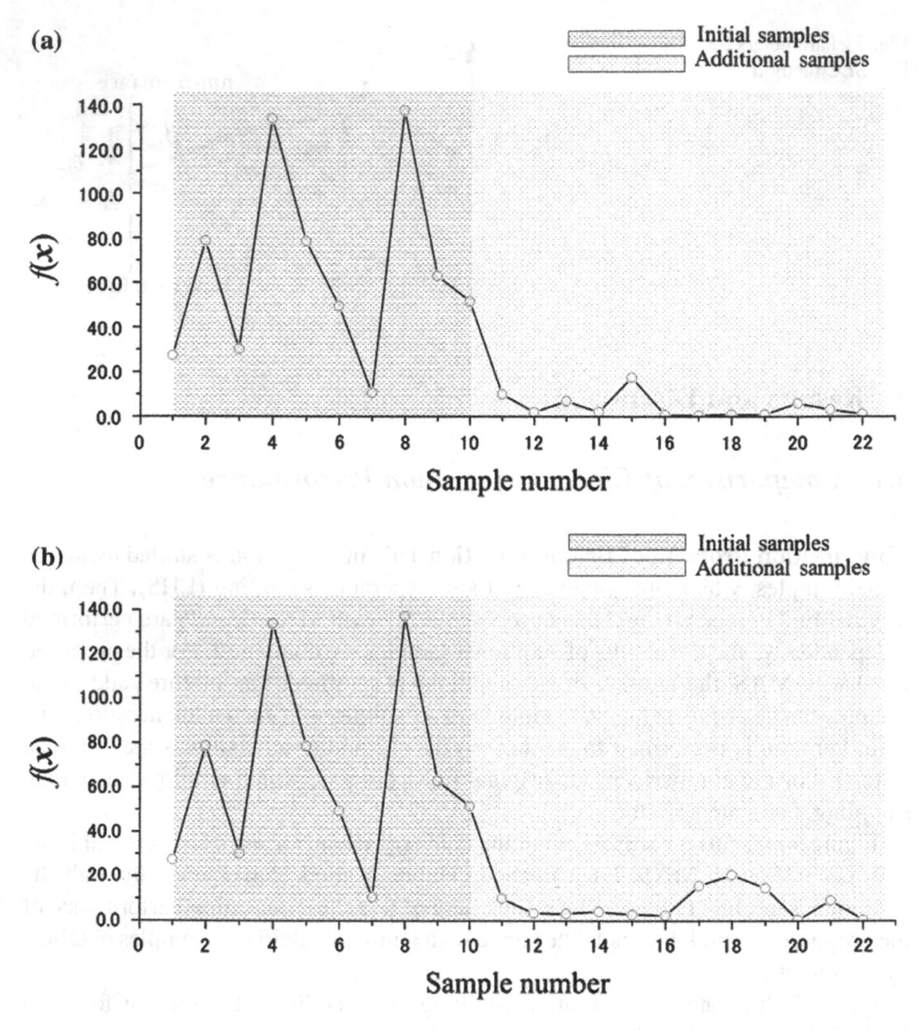

Fig. 6 Comparison of convergence histories in the minimization of the Branin function. **a** Original EGO. **b** EGO with MAS

Drag Minimization Problem of Airfoil This investigation is started using 89 initial samples obtained by Latin hypercube sampling (LHS). Then, the original EGO process (Fig. 1) and the proposed EGO with MAS (Fig. 4) are performed independently, and the results of additional samples are compared. For the proposed EGO with MAS, the number of sub-iterations is set three; that is, three additional samples can be acquired for the evaluations of exact solutions in each main iteration. The main iteration is performed seven times, so that 21 additional samples are obtained in total. For the comparison, the original EGO process using sequential sampling is iterated 12 times.

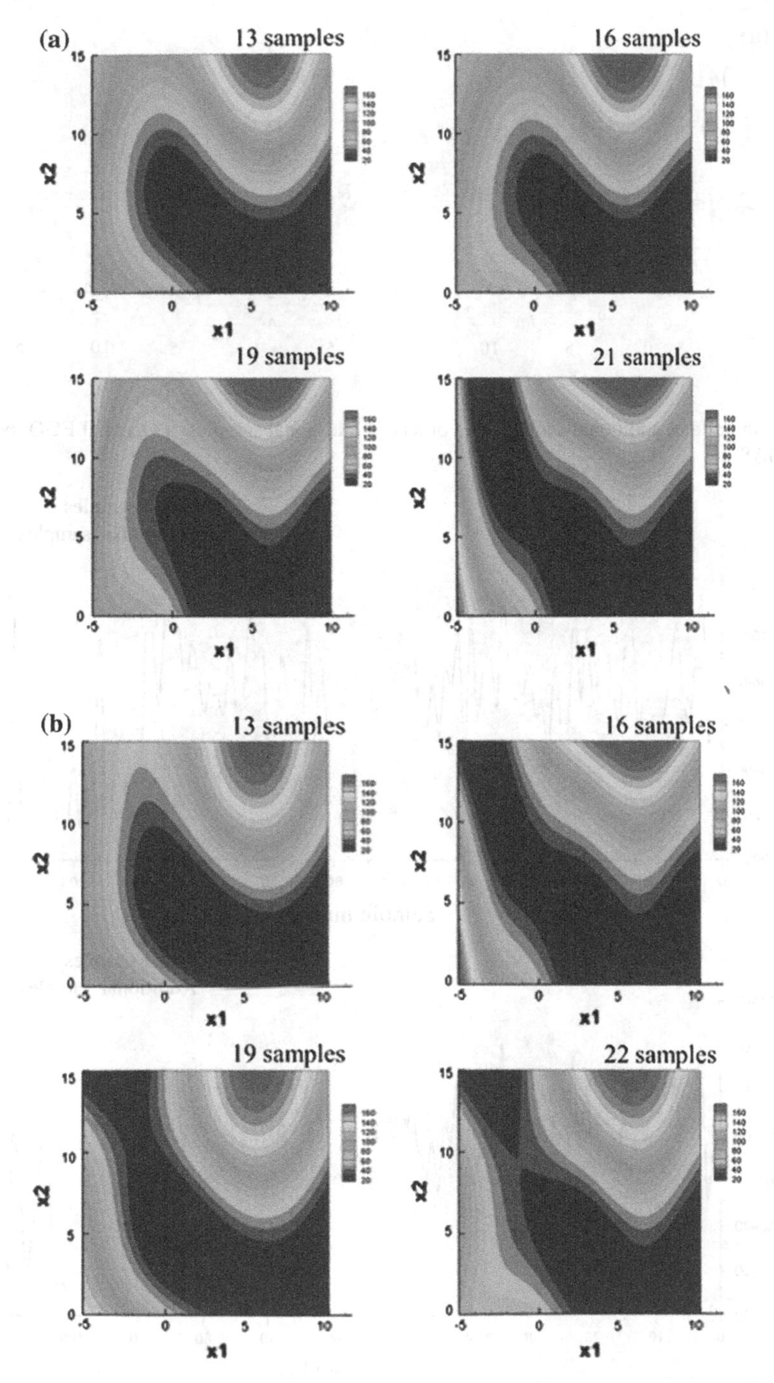

Fig. 7 Variations of the shape of the surrogate model for additional samples for the minimization problem of the Branin function. **a** Function shape for the original EGO. **b** Function shape for the proposed method (EGO with MAS)

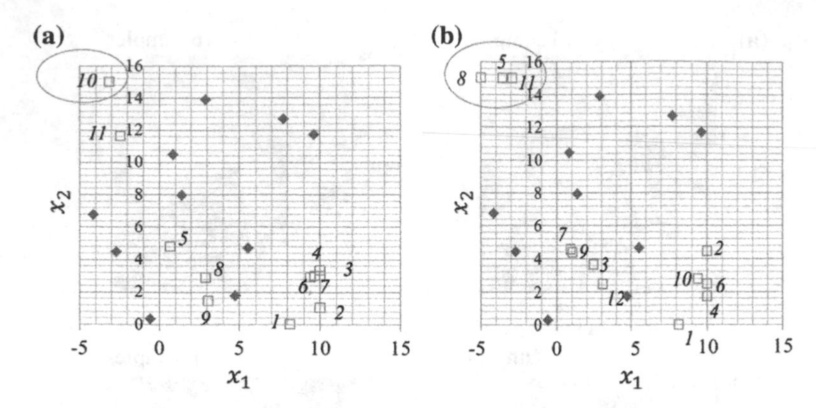

Fig. 8 Initial and additional samples and orders of additional samples. **a** Original EGO. **b** EGO with MAS

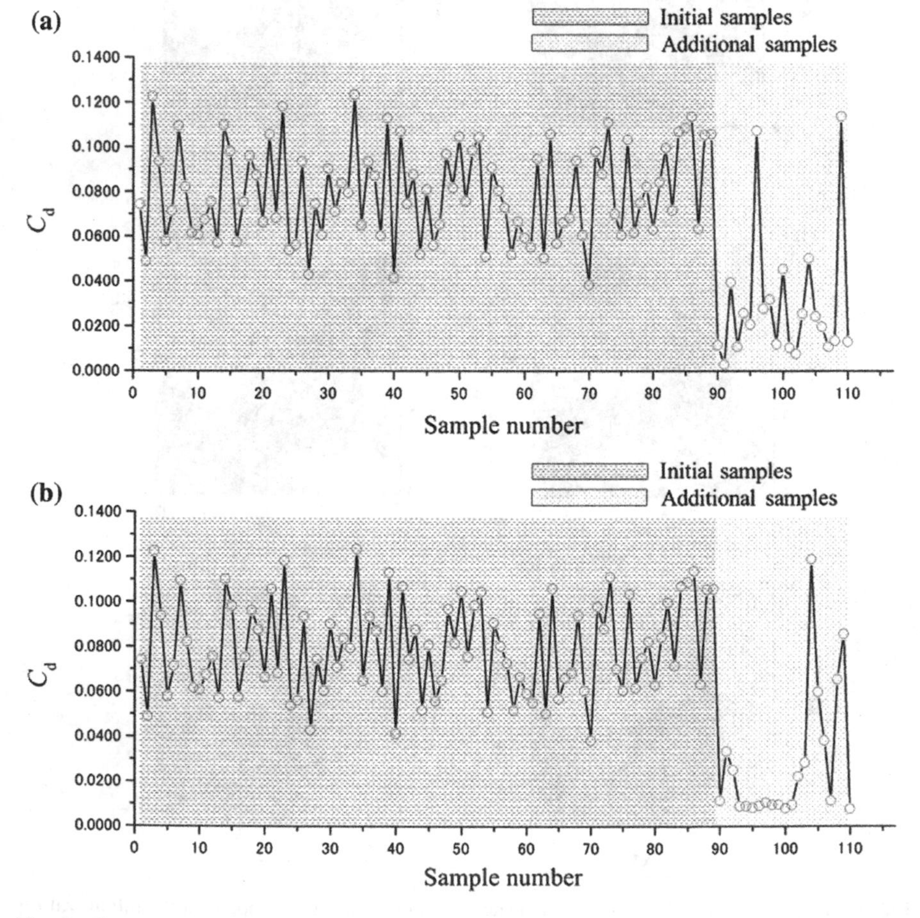

Fig. 9 Comparison of convergence histories in the drag minimization of airfoil. **a** Original EGO. **b** EGO with MAS

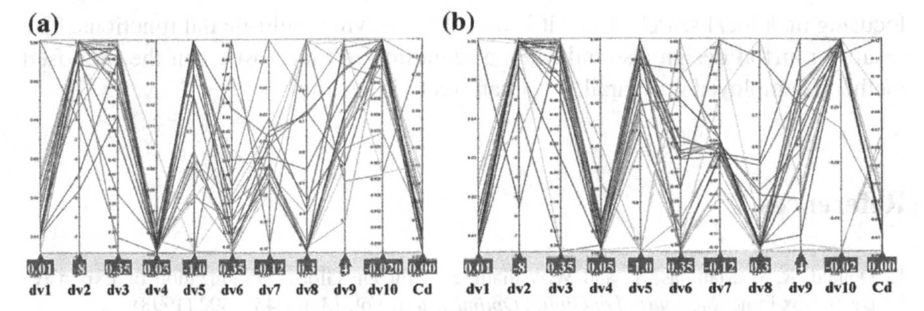

Fig. 10 Design problem visualization by PCP colored by C_D. **a** Original EGO. **b** EGO with MAS

Figure 9 presents a comparison of the convergence histories of the original EGO and the EGO with MAS. Both methods exhibit a similar convergence through the additional sampling process. This result suggests that the proposed method is applicable to this real-world problem, with an increased computational efficiency.

Figure 10 presents a comparison of the visualizations of the design problem for each method using a parallel coordinate plot (PCP). modeFrontier™version. 4.2.2 was used for this visualization. According to this comparison, few differences can be observed in the design spaces. This result suggests that an almost equal global exploration was performed by each method.

The CFD employed in this investigation requires about four minutes.[1] Therefore, in this study the original EGO obtaining the additional samples sequentially takes 53 min to obtain 21 additional samples. On the other hand, the proposed method employing the three sub-iterations takes 28 min to obtain 21 additional samples in the parallel evaluation environment. Thus, it is suggested that the proposed method improves the efficiency of the original EGO when a parallel evaluation environment is employed.

6 Conclusions

In this paper, a multiple additional sampling method was proposed for efficient global optimization (EGO) in order to make use of a parallel evaluation environment. The proposed method was developed by including sub-iterations that can obtain multiple possible additional samples to improve the surrogate model. The proposed method was investigated by solving the Branin function minimization problem and a drag minimization problem for airfoil. The results were compared with the original EGO method.

According to the investigation, the proposed method can achieve global optimization as effectively as the original EGO. In addition, the proposed method can avoid

[1]Using a computer with Intel®Core™i7-3770.

focusing on a local search. Thus, it is useful for solving multi-modal functions. The result for airfoil design also indicated a reduction in time cost when the proposed method is employed in a parallel evaluation environment.

References

1. Donald, R. J., Matthias, S., and W.William, J.: Efficient Global Optimization of Expensive Black-Box Function, *Journal of Global Optimization*, Vol. 13, pp. 455–492 (1998)
2. Ito, Y. and Nakahashi, K.: Direct surface triangulation using stereolithography data, *AIAA Journal*, Vol. 40, No. 3, pp. 490–496 (2002)
3. Ito, Y. and Nakahashi, K.: Surface triangulation for polygonal models based on CAD data, *International Journal for Numerical Methods in Fluids*, Vol. 39, pp. 75–96 (2002)
4. Jeong, S., Murayama, M., and Yamamoto, K.: Efficient Optimization Design Method Using Kriging Model, *Jounal of Aircraft*, Vol. 42, No. 2, pp. 413–420 (2005)
5. Kanazaki, M., Obayashi, S., and Nakahashi, K.: Exhaust Manifold Design with Tapered Pipes Using Divided Range MOGA, *Engineering Optimization, Taylor & Francis*, Vol. 36, No. 2, pp. 149–164 (2004)
6. Kitagawa, Y., Kitagawa, K., Nakamiya, M., Kanazaki, M., and Shimada, T.: Multi-Stage Hybrid Rocket Conceptual Design for Mi-cro-Satellites Launch using Genetic Algorithm, *Transactions of the Japan Society for Aeronautical and Space Sciences*, Vol. 55, No. 4, pp. 229–236 (2012)
7. Obayashi, S., Sasaki, D., Takeguchi, Y., and Hirose, N.: Multiobjective Evolutionary Computation for Supersonic Wing-shape Optimization, *Evolutionary Computation, IEEE Transactions on*, Vol. 4, No. 2, pp. 182–187 (2005)
8. Sobieczky, H.: Parametric Airfoils and Wing, *Notes on Numerical Fluid Mechanics, Viewing*, Vol. 68, pp. 71–88 (1998)

Dynamic Job Shop Scheduling Under Uncertainty Using Genetic Programming

Deepak Karunakaran, Yi Mei, Gang Chen and Mengjie Zhang

Abstract Job shop scheduling (JSS) is a hard problem with most of the research focused on scenarios with the assumption that the shop parameters such as processing times, due dates are constant. But in the real world uncertainty in such parameters is a major issue. In this work, we investigate a genetic programming based hyper-heuristic approach to evolving dispatching rules suitable for dynamic job shop scheduling under uncertainty. We consider uncertainty in processing times and consider multiple job types pertaining to different levels of uncertainty. In particular, we propose an approach to use exponential moving average of the deviations of the processing times in the dispatching rules. We test the performance of the proposed approach under different uncertain scenarios. Our results show that the proposed method performs significantly better for a wide range of uncertain scenarios.

1 Introduction

Most research on job shop scheduling use a deterministic model [20]. A deterministic model for a scheduling environment assumes that the shop parameters viz. processing times are constant throughout the realization of a schedule. However, in practice, the job shop environment always has uncertainty which makes scheduling a challenging and difficult task [18]. Handling uncertainty during scheduling is of practical importance. In *dynamic* job shop scheduling, the information about jobs is unknown

D. Karunakaran (✉) · Y. Mei · G. Chen · M. Zhang
School of Engineering and Computer Science, Victoria University of Wellington,
Wellington, New Zealand
e-mail: deepak.karunakaran@ecs.vuw.ac.nz

Y. Mei
e-mail: yi.mei@ecs.vuw.ac.nz

G. Chen
e-mail: aaron.chen@ecs.vuw.ac.nz

M. Zhang
e-mail: mengjie.zhang@ecs.vuw.ac.nz

G. Leu et al. (eds.), *Intelligent and Evolutionary Systems*,
Proceedings in Adaptation, Learning and Optimization 8,
DOI 10.1007/978-3-319-49049-6_14

before their arrival. In this work, we consider the dynamic job shop scheduling problem with uncertain processing times.

There are many sources of uncertainty which can affect scheduling. A review of the literature shows that [7] uncertainty in processing times, machine breakdowns, fluctuating arrival rates of orders, modification of due dates, cancellation or modification of orders and uncertain arrival time of raw materials have been considered. In particular, a lot of focus is given to uncertainty in processing times. This could be attributed to the fact that variability in processing times has an impact on almost all scheduling objectives e.g. makespan, mean flowtime, tardiness etc.

The methodologies to tackle uncertainty depend on the description of uncertainty. A review of the literature shows that the three widely used methods to describe uncertainty [20] are: (1) *bounded form*, (2) *probability description* and (3) *fuzzy description*. The probability description is frequently used in the literature and is used for this work.

Broadly, the approaches to scheduling under uncertainty are classified as *preventivescheduling* and *reactive scheduling*. Preventive scheduling generates the schedules before the uncertain event by employing historical information to determine uncertainty parameters (say in the form of a stochastic distribution). Preventive scheduling methods are further classified into stochastic scheduling [1], robust optimization [18], fuzzy programming [8] and sensitivity analysis [25]. Reactive scheduling modifies the generated schedule in the case of an uncertain event. The literature shows a number of techniques for reactive scheduling, for example, using Mixed Integer Linear Programming [29], but most of the works are based on using dispatching rules. Scheduling using dispatching rules, in particular, have been shown to be a good approach for dynamic job shop scheduling [24]. Our work focuses on using dispatching rules for reactive scheduling.

The literature survey shows that a lot of research works [6, 9, 19, 21, 23] have proposed approaches for using dispatching rules for scheduling under uncertainty. For instance, in [23], Matsuura et al. compare performance of dispatching rules with re-sequencing and switching as a scheduling approach under different types of uncertainty and show that dispatching rules perform better in a dynamic environment with rush jobs and specification change (number of machines a job needs to visit). In the work by Lawrence et al. [19], they show that when the uncertainty in processing times increases, the quality of schedules generated by dispatching rules is as good as those generated by computationally expensive optimization techniques viz. branch and bound. Several other works have considered dispatching rules and scheduling heuristics in reactive scheduling [13, 16, 17]. Pinedo and Weiss [27] show that for a certain class of scheduling problems on parallel machines "largest variance first" policy minimizes flow time and makespan. This has been considered as a dispatching rule in [19]. Later, Pinedo [26] shows that "smallest variance first" minimizes the expected makespan and total expected completion time for a certain class of stochastic batch scheduling problems.

Design of dispatching rules is challenging and requires rigorous experimental validation. Recently genetic programming has been used to automatically evolve dispatching rules [3, 15]. In these methods, the composite dispatching rules are evolved as genetic programs and are shown empirically to be more effective and flexible than simple dispatching rules and manually generated dispatching rules [24]. Similarly, Vazquez-Rodriguez and Ochoa [31] modified the dispatching rule under an existing heuristic using genetic programming for solving permutation flow shop scheduling problem. In a related work by Hunt et al. [14], they evolved "less-myopic" dispatching rules based on genetic programming. Yin et al. [34] learn genetic programming based predictive scheduling heuristics for stochastic machine failures. The genetic programming based hyper-heuristic approach to evolving dispatching rules has shown promise recently, but its application to evolve dispatching rules suitable for a stochastic environment is still nascent.

Therefore, considering the difficult nature of the job shop scheduling problem under uncertainty, creating dispatching rules which are automatically designed using genetic programming is a key approach. With the flexible representation of the genetic programs, the evolved composite dispatching rules are expected to produce better quality schedules under uncertainty. Furthermore, to leverage the advantages of genetic programming, it is a good direction of research to investigate a new terminal set for genetic programming to consider the randomness in processing times and other uncertain parameters of the scheduling problem.

In this work, we consider the problem of job shop scheduling under uncertain processing times. In particular we propose the exponential moving average (*EMA*) of the deviation from processing times as a new terminal for genetic program. The overall goal is to present a GPHH method which is able to improve the performance of scheduling under uncertain processing times pertaining to different scenarios. The primary objective is evolving dispatching rules with the *EMA* terminal and verify that the rules perform significantly better against current GPHH approaches with mean flow time as the scheduling objective. The secondary objective is to show that appropriate training instances are required for evolving dispatching rules which can perform better under different levels of uncertainty.

2 Background

2.1 Job Shop Scheduling

We briefly introduce the dynamic job shop scheduling problem (DJSS). In most of the works [14, 24], the arrival of the jobs to the shop is assumed to follow a Poisson process. Also the general assumptions are: (1) no recirculation of jobs, (2) no preemption, (3) no machine failure, (4) no alternate routing and (5) zero transit time between machines. No prior information about the jobs is known before their arrival. We follow these assumptions in our work.

Each job j arriving at the shop has n_j operations and the sequence of these operations is O_j. Here, $n_j \in \{1, \dots, n\}$ and $O_j = (o_{j,1} \rightarrow o_{j,2} \rightarrow, \dots, o_{j,n_j})$ is the predefined route. The ith operation of the job j is $o_{j,i}$ and $p(o)$ is the expected processing time of the operation o. In practice, the realized processing time may vary from the expected value. $p'(o)$ is defined as the *realized* processing time. Furthermore, it is practical to assume that $p'(o) \geq p(o)$ and the deviation from the processing time follows some probability distribution. d_j and r_j are the due dates and arrival dates (released immediately to shop) respectively for the job j. $m(o)$ is the machine at which the operation o is processed. Let C_j be the completion time of the job. Then the total flowtime is defined as

$$F = \sum_j (C_j - r_j)$$

The jobs arrive at the shop continuously and the first operation of each job is queued at the first machine in the job's route. A dispatching rule (DR) is used to assign priority values to each operation queued on a machine. The operation which is first in the priority queue of that machine is processed as soon as the machine is free. Once the operation is completed, the next operation in the job's route is enqueued to its corresponding machine. This continues till the entire job is completed.

Depending upon the scheduling objectives, different DRs are used in the job shops. Some of the popular DRs for dynamic job shop scheduling problem are *FIFO* (first in first out), *SPT* (shortest processing time), *COVERT* (cost over time) [30], *ATC* (apparent tardiness cost) [32], etc.

2.2 Genetic Programming Based Hyper Heuristics (GPHHs)

Hyper-heuristics are a set of approaches which automate the design of heuristics to solve hard problems, particularly combinatorial optimization problems [4]. Hyper-heuristic search is performed in the heuristic space rather than the solution space.

Genetic programming has been shown to be a good method for hyper-heuristic search [5]. Moreover, a genetic program is an executable data structure which is conducive to represent a dispatching rule. It has been shown to outperform other representations [3] viz. linear and neural networks representations. It is possible to vary the depth of the trees and for the domain-experts to easily incorporate useful functions and terminals. Furthermore, it is possible to use multi-population systems to create genetic programs which incorporate diverse characteristics [24] in dispatching rules. There are many more examples from literature which use GPHH to develop composite dispatching rules. For example, Tay and Ho [12] evolved scalable and flexible dispatching rules for multi-objective flexible job shop problem. Hildebrandt et al. [11] use GPHH to evolve dispatching rules which generalize well across varying scenarios in the job shop. GPHH has been used to evolve dispatching rules in dynamic environments where the arrival of jobs is uncertain and follows a particular probability distribution [24]. Motivated by the success of GPHH approach in dynamic scheduling environment, our GPHH approach aims at evolving dispatching rules which are suitable for DJSS under uncertainty in processing times.

3 Proposed Method

We present an approach to evolving dispatching rules which perform well under uncertain processing times compared to the current GPHH approaches. In particular, we investigate a new terminal for genetic programming which estimates the information about uncertainty in processing times. We use *exponential moving average* as one of the candidates for the terminals in the genetic programs and expect to capture the information about uncertainty in processing times of the operations. We use ECJ22 [22] for evolving the dispatching rules; the dynamic job shop scheduling simulation was done using a discrete-event simulation system [10]. The objective of scheduling is to minimize total flowtime of all jobs, which is a basic and frequently considered objective in job shops.

3.1 Simulation Model with Uncertainty

We assume that the uncertainty in the processing time of a particular job is associated with a probability distribution. This probability distribution may change from one job to another. This is similar to many other studies in related work e.g. [28, 33]. Rai et al. [28] consider scheduling in printing industry, where operator skills and job characteristics are main sources of variation in processing times. So it is reasonable to assume dissimilar uncertainty distributions for different types of jobs. Akker et al. [33] consider processing times with a deterministic component and a random disturbance, which is identically distributed for each job. We model the uncertainty using the gamma distribution, which is widely used to model parameters that are required to be positive or skewed, and have been used to model uncertainty e.g. [19]. It is a continuous probability distribution with two parameters, *shape* ($\alpha \in \mathbb{R}^+$) and *scale* ($\beta \in \mathbb{R}^+$). In this work, we define the deviation from the processing time ($p(o)$) as δ, such that

$$p'(o) = p(o) + p(o) \times \delta$$

where, δ follows a gamma distribution and different jobs are assigned gamma distribution parameters viz. *shape* (α) and *scale* (β) from a set. In particular, we use $\alpha = 1$ and $\beta \in \{0.1, 0.2, 0.3, 0.4, 0.5, 0.6, 0.8, 1.2\}$ in our experiments. The choice of $\alpha = 1$ is motivated by the evidence from literature that the information contained in standard deviation of uncertain processing times is useful for scheduling [19, 27]. For a gamma distribution, the standard deviation is equal to $\alpha \times \beta^2$ and the mean is equal to $\alpha \times \beta$. If we assign $\alpha = 1$, then the mean and the standard deviation of this gamma distribution are β and β^2 respectively. Note that when $\alpha = 1$, the distribution essentially becomes exponential, which has been frequently used to model uncertain processing times. Since the mean and the standard deviation have a simple relation, we expect that the evolved rules will be able to estimate the uncertainty information. This estimation is done using exponential moving average as explained below (Sect. 3.2).

3.2 Exponential Moving Average (EMA) Terminal

The moving average is a rolling mean over different subsets of data. In exponential moving average the data points are weighted such that for the older data the weight keeps decreasing. It is presented in the equation below, where $\bar{\delta}_i$ is the exponential moving average at step $i(>0)$ and Y_i is the new data. κ is a constant smoothing factor.

$$\bar{\delta}_i = \kappa \times Y_i + (1 - \kappa) \times \bar{\delta}_{i-1}$$

According to literature, $\kappa \leq 0.3$ is considered good. We used $\kappa = 0.2$ in our experiments, and did not observe any difference in performance with the values in $\{0.1, 0.3\}$.

In our work, $\bar{\delta}$ is evaluated for every job and for a job j,

$$Y_i = \delta_{j,i} = p'(o_{j,i})/p(o_{j,i}) - 1$$

Therefore for ith operation of the job j,

$$\bar{\delta}_{j,i} = \kappa \times \delta_{j,i} + (1 - \kappa) \times \bar{\delta}_{j,i-1}$$

We assigned $\bar{\delta}_{j,0} = \delta_{j,1}$ at the beginning. Our aim, as mentioned earlier, is to obtain information about uncertainty and use it in the dispatching rule. The exponential moving average, which is maintained for every job, captures the expected deviation from the processing times of operations that are already completed. We expect that the GPHH approach will evolve programs which could utilize the information from the *EMA* terminal to create better schedules.

3.3 Genetic Programming System

Table 1 lists all the functions and terminals which we use to construct a genetic program. The function *if* takes three arguments; if the first argument is larger than 0 then it returns the second argument else it returns the third one. The protected division returns 1 if the second argument is 0. The other functions are self-explanatory. A population size of 1024 is used with the number of generations set to 50. A maximum depth of 8 was considered for the genetic programs. The crossover, mutation and elitism rates are set as 0.85, 0.1 and 0.05 respectively [24].

In order to compare the performance of exponential moving average approach, for the two training configurations mentioned before, we evolve dispatching rules using all of these terminals and compare them with dispatching rules obtained by GPHH without *EMA* as the terminal.

Table 1 Function and Terminal sets for genetic programs

Function set	Meaning
+	Addition
−	Subtraction
*	Multiplication
/	Protected division
Max	Maximum
Min	Minimum
If	Conditional
Terminal set	Meaning
DD	Due date of job
PR	Processing time of operation
RO	Remaining operations for job
RJ	Ready time of job
RT	Remaining processing time of job
RM	Ready time of machine
ERC	Ephemeral random constant
EMA	Exponential moving average ($\bar{\delta}$) of processing time deviation

4 Experimental Results

In this section, we describe the results from our experiments. First, we briefly describe the simulation configuration. The instances of dynamic job shop scheduling problem are created randomly using a discrete event simulation system. We use a configuration with ten machines, eight operations per job, with the processing times sampled uniformly at random from the range $[1, 49]$. We use this configuration because it has been used in previous works [24]. The stochastic arrival of the jobs at the shop follows a Poisson process with a rate λ. λ is a function of the utilization factor [2]; we assign 0.85 as the *expected* utilization factor. The first 500 jobs are considered as the warm-up and the next 2000 jobs are used for analysis, with new jobs arriving at the shop till the 2500th job is completed. If we use $\lambda > 0.85$, by the time the 2500th job is finished, the number of jobs queued on the machines is very high and the simulation runs for a very long time. This is because the *realized* processing times are higher and consequently the *realized* utilization is also high.

Training Since we consider different jobs to be associated with different probability distributions, we use two configurations for training by varying the number of job types.

- In the first case, every job on arrival is assigned gamma distribution parameters as $\alpha = 1,\ \beta \in \{0.1, 0.6\}$ with equal probability i.e. 50 % of the jobs are associated with $\{\alpha = 1, \beta = 0.1\}$ and the rest with $\{\alpha = 1, \beta = 0.6\}$ (Fig. 1). Consequently,

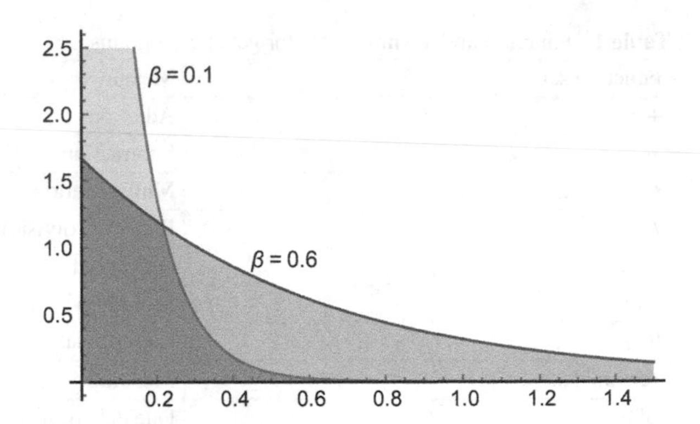

Fig. 1 Gamma Distributions, $\alpha = 1$

the values of δ for an operation are sampled from the assigned distribution of its job. Thus, a job is associated with either of the *two* levels of uncertainty.

- In the second case, we consider *five* types of jobs, such that the gamma distribution parameters are $\alpha = 1$, $\beta \in \{0.1, 0.2, 0.3, 0.4, 0.5\}$ with equal probability. In other words, 20 % jobs are associated with each gamma distribution arising out of each of the five possible values of $\beta \in \{0.1, 0.2, 0.3, 0.4, 0.5\}$.

Our primary objective is to test the performance of our proposed genetic programming representation which uses exponential moving average as a terminal. As mentioned earlier, we use two training sets, and for each set we evolve dispatching rules with and without the proposed terminal. Therefore, we get four sets of dispatching rules. We denote the four sets as $\mathcal{D}_2^{\bar{\delta}}$, $\mathcal{D}_5^{\bar{\delta}}$, $\mathcal{D}_2$ and $\mathcal{D}_5$, where the subscripts denote the number of types of jobs (with respect to uncertainty level) used. The rules with superscript $\bar{\delta}$ are evolved using exponential moving average as a terminal. Each set consists of 30 rules, which is considered as a minimum number of samples required for statistical tests.

Testing The experiment is conducted on a number of different configurations. In particular, we use different gamma distributions by varying the *scale* parameter (β). We also vary the proportion of the number of jobs associated with each distribution. These different configurations are shown in the Table 2. Each test set consists of

Table 2 Test configurations

Test-set	Scale (β)	#Job-ratio
I	{0.1, 0.6}	1 : 1
II	{0.1, 0.6}	2 : 1
III	{0.1, 0.6}	3 : 1
IV	{0.1, 0.3, 0.6}	1 : 1 : 1
V	{0.1, 0.3, 0.6, 0.8}	1 : 1 : 1 : 1
VI	{0.1, 0.2, 0.3, 0.4, 0.5}	1 : 1 : 1 : 1 : 1
VII	{0.1, 0.3, 0.6, 0.8, 1.2}	1 : 1 : 1 : 1 : 1

30 problem instances. The second column in this table shows the scale parameter values of the gamma distributions, which the jobs in a problem instance are assigned. The column '#job-ratio' shows the ratio of the number of jobs which are assigned a particular gamma distribution. For example, in row 2, corresponding to the test set-II, out of every three jobs, *two* have uncertainty represented using a gamma distribution scale parameter of 0.1 and *one* has scale parameter of 0.6.

Our aim is to verify if the exponential moving average i.e. the EMA terminal, is able to improve the performance of scheduling (Fig. 2). While the dispatching rules are evolved using a training set with just two configurations, the test configurations are more comprehensive. Therefore, improvement of performance, if any, could indicate more practical applicability of evolved rules. We test the trained dispatching rules over different test sets and present the box-plots. The Wilcoxon-rank-sum test with significance level of 0.05 is used to test the statistical significance of the result.

We present our findings in the Figs. 3 and 4. Though we conducted our experiments on 30 problem instances in each test set, we have presented box-plot comparisons only for 9 of them due to space limitation. We observe similar pattern for other instances. The results for the 30 problem instances are mentioned below each plot in the form of triplets. The first value of the triplet is equal to the number of test instance for which the performance of proposed method is significantly 'better', second value equals the number of instances when it is significantly the 'same'. And the third value is equal to number of instances with significantly poor performance; thus the sum of the three values is 30 for each test set. The x-axis denotes the #problem-instance which is the *id* of the problem instance; and each of the adjacent box plot pairs is denoted by same *id* #problem-instance on the x-axis. The statistically significant test result is indicated by a colored box-plot. '*' is added as a superscript to the corresponding #problem-instance on the x-axis as well.

We split the description of our results into *three*. Firstly we discuss the performance of the dispatching rules obtained by training on 2 job types i.e., comparing $\mathcal{D}_2^{\bar{\delta}}$ with $\mathcal{D}_2$ and then with 5 job-types i.e., $\mathcal{D}_5^{\bar{\delta}}$ with $\mathcal{D}_5$. In order to have a holistic analysis, we also compare $\mathcal{D}_2^{\bar{\delta}}$ with $\mathcal{D}_5^{\bar{\delta}}$ and to show the influence of different training sets on the performance of GP and the results of some of the test sets are shown.

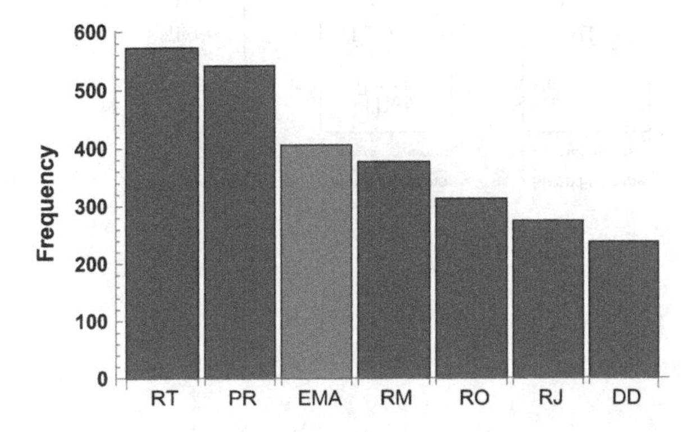

Fig. 2 Frequency of terminals

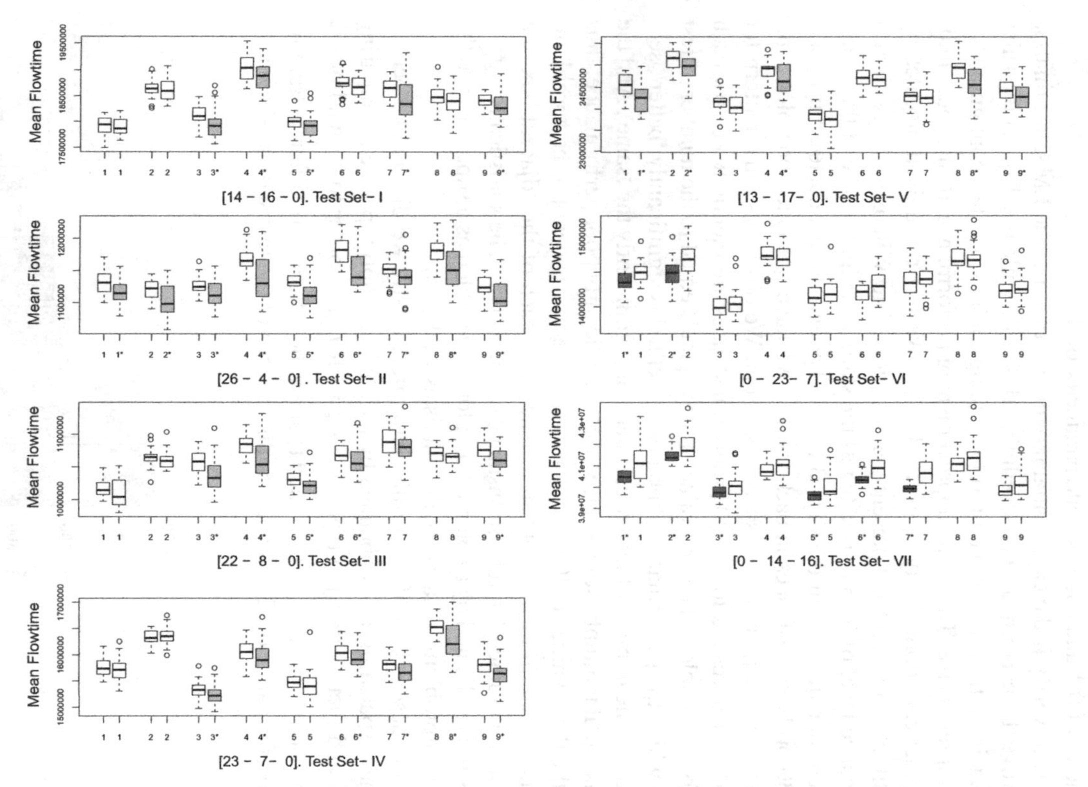

Fig. 3 $\mathcal{D}_2$ vs $\mathcal{D}_2^{\bar{\delta}}$ on different Test Sets. The box plots comparing 9 problem instances out of a total of 30 are shown for each test set. Each pair of box plots is associated with one test instance as indicated on x-axis. The labels below each figure indicate the index of the test instances (out of 30). For each pair, the box-plot to the right represents the evolved rules with *EMA* terminals. The box plots corresponding to significantly better result (Wilcoxon test) are colored: *yellow* when $\mathcal{D}_2^{\bar{\delta}}$ is significantly better and *red* when $\mathcal{D}_2$ is better. The triplets under each plot show number of instances which are significantly better, same and poor respectively [better—same—poor]

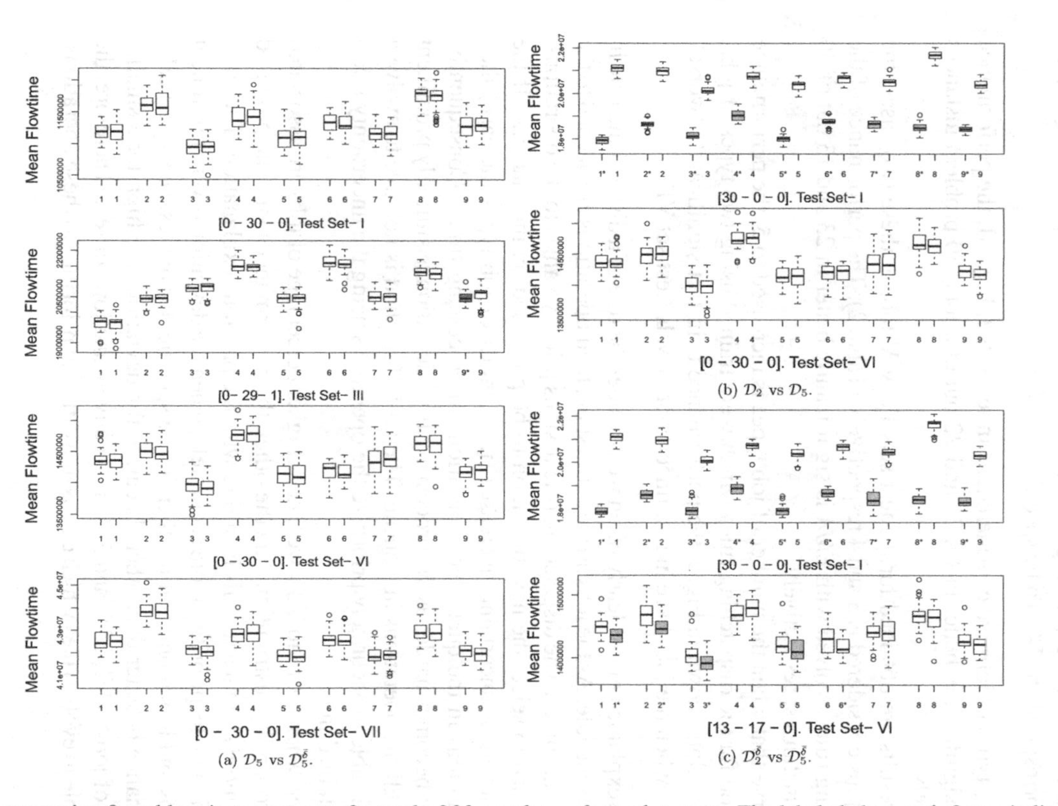

Fig. 4 The box plots comparing 9 problem instances out of a total of 30 are shown for each test set. The labels below each figure indicate the index of the test instances (out of 30). Each pair of box plots is associated with one test instance as indicated on x-axis. For Fig. a, the box-plot to the *right*, in each pair, represents the evolved rules with *EMA* terminals. For Fig. (b, c), the box-plot to the *left*, in each pair, represents the evolved rules with 2 job types. The box plots corresponding to significantly better result (Wilcoxon test) are colored. The triplets under each plot show number of instances which are significantly better, same and poor respectively [better—same—poor]

4.1 Performance Comparisons: $\mathcal{D}_2^{\bar{\delta}}$ vs $\mathcal{D}_2$

In this set of comparisons, between $\mathcal{D}_2^{\bar{\delta}}$ and $\mathcal{D}_2$, shown in the box plots of Fig. 3, our primary observation is that the terminal EMA is able to evolve rules which perform better across most of the test configurations. For all pairs of box plots in Fig. 3, the right one corresponds to $\mathcal{D}_2^{\bar{\delta}}$. To be specific:

- When two types of jobs are considered, as in test sets I, II & III, the performance of $\mathcal{D}_2^{\bar{\delta}}$ is significantly better in 14, 26 and 22 out of a total 30 problem instances respectively.
- Similar results are obtained for the test sets IV & V too. In these two cases, the types of jobs considered is 3 and 4 respectively (Table 2). The performance of the dispatching rules evolved with *EMA* is significantly better in 23 and 13 out of 30 problem instances, respectively.
- We observe that when the number of job types is increased to 5, the performance is not good for the dispatching rules which were trained using two types of jobs. In the box plots shown in Fig. 3, the performance is in fact significantly 'worse', especially when the jobs have higher uncertainty levels (test set VII).

This can be explained more clearly when we consider the role of a dispatching rule in creating the schedule. As the jobs arrive at the shop, their operations are queued to the appropriate machines. The dispatching rule assigns priority values to the operations, which are used to select the next operation to be processed on the machine. If the number of levels of uncertainty increases (i.e. number of job types increase) then all the operations in the queue have variations in processing times. Consequently, the problem becomes harder as the processing time deviation is similarly prominent throughout all the operations in a queue. The GPHH method is not able to evolve a rule which is able to accurately prioritize the operations using the uncertainty information contained in $\bar{\delta}$.

Moreover, when the ratio is skewed, i.e. a higher percentage of jobs are associated with one probability distribution than the other, the performance using the proposed terminal is more significant. This is in line with the previous explanation. The rules are able to assign the priorities more accurately, when the deviation is very prominent in some jobs and less prominent in other. To be specific, for the test sets II, III &IV the performance is better than others because, the deviations are high for a smaller percentage of jobs; 33 %, 25 % and 33 % respectively and low for rest. Therefore, the GPHH is able to evolve rules which can assign different priorities to these operations.

4.2 Performance Comparisons: $\mathcal{D}_5^{\bar{\delta}}$ versus $\mathcal{D}_5$

In this set of comparisons as shown in Fig. 4a, our primary observation is that for dispatching rules, which are trained on 5 types of jobs, the performance on test cases shows no significant difference to either test cases with lower (two) or higher (five) number of job types.

- $\mathcal{D}_5^{\bar{\delta}}$ performs significantly 'worse' in only one problem instance of test set III, out of a total of 30 problem instances. On other test sets, the performance is significantly same (including on test sets II, IV & V which are not shown due to space limitation).

In other words, the performance of $\mathcal{D}_5^{\bar{\delta}}$ is similar on almost all test set. Our explanation to this behavior of dispatching rules is similar to the one given before. The scheduling problem becomes more difficult when the number of levels of uncertainty assigned to the different jobs increase. Also, this result points to the fact that $\mathcal{D}_2^{\bar{\delta}}$ suffers from over-fitting problem when presented with problems with *five* levels of uncertainties in jobs as test sets.

4.3 Performance Comparisons: $\mathcal{D}_2^{\bar{\delta}}$ versus $\mathcal{D}_5^{\bar{\delta}}$ and $\mathcal{D}_2$ versus $\mathcal{D}_5$

We also cross-compare the test results between the set of evolved rules which are trained on different configurations. The results are shown as box plots in the Fig. 4b, c. For these pairs of box plots, the left one always corresponds to rules trained with 2 job types.

- For test set I, the results show that there is a big drop in performance when the rules trained on another configuration is tested on a different configuration. This is an expected result.
- Similarly, for test set VI, though the results are not that apparent as in the previous case, the rules in $\mathcal{D}_5^{\bar{\delta}}$ perform significantly better in 13 test instances. Though in the case of $\mathcal{D}_5$ the performance is significantly same as that of $\mathcal{D}_2$.

  This shows the significance of choosing proper training instances depending on the level of uncertainty in processing times.

4.4 Analysis of Dispatching Rules

Now we analyze the evolved rules to make more sense of our observations. Firstly, we count the number of each terminal in the set of 30 evolved rules from $\mathcal{D}_2^{\bar{\delta}}$. The bar chart is shown in Fig. 2. The prominence of the terminal EMA is an indicator that the terminal is useful. Furthermore, we counted the number of occurrences of pairs of the terminal EMA and of PR, RT. These pairs are highlighted (in bold) in the example of a dispatching rule below (Listing 1.1). These terminals are chosen because the EMA value is expected to combine with terminals related to processing times. We found that among the 30 rules, they occurred 92 times (multiple occurrences within same rule). Moreover, we calculated the frequency of these combinations for the dispatching rules in $\mathcal{D}_5^{\bar{\delta}}$. We found that the frequency is 61, which indicates that for a configuration with five job types, the effect of EMA is less prominent.

An example of one of the best evolved dispatching rules is presented below. We can see that the occurrences of the terminal EMA and its combination with the terminals related to processing times is frequent.

Listing 1.1 One of the best evolved rules

```
(* (Min (* RT EMA) (- (/ 0.584 RO) (+ (Max 0.724 RT)
(+ (* (* (If (+ (Max EMA 0.219) (- RO DD)) (Max
(+ EMA(- RT RO))( If (+ 0.990 (* RT EMA))
0.341 (/ 0.956 (+(/ 0.584 RO) (Max(*RT EMA) (/ 0.584 RO)
))))) 2)(+(+ EMA(- RT RO)) (Min (Min (*EMA PR) (Max RJ
RM)) (- RT RO)))) (+ (Min RO 0.321) (Min RO 0.321))) (Max
(* RT EMA) (/ 0.584 (If (+ (Max (*RT EMA)
(/ 0.584 RO)) (- RO DD)) (Max 0.724 RT) (+ 2 ))))))))
(Max RO PR))
```

5 Conclusions

In this paper, we focused on evolving dispatching rules using a GP-based hyper heuristic approach for dynamic job shop scheduling problem under uncertain processing times. In particular, we introduced a new terminal which captures the information about uncertainty in processing times to evolve suitable dispatching rules. The terminal computes the exponential moving average of deviation in processing times. We considered different uncertainty levels in processing times and considered different ratios of jobs pertaining to these levels.

The primary conclusion is that, the inclusion of *EMA* terminal in the GPHH approach could evolve better dispatching rules with total flow time as the scheduling objective when the number of uncertainty levels is up to *four*. When the number of uncertainty levels is *five*, the performance of the proposed terminal shows a decline. We conclude that up to *four* levels of uncertainty the proposed terminal is able to assign more accurate priorities to operations queued on a machine. But beyond that, the processing time variations are prominent across all the queued operations and consequently the *EMA* terminal is not effective.

Furthermore, our observations concur with the fact that different job shop scenarios require specific dispatching rules. In particular, when the following parameters in the job shop are varied: (1) the number of different levels of uncertainty and (2) the ratio of jobs pertaining to the different levels of uncertainty; different sets of dispatching rules are required to maintain schedule quality.

In our future work, we will investigate better estimation techniques to evolve dispatching rules which perform well even with higher number of levels of uncertainty. We will also consider evolving dispatching rules which perform well under more types of uncertainty e.g. variation in due dates.

References

1. Balasubramanian, J., Grossmann, I.: Approximation to multistage stochastic optimization in multiperiod batch plant scheduling under demand uncertainty. Industrial & engineering chemistry research 43(14), 3695–3713 (2004)
2. Bhat, U.N.: An introduction to queueing theory: modeling and analysis in applications. Birkhäuser (2015)
3. Branke, J., Hildebrandt, T., Scholz-Reiter, B.: Hyper-heuristic evolution of dispatching rules: A comparison of rule representations. Evolutionary computation 23(2), 249–277 (2015)
4. Burke, E.K., Gendreau, M., Hyde, M., Kendall, G., Ochoa, G., Özcan, E., Qu, R.: Hyper-heuristics: A survey of the state of the art. Journal of the Operational Research Society 64(12), 1695–1724 (2013)
5. Burke, E.K., Hyde, M.R., Kendall, G., Ochoa, G., Ozcan, E., Woodward, J.R.: Exploring hyper-heuristic methodologies with genetic programming. In: Computational intelligence, pp. 177–201. Springer (2009)
6. Calleja, G., Pastor, R.: A dispatching algorithm for flexible job-shop scheduling with transfer batches: an industrial application. Production Planning & Control 25(2), 93–109 (2014)
7. Davenport, A.J., Beck, J.C.: A survey of techniques for scheduling with uncertainty. Unpublished manuscript. Available from http://tidel.mie.utoronto.ca/publications.php (2000)
8. Fortemps, P.: Jobshop scheduling with imprecise durations: a fuzzy approach. IEEE Transactions on Fuzzy Systems 5(4), 557–569 (1997)
9. Gao, K.Z., Suganthan, P.N., Tasgetiren, M.F., Pan, Q.K., Sun, Q.Q.: Effective ensembles of heuristics for scheduling flexible job shop problem with new job insertion. Computers & Industrial Engineering 90, 107–117 (2015)
10. Hildebrandt, T.: Jasima – an efficient java simulator for manufacturing and logistics. http://code.google.com/p/jasima (2012)
11. Hildebrandt, T., Heger, J., Scholz-Reiter, B.: Towards improved dispatching rules for complex shop floor scenarios: a genetic programming approach. In: Proceedings of the 12th annual conference on Genetic and evolutionary computation. pp. 257–264. ACM (2010)
12. Ho, N.B., Tay, J.C.: Evolving dispatching rules for solving the flexible job-shop problem. In: 2005 IEEE Congress on Evolutionary Computation. vol. 3, pp. 2848–2855. IEEE (2005)
13. Huercio, A., Espuna, A., Puigjaner, L.: Incorporating on-line scheduling strategies in integrated batch production control. Computers & chemical engineering 19, 609–614 (1995)
14. Hunt, R., Johnston, M., Zhang, M.: Evolving less-myopic scheduling rules for dynamic job shop scheduling with genetic programming. In: Proceedings of the 2014 conference on Genetic and evolutionary computation. pp. 927–934. ACM (2014)
15. Jakobović, D., Jelenković, L., Budin, L.: Genetic programming heuristics for multiple machine scheduling. In: Genetic Programming, pp. 321–330. Springer (2007)
16. Janak, S.L., Floudas, C.A., Kallrath, J., Vormbrock, N.: Production scheduling of a large-scale industrial batch plant. ii. reactive scheduling. Industrial & engineering chemistry research 45(25), 8253–8269 (2006)
17. Kanakamedala, K.B., Reklaitis, G.V., Venkatasubramanian, V.: Reactive schedule modification in multipurpose batch chemical plants. Industrial & engineering chemistry research 33(1), 77–90 (1994)
18. Kouvelis, P., Yu, G.: Robust discrete optimization and its applications, vol. 14. Springer Science & Business Media (2013)

19. Lawrence, S.R., Sewell, E.C.: Heuristic, optimal, static, and dynamic schedules when processing times are uncertain. Journal of Operations Management 15(1), 71–82 (1997)
20. Li, Z., Ierapetritou, M.: Process scheduling under uncertainty: Review and challenges. Computers & Chemical Engineering 32(4), 715–727 (2008)
21. Liu, K.C.: Dispatching rules for stochastic finite capacity scheduling. Computers & industrial engineering 35(1), 113–116 (1998)
22. Luke, S.: Essentials of metaheuristics. Lulu Com (2013)
23. Matsuura, H., Tsubone, H., Kanezashi, M.: Sequencing, dispatching and switching in a dynamic manufacturing environment. The International Journal of Production Research 31(7), 1671–1688 (1993)
24. Nguyen, S.: Automatic design of dispatching rules for job shop scheduling with genetic programming (2013)
25. Penz, B., Rapine, C., Trystram, D.: Sensitivity analysis of scheduling algorithms. European Journal of Operational Research 134(3), 606–615 (2001)
26. Pinedo, M.: Stochastic batch scheduling and the "smallest variance first" rule. Probability in the Engineering and Informational Sciences 21(04), 579–595 (2007)
27. Pinedo, M., Weiss, G.: The largest variance first policy in some stochastic scheduling problems. Operations Research 35(6), 884–891 (1987)
28. Rai, S., Duke, C.B., Lowe, V., Quan-Trotter, C., Scheermesser, T.: Ldp lean document production-or-enhanced productivity improvements for the printing industry. Interfaces 39(1), 69–90 (2009)
29. Rodrigues, M., Gimeno, L., Passos, C., Campos, M.: Reactive scheduling approach for multipurpose chemical batch plants. Computers & chemical engineering 20, S1215–S1220 (1996)
30. Salvendy, G.: Handbook of industrial engineering: technology and operations management. John Wiley & Sons (2001)
31. Vazquez-Rodriguez, J.A., Ochoa, G.: On the automatic discovery of variants of the neh procedure for flow shop scheduling using genetic programming. Journal of the Operational Research Society 62(2), 381–396 (2011)
32. Vepsalainen, A.P., Morton, T.E.: Priority rules for job shops with weighted tardiness costs. Management science 33(8), 1035–1047 (1987)
33. Van den Akker, M., Hoogeveen, H.: Minimizing the number of late jobs in a stochastic setting using a chance constraint. Journal of Scheduling 11(1), 59–69 (2008)
34. Yin, W.J., Liu, M., Wu, C.: Learning single-machine scheduling heuristics subject to machine breakdowns with genetic programming. In: Evolutionary Computation, 2003. CEC'03. The 2003 Congress on. vol. 2, pp. 1050–1055. IEEE (2003)

Similarity Analysis of Survey on Employment Trends in Japan

Masao Kubo, Hiroshi Sato, Akihiro Yamaguchi and Yuji Aruka

Abstract This paper analyzes the characteristics of employment in Japan from open data. Applying methods of AI research field to Survey on Employment Trends in Japan as the way to understand a trend of economic activity is proposed.

Keywords Open data · Complex network analysis · Machine learning

1 Introduction

Adequate agent models of a social simulation is one of critical issue for obtain a good findings. We will want to build a good employment simulator in future and a lot of hypotheses about employers and employees will be required. Although a field survey is a good approach to introduce plausible assumptions for them, it has been difficult to collect valuable information about employment because this kind of information should be kept secret on the both side. Therefore this trait may make researches of this field difficult. Employment and monetary dynamics are major factors of macroscopic economics but the monetary dynamics is much more studied so far. (For example, Input-output table, http://www.oecd.org).

M. Kubo (✉) · H. Sato · A. Yamaguchi · Y. Aruka
National Defense Academy of Japan,
Fukuoka Institute of Technology, Fukuoka, Japan
e-mail: masaok@nda.ac.jp
URL: http://www.nda.ac.jp/cs/staff/masaok.html

H. Sato
e-mail: hsato@nda.ac.jp

A. Yamaguchi
e-mail: aki@fit.ac.jp

Y. Aruka
e-mail: aruka.tamacc@chuo-u.ac.jp

M. Kubo · H. Sato · A. Yamaguchi · Y. Aruka
Chuo University, Tokyo, Japan

G. Leu et al. (eds.), *Intelligent and Evolutionary Systems*,
Proceedings in Adaptation, Learning and Optimization 8,
DOI 10.1007/978-3-319-49049-6_15

Applying an analysis method of AI research field to employment data of open data as the way to understand a trend of economic activity is proposed. Recently, government, local administration, and public entity offer their data in public. It will promote that citizens understand their authority's activity more clearly. Also the open data makes activity of research and business more efficient and consistent. On the other hand, the privacy of the data which the authorities handle should be kept secret. Therefore it seems that data without any modification is not published but its statical data is released in public. The modified data is not convenient in sometimes so that a user needs some inferences and estimations for knowing what it wants to know. Complex network analysis, machine learning, and other artificial intelligence techniques are expected to be prospective.

In this paper, the data of Survey on Employment Trends in Japan is discussed, which is a biannual report published by Health, Labor and Welfare Ministry of Japan. This reports trends of employment and unemployment about labor's sex, age, previous job, and reason why they choose the job and so on.

An index, increasing rate is introduced which is (employment rate–separation rate) of a sector. The similarity of the increasing rate is visualized and we show the behavior of the rate of all of the sectors are very similar but sectors in Japan can be divided into 2 types: sectors of the first type influence each other and form a large group. A sector in the second is independent: the behavior of the increasing rate is not similar to other sectors.

2 Annual Data of Survey on Employment Trends in Japan

There are a lot of open data about employment. For example, input-output table is such a data because this is a primary information of employee. In this paper, annual data of survey on Employment Trends in Japan is used (http://www.mhlw.go.jp/toukei/list/9-23-1.html). This survey mainly discusses employment from a perspective of labor. Health, Labor and Welfare Ministry of Japan takes a survey every 6 months. This questionnaire survey reports accession rate and separation rate of a sector which is a division of Japanese industry. The reports from 2000 to 2014 are utilized here and 55 sectors are introduced in total.

Table 1 shows the sectors in 2013. These sectors are a part of the following 16 industries based on the Japan Standard Industrial Classification, namely, Mining and quarrying of stone and gravel; construction; manufacturing; electricity, gas, heat supply and water; information and communications; transport and postal activities; wholesale and retail trade; finance and insurance; real estate and goods rental and leasing; scientific research, professional and technical services; accommodation, eating and drinking services; living-related and personal services and amusement services (excluding housework services); education and learning support; medical, health care and welfare; compound services; services not elsewhere classified (other unclassified) (excluding foreign public services). Note that this survey does not handle agricultural industry and fishery industry.

Table 1 Sample of sectors in this analysis

Industries covered	Mining and quarrying of stone and gravel	Construction
Manufacturing	Manufacture of beverages, tobacco and feed	Manufacture of textile mill products
Manufacture of lumber and wood products, except furniture	Manufacture of furniture and fixtures	Manufacture of pulp, paper and paper products
Printing and allied industries	Manufacture of chemical and allied products	Manufacture of petroleum and coal products
Manufacture of rubber products	Manufacture of plastic products, except otherwise classified	Manufacture of rubber products
Manufacture of ceramic, stone and clay products	Manufacture of iron and steel	Manufacture of non-ferrous metals and products
Manufacture of fabricated metal products	Manufacture of general-purpose machinery	Manufacture of production machinery
Manufacture of business oriented machinery	Electronic parts, devices and electronic circuits	Manufacture of electrical machinery, equipment and supplies
Manufacture of information and communication electronics equipment	Manufacture of transportation equipment	Miscellaneous manufacturing industries, Manufacture of leather tanning, leather products and fur skins
Electricity, Gas, Heat supply and Water	Information and communications	Transport and postal activities
Wholesale and Retail trade	Wholesale	Retail trade
Finance and Insurance	Real estate and goods rental and leasing	Scientific research, professional and technical services
Accommodations, eating and drinking services	Living-related and personal services and amusement services	Services for amusement and hobbies
Education, learning support	Medical, health care and welfare	Medical and other health services
Social insurance and social welfare	Compound services	Services, N.E.C
Automobile maintenance services? CMachine, ETC. repair services, except otherwise classified	Miscellaneous business services	

A classification of data of this report in 2014 is shown below (http://www.mhlw.go.jp/english/database/db-l/dl/employment_trends_2014_outline.pdf):

1. Establishment Survey

    (a) Number of regular employees in the enterprise overall

   (b) Changes in regular employees by gender and type of employment
   (c) Number of regular employees by gender, age and type of employment
   (d) Number of regular employees and unfilled job vacancy by occupation and type of employment

2. Hired Employee Survey
   (a) Individual attributes
   Gender, age, highest level of education
   (b) Hiring
   Hiring route, type of employment, occupation, address prior to hiring
   (c) Previous employment
   Industry, occupation, status in employment, length of separation, size of enterprise, reason

3. Separated Employee Survey
   (a) Individual attributes
   Gender, age, highest level of education
   (b) Employment immediately before separation
   Type of employment, occupation, length of service, reason for separation

3 Increasing Rate and Analysis of This Index

In this section, increasing rate is introduced which is *accession rate–separation rate*. The benefit of this index will be shown by a simple similarity analysis.

Figure 1 shows the accession rate and separation rate of the 44 sectors in 2014. *Accession rate* is a ratio of the number of new employees of a sector and *separation rate* is a ratio of the number of employees who leave their previous sector. The x axis means sector. 11 sectors (55–11) are not used in this year's report. Roughly speaking,

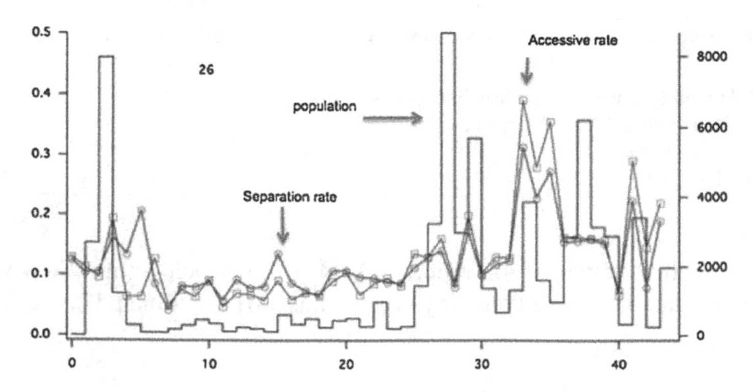

Fig. 1 The result of the survey on employment trends in Japan in 2014

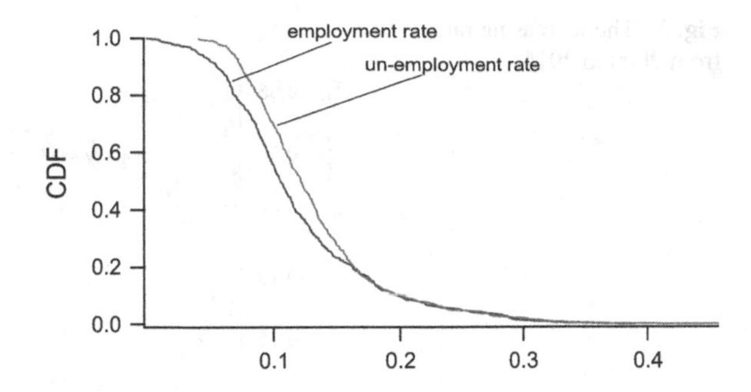

Fig. 2 The cumulative density distribution of the employment and separation rate from 2000 to 2014

accession rate and separation rate of a sector is almost equal but the magnitude and the (+, −) sign of the difference is different among sectors. For example, sector 28s accession rate and separation rate is equal so that the number of employee of this section is stable in spite that a part of employee is exchanged. Sector 35s separation rate is nearly 30 % but its accession rate is about 40 % so the number of employee of this sector is increasing in spite of this large separation rate. Figure 1 suggests that accession rate and separation rate should be treated as a pair.

Next the frequency of the both rates is verified. Figure 2 illustrates the cumulative density distribution of the both rates of all sectors from 2000 to 2014. The curve of separation rate starts about 4 % so at least 4 % of employees quit their job. Also the probability of both rates when they are smaller than 20 % is different. There is no difference if these rates are larger than this percentage. Therefore the sign of (accession rate–separation) is also important even if the difference is small.

3.1 *Introduction of Increasing Rate*

We introduce an index, increasing rate $z_{i,t}$ as follows. Increasing rate $z_{i,t}$ is

$$z_{i,t} = a_{i,t} - s_{i,t} \tag{1}$$

where i is the sector number and t is year. $a_{i,t}$ is the accessing rate and $s_{i,t}$ is the separation rate.

3.2 *Increasing Rate of This Period*

Figure 3 shows the progress of the increasing rate in the period. The x axis indicates a year. Each graph represents a time series of increasing rate $z_i = \{z_{i,t}\}$. Roughly speaking, these graphs synchronize but there are small differences. For example, after the Bankruptcy of Lehman Brothers the increasing rate of the almost of all

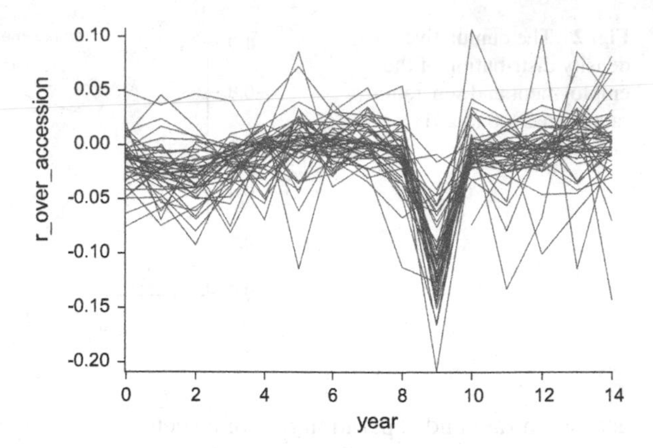

Fig. 3 The increasing rate from 2000 to 2014

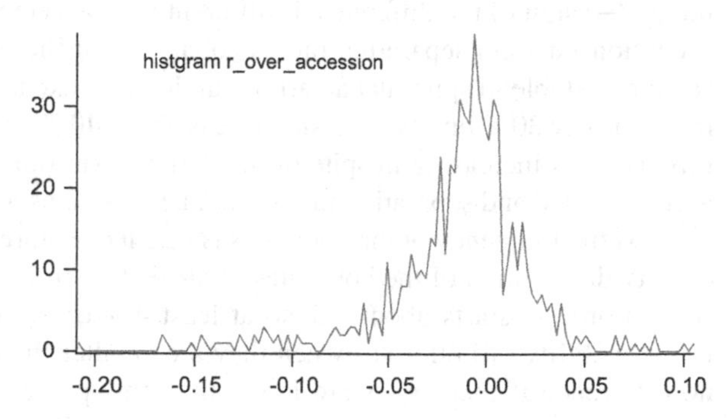

Fig. 4 Histogram of the increasing date from 2000 to 2014

sectors at $t = 2009$ are decreased but the magnitude is different. Also, in 2005, the sectors are equally divided into a group of positive z and a group of negative z.

The histogram of this increasing rate is shown by Fig. 4. The average is -0.013 and the standard deviation is 0.037. This seems to be a Gaussian distribution.

3.3 *Correlation of Increasing Rate*

Next, correlation of increasing rate of sectors are investigated. Pearson's correlation coefficient is adopted.

$$r_{i,j,k} = \frac{\sum_{i=1}^{n}(z_{i,t} - \bar{z}_i)(z_{j,t-k} - \bar{z}_j)}{((\sum_{i=1}^{n}(z_{i,t} - \bar{z}_i)^2)(\sum_{i=1}^{n}(z_{j,t-k} - \bar{z}_j)^2))^{1/2}} \quad (2)$$

where $k \geq 0$ is a time delay. Figure 5 shows the correlation matrix $r_{i,j,k=0}$ when $k = 0$. A black point indicates $r_{i,j} = -1$ and a yellow means $r_{i,j} = 1$. Simply speaking, almost of all sectors are positively correlated. However, for example, there are some negatively correlated sectors around sector 30–40.

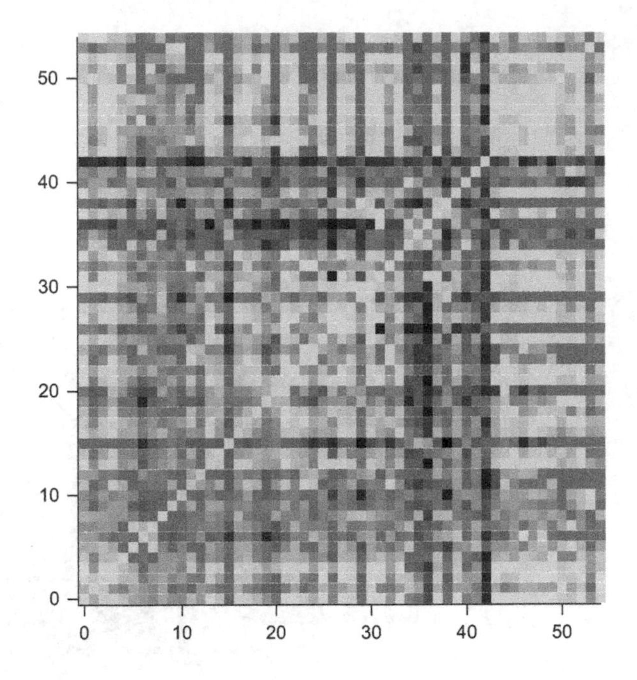

Fig. 5 Correlation matrix of increasing rate of sectors

3.4 Correlation Graph of Increasing Rate

Next, a correlation graph of the increasing rate is generated from the correlation matrix. This matrix is too complicated to understand its structure because this includes many strong negative relations. First weak correlations are omitted and only strong correlations are illustrated as link of graph. Figure 6 shows the correlation graph with only links which $|r_{i,j,k=0}| \geq 0.7$ which is the largest value still all of nodes are connected. A node represents a sector and a blue link is a positive strong relation and a red link indicates a strong negative link.

1 large group of sectors which have positive link each other is found. A few sectors have small correlation to others, for example, "medical, health care and welfare" sector have only 1 strong correlation link to "manufacture of general purpose machinery". It is reasonable because "medical, health care and welfare" sector uses products of "manufacture of general purpose machinery" and this is a kind of relation of a stream of production network [1].

Next, the correlation with time delay is calculated. By changing time delay $k = \{0, \ldots, 9\}$ we expect that a strong correlation with sectors of the past is detected. Figure 7 shows the result. The x axis is k. The y axis indicates the total correlation r_k which is calculated by

$$r_k = \sum_{i=0}^{n} \sum_{j=0}^{n} |r_{i,j,k}| \tag{3}$$

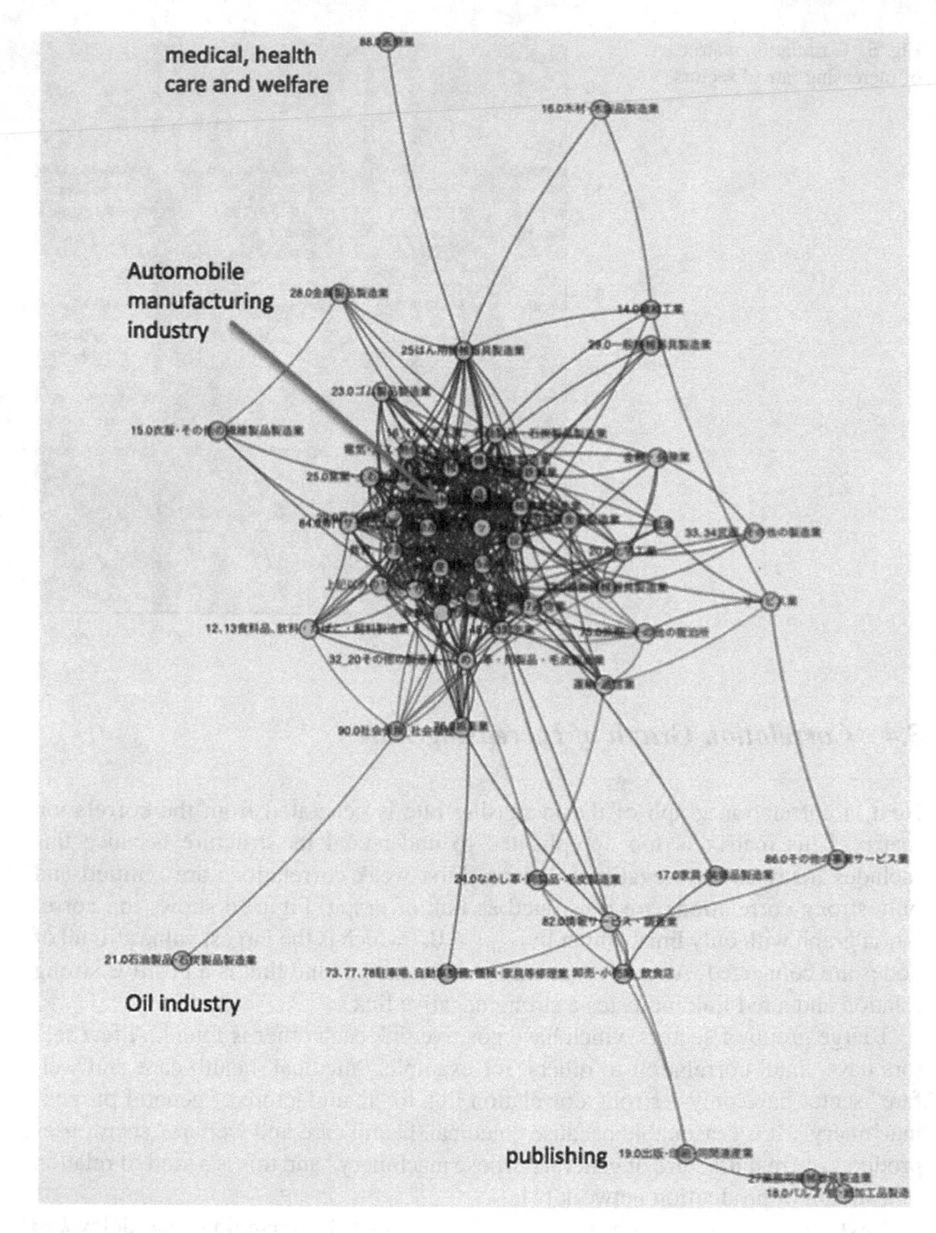

Fig. 6 Graphic representation of the correlation matrix

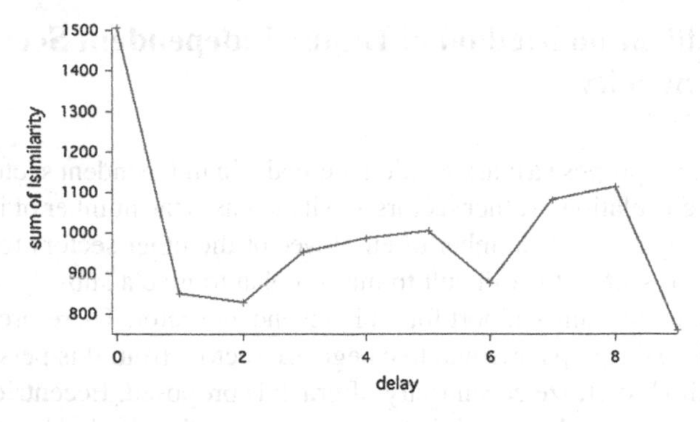

Fig. 7 Time delay and correlation

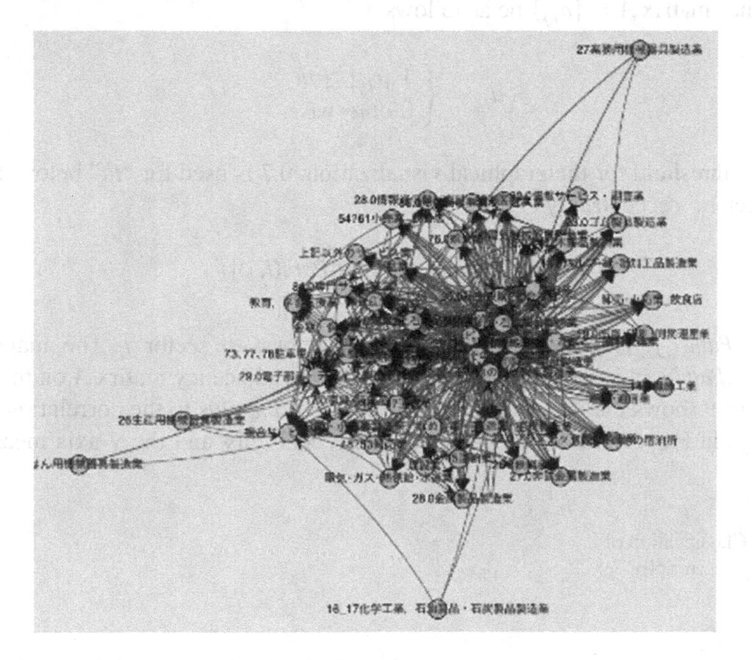

Fig. 8 The correlation graph when $k = 8$

$r_{k=8}$ is a highest correlation except $k = 0$. It suggests that Japanese industry makes a plan of employment in about 1 decade long perspective. Figure 8 illustrates the correlation graph of this delay. There seems to be 2 groups which have positive and negative influence respectively. The analysis of this graph is a future subject.

By this series of experiments we think that the increasing rate is a reasonable index for analysis of employment of the survey. In the correlation graph in $k = 0$ there is 1 large group and a few independent sectors, for example, "medical, health care and welfare" and "oil industry". Oil industry may strongly depend on oil price. Also we suppose it is natural that "medical" industry has not strong relations to other sectors.

4 A Classification Method of Highly Independent Sectors by Eccentricity

In this section, we propose a classification method of a independent sector. If a sector has a negative correlation to other sectors and it increases the number of its employee, this increase may force the number of employee of the other sectors to be reduced. In such a case of sector it is difficult to make a plan to give a subsidy. On the other hand it is easy to give any support for an independent sector. Therefore we suppose that it is worth to develop a method to categorize sectors from this perspective.

Here a method to utilize eccentricity of graph is proposed. Eccentricity of graph is an index representing how far it is from its center of its graph. Now let a graph's adjacency matrix $A = \{a_{i,j}\}$ be as follows

$$a_{i,j} = \begin{cases} 1 \ |r_{i,j}| \geq th \\ 0 \ otherwise. \end{cases} \tag{4}$$

th is the threshold for the graphical visualization. 0.7 is used for "*th*" below. In this case sector *i*'s eccentricity ec_i is

$$ec_i = \max_j(|ShortestPath(i,j)|) \tag{5}$$

ShortestPath(*i*,*j*) is the shortest path from sector *i* to sector *j*. The magnitude |*ShortestPath*(*i*,*j*)| is the sum of the element of the adjacency matirx *A* on the path.

Figure 9 shows the classification result that Eq. 5 applys to the correlation graph of $r_{i,j,k=0}$ in Fig. 6. The x axis indicates the eccentricity and the y axis means the

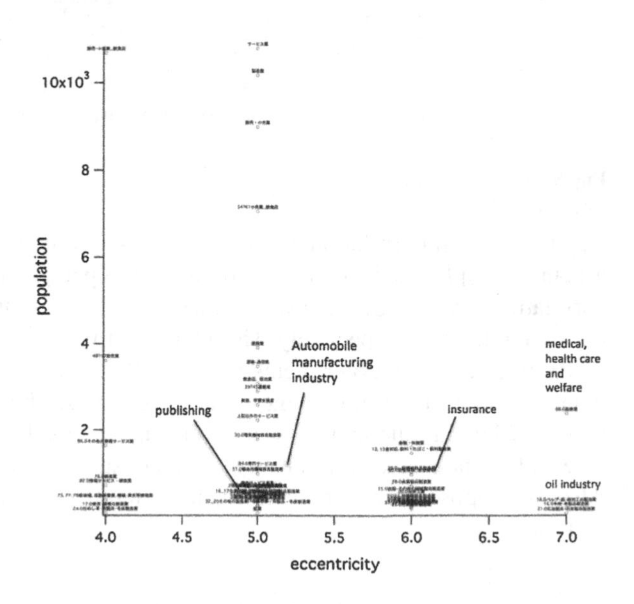

Fig. 9 Classification of sector by eccentricity

number of employee of the sector. The eccentricity of "automobile", "publishing", and "eating" industries is "5.0" which is minimum. On the other hand the eccentricity of "medical" and "oil industry" is "7.0" which is the largest. This index can classify sectors adequately.

5 Conclusion

In this paper, we investigate the survey on employment trends in Japan. This survey is an open data published by Health, Labor and Welfare Ministry of Japan. We introduce an index, increasing rate and the similarity of 55 sectors is analyzed. Consequently we found that there is 1 large group of sectors which are correlated each other and a few independent sectors. Finally we propose the method to classify the independent sectors and its performance is verified.

Reference

1. Carvalho, V.M.: From micro to macro via production networks. Journal of Economic Perspectives 28(4), 23–48 (2014)

On Deriving a Relationship Between Complexity and Fidelity in Rule Based Multi-agent Systems

Erandi Lakshika and Michael Barlow

Abstract Understanding the relationship between model complexity and fidelity in simulations is particularly important as increased complexity is one of the major cost drivers of any simulation. In this work we explore the relationship between complexity and fidelity in simple rule based multi-agent systems by employing a multi-objective evolutionary framework in two problem domains: (1) Simulation of conversational group dynamics (2) Simulation of sheepdog herding dynamics. Firstly, a new complexity measure is introduced to characterise complexity of the multi-agent systems. Thereafter the interplay between complexity and fidelity is analysed and the relationship is derived empirically by fitting the obtained data into functions that can describe the relationship in a compact and meaningful manner. This empirical study will be useful to develop theoretical understandings of the complexity and fidelity trade-off in multi-agent based simulations and the approach may be generalised to other simulation types.

Keywords Multi-agent simulations · Trade-off · Complexity · Fidelity · Multi-objective optimisation · Complex behaviours · Simulation · Model

1 Introduction

The importance of understanding the relationship of simulation model complexity and fidelity (measured in various forms) has long been discussed in the literature [1–5]. These works stress the importance of choosing the best model in terms of appropriate level of detail and complexity considering the objectives of the modelling process and available resources. This is particularly important because choosing a simple model may not satisfy the simulation objectives whereas a complex model could suffer resource constraints. Despite the importance of this relationship

---

E. Lakshika (✉) · M. Barlow
School of Engineering & IT, University of New South Wales, Canberra, Australia
e-mail: erasuru@gmail.com

G. Leu et al. (eds.), *Intelligent and Evolutionary Systems*,
Proceedings in Adaptation, Learning and Optimization 8,
DOI 10.1007/978-3-319-49049-6_16

of model complexity and fidelity, little theoretical or empirical work has been performed to elucidate the relationship.

Multi-agent based simulations are an effective form of simulation which facilitate modelling the complex behaviours of real world entities through simple rules codified into the agents. The decentralised nature, autonomous operation and ability to capture complex macro level emergent phenomena through micro level interactions make them desirable in simulating complex systems that are otherwise difficult to understand or simulate. Therefore, multi-agent based simulations are being extensively used in simulating complex systems and behaviours in numerous application areas ranging from engineering, biology, medicine to economics, social sciences, psychology and artificial intelligence. However, when it comes to multi-agent based simulations (referred as multi-agent systems in this paper interchangeably), little theoretical or empirical work has been undertaken to understand the complexity-fidelity interplay.

Previous work by the authors analysed the interplay of complexity and fidelity in multi-agent systems in three problem domains—simulation of conversational group dynamics, sheepdog herding dynamics and lane merge traffic dynamics—using a multi-objective evolutionary framework [5]. In particular, the work analysed the properties of the resulted Pareto frontiers [6] by simultaneously optimising the two objectives complexity and behaviour fidelity (look and feel of the resulted multi-agent behaviours) based on multiple individual complexity measures: rule count, parameter count and cyclomatic complexity [7]. Despite being abstract, the measures were highly useful in understanding the complexity fidelity trade-off in the three multi-agent systems that were analysed. In this work, we combine the two complexity measures rule count (number of rules in the multi-agent system) and parameter count (number of parameters codified in the rules used in the multi-agent system) in a novel fashion, overcoming the resolution limitations of using each measure alone and better understanding the complexity-fidelity trade-off. The resulted non-dominated solutions [6] are then used to describe the relationship of complexity and fidelity in a more compact and meaningful manner by finding the equation of the curve of best fit. This empirical understanding is useful in developing theoretical relationships between simulation model complexity and fidelity, not only in multi-agent based simulations, but also in other types of simulations.

2 Related Work

Often the guidelines for selecting the best simulation model is based on Occam's (Ockham's) razor- "*entities should not be multiplied beyond necessity*" [1, 2, 8]. A complex model is not preferred over a simpler model when both can satisfy the same objectives. However, the guidelines are vague and there is no clear methodology to understand model performance as the complexity increases. The notional relationship between model accuracy and level of detail (which is also referred to as complexity) presented in [8] shows that beyond a certain point little can be gained by adding more and more complexity to the model. Also, it states that 100 % accuracy is unattainable because it is impossible to know the exact detail of the real system. Further, beyond a certain point (of increasing complexity) the accuracy will decrease because no data or information is available for such a detailed model. However, the relationship is rather notional and no extensive experiments were reported.

The agent based approach has opened up new frontiers to model complex situations, advancing studies in the areas of physical, natural, social, military, information sciences and engineering [9]. According to [9], these trends showed that simulation model fidelity and complexity are expected to increase continuously and rapidly in the coming decades. However, the statement is broad and not based on any experimental outcome. Although, the growing range of applications and simulation requirements have imposed needs for high fidelity simulations, it is unlikely that the complexity of a system could be increased exponentially without an upper bound.

One of the main reasons for the continuing possibility of an increase in the complexity of simulations is the advancements in computational resources [2]. Moore's law [10] stated that the number of transistors per square inch on integrated circuits had doubled every year (later this became every two years) since the integrated circuit was invented and the trend will continue in the foreseeable future. This predicted trend has continued for about the last 50 years [11] resulting in corresponding advancements in computational power. However, this exponential rise will not continue indefinitely due to both economic limits and technology limits [11]. Therefore, it is unlikely that the simulation model complexity and fidelity can be increased without an upper limit. For example, Level-of-Detail AI (LOD AI) research promotes a simplification approach for characters in virtual environments, based on different criteria such as distance to the camera, in order to save computational power [12].

When it comes to multi-agent systems and agent based systems, empirical or theoretical studies on the relationship of model complexity and fidelity are particularly lacking. Even the existing works focus on specific application areas such as air traffic simulations [4]. However, given that multi-agent systems are used in many application domains, understanding the model complexity-fidelity interplay is vital. This is not only because increasing fidelity comes at the cost of complexity and computational cost, but also because it is one of the major cost drivers of any simulation [13].

In the context of this work complexity is defined as "*measures of those properties of a model that impose more resource requirements and consumption in terms of computation, design, testing or maintenance time and effort*" [5] and fidelity is defined as "*a measure of the perceived (by humans) naturalness and realism of the behavioural dynamics occurring in the virtual environment*" [5]. Previous work by the authors introduced a multi-objective evolutionary framework to explore the interplay of model complexity and simulation fidelity in the cases where fidelity depends on the quality of the emergent behaviours of the multi-agent systems [5]. This framework was proven to be successful in understanding the trade-off of model complexity and fidelity in three problem domains—simulation of conversational group dynamics, sheepdog herding dynamics and lane merge traffic dynamics. Firstly, three machine learning systems were trained using human evaluations of the quality of the behaviours in the three domains as well as the features used by humans to determine their scores. The ability of this machine learning approach to recreate human judgement was validated in [14]. These machine learning systems were integrated to the multi-objective evolutionary framework in order to automatically score the fidelity of the resulted behaviours at varying complexity levels. The three multi-agent systems consist of rules and their parameters (including the weights by which they are combined) which determine the complexity of the agents (the model). Four complexity measures—rule count, parameter count, cyclomatic complexity algorithmic level and cyclomatic complexity implementation level—were used to measure model complexity. Separate experiments were conducted for each problem domain by employing the Non-dominated Sorting Genetic Algorithm II (NSGA II) [15], simultaneously optimising the two conflicting objectives complexity and fidelity. Analysis of the solutions in the Pareto-frontier showed that initially rapid fidelity increases can be observed with increased complexity, however, beyond a certain point increasing complexity resulted in diminishing returns in fidelity.

Using the above mentioned complexity measures alone has a number of shortcomings. For example, the rule count complexity measure simply accounts for the number of rules in the multi-agent system without considering the parameters codified into the rules. Intuitively when a particular rule has more parameters that rule becomes more complex than a rule that has fewer parameters. On the other hand, the parameter count complexity measure simply accounts for the number of parameters in the multi-agent system without considering the relative importance of each parameter. For example, when rules are combined with their relative weightings, the impact of the parameters codified into a higher weighted rules are larger than the impact created by a parameter within a lower weighted rule. In this work a new complexity measure is introduced in order to address these shortcomings by considering the rule count, parameter count as well as the relative weighting of the rules. Further, the relationship of model complexity and fidelity will be analysed by fitting the non-dominated solutions to a function that can describe the relationship in a meaningful manner in two case studies, namely a conversational group dynamics simulation and a sheepdog herding simulation.

3 Case Studies

3.1 Multi-agent Based Conversational Group Dynamics Simulation

The first case study is based on a simple rule based multi-agent simulation of people engaged in a standing conversational group. The multi-agent system simulates the dynamics of conversational groups when new people join a conversational group as well as when participants leave a conversational group. 29 agents organised as 1 group of 8 agents, 2 groups of 6 agents, 1 group of 4 agents, 1 group of 3 agents, and

Table 1 The pool of rules of the conversational agents with their parameters

Rule	Parameter	Conceptual function
Keep Personal Distance - A	minPersonalDistance maxPersonalDistance	If another agent is within the personal distance move away
Keep Personal Distance - B	inertiaThreshold	If total repulsion forces is above the *inertiaThreshold* move away
Keep Centre of the Conversation	Tolerance	Keep the distance to the center of the group within a *tolerance*
Keep Visibility	wideArc *x%*	If not *x%* of agents are visible within a *wideArc* move away from center
	narrowArc *y%*	If more than *y%* of agents are visible within a *narrowArc* move towards center
Keep Distance to the Nearest Neighbour-A	thresholdDistance	Move towards the average position of the neighbours within a *thresholdDistance*
Keep Distance to the Nearest Neighbour-B	toleranceDistance	Move towards the furthest of the nearest neighbours within a *toleranceDistance*

another group of 2 agents have been simulated with the same scenario of two agents leaving groups and joining two other groups [5, 14]. The rule space comprised of 4 rules shown in Table 1. The conceptual rules, Keep Personal Distance and Keep Distance to the Nearest Neighbours (KDNN) have two variants (indicated as A and B) based on the choice of parameters and implementation, however in a given simulation only one implementation (either A or B) can be active. Rules are combined using a simple linear approach based on the weights, i.e.: the agents' movement in each simulation tick is determined by the weighted average of the active rules within a simulation.

3.2 Multi-agent Based Sheepdog Herding Simulation

The second case study is based on a sheepdog herding dynamics simulation [5, 16]. This simulation follows a simple rule based multi-agent approach to simulate the dynamics occurring when a flock of three sheep is herded by a herding dog. The flock and the dog are initially placed on the opposite ends of a circular arena. The dog then approaches the flock and herds them back to the dog's starting position as quickly and as directly as possible. The sheep in the flock apply the same set of

Table 2 The pool of rules of the sheep agents with their parameters

Rule	Parameter	Conceptual function
Separation	Separation Distance	If within *Separation Distance* move away
Alignment		Follow velocities of other sheep agents
Cohesion		Keep the centre of the herd
Repel from the dog-A	Speedup Parameter Repulsion Threshold	If total repulsion force is above *Repulsion Threshold* move away based on *Speedup Parameter*
Repel from the dog-B	Speedup Parameter Activation Distance	If dog within *Activation Distance* move away based on *Speedup Parameter*
Repel from the Walls	Activation Threshold	If closer to walls than *Activation Threshold* move away
Wandering	Wandering Angle Activation Threshold Speedup Parameter	Add variability to the motion within *Wandering Angle* based on *Activation Threshold* & *Speedup Parameter*

rules from a pool of 6 rules shown in Table 2.[1] These rules have different numbers of parameters that need to be tuned for realistic dynamics to occur. The rules are combined using a simple linear combination based on their weights.

4 Methodology

In order to analyse the relationship between simulation fidelity and complexity the same multi-objective evolutionary framework introduced in [5] is applied in this work. A machine learning algorithm trained based on bootstrapped human aesthetic judgement was employed to automatically score the fidelity of the simulations. In depth details of the machine learning systems can be found in [14, 16]. The following summarises the machine learning approach followed in training the sequential minimal-optimisation algorithm for learning a support vector regression model (SMOreg) [17] implemented in the Weka machine learning software [18].

Firstly, a representative set of simulations from each problem domain was selected by varying the number of active rules, their weights and parameter values. Thereafter human participants were recruited on a voluntary basis to determine the aesthetic quality of the behaviours. 33 and 20 volunteers were recruited to evaluate the

[1]The *repel from the dog* rule has two different implementations based on the parameter choices and these two implementations are mutually exclusive in a given simulation.

conversational group dynamics simulations and sheepdog herding dynamics respectively. In each experiment (one experiment per problem domain), firstly the participants were shown videos and pictures of scenarios in real life and asked to list the feature that they thought as important to determine the realism (fidelity) of simulations of such scenarios. Thereafter, they were shown simulations (each simulation as a video) in a randomised order and asked to score the simulation for their realism in a range of 0 to 9 (0 being not realistic at all, and 9 being as realistic as the real world). Their reason for the score was also obtained as a written response.

Based on the scores provided by the human volunteers and the features they listed the SMOreg machine learning algorithm was trained to automatically evaluate the unseen simulations. Based on leave one-out-cross validation results for 1295 training samples, the machine learning system trained for the conversational group dynamics model showed a 0.88 correlation between the predicted and the actual scores and a 0.86 root mean squared error (in a range of 0–9) whereas for the sheepdog herding dynamics model the correlation was 0.98 and the root mean squared error was 0.49 for 420 training samples. Pilot experiments were used to validate the machine learning scores with the human scores as explained in [14, 16].

A series of complexity measures that can be used to characterise the model complexity of multi-agent systems is discussed in [5]. In this work, a new complexity measure is introduced in order to better understand the complexity fidelity tradeoff by combining two complexity measures used in [5] in order to overcome the limitations. The number of rules used to codify the agents has a significant impact on the complexity of the multi-agent system because an agent codified with two rules requires more computational resources than an agent codified with a single rule. As these rules are sensitive to the parameters encoded within the rule, rules with more parameters require more resources for calibrating them for desired behaviours. Similarly, if a particular rule has a higher weight, its parameters arguably have a higher impact on the resulted behaviours. As such using either rule count or parameter count in isolation conceals some aspects of complexity which is multi-dimensional. The complexity measure that is used to examine the relationship of model complexity and fidelity in this work is defined in Eq. (1) where; R is the number of rules agents are codified with (e.g.: if the sheepdog behaviours are codified with the rules separation and cohesion, $R = 2$), i = ith rule in the simulation, w_i = weight of the ith rule and p_i = number of parameters in the ith rule (for example, if the sheepdog behaviours are codified with the rules separation and cohesion $p_1 = 1, p_2 = 0$).

$$Complexity = R + \sum_{i=1}^{R} .w_i.p_i \tag{1}$$

NSGA II [15] was employed to understand the relationship of model complexity and fidelity by simultaneously optimising the two conflicting objectives low complexity and high fidelity. The machine learning systems described above were employed to determine the fidelity of the simulations generated at the varying complexity levels. A population size of 100 was used for the conversational group dynamics model and a population size of 30 was used for the sheepdog herding dynamics model. The chromosome was encoded with both Boolean and real values. Boolean genes specified if a rule is active or not while real value genes were used to represent rule weights and parameters. Mutation rates of 0.1 and 0.01 were applied for the sheepdog herding model and conversational group dynamics model respectively. A crossover rate of 0.9 was applied for both models and 7 evolutionary runs for each model were conducted. Since no significant differences between the runs were observed 7 evolutionary runs were sufficient to understand the properties of the resulted non-dominated solutions. All the evolutionary parameters were determined based on pilot studies.

5 Experimental Results

Figure 1a shows all non-dominated solutions generated by the 7 evolutionary runs for the conversational group dynamics domain and Fig. 1b demonstrates all non-dominated solutions generated by the 7 evolutionary runs for the sheepdog herding dynamics domain. Generally, in both problem domains the shape of the Pareto-frontier remains the same showing initial rapid fidelity increases as complexity increases and thereafter diminishing returns. This confirms the results presented in [5]. One of the main differences observed between the Pareto-frontiers obtained in the two problem domains is the initial starting point which varies due to the unique features in each problem domain and their starting configuration. Since it is difficult to analyse all the non-dominated solutions, the next sections analyse the centroid solutions from important regions in the Pareto-frontier.

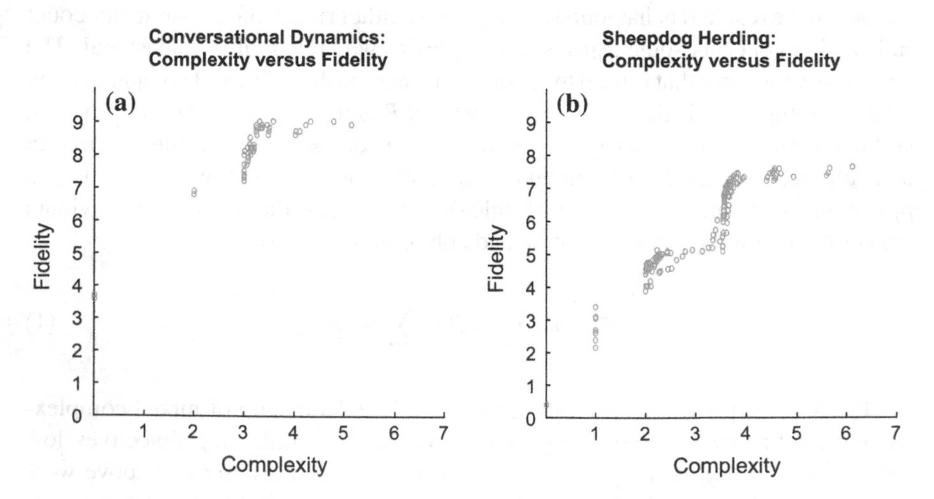

Fig. 1 Complexity versus fidelity: **a** Conversational group dynamics. **b** Sheepdog herding dynamics

In the conversational group dynamics domain, the first non-dominated solution (complexity = 2, fidelity ≅ 6.8) is the rule KDNN (both versions were present in the Pareto-frontier). The next important combination is the rules Keep Centre of the Conversation and KDNN-B (complexity = 3, fidelity ≅ 7.5). Thereafter, the fidelity gain is marginal as the complexity increases. The solutions clustered around fidelity 8–8.5, show a complexity of around 3.3 with the two active rules being Keep Visibility and KDNN-B with a significantly higher weight for the rule KDNN-B. The cluster around fidelity 8.5–9 show a complexity of around 4.3 and consists of rules Keep Centre of the Conversation, Keep Visibility and KDNN-B with approximate relative weightings of 0.4, 0.4, 0.2 respectively. The most complex solution in the Pareto-frontier consists of the rules Keep Personal Distance-A, Keep visibility and KDNN-B. Relative weightings between the rules are 0.2, 0.3 and 0.5 respectively. The results are different from the non-dominated solutions obtained by optimising the most abstract rule count complexity measure [19, Chap. 7], however similar to the non-dominated solutions generated by the parameter count complexity measure [19, Chap. 7].

Unlike the conversational group dynamics model, the sheepdog herding dynamics model has rules with no parameters. The first cluster of the non-dominated solutions consists of the rule Cohesion[2] with a complexity of 1 which plays a significant role in generating flocking dynamics. The centroid solutions around complexity 2 consist of the rule combination Cohesion and Repel from the dog-A. The analysis of the solutions clustered between complexity 2 and 3, shows that as the weight for the Repel from the dog rule increases fidelity also increases. This is due to the fact that the rule Repel from the dog is the prominent rule which makes the herding dynamics occur. The centroid solutions clustered around complexity 3.5–4 consist of the three rules Cohesion, Separation and Repel From the Dog. Similarly to the previous cluster, fidelity increases as the weight for the rule Repel from the dog increases. After complexity reaches 4, the fidelity gain becomes marginal. The centroid solutions in the cluster around complexity of 4.5 consist of the rule combination Cohesion, Separation, Repel From the Dog and Alignment. The most complex solutions consist of the rules Cohesion, Separation, Repel From the Dog, Alignment and Wandering, however there was no significant fidelity improvement that could be observed by adding the extra rule. The results are quite similar to the results obtained by using the rule count and parameter count complexity measures in isolation [19, Chap. 7] except that when using the rule count complexity measure the first non-dominated solution was the rule Repel From the Dog and there were no 5-rule solutions in the Pareto-front. When applying the parameter count complexity measure, the extra rule added after the rapid fitness increase was the rule Repel from the Walls instead of Wandering. Generally, after reaching a certain complexity level to get important dynamics of the simulation, the addition of extra rules only added insignificant fidelity improvements.

---

[2]Fidelity values show minor variations between the runs due to different seeds fed into the starting configuration.

Table 3 Curve fitting: conversational group dynamics model

Function	Fit	RMSE	R^2
Exponential decay	$F = 10.7(1 - 0.7e^{-0.3x})$	0.38	0.89
Logistic growth	$F = 9.7/(1 + 1.7e^{-0.7x})$	0.38	0.90
Richard's curve	$F = 8.8/(1 + 212.3e^{(-52.0(x-3.3))})^{(1/212.3)}$	0.31	0.93

Table 4 Curve fitting: sheepdog herding dynamics model

Function	Fit	RMSE	R^2
Exponential decay	$F = 10.4(1 - 1.0e^{-0.3x})$	0.46	0.89
Logistic growth	$F = 8.2/(1 + 4.6e^{-0.8x})$	0.45	0.90
Richard's curve	$F = 8.4/(1 + 0.4e^{(-0.7(x-1.6))})^{(1/0.4)}$	0.45	0.90

In order to describe the data presented in Fig. 1 in a compact and meaningful manner that captured the key attributes (such as shape) of the relationship, a series of functions were considered by looking at the shape of the data distribution (in order to define fidelity $F_{(x)}$ as a function of complexity x). Obviously, the relationship does not appear to be linear and such a linear curve fitting is not useful to understand the relationship. An exponential curve $F_{(x)}$ increases without a bound as x increases which is not the case in the data obtained. A log curve in the form $f_{(x)} = a + b.log(r + x)$ would be appropriate to represent the decay of growth, however, such a curve does not have an upper-bound.

After considering a series of functions the most appropriate functions were exponential decay in the form $F_{(x)} = a(1 - be^{-r.x})$ (upper-bound represented by a), a logistic growth function in the form $F_{(x)} = c/(1 + a.e^{-r.x})$ (upper-bound represented by c, growth rate represented by r) or a more generalised logistic growth function know as the Richard's curve [20] as shown in Eq. (2) where c is the upper-bound, r is the growth rate, x_0 is the point of inflection and a is a parameter that determines the asymmetry of the curve. According to the Goodness of fit (R^2) and Root Mean Squared Error (RMSE) shown in Table 3 the Richards curve appears to be the most appropriate function to describe fidelity as a function of complexity in the conversational group dynamics model ($R^2 = 0.93$ and RMSE $= 0.31$), however both logistic function and the Richard's curve showed similar performance (See Table 4) for the sheepdog herding dynamics model ($R^2 = 0.90$ and RMSE $= 0.45$).

$$F_{(x)} = c/(1 + ae^{-r(x-x_0)})^{1/a} \tag{2}$$

Figures 2 and 3 demonstrate Richard's curve fittings for the data obtained in the conversational group dynamics domain and sheepdog herding dynamics domain respectively. Both curves capture an initial rapid fitness increases in response to increased complexity, beyond a certain point increased complexity only resulted in marginal fidelity improvements. In the conversational group dynamics domain, the

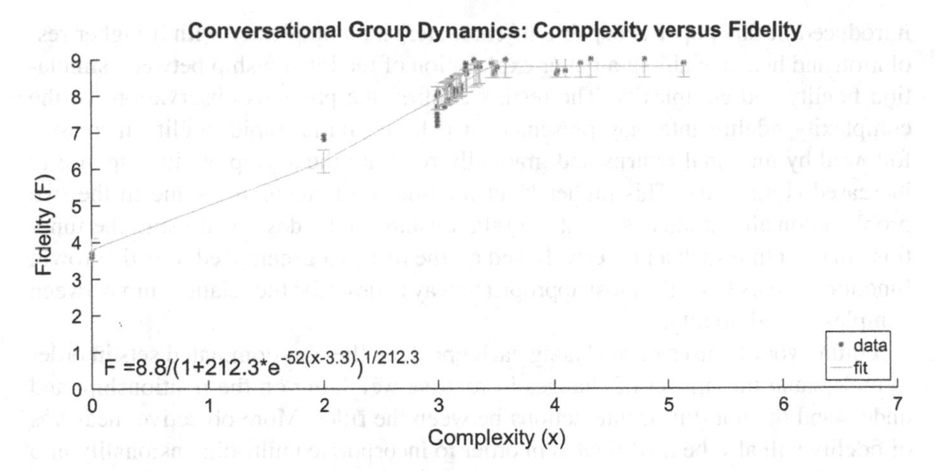

Fig. 2 Richard's curve fitting: conversational group dynamics

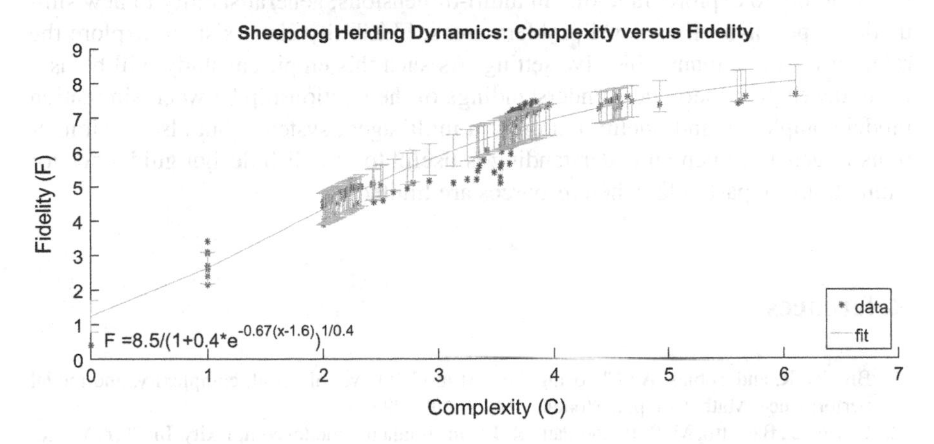

Fig. 3 Richard's curve fitting: sheepdog herding dynamics

asymptote was nearly 9 and the point of inflection was about 3.3 units of complexity whereas in the sheepdog herding dynamics domain the asymptotes was nearly 8.4 and the point of inflection was about 1.6 units of complexity. It appears that the relationship between simulation fidelity and complexity can be described by the curve types that can model diminishing returns, however the growth rate, y-intercept and the point of inflection depend on the unique properties of the problem domains.

6 Conclusion and Future Work

In this paper a new complexity measure to measure complexity of the rule and parameter space of multi-agent systems was introduced. In contrast to the discrete complexity measures presented in the previous work [5], the novel complexity measure introduced in this paper is capable of characterising complexity with a higher resolution and hence enabling a better exploration of the relationship between simulation fidelity and complexity. The results confirm the previous observations of the complexity fidelity interplay presented in [5], i.e.: initial rapid fidelity increases, followed by marginal returns and gradually reaching the asymptote in response to increased complexity. This higher level relationship remains the same in the two problem domains studied, so that the relationship can be described using the functions that capture such a property. Based on the results a generalised logistic growth function appears to be the most appropriate way to describe the relationship between complexity and fidelity.

Future work focuses on analysing each epoch on the non-dominated sets in order to understand the impact of changes in relative weighting on the relationship, and understand the non-linear interactions between the rules. More objective measures of fidelity will also be investigated in order to incorporate multi-dimensionality into the relationship. The power of the multi-objective framework used in this work is its capability to explore trade-offs in multi-dimensions, generalisability to new simulation types, as well as novel problem areas. Flexibility also exists to explore the relationships in a many-objective setting. As such this empirical study will be useful in developing theoretical understandings of the relationship between simulation model complexity and fidelity not only in multi-agent systems, but also in simulations in general. Such an understanding is useful to establish design guidelines for simulations; in particular when resources are limited.

References

1. Brooks, R. and Tobias, A.: Choosing the best model: Level of detail, complexity, and model performance. Math. Comput. Model, 24(4), 1–14 (1996)
2. Chwif, L., Barretto, M. R. P., and Paul, R. J.: On simulation model complexity. In: 2000 Winter Simulation Conference, volume 1, pp. 449–455. IEEE (2000)
3. Robinson, S.: Tutorial: Choosing what to model—conceptual modeling for simulation. In: 2012 Winter Simulation Conference (WSC), pp. 1–12. IEEE (2012)
4. Amin, R., Tang, J., Ellejmi, M., Kirby, S., and Abbass, H. A.: Trading-off simulation fidelity and optimization accuracy in air-traffic experiments using differential evolution. In: 2014 IEEE Congress on Evolutionary Computation, pp. 475–482. IEEE. (2014)
5. Lakshika, E., Barlow, M., and Easton, A.: Understanding the interplay of model complexity and fidelity in multi-agent systems via an evolutionary framework. IEEE Trans. Comput. Intell. AI in Games, doi:10.1109/TCIAIG.2016.2560882, IEEE (in press)
6. Konak, A., Coit, D. W., and Smith, A. E.: Multi-objective optimization using genetic algorithms: A tutorial. Reliab. Eng. Syst. Safe., 91(9), 992–1007 (2006)
7. McCabe, T. J.: A complexity measure. IEEE Trans. Softw. Eng., 2(4), 308–320 (1976)
8. Robinson, S.: Conceptual modelling for simulation part I: definition and requirements. J. Oper. Res. Soc., 59(3), 278–290 (2007)
9. Yilmaz, L., ren, T., and Aghaee, N.-G.: Intelligent agents, simulation, and gaming. Simul. Gaming, 37(3), 339–349 (2006)
10. Moore, G. E.: Cramming More Components onto Integrated Circuits. Electronics, 38(8), 114–117 (1965)

11. Mack, C. A.: Fifty Years of Moores Law. IEEE Trans. Semicond. Manuf., 24(2), 202–207 (2011)
12. Brom, C., Ser'y, O., and Poch, T.: Simulation level of detail for virtual humans. In: 7th international conference on Intelligent Virtual Agents, pages 1–14. Springer (2007)
13. Gross, D. C. et al.: Report from the fidelity implementation study group. In: Fall Simulation Interoperability Workshop Papers. (1999)
14. Lakshika, E., Barlow, M., and Easton, A. : Fidelity and complexity of standing group conversation simulations: A framework for the evolution of multi agent systems through bootstrapping human aesthetic judgments. In: 2012 IEEE Congress on Evolutionary Computation (CEC), pp. 1044–1051 (2012)
15. Deb, K., Pratap, A., Agarwal, S., and Meyarivan, T.: A fast and elitist multiobjective genetic algorithm: Nsga-II. IEEE Trans. Evol. Comput., 6(2):182–197 (2002)
16. Lakshika, E., Barlow, M., and Easton, A.: Evolving high fidelity low complexity sheepdog herding simulations using a machine learner fitness function surrogate for human judgement. In: Australasian Joint Conference on Artificial Intelligence, pp. 330–342. Springer (2015)
17. Smola, A. J. and Scholkopf, B.: A tutorial on support vector regression. Stat. Comput., 14(3):199–222 (2004)
18. Witten, I. H. and Frank, E.: Data Mining: Practical Machine Learning Tools and Techniques, Second Edition (Morgan Kaufmann Series in Data Management Systems). Morgan Kaufmann Publishers Inc., San Francisco, CA, USA (2005)
19. Lakshika, E.: An exploration of the complexity-fidelity trade-off in multi-agent systems via an evolutionary framework. Phd thesis, Engineering & Information Technology, University of New South Wales, Canberra, Australia (2014)
20. Richards, F. : A flexible growth function for empirical use. J. Exp. Bot., 10(2):290–301 (1959)

Feature Construction Using Genetic Programming for Figure-Ground Image Segmentation

Yuyu Liang, Mengjie Zhang and Will N. Browne

Abstract Figure-ground segmentation is a process of separating regions of interest from unimportant backgrounds. For complex images (e.g. images with high variations), feature construction (FC) is necessary, which can produce high-level features, to help achieve accurate segmentation performance. Genetic programming (GP) is considered as a well-suited FC technique, which is employed for the first time to build FC methods that aim to improve the segmentation performance in this paper. One filter GP method (FGP), in which a novel entropy based fitness function is developed, and one embedded GP method (EGP), in which the error rate is used as the fitness function, are proposed. The single constructed feature and the combined features (the constructed feature + original features) are tested on two standard image datasets with high variations, i.e. Weizmann and Pascal datasets. Compared with the original features extracted by existing feature descriptors, both methods can construct useful features from the original ones with the combined features improving the segmentation performance on both datasets generally. Moreover, EGP is more efficient and perform better than FGP.

Keywords Feature construction · Genetic programming · Fitness function · Figure-ground image segmentation

---

Y. Liang (✉) · M. Zhang · W.N. Browne
School of Engineering and Computer Science, Victoria University of Wellington,
P.O. Box 600, Wellington 6140, New Zealand
e-mail: yuyu.liang@ecs.vuw.ac.nz

M. Zhang
e-mail: mengjie.zhang@ecs.vuw.ac.nz

W.N. Browne
e-mail: will.browne@ecs.vuw.ac.nz

G. Leu et al. (eds.), *Intelligent and Evolutionary Systems*,
Proceedings in Adaptation, Learning and Optimization 8,
DOI 10.1007/978-3-319-49049-6_17

1 Introduction

Feature construction (FC) is the process of transforming a set of original features to a new set of more powerful features [1]. FC is crucial for many image analysis tasks, such as image recognisation, classification and segmentation [2]. Considering figure-ground image segmentation as an example; it is the process of separating regions of interest from backgrounds, thus producing binary result images. As there exist images with high variations, e.g. containing cluttered backgrounds or varying object shapes, a good feature space is necessary to capture difference between object pixels and background ones, which is the prerequisite to achieve accurate segmentation performance. Even though many image descriptors exist, e.g. Gabor filters, they are hand-crafted techniques and often do not match the underlying patterns in the image, thus can not provide effective feature sets [3]. Moreover, the optimal feature representation is often not clear for a given problem, and it is infeasible to try all possible features [3]. Therefore, it is important to conduct feature construction on the features extracted by existing descriptors to produce high-level features with more identifying information of objects.

According to how the constructed new features are evaluated in the feature construction process, existing methods can be divided into wrapper, filter and embedded approaches [4]. Wrapper methods use an inductive algorithm, e.g. classifiers, to evaluate the new features; while filter methods depend on general characteristics of training data rather than the feedback of an inductive algorithm. The embedded approach is similar to the wrapper approach, which is directed by an inductive algorithm for the feature evaluation, but the inductive algorithm is the learning algorithm itself. Therefore, embedded methods construct features and build a learning model in one step, while filter and wrapper methods treat them as two separate steps.

GP has been applied to feature construction by existing works [1, 5–7]. Guo et al. [6, 7] develop two filter FC methods using GP for identification problems. In work [6], the Fisher criterion is applied as the fitness measure, and the created features are tested by artificial neural networks (ANNs) and support vector machines (SVMs) for the identification of six bearing conditions. In work [7], a modified Fisher linear discriminant analysis (MFLDA) is developed as the fitness function. Multi-layer perceptrons and SVMs are used to measure the constructed features. In both works [6, 7], the constructed features improve classification performance compared with the original features. Bishop et al. [1] presents a wrapper GP based method for image classification. The fitness function is the classification accuracy. The constructed features improve the classification results significantly over the manually designed features, e.g. features that measure the image aesthetic value in Datta's work [8]. Ahmed et al. [5] propose an embedded GP based method for classification and biomarker identification of mass spectrometry data. The fitness function combines Fisher criterion and p-values between different classes. Results show that the constructed high-level features outperform the original feature set in the classification performance and biomarker detection rate. The sucess of these existing works lies in GP's ability of constructing and searching for more discriminative features, and the proper settings of GP (e.g. an effective fitness function).

Even though promising results have been achieved by existing works, there are still several issues in current GP based FC methods [1, 5–7]. Firstly, existing approaches have been tested mainly on detection and classification problems, other tasks (e.g. segmentation problems) have not been extensively investigated. Secondly, the GP-based methods often face the problem of high computational costs due to the fitness evaluation of each individual (solution). Considering wrapper methods are normally more computationally intensive, filter and embedded GP methods will be investigated for feature construction in this work. Thirdly, it is still challenging to design an effective fitness function to measure the goodness of the constructed features, especially for filter methods.

1.1 Goals

This paper employs GP for the first time to construct high-level features from low (raw features, e.g. intensity)/middle-level features (processed features, e.g. Gabor features) extracted by existing descriptors for segmentation tasks on complex images (e.g. images with high variations). One filter GP method (FGP) and one embedded GP method (EGP) are investigated to evolve feature construction programs. In FGP, a novel entropy based fitness function is designed. The constructed features are expected to improve the segmentation performance. Specific objectives are shown as follows.

1. whether the constructed features can outperform the original features for the given segmentation tasks;
2. which one of the proposed methods, FGP or EGP, can construct more effective features;
3. how to interpret the constructed features to reveal their effectiveness in distinguishing object and background pixels.

As the background to this work has been covered above, Sect. 2 introduces the baseline methods. Section 3 describes the two new GP based feature construction methods. In Sect. 4, the experiment preparation is described. Section 5 discusses the results. Conclusions and future work are shown in the Sect. 6.

2 Baseline Methods

2.1 Pixel Classification Based Segmentation with Feature Construction

Figure 1 displays the framework of the pixel classification based figure-ground segmentation, which includes the process of feature construction. Firstly, an equal number of object and background pixels are captured from training images. Low/middle-level features are extracted for these pixels along with their class labels (class *object* or *background*) to form the training set. Accordingly, the same type of features are extracted for each pixel in test images. Secondly, the training set is input to a proposed GP based method to evolve feature construction functions, among which the single best function is returned (details are described in Sect. 3). Based on this feature construction function, each sample (a feature vector) in the training set and the test set is converted to a single new feature. This single constructed feature f_c

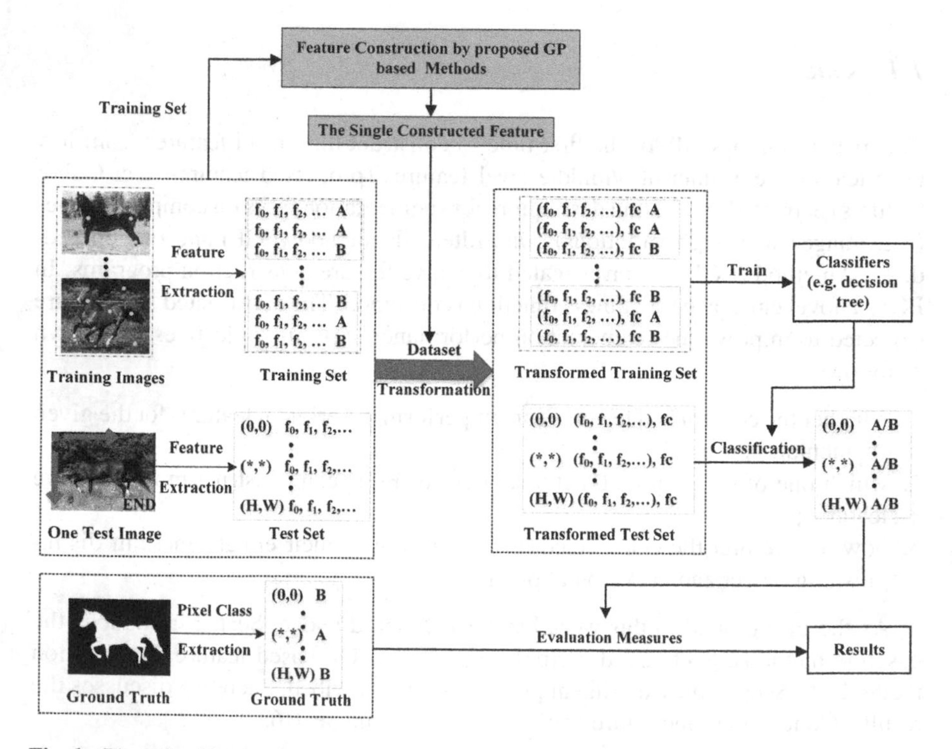

Fig. 1 Pixel classification based figure-ground segmentation with feature construction (f_0 is a feature with 0 as its index and so forth; (f_0,f_1,f_2, ...) is a feature vector extracted by existing feature descriptors; f_c is the single constructed feature; A and B represent class *object* and *background* respectively; (0, 0), (*, *) and (H, W) represent the pixel coordinates; H and W are the test image's height and width)

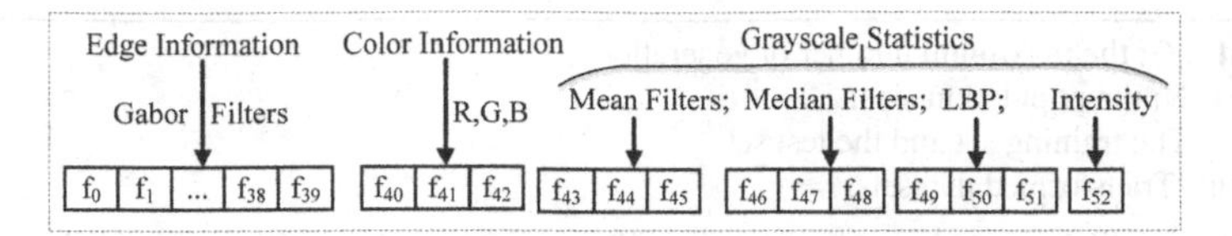

Fig. 2 Low/middle-level features (f_0 is a feature with 0 as its index and so forth)

can be used by itself and can also be combined with the original features to form the transformed training and test sets. Thirdly, the transformed training set is employed to train a standard classifier, e.g. decision tree, which is then applied to classify the transformed test set. Based on certain evaluation measures and the ground truth, the segmentation results are generated.

2.2 Low/Middle-Level Features

Existing feature descriptors are used to extract low or middle-level features, which are then input into the proposed FC methods to construct high-level features. In this paper, several standard feature descriptors are employed, which can extract three general types of image features, i.e. edge, colour and grayscale statistics features. As shown in Fig. 2, there are Gabor filters to extract edge information; R, G, B colour features; mean filters, median filters, local binary patterns (LBP) and intensity values to provide grayscale statistics information. All the extracted features are normalized into the value range of [0, 1].

3 Novel GP Based Feature Construction Methods

This paper employs GP to construct a single high-level feature from the low/middle-level features based on the best individual in each GP run. Even though a GP individual (tree) has many sub-branches, only the output of the root node is considered for the single feature construction, as it can contain multiple feature combinations. Algorithm 1 shows the process of GP based feature construction. The parts in blue colour are the standard procedures in GP algorithm. The parts in red colour are introduced to GP to conduct feature construction, which are mainly about two aspects, i.e. the fitness function to evaluate the generated single features in the evolutionary process and the normalization of the best feature to [0,1]. As the original features are in [0,1], the feature normalization is necessary to make the constructed feature compatible with the original features.

```
input : G: the maximum number of generations;
        N: the population size;
        The training set and the test set.
output: Transformed datasets.
1  Create an initial population at iteration zero (P_0) of GP trees using the Ramped half-and-half
   method;
2  g ← 0
3  while g < G and the ideal individual (the individual with fitness value == 0.0) is not found
   do
4      Create a child population P_{g+1} from P_g:
5      begin
6          Evaluate each individual of P_g based on the proposed fitness function and assign it
           a fitness value;
7          for i = 0 to N-1 do
8              Select parent(s) by tournament selection from P_g;
9              Generate children by mutation, crossover or elite operation;
10             Add the children to P_{g+1};
11         end
12     end
13     g ← g + 1;
14 end
15 Get the best-so-far individual (feature construction function);
16 for each sample in the training set and the test set do
17     Feed the sample to the feature construction function;
18     Get the output of the function;
19     Convert the output to a value within [0,1] based on the sigmoid function;
20     The transformed value is regarded as the constructed feature for this sample.
21 end
22 Return the transformed training set and the transformed test set.
```

Algorithm 1: Pseudo-code of GP based feature construction method (GP procedure is in blue; feature construction related parts are in red).

Both the filter and the embedded GP methods are investigated for feature construction in this paper. The major difference between the filter and the embedded methods lies in the fitness function, which is described in Sect. 1. A novel fitness function is designed for the filter GP method; while the classification error rate is used as the fitness function for the embedded GP method. They will be described in detail in Sects. 3.1 and 3.2.

Considering that the constructed features may contain outliers, data normalization (or feature scaling) is required. Since the sigmoid function is monotonic and can transform numeric values into the range of [0, 1], it can conduct feature normalization while preserving the distribution of original data. Therefore, the sigmoid function is used to normalize the constructed features.

3.1 A Novel Fitness Function for the Filter GP Method

As this work transforms the segmentation task to a pixel classification based problem, a "good" feature should be the one that can separate pixels of two classes, i.e. class *object* and *background*, by projecting them in the feature space. As shown in Fig. 3a, pixels from two classes are commonly overlapping. For example, the interval (the range of feature values) of class *background* contains object pixels. Since the less overlapping (or less impure) a class interval means better separation, the goodness of constructed features can be reflected by the impurity of the class intervals [9]. Therefore, a novel fitness function is proposed for the filter GP method, which aims to minimize the impurity of the class intervals. Considering that there may exist outliers, which are noisy samples, 0.5 % (determined empirically) of its total pixels on both minimum and maximal edges are removed from each class interval (illustrated in Fig. 3b).

As a popular entropy measure, Shannon entropy [10] is employed in this work to measure the impurity of the class intervals. Shannon entropy is described as follows. Given discrete probability distributions $\vec{p} = p_1, p_2, ..., p_N$, where the probabilities are all non-negative numbers and $\sum_{i=1}^{N} p_i = 1$, the Shannon entropy of $\vec{p}$ is shown in Eq. 1.

$$H(\vec{p}) = -\sum_{i=1}^{N} p_i * lnp_i. \tag{1}$$

Based on the Shannon entropy, the impurity of one class interval can be calculated by Eq. 2, where C is the set of class labels, Ic represents the interval in the feature space for a specific class c ($c \in C$), and $H(Ic)$ means the Shannon entropy of the interval Ic. Specifically, given a set of training samples (feature vectors with class labels), they are transformed based on the evolved feature construction func-

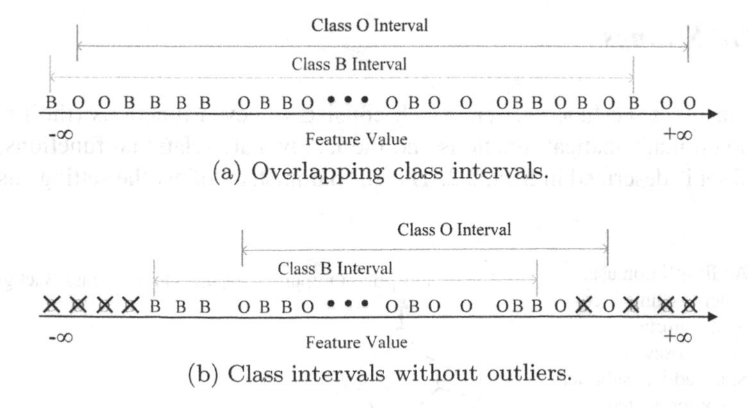

(a) Overlapping class intervals.

(b) Class intervals without outliers.

Fig. 3 Pixel distribution in the feature space (O and B represent class *object* and *background* respectively; × means removing the corresponding pixels)

tion. Then in the constructed feature space, for a certain class c, the class interval Ic can be determined. $NPIXS_{Ic}$ refers to the number of pixels that fall into the interval Ic; while $NPIXS_{i;Ic}$ refers to the number of pixels whose class labels are class i ($i \in C$) and who fall into the interval Ic.

$$H(Ic) = -\sum_{i \in C} p(NPIXS_{i;Ic}|NPIXS_{Ic}) * ln\, p(NPIXS_{i;Ic}|NPIXS_{Ic}). \qquad (2)$$

As in this paper the problems are binary classification problems, we define $C = \{c_1, c_2\}$, $Ic = \{Ic_1, Ic_2\}$. The fitness function is defined as the mean of the two class intervals' entropies. A lower fitness value means a better separation of two classes, which suggests a better constructed feature.

$$fitness = \frac{H(Ic_1) + H(Ic_2)}{2}. \qquad (3)$$

3.2 A Fitness Function for the Embedded GP Method

A GP solution can be used as not only a feature construction function, but also a classifier, the commonly-used error rate based on the evolved solution is selected as the fitness function for the embedded GP method. Specifically, Fig. 4 shows an evolved solution, which can be used as a binary classifier by comparing the output of its root node with a threshold 0 (other thresholds are also suitable). If the output is negative, the pixel is categorized as class *object*; otherwise class *background*. Based on the classification results of training samples, $error\ rate = \frac{\text{The number of wrongly classified samples}}{\text{The total number of training samples}}$ can be generated.

3.3 GP Settings

Table 1 displays the function set, which consists of seven functions (the first five are standard mathematical functions and the last two are relational functions). The terminal set is described in Sect. 2.2. The GP parameters follow the settings used by

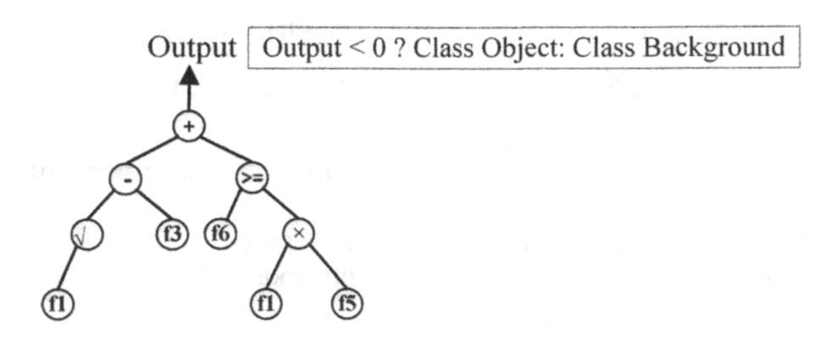

Fig. 4 A GP solution used as a classifier (assume there are six input features $f_1, f_2, \ldots, f_6$, and seven operators, i.e. add +, subtract −, multiply ×, protected divide %, >=, <=, $\sqrt{}$)

Table 1 Function set

Function name	Definition	Function name	Definition
$+(a_1, a_2)$	$a_1 + a_2$	$-(a_1, a_2)$	$a_1 - a_2$
$*(a_1, a_2)$	$a_1 * a_2$	$\%(a_1, a_2)$	$\begin{cases} a_1/a_2 & \text{if } a_2 != 0 \\ 0 & \text{if } a_2 == 0 \end{cases}$
$\mathrm{sqrt}(a_1)$	$\sqrt{\lvert a_1 \rvert}$	$>(a_1, a_2)$	$\begin{cases} a_1 & \text{if } a_1 > a_2 \\ a_2 & \text{if otherwise} \end{cases}$
$<(a_1, a_2)$	$\begin{cases} a_1 & \text{if } a_1 < a_2 \\ a_2 & \text{if otherwise} \end{cases}$		

Koza [11] except for the population size and reproduction operators. The population size in this work is set to 512 (1024 set by Koza), as it is sufficient to solve the related problems. Moreover, crossover, mutation and elitism are used as reproduction operators (rates are 0.89, 0.1 and 0.01 respectively), which are set empirically.

4 Experiment Preparation

4.1 Datasets

Two standard datasets, the Weizmann horse dataset [12] and Pascal VOC2012 (VOC, Visual Object Classes) [13], are employed in this paper. Both datasets consist of images with high variations, which are considered as difficult segmentation tasks. Figure 5 displays several examples, along with the ground truth images (object in white and background in black colour). Specifically, the Weizmann dataset has 328 horse images with varying horse positions. The average size of the Weizmann images

Fig. 5 Example images

is around 250 × 200 pixels. There are 178 aeroplane images in the Pascal dataset, and their average size is around 500 × 350 pixels.

4.2 Evaluation Measures

The F_1 score (Eq. 4) is applied to evaluate the segmentation results, which reaches the worst at 0 and the best at 1. In Eq. 4, *TP*, *TN*, *FP* and *FN* stand for true positives, true negatives, false positives and false negatives respectively across the pixels of all test images. Therefore, the F_1 score is actually the average across the test images.

$$\begin{aligned} F_1 &= 2 * Precision * Recall/(Precision + Recall) \\ Precision &= TP/(TP + FP) \\ Recall &= TP/(TP + FN) \end{aligned} \quad (4)$$

4.3 Experiment Design

In the Weizmann dataset, two thirds of total images are used as training images (218 images), from each of which 20 samples (10 from object/background pixels respectively) are extracted. The remaining 110 images are for testing. In Pascal dataset, 88 images are for training and 90 images are for testing, which is suggested by Everingham et al. [13]. Since the average size of Pascal images is much larger than that of Weizmann images, 50 samples (25 from object/background pixels respectively) are extracted from each training image. Five standard classifiers, i.e. J48, Naïve Bayes (NB), PART, random forest (RF) and Support Vector Machines using Sequential Minimal Optimization (SMO), are selected from the Weka package to measure the constructed features. All GP related experiments conduct 30 independent runs, and the results are the average of those based on 30 single best solutions.

5 Result

This section presents the training and test results in Table 2, and provides analyses in Sects. 5.1 and 5.2. The comparison between features constructed by the proposed methods with the original features are based on a statistical significance test technique: the one sample median test [14] at the significance level 5 %, which can test whether the median of a sample set of values is the same as a hypothesized value. Since the proposed methods are indeterministic, thus they are run multiple times producing multiple independent feature sets; while there is only one original feature set. The one sample median test is suitable to test whether the median of results based

Table 2 Performance of different feature sets (Weiz and Pasc mean Weizmann and Pascal datasets respectively; #F means the number of features; Ori is the original feature set; CF and Combined stands for the single constructed feature and the combined feature set with adding the constructed feature to the original features respectively; ↑, ↓ or = mean significantly better, worse or similar, compared with the results using the original features only)

Dataset	Method (#F)	Training accuracy (%)					Test F_1 score				
		J48	NB	PART	RF	SMO	J48	NB	PART	RF	SMO
Weiz	Ori (53)	98.07	72.53	84.36	100.00	82.43	0.530	0.568	0.555	0.621	0.651
	FGP	68.99↓	64.62↓	68.93↓	73.97↓	60.45↓	0.510↓	0.457↓	0.509↓	0.480↓	0.397↓
	CF(1)	±3.60	±6.42	±3.56	±3.68	±9.18	±0.005	±0.014	±0.005	±0.006	±0.028
	FGP	96.73↓	72.76=	85.42=	100.00=	82.18↓	0.544↑	0.561↓	0.559↑	0.626↑	0.648↓
	Combined (54)	±1.09	±0.46	±2.36	±0.00	±0.23	±0.003	±0.001	±0.005	±0.002	±0.000
	EGP	77.06↓	75.37↑	77.06↓	79.69↓	75.84↓	0.584↑	0.581↑	0.584↑	0.544↓	0.585↓
	CF (1)	±1.28	±2.62	±1.28	±1.22	±2.24	±0.005	±0.005	±0.005	±0.006	±0.004
	EGP	95.84↓	73.31↑	83.93=	100.00=	82.34=	0.556↑	0.566↑	0.565↑	0.627↑	0.649↓
	Combined (54)	±1.74	±0.29	±2.25	±0.00	±0.21	±0.006	±0.001	±0.010	±0.003	±0.001
Pasc	Ori (53)	97.23	80.99	92.22	100.00	84.82	0.438	0.466	0.452	0.506	0.503
	FGP	73.27↓	66.82↓	73.25↓	74.11↓	63.76↓	0.396↓	0.311↓	0.397↓	0.387↓	0.271↓
	CF (1)	±9.78	±10.01	±9.76	±10.51	±12.82	±0.016	±0.024	±0.017	±0.017	±0.050
	FGP	96.60=	80.71↓	89.19↓	100.00=	84.86↑	0.452↑	0.479↑	0.476↑	0.520↑	0.510↑
	Combined (54)	±1.15	±0.38	±1.69	±0.01	±0.11	±0.002	±0.001	±0.003	±0.000	±0.000
	EGP	84.01↓	79.89=	84.01↓	85.59↓	81.48↓	0.483↑	0.476↑	0.483↑	0.465↓	0.485↓
	CF (1)	±0.54	±4.22	±0.54	±1.06	±3.80	±0.001	±0.007	±0.001	±0.004	±0.006
	EGP	96.42=	81.11↑	88.28↓	100.00=	85.08↑	0.461↑	0.481↑	0.467↑	0.521↑	0.516↑
	Combined (54)	±1.42	±0.18	±1.51	±0.00	±0.33	±0.002	±0.000	±0.003	±0.001	±0.001

on a set of feature sets is significant different from the result based on the original feature set.

5.1 Training Performance

According to Table 2, the training performance of single constructed features produced by both FGP and EGP is generally worse than that of the original features on both datasets. As the single constructed feature is only one feature, it may not contain sufficient distinguishing information of object/background pixels, which leads to worse training performance than the original features. The combined feature sets, which consists of the original features and the constructed feature, achieve varied performance compared with the original features. For example, compared with original features on Pascal dataset, the FGP combined features perform better for classifier SMO; worse for classifiers NB and PART; while similar for classifiers J48 and RF. Even though the combined feature sets cannot generally outperform the original features in training accuracy, they have generally better test performance on both datasets than the original features (based on the analyses in Sect. 5.2), which means that the combined feature sets are more robust than the original features.

Table 3 displays the CPU time cost in the feature construction (or training) process of FGP and EGP. The values in this table are the average of 30 independent runs. Based on the Mann-Whitney U test (suitable for testing the median of two equal sized sample sets) [15] at the significance level 5 %, EGP is significantly faster than FGP in constructing a single best feature on both datasets, which indicates that the fitness evaluation in EGP is more efficient than that in FGP.

5.2 Test Performance

For FGP, Table 2 shows that the test results based on the single constructed feature (CF) are significantly worse than those based on the original features on both Weizmann and Pascal datasets. However, the combined feature set (Combined), which adds the single constructed feature into the original features, can generally increase the standard classifiers' performance. Specifically, on the Weizmann dataset, the performance based on the FGP combined feature set increases for classifiers J48, PART

Table 3 Time cost for feature construction (second)

Method	Weizmann dataset	Pascal dataset
FGP	11.419 ± 1.721	13.439 ± 2.500
EGP	9.609 ± 1.990	10.715 ± 1.879

and RF, except for classifiers NB and SMO, whose results are similar to those of the original features. On the Pascal dataset, all the five classifiers achieve higher F_1 scores using FGP combined features than those using original features. This indicates that the single feature constructed by FGP contains useful distinguishing information of the objects and backgrounds in test images, but the information is limited and cannot completely replace that contained in the original features; therefore, the combined features including both original features and the single constructed feature perform better than the original features.

For EGP, Table 2 shows promising results. The single constructed feature of EGP can help several classifiers achieve better results than the original features. For example, J48, NB and PART using only one constructed feature produce significantly better results on the Weizmann dataset and the Pascal dataset than those based on the original 53 features. Even though the EGP constructed feature cannot improve all the classifiers' performance, the combined feature set of EGP helps improve most classifiers' test performance, except for SMO on the Weizmann dataset, which holds a similar result to that of the original features. One interesting point is that the single EGP constructed feature outperforms the combined features for certain cases, e.g. for classifiers J48, NB and PART on Weizmann dataset, and for classifiers J48 and PART on Pascal dataset. The observation suggests that EGP can construct powerful features, which can replace the original features for the given segmentation tasks based on certain classifiers.

When comparing FGP and EGP, EGP are generally better than FGP in both the single constructed feature and the combined features. For example, on Weizmann dataset, the five classifiers using the single FGP constructed feature achieve 0.510, 0.457, 0.509, 0.621 and 0.651 in F_1 score respectively; while all the five classifiers using the EGP constructed feature increase in F_1 score by 0.074, 0.124, 0.075, 0.064 and 0.188 respectively. The EGP combined feature sets also generally outperform the FGP combined feature sets on both datasets. The analyses reflect that EGP is better than FGP in constructing effective features for given segmentation tasks.

6 Conclusions

This paper developed two GP based feature construction methods for figure-ground image segmentation, i.e. one filter GP method (FGP) and one embedded GP method (EGP). This is the first time to employ GP to construct image features for figure-ground segmentation tasks. Moreover, a novel entropy based fitness function was designed for FGP; while EGP used GP evolved solutions as classifiers to generate the error rate as the fitness function. The single constructed feature and the combined features (the constructed feature + original features) were compared with the original features by five standard classifiers, i.e. J48, NB, PART, RF and SMO, based segmentation tasks on two standard image datasets, i.e. Weizmann and Pascal datasets.

The combined (constructed and original) features of both FGP and EGP perform generally better than original features for the given segmentation tasks, which indicates that the proposed methods can construct useful features. Specifically, the combined features of EGP help the standard classifiers achieve higher F_1 scores than FGP combined features. Moreover, EGP is more efficient than FGP in feature construction, as the fitness evaluation in EGP cost less time than that in FGP.

This paper works on the single feature construction. As the single constructed feature is often not sufficient and original features are also needed to form combined features in order to achieve better results than the original features, we will consider multiple feature construction of the single constructed features to capture diverse aspects of image segmentation in the future.

References

1. Bishop, A., Ciesielski, V., Trist, K.: Feature construction using genetic programming for classification of images by aesthetic value. In: Evolutionary and Biologically Inspired Music, Sound, Art and Design, pp. 62–73. Springer (2014)
2. Lillywhite, K., Lee, D.J., Tippetts, B., Archibald, J.: A feature construction method for general object recognition. Pattern Recognition 46(12), 3300–3314 (2013)
3. Sondhi, P.: Feature construction methods: a survey. sifaka. cs. uiuc. edu 69, 70–71 (2009)
4. Poli, R.: Genetic programming for feature detection and image segmentation. In: Evolutionary Computing, pp. 110–125. Springer (1996)
5. Ahmed, S., Zhang, M., Peng, L., Xue, B.: Multiple feature construction for effective biomarker identification and classification using genetic programming. In: Proceedings of the 2014 conference on Genetic and evolutionary computation. pp. 249–256. ACM (2014)
6. Guo, H., Jack, L.B., Nandi, A.K.: Feature generation using genetic programming with application to fault classification. Systems, Man, and Cybernetics, Part B: Cybernetics, IEEE Transactions on 35(1), 89–99 (2005)
7. Guo, H., Nandi, A.K.: Breast cancer diagnosis using genetic programming generated feature. Pattern Recognition 39(5), 980–987 (2006)
8. Datta, R., Joshi, D., Li, J., Wang, J.Z.: Studying aesthetics in photographic images using a computational approach. In: Computer Vision–ECCV 2006, pp. 288–301. Springer (2006)
9. Neshatian, K.: Feature manipulation with genetic programming (2010)
10. Shannon, C.E.: A mathematical theory of communication. ACM SIGMOBILE Mobile Computing and Communications Review 5(1), 3–55 (2001)
11. Koza, J.R.: Genetic programming: on the programming of computers by means of natural selection, vol. 1. MIT press (1992)
12. Borenstein, E., Ullman, S.: Combined top-down/bottom-up segmentation. Pattern Analysis and Machine Intelligence, IEEE Transactions on 30(12), 2109–2125 (2008)
13. Everingham, M., Eslami, S.A., Van Gool, L., Williams, C.K., Winn, J., Zisserman, A.: The pascal visual object classes challenge: A retrospective. International Journal of Computer Vision 111(1), 98–136 (2014)
14. The 1-sample median test – analysis of a single quantitative variable. Available at http://psych.unl.edu/psycrs/handcomp/hc1median.pdf
15. Ruxton, G.D.: The unequal variance t-test is an underused alternative to student's t-test and the mann–whitney u test. Behavioral Ecology 17(4), 688–690 (2006)

Estimation of Distribution Algorithms with Graph Kernels for Graphs with Node Types

Kenta Maezawa and Hisashi Handa

Abstract We have proposed a novel Estimation of Distribution Algorithms with graph kernels (EDA-GK). By using the graph kernels, we can search for solutions in a feature space. The use of the graph kernel can eliminate the effect of the ruggedness of genotype-phenotype mappings of evolutionary algorithms. In this paper, we extend the EDA-GK to cope with graphs with node types. In order to achieve this, the histogram in the kernel density estimation function is separated into several subhistograms for modes of inter-types, and for nodes of each type. Experimental results on the Edge-Max problems show the effectiveness of the proposed method.

Keywords Estimation of distribution algorithms · Graph kernels · Kernel density estimation

1 Introduction

Estimation of Distribution Algorithms (EDAs) are a class of Evolutionary Algorithms, which use the probabilistic models to generate offspring [1]. Hence, the EDAs do not use genetic operators such as crossovers and mutations. In our previous studies, the EDAs with graph kernels (EDA-GK) have been proposed [2]. In the EDA-GK, graph kernels are employed to estimate probabilistic distributions of graphs by means of the kernel density estimation.

The difficulties of solving graph-related problems are described as follows: Suppose that, as individual representation, the adjacency matrix is employed. In this example, suppose that we have only the interests of the topology of graphs for fitness calculations. In Fig. 1a and b, the genotype of individuals, i.e., the adjacency matrices, are described in the upper row of these figures. The phenotype of individuals, i.e., graphs, are depicted in the lower row. The fitness of these individuals is the same if only the typologies are affected to the fitness calculations. On the other hand,

K. Maezawa · H. Handa (✉)
Kindai University, Higashi-Osaka, Japan
e-mail: handa@info.kindai.ac.jp

G. Leu et al. (eds.), *Intelligent and Evolutionary Systems*,
Proceedings in Adaptation, Learning and Optimization 8,
DOI 10.1007/978-3-319-49049-6_18

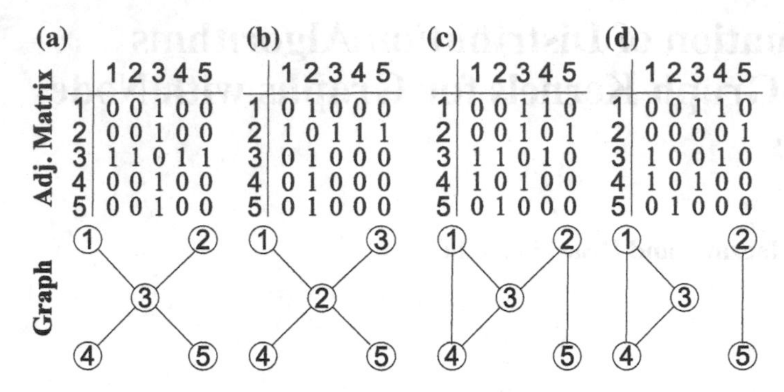

Fig. 1 Difficulties of evolutionary algorithms in graph-related problems

in the case of Fig. 1c and d, the genotype is almost the same: The difference among them is an edge between nodes 2 and 3. The phenotype, however, are quite differ: The graph of Fig. 1d consists of two sub-graphs. Therefore, the genotype-phenotype mappings in evolutionary algorithms for solving graph-related problems are rugged so that the difficulties of solving the graph-related problems are caused. The graph kernels explained in Sect. 2 might address this difficulties.

The EDA-GK proposed by us is an extension of the Estimation of Distribution Algorithms, which tackle to the graph-related problems. The EDA-GK uses graphs to represent individuals. In order to estimate the distribution of selected individuals, the kernel density estimation with graph kernels are employed. The original paper of the EDA-GK copes with simple graphs. In this paper, we extend the EDA-GK for graphs with typed nodes. The nodes of the graphs used in this paper are associated with one of several types. This extension is useful for applying the proposed method to scene recognitions from image, synthesis of digital circuits, graph mining problems of the web, and so on.

Related works are summarized as follows: Due to the flexibility of Evolutionary Algorithms, graph-related problems are potential application area of EAs. However, we do not have many EA applications for graph-related problems because of the time-consuming nature of the problems. Chicano and Alba propose ACOhg (Ant Colony optimization for huge graphs) in the soft engineering area, i.e., model verification [3, 4]. McDermott and O'Reilly employ graphs for music generation [5]. In addition, the notion of graph is incorporated into some evolutionary algorithms [6–8].

The organization of the remainder of the paper is as follows: Sect. 2 mention about the shortest-path distance kernel which is used in this paper as a graph kernel of the proposed methods. The original EDA-GK is explained in Sect. 3. An extension to graphs with node types of the proposed method is introduced in Sect. 4. Experimental results on Edge-Max problems are shown in Sect. 5.

2 Graph Kernels

Machine Learning algorithms based upon kernel methods have attracted much attention. The notion of kernel functions has been extended to cope with graphs [9]. This paper employs the shortest-path graph kernel proposed by Borgwardt and Keiegel [10]. The followings explain the calculation procedure of the shortest-path graph kernel:

1. The Floyd-Warshall algorithm are used to calculate all the pairs-shortest-paths in two graphs $G = (V, E)$, and $G' = (V', E')$.
2. Then, the shortest-path kernel is defined by comparing all the pairs of the shortest path lengths among nodes in G, and G':

$$k(G, G') = \sum_{v_i, v_j \in V} \sum_{v'_i, v'_j \in V'} k_{length}(d(v_i, v_j), d(v'_i, v'_j)), \tag{1}$$

where $d(v_i, v_j)$ denotes the path distance of the shortest path between nodes v_i, v_j. k_{length} is a kernel for comparing the lengths of two shortest paths:

In the case of a delta kernel [10],

$$k_{length}(d(v_i, v_j), d(v'_i, v'_j)) = \begin{cases} 1 & \text{if } d(v_i, v_j) = d(v'_i, v'_j) \\ 0 & \text{otherwise.} \end{cases}$$

Let $h(G)$ be a histogram of distances for all the shortest paths in a graph $G = (V, E)$. Each bin of the path distance D of $h(G)$ can be written as follows:

$$h(D, G) = |\{v_i, v_j \in V | d(v_i, v_j) = D\}|.$$

In the case of the delta kernel, the shortest-path kernel $k(G, G')$ can be rewritten as follows:

$$k(G, G') = \sum_D h(D, G) \cdot h(D, G'). \tag{2}$$

Figure 2 illustrates how the shortest-path graph kernels work. The graphs in Fig. 2 are the same as the ones in Fig. 1. "Dist. Matrix" in Fig. 2 represents the distance matrices of the shortest-path between corresponding nodes. "Histogram" means the histogram of distances $h(G)$ mentioned in the previous paragraph. Numbers at the left hand and the right hand of the colon means the distance D and the bin $h(G, D)$, respectively. The rotated/reflected solutions, i.e., the graphs in Fig. 2a and b, indicate the same histograms. Meanwhile, the solutions in Fig. 2c and d, which are similar in the adjacency matrix representation, show quite different histograms.

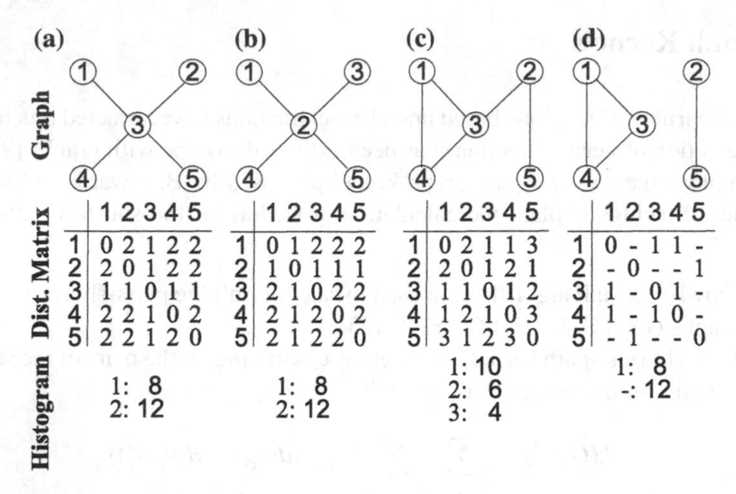

Fig. 2 Shortest-path graph kernels: the graphs with the same phenotype as in (**a**) and (**b**) show the same histogram. Meanwhile, the graphs with the similar genotype but with different phenotype show quite different histogram

3 Estimation of Distribution Algorithm with Graph Kernels

This section explains Estimation of Distribution Algorithm with graph kernels (EDA-GK) proposed by us [2]. The procedure of the EDA-GK is depicted in Fig. 3, which is quite similar to the one of the conventional EDAs. One of the differences with the conventional EDAs is the use of graphs for the individual representation. Because of this difference, the estimation phase and the sampling phase are also different. Therefore, "individual representation," "estimation," and "sampling" are separately described in the following subsections. At the same as in the conventional EDAs, we assume that fitness is evaluated by user-defined fitness functions.

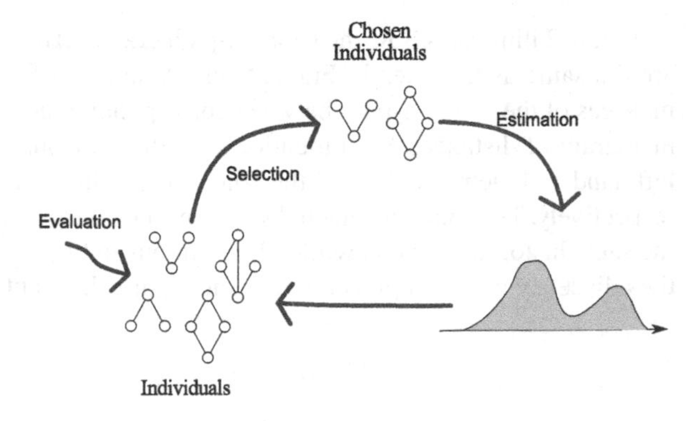

Fig. 3 Procedure of the estimation of distribution algorithms with graph kernels (EDA-GK)

3.1 Individual Representation

The proposed method employs the direct coding for representing individuals so that a graph indicates an individual. For an implementation issue, we can use either of the adjacency matrix data structure or the adjacency list data structure. The proposed method does not manipulate graphs by using crossover and mutation operations but requires the data structures to add/remove edges and to calculate all the shortest-path distances. Hence, both the data structures can be used. We can use the adjacency matrix for small/dense problems and the adjacency lists for large/sparse problems. For ease of implementation, we could use public domain software of graph libraries.

3.2 Estimation of Individual Distributions

Suppose that N denotes the number of the selected individuals at each generation, and the selected individuals are represented by $G_1, G_2, \ldots, G_N$. A kernel density function $f(G)$ can be defined as follows:

$$f(G) = \frac{1}{N \cdot p_n} \sum_i k(G, G_i) \tag{3}$$

where k denotes the kernel function in Eq. (2), and p_n stands for a normalizing parameter which normalizes the magnitude of the kernel function to 1. By using Eq. (2), Eq. (3) can be rewritten as follows:

$$\begin{aligned} f(G) &= \frac{1}{N \cdot p_n} \sum_i \sum_D h(D, G) \cdot h(D, G_i) \\ &= \sum_D h(D, G) \cdot \frac{1}{N \cdot p_n} \sum_i h(D, G_i) \end{aligned}$$

Therefore, the probability density estimation for the selected individuals in the EDA-GK is to sum up $h(D, G_i)$ of all the selected individuals for each possible distance D. Hence, the procedure of this is summarized as follows:

1. The distance matrices of shortest-path for all the selected individuals are calculated.
2. A histogram of distances for all the selected individuals is constituted.

3.3 Sampling of New Individuals

The following sampling method is used for the EDA-GK.

1. A graph G' is copied from one of the selected individuals.
2. Two nodes are selected from the graph G'.
3. An edge is added if there is no edge between the selected nodes. Otherwise, an edge is removed.
4. The new graph G' is accepted if $f(G') > f(G)$. Otherwise the new graph G' is accepted with the following probability $P(G')$:

$$P(G') = \frac{f(G')}{f(G) + f(G')},$$

where the function f is the kernel density function in the previous section[1]
5. If the graph G' is accepted in the previous step, $G \leftarrow G'$.
6. Go back to Step (2) until the number of iteration reaches a predefined parameter N_i.

Note that the sampling method in this paper is improved from the one in [2]. In the previous study, the sampling process is biased so that the Edge-Max problems were easier than Edge-Min problems for the EDA-GK, where the Edge-Max problems are to find out the complete graph, and the Edge-Min problems are to find out the empty graph. By using sampling method in this paper, there is no bias so that the results of both problems are almost the same.

4 Extension of EDA-GK for Node Types

4.1 Graphs with Node Types

This section introduces a belief definition of graphs with node types. Firstly, we define a graph G as follows:

$$G = (V, E),$$

where $V = \{v_1, v_2, \ldots, v_l\}$ and $E = \{e_1, e_2, \ldots, e_m\}$ denote a set of nodes and a set of edges, respectively. The original EDA-GK can cope with simple graphs [2], where there are no types of nodes such that we do not have to distinguish between nodes, there are no types of edges too.

In this paper, the type is associated to each of nodes: nodes are separated into groups of nodes V_i:

$$V = V_1 \cup V_2 \cup \cdots \cup V_k$$

Each node must be a member of any one of these groups.

[1]For the extended version of the EDA-GK, i.e., the proposed method in this paper, Eq. (5) is used for the calculation of the function f in Eq. (3).

$$V_i \cap V_j = \emptyset \quad \text{for all } i, j$$

For instance, suppose that we would like to represent a chemical compound H_2O as a graph with node type. There is a two groups of nodes $V_1 = \{H, H\}$ and $V_2 = \{O\}$. We do not have to distinguish between two hydrogen atoms but we have to distinguish between an oxygen atom and a hydrogen atom.

4.2 *Kernel Density Function*

The modification for the extension of node types is individual representation so that each node should store type information. This modification only affects to representation of the kernel density function of the EDA-GK. We can distinguish nodes in different groups so that Eq. (1) is rewritten as follows:

$$k_{nt}(G, G') = \sum_k \sum_l \sum_{v_i \in V_k, v_j \in V_l} \sum_{v'_i \in V'_k, v'_j \in V'_l} k_{length}(d(v_i, v_j), d(v'_i, v'_j)), \tag{4}$$

where $d(v_i, v_j)$ denotes the path distance of the shortest path between nodes v_i, v_j. These appended summations in Eq. (4), i.e., $\sum_k \sum_l$, mean the shortest-path distance matrix in the original graph kernel are separated into several regions. Therefore, total calculation costs in the graph kernels are not increased so much. However, we need that the histogram is separated into several sub-histograms for inter-types, and for each type as in Fig. 4.

Let $h_{k,l}(G)$ be the sub-histogram of inter-types for all the shortest paths between nodes in V_k and V_l:

$$h_{k,l}(D, G) = |\{v_i \in V_k, v_j \in V_l | d(v_i, v_j) = D\}|.$$

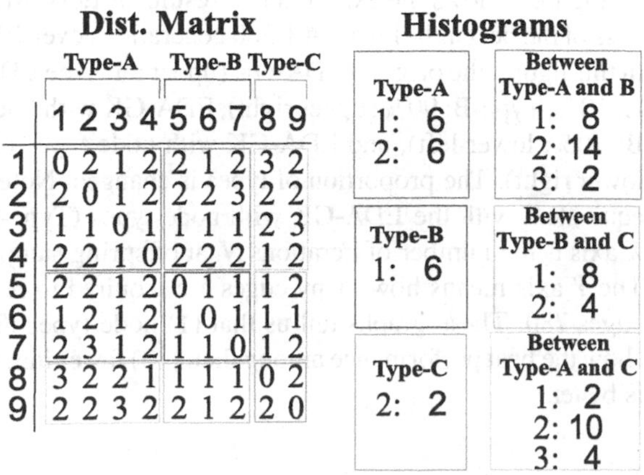

Dist. Matrix

	Type-A				Type-B			Type-C	
	1	2	3	4	5	6	7	8	9
1	0	2	1	2	2	1	2	3	2
2	2	0	1	2	2	2	3	2	2
3	1	1	0	1	1	1	1	2	3
4	2	2	1	0	2	2	2	1	2
5	2	2	1	2	0	1	1	1	2
6	1	2	1	2	1	0	1	1	1
7	2	3	1	2	1	1	0	1	2
8	3	2	2	1	1	1	1	0	2
9	2	2	3	2	2	1	2	2	0

Histograms

Type-A
- 1: 6
- 2: 6

Type-B
- 1: 6

Type-C
- 2: 2

Between Type-A and B
- 1: 8
- 2: 14
- 3: 2

Between Type-B and C
- 1: 8
- 2: 4

Between Type-A and C
- 1: 2
- 2: 10
- 3: 4

Fig. 4 Constitution of sub-histograms: there are three types for instance. There are six sub-histograms: three histograms for each type and three histograms for inter-types

As in the previous section, the shortest-path kernel $k_{nt}(G, G')$ for node types can be rewritten as follows:

$$k(G, G') = \sum_k \sum_l \sum_D h_{k,l}(D, G) \cdot h_{k,l}(D, G'), \qquad (5)$$

where $h_{k,l}(D, G)$ denotes the bin the distance k of the histogram of node group k and l in graph G.

5 Experiments

This paper shows the experimental results on the Edge-Max problems. Edge-Max problems can be regarded as a graph edition of the One-Max problems of binary Genetic Algorithm researches. The target of the Edge-Max problems is to find out the complete graph, where all the pairs of nodes have an edge. The fitness f_{emax} of the Edge-max problems is defined as follows:

$$f_{emax}(G) = (\text{the number of edges in } G),$$

where $f_{emax}(G)$ should be maximized. For the proposed method, i.e., the EDA-GK with node types, two types are introduced: Type-A, and Type-B. These types are not affected for fitness calculations.

The number of selected individuals N is set to be half of the number of individuals M. Plus and Comma selection method is used, μ is set to be 100 and λ is set to be 200. The number of iterations N_i at the sampling explained in Sect. 3.3 is set to be ether of 1, 2, 3, 5, 7, 10, 15, 20, 30, or 50. We examined the Edge-max problems with 40 nodes. Hence the optimal value for these problems is 780. The number of generations is set to be 1,500. For comparison, one-max problems of 780 bits are examined for a few conventional EDAs, UMDA and ECGA.

Figure 5 shows the experimental results of Edge-Max problems: the averaged fitness of the best individuals at 1,500 generations over 10 runs. Four graphs are located in this figure: the original EDA-GK (upper left), the EDA-GK with node types (Type-A: 10 %, Type-B: 90 %; upper right), EDA-GK with node types (Type-A: 30 %, Type-B: 70 %; lower left), and EDA-GK with node types (Type-A: 50 %, Type-B: 50 %; lower right). The proportion of types is changed. Note that the original EDA-GK is equivalent with the EDA-GK with node types (Type-A: 100 %, Type-B: 0 %). The X axis is the number of iterations N_i at offspring sampling as described in Sect. 3.3. The Y axis means how many edges are required to reach to the optimal number of edges 780. These graphs tell us that (1) node types (Type-A: 50 %, Type-B: 50 %) show the best performance among them, (2) fewer iterations N_i at offspring sampling is better.

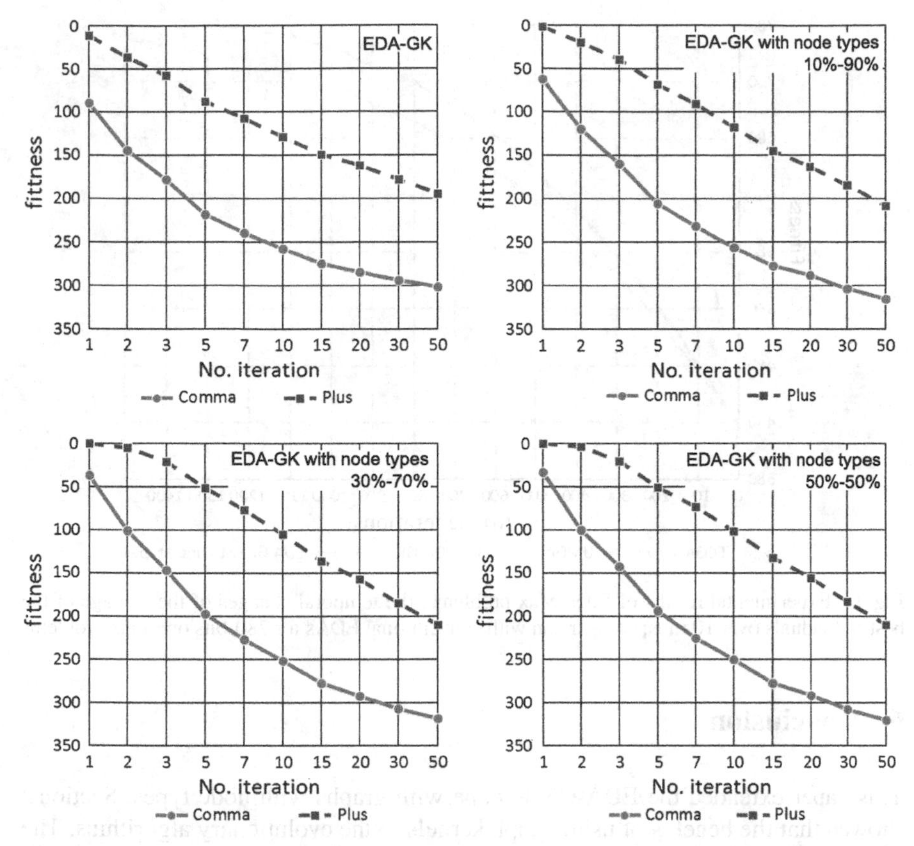

Fig. 5 Experimental results of Edge-Max problems: the average of the best individuals over 30 runs at 1,500 generations: results of the original EDA-GK (*upper left*); EDA-GK with node types (Type-A: 10 %, Type-B: 90 %; *upper right*); EDA-GK with node types (Type-A: 30 %, Type-B: 70 %; *lower left*); EDA-GK with node types (Type-A: 50 %, Type-B: 50 %; *lower right*)

Figure 6 shows the experimental results of Edge-Max problems: the temporal changes of the average of the best individuals over 10 runs. This graph compares the proposed methods with conventional EDAs: UMDA, and ECGA. The problems for the conventional EDAs are 780-bits One-Max problems, which are equivalent to 40 nodes Edge-Max problems. The EDA-GKs show the better performance than the conventional EDAs. This comparison would not be fair comparisons because the Edge-Max problems are on graphs. However, this graph elucidates that the EDA-GKs show the sufficient performance.

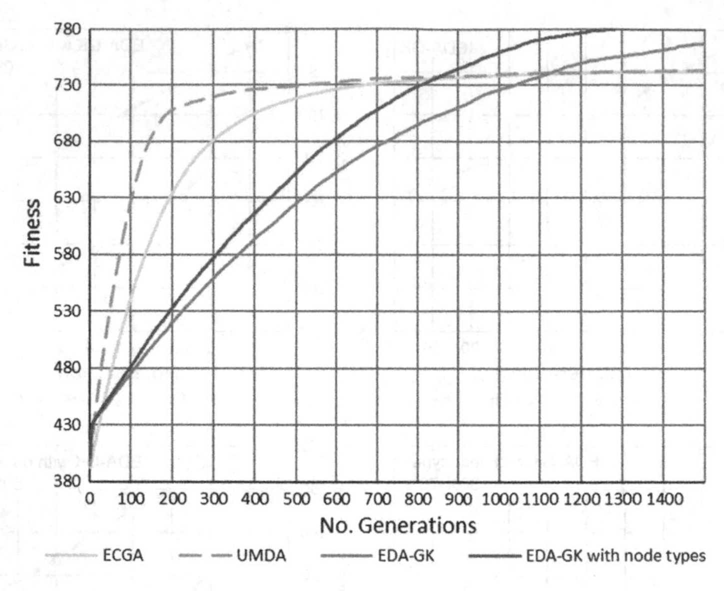

Fig. 6 Experimental results of Edge-Max problems: the temporal changes of the average of the best individuals over 10 runs; comparison with conventional EDAs for 780-bits one-max problems

6 Conclusion

This paper extended the EDA-GK to cope with graphs with node types. Section 2 showed that the benefits of using graph kernels to the evolutionary algorithms. The proposed mechanism of coping with node types is that the histogram of the kernel density estimate function is separated into several sub-histograms for inter-types, and for each type. Experimental results on the Edge-Max problems show the effectiveness of the EDA-GK with node types. Although the Edge-Max problems have no characteristics of node types, the EDA-GK with node types works well. It would be caused by the separation of the histogram in the kernel density estimation function.

Future works are summarized as follows: In this paper, we only examined simple problems such that adding/removing edges are required. The proposed method can easily extend to adding/removing nodes. This extension will be useful for applying graph mining problems. Another extension is for graphs with weighted edges.

Acknowledgments This work was partially supported by the Grant-in-Aid for Young Scientists (B) and the Grant-in-Aid for Scientific Research (C) of MEXT, Japan (23700267, 26330291).

References

1. P. Larrañaga, and J.A. Lozano, *Estimation of Distribution Algorithms*, Kluwer Academic Publishers, 2003
2. H. Handa. Use of graph kernels in Estimation of Distribution Algorithms. In *Proc. of 2012 IEEE Congress on Evolutionary Computation (CEC)*, pages 1–6, 2012.
3. E. Alba, and F. Chicano, *ACOhg: Dealing with huge graphs*, Proc. the 2007 ACM Genetic and Evolutionary Conference, pp. 10–17, 2007.
4. F. Chicano, and E. Alba, *Ant colony optimization with partial order reduction for discovering safety property violations in concurrent models*, Information Processing Letters, Vol. 106, No. 6, pp. 221–231, 2008.
5. J. McDermott, U.-M. O'Reilly, *An executable graph representation for evolutionary generative music*, Proc. the 2011 ACM Genetic and Evolutionary Conference, pp. 403–412, 2011.
6. T.E. Lewis, and G.D. Magoulas, *Strategies to minimise the total run time of cyclic graph based genetic programming with GPUs*, Proc. the 2009 ACM Genetic and Evolutionary Conference, pp. 1379–1386, 2009.
7. S. Shirakawa, and T. Nagao, *Graph structured program evolution with automatically defined nodes*, Proc. the 2009 ACM Genetic and Evolutionary Conference, pp. 1107–1115, 2009.
8. S. Mabu, K. Hirasawa, and J.Hu, *A Graph-Based Evolutionary Algorithm: Genetic Network Programming (GNP) and Its Extension Using Reinforcement Learning*, Evolutionary Computation, Vol. 15, No. 3, pp. 369–398, 2007.
9. H. Kashima, K. Tsuda, and A. Inokuchi, *Marginalized Kernels Between Labeled Graphs*, Proc. 20th International Conference on Machine Learning, pp. 321–328, 2003.
10. K.M. Borgwardt, and H.-P. Kriegel, *Shortest-path kernels on graphs*, Proc. 5th International Conference Data Mining, 2005.

Generating Hub-Spoke Network for Public Transportation: Comparison Between Genetic Algorithm and Cuckoo Search Algorithm

Takahiro Majima, Keiki Takadma, Daisuke Watanabe and Mitujiro Katuhara

Abstract Scheduled transportation service is a proper system for mass transportation and it is adopted by wide range of transportation modes, such as railway, airline, maritime container shipping and bus. The providers of the service are required to organize effective routes and networks. This paper tackles the problem combining two problems. One is generating Public Transit Network (PTN) as one of the scheduled transportation services. The other is network hub location problem to find out the effective position of the node as a hub station. The method generating PTN is based on a growing network model and the method for the hub location problem is based on a genetic algorithm and a cuckoo search algorithm. This method can find out effective position of the hub node and transportation line network simultaneously. In this framework, this paper reports the comparison result between the genetic algorithm and the cuckoo search algorithm for the hub location problem.

Keywords Public transportation network · Growing network model · Complex network · Genetic algorithm · Cuckoo search algorithm

---

T. Majima (✉)
National Maritime Research Institute, 6-38-1, Tokyo, Mitakashi, Japan
e-mail: majy@nmri.go.jp

K. Takadma
The University of Electro-Communications, Tokyo, Chufu 1-5-1 Chufugaoka, Japan
e-mail: keiki@inf.uec.ac.jp

D. Watanabe
Tokyo University of Marine Science and Technology,
2-1-6 Etchujima, Tokyo, Koto-ku, Japan
e-mail: daisuke@kaiyodai.ac.jp

M. Katuhara
Socio Tech Data, Tokyo, Japan
e-mail: kat-151@mail.bbexcite.jp

G. Leu et al. (eds.), *Intelligent and Evolutionary Systems*,
Proceedings in Adaptation, Learning and Optimization 8,
DOI 10.1007/978-3-319-49049-6_19

1 Introduction

Scheduled transportation service is a proper system for mass transportation and it is adopted by wide range of transportation modes, such as railway, airline, maritime container shipping and bus. The providers of the service are required to organize effective routes and networks. This paper tackles the problem combining two problems. One is generating Public Transit Network (PTN) as one of the scheduled liner services. The other is network hub location problem to find out the optimized position of a hub station. The method generating PTN is based on a growing network model and the method for hub location problem is based on a genetic algorithm in our previous method [1]. It was confirmed that the method combining above algorithms can out put proper position of the hub and transportation line network simultaneously. To pursue more efficient algorithm, a cuckoo search algorithm is applied for the hub location problem in this paper and it is compared with the Genetic Algorithm in our original method.

A large number of the bio-inspired optimization algorithms are reported recently. In that algorithms, the cuckoo search algorithms is appropriate for optimizing the hub node in the PTN generation problem, because it relies on the Lévy flight. Lévy distribution has fat tail distributions, in which the probability of random variables far from its mean value is larger than that of the normal distributions. It implies that the searching space is focused around a certain position in most case, but sometime it jumps. This characteristic is effective to search in a large space without trapping into the local optimums and This paper reports the result of the comparison between the genetic algorithm and the cuckoo search algorithm for the PTN generation.

2 Method

In this paper, the nodes in networks represent bus stops or stations. The position and demand of the nodes, street network, vehicle speed and vehicle capacity are given. The destination node is limited to only one node represented by a 'hub node'. The evaluation function similar to the following equation is applied frequently to the PTN generation problem [2].

$$\min Z = Z_1 + Z_2 = \sum_{i \in ST} T_i D_i + w \sum_{k \in BL} B_k \tag{1}$$

where, ST is a set of nodes (bus stops or station). BL is a set of lines. D_i is a demand at node i(number of passengers from node i to the hub node), T_i is a travel time from node i to the hub node composed of the moving time, Tm_i and the expected waiting time, Tw_i. B_k is a number of vehicle deployed into the line k. w is a control parameter (in this paper, w is fixed to 10). The passenger's total travel time represents the passenger's cost, whereas the vehicle number represents the cost of transportation company. The destination of passengers is the hub node which position is determined by two algorithms, Genetic Algorithm and Cuckoo Search Algorithm in Sect. 2.2.2 Thus optimizing process to find out the best position of the hub node reduces the evaluation value, Z.

The method described below separates two processes. One method produces transit routes of the transportation line with an improved growing network model. Second method with location optimization algorithm finds the position of the hub node.

2.1 Growing Network Model

Our model is inspired by Gastner [3]. In this model, after determining a root node, another node is added to the sub-network including the root node with linking the node at each evolution step. The node added to the sub-network is selected according to the ascending order of the distance to the root node. Through such a node addition, the sub-network spreads outward from the root node. In this paper, the model was improved to deal with the transportation line. Figure 1 shows the process of the growing line network.

In this figure, the newly added node (i.e. the node 3 in Fig. 1) is defined as the 'target node', while the nodes in the sub-network (i.e. the node 0, 1, and 2 in Fig. 1) are defined as the 'candidate node'. The target node is selected in the ascending order of the distance between the hub node and the target node.

The link connects between the target node and one of the candidate nodes. The route of the line starting from the target node is same to that of the already existing line besides the newly added link. The evaluation of the line starting form the target node is calculated for all candidate nodes according to Eq. 1. (note that this calculation is limited within the sub-network.). After calculation of such an evaluation, the candidate node with the smallest evaluation value is selected and is linked with the target node. This process repeats until all nodes are connected to the hub node. Note that the vehicle number, the waiting time and the connection type should be determined to calculate the evaluation value. Following two sections explain how to determine them.

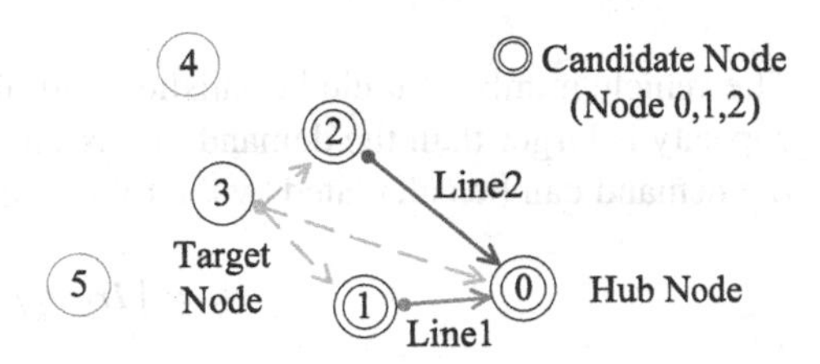

Fig. 1 Schematic of growing line network model

Fig. 2 Schematic of two connection types

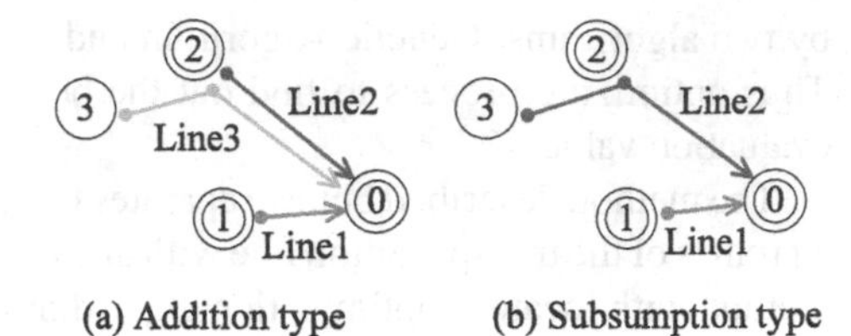

2.1.1 Connection Type and Waiting Time

If the candidate node is a terminal of the existing lines, there are two types of connection can be considered as illustrated in Fig. 2.

The node 2 in Fig. 2 is the terminal of the line 2 and it is the candidate node for the target node 3. In case of the 'addition type', the line 3 for the newly added target node (i.e. the node 3) is generated. In case of the 'subsumption type', the new line 3 subsumes the already existing line 2. Thus, the evaluation value is calculated not only for the candidate node but also for the connection type.

The expected waiting time at a node where more than one bus lines stop can be calculated by the following equation [4].

$$Tw_i = t_1 \left\{ \frac{1}{2} + \sum_{r=1}^{N-1} \frac{(-1)^r t_1^r}{(r+1)(r+2)} \sum_{j_1=2}^{N-r+1} \sum_{j_2=j_1+1}^{N-r+2} \cdots \sum_{j_r=j_{r-1}+1}^{N} \frac{1}{t_{j1} t_{j2} \cdots t_{jr}} \right\} \tag{2}$$

where, N is the number of lines stopping at the node i, t_1 is the smallest headway (frequency) in the set of t_j representing the headway of bus line j. Furthermore, the headway for each line is calculated by the averaged value of its round trip time and the vehicle number.

2.1.2 Vehicle Number

The vehicle number should be satisfied with the condition where the transportation capacity is larger than the demand of passengers. The number of vehicle satisfying the demand can be calculated by the following equation.

$$B_k = \lceil Tr_k D_k / B_{capa} \rceil \tag{3}$$

where, Tr_k is a round trip time of the target node k. B_{capa} is a capacity of one vehicle.

However, in the case of subsumption type in Fig. 2, Eq. 3 does not satisfy the condition of the demand, because the round trip time of the subsumed line changes. In this paper, the following equation is employed for the subsumption type.

$$B_k = \lceil Tr_k D_k / B_{capa} + B_j Tr_k / Tr_j \rceil \quad (4)$$

where, subscript k and j represent the target node and the candidate node, and Tr_j, B_j are the round trip time of the node j and the vehicle number of the subsumed line at the node j. Equation 4 satisfies the demand at the target node and does not deteriorate the frequency of the vehicle stopping at the candidate node.

An example of the growing line network is illustrated in Fig. 3. Further information on the characteristic of the growing line network model is summarized in a paper [5].

2.2 *Hub Location*

After the publication of the paper by O'Kelly [6], a number of methods on the hub location problem have been reported [7]. In that reports, some methods output hub-spoke network similar to the topology in this paper [8, 9]. (It is not supposed that the all links directly connect the hub node and the other nodes.) However they lack the concept of the transportation line (route and frequency of line).

The performance of the following two algorithms, genetic algorithm and cuckoo search algorithm, are compared for the hub location problems of the PTN generation.

2.2.1 Genetic Algorithm

The position of the hub node (only one destination node) relied on a standard genetic algorithm (here in after referred to GA). The chromosome encodes the position on the xy plane of the hub node with length of 20 digits binary string (10 digits for each x, y position). Population size is 25 (Initial arrangement on the plane is 5 × 5). Fitness of the chromosome is inverse of the evaluation value computed by Eq. 1. Crossover rates from 0.2 to 1.0 and position-based crossover is employed. Mutation rate ranges from 0.01 to 0.05. Selection of parents is wheel roulette strategy with respect to the fitness. Deletion of chromosome is greedy strategy with respect to the fitness.

2.2.2 Cuckoo Search Algorithm

The cuckoo search algorithm (here in after referred to CS) was inspired by the brooding parasitism of some cuckoo species [10]. It is reported that CS is widely adopted by the research domains requiring optimization [11]. The location searching process

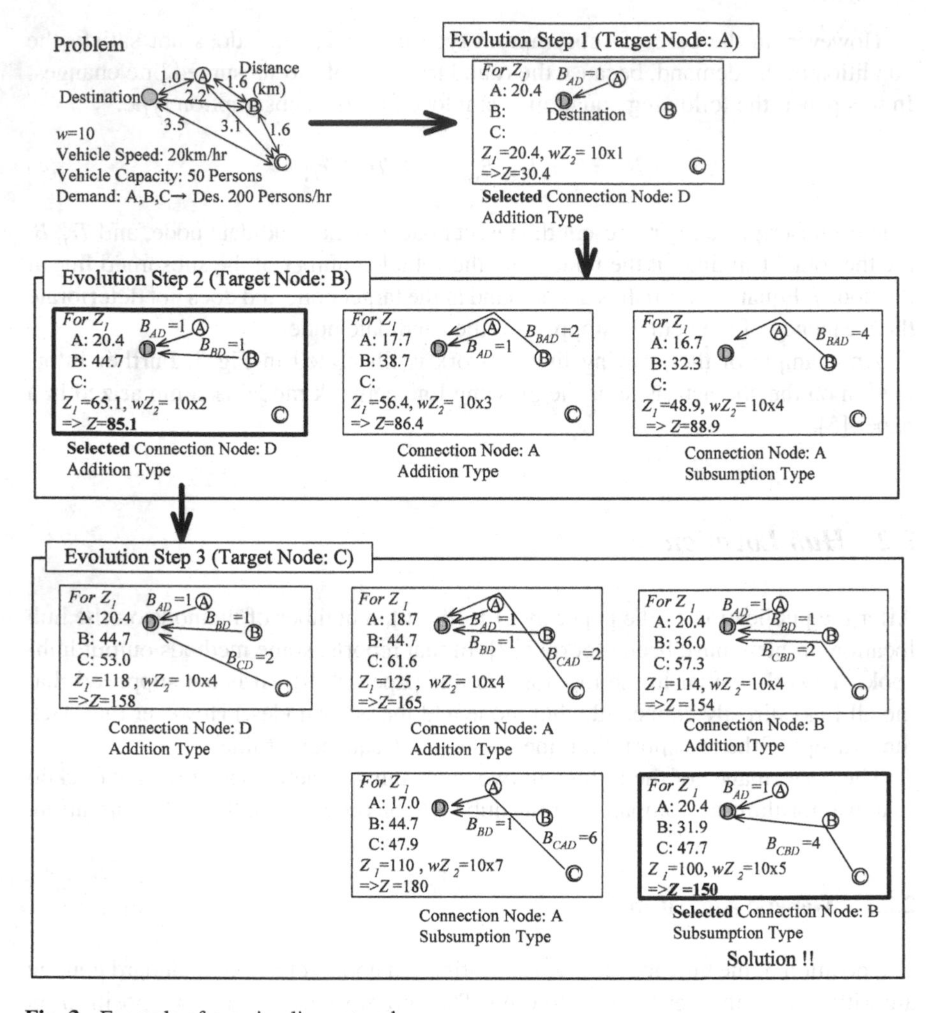

Fig. 3 Example of growing line network

of the CS relies on the Lévy flight that follows the power law distribution in the range of the large random variables. The advantages of the CS are as follows.

- The Lévy flight allows to conduct global and local search effectively.
- Number of the setting parameter is small.
- The solution variables are not discrete values but real numbers.

In this paper, CS is applied to searching the optimum hub node location. Figure 4 shows the pseudo code of the CS algorithm.

The position of the nest is updated by the following equation.

$$\mathbf{x_i}^{(t+1)} = \mathbf{x_i}^{(t)} + \alpha \oplus L\acute{e}vy(\lambda) \tag{5}$$

```
Objective function f(x)
Generate initial population of  n  host nests
   while( t < MaxGeneration) or (stop criterion)
      Get a cuckoo randomly/generate a solution by Levy flights
            and then evaluate its quality / fitness F_i
      Choose a nest among  n  (say,  j ) randomly
      if(F_i > F_j),
         Replace  j  by the new solution
      end
      A fraction(p_a) of worse nests are abandoned
          and new ones/solutions are built/generated
          (with the Levy flight from the position of the best solution)
      Keep best solutions and find the current best
   end while
Post Process results and visualization
```

Fig. 4 Pseudo code of the Cuckoo Search

where, the product $\oplus$ means entry-wise multiplications. α is the control parameter for the step length. Using Mantegna algorithm, the step length s following Lévy distribution can be obtained from

$$s = \frac{u}{|v|^{1/\beta}} \, , \; \beta \equiv \lambda - 1 \tag{6}$$

where, u and v are drawn from normal distributions.

$$u \sim N\left(0, \sigma_u^2\right) \, , \; v \sim N(0, 1) \tag{7}$$

$$\sigma_u = \left[\frac{\Gamma(\beta + 1) \sin\left(\frac{\pi\beta}{2}\right)}{\Gamma\left(\frac{\beta+1}{2}\right) \beta 2^{\left(\frac{\beta-1}{2}\right)}} \right]^{\frac{1}{\beta}} \tag{8}$$

Here, Γ is the Gamma function.

The number of the host nest is 25 which is same to the population number of the GA. The fitness of the host nest is inverse of the evaluation value computed by Eq. 1. The fraction of abandoned nests is set to $p_a = 0.25$. The exponent of the power law distribution is set to $\lambda = 1.5$. (Lévy distribution approaches to the power law distribution in the range of the large random variable (i.e. step length).) The control parameter for the step length is set to $\alpha = 0.01$;

3 Application

3.1 Simple Network

Firstly, the function of the two algorithms to find an optimal position of the hub node is confirmed by applying to a simple network. Figure 5 shows the condition of the simple network problem. Twelve nodes are arranged on a circle with a diameter of 5 km. The infrastructure network (street network) is assumed to be complete graph. Each node has demand of 200 (person/h). (It is allowed that the generated hub node has direct links to the other 12 nodes.) The capacity and speed of the vehicle are 50 (persons) and 20 km/h respectively.

Figure 6 shows the comparison of the history of the maximum fitness in the populations between GA and CS. In the figure (a), GA with larger crossover rate outputs better solution. In the figure (b), mutation rate of the GA does not affect to the quality of the final solution. But, the fitness by the mutation rate set to 1.0 needs more steps to reach the final solution than that of mutation rate set to 0.8. Although the final solution is the same quality between GA (Pc ≥ 0.8) and CS, it is clear that the CS finds the best solution earlier than GA.

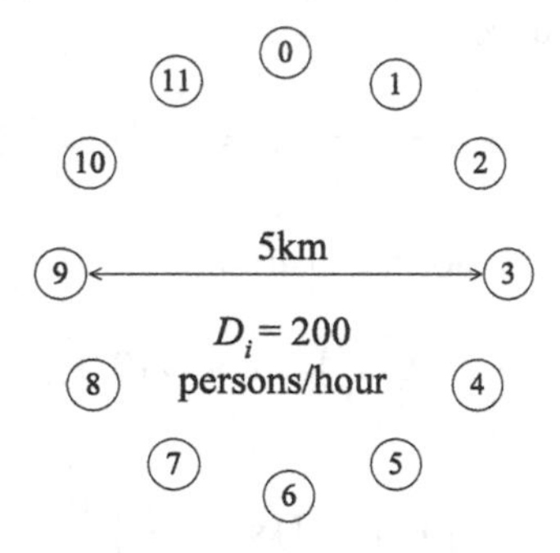

Fig. 5 Simple network problem

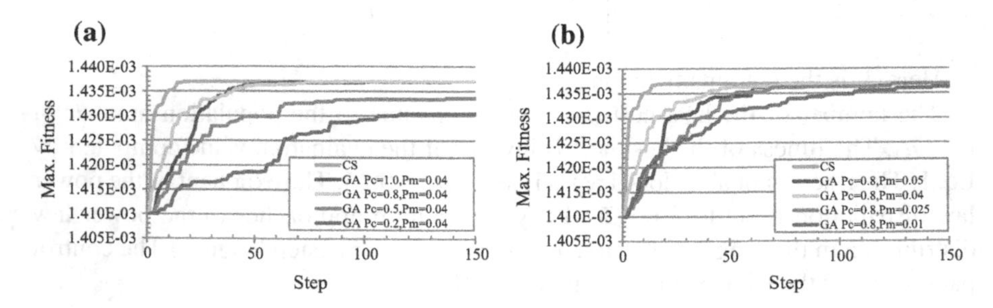

Fig. 6 Evolution history of maximum fitness averaged on 10 trials (Pc: Crossover Rate, Pm: Mutation Rate). **a** Pm0.04, **b** Pc = 0.8

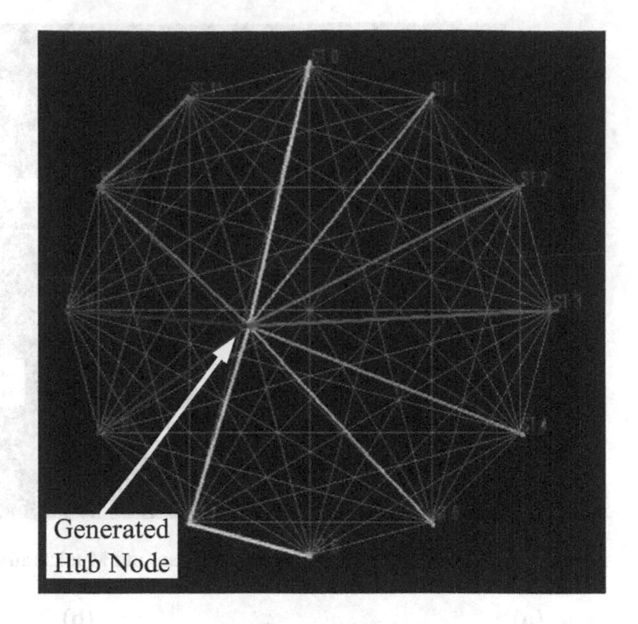

Fig. 7 Optimized hub location and route chart. Different color means different line. *Gray line* is street network

Figure 7 shows the hub location and he route chart of the best solution. The line number of the best solution is 10. The vehicle number is 20. The total travel time is 495 h, in which moving time is 324 h and waiting time is 171 h. Although the demand pattern is homogeneous, the hub location is not center of the circle. The number of vehicle in the Eq. 1 is integer and this constraint causes asymmetrical route pattern. Furthermore, it is noteworthy that routes of the lines go through several nodes before reaching to the hub node. This route pattern distinguishes this method from the other methods for the hub location problems.

3.2 *Real Street Network*

Figure 8 is the street network and bus stops in the southern area of the Mitaka railway station in Tokyo. Fifty bus stops have passengers going to the Mitaka station. Total demand during rush hours is assumed to be one-third for one day, about 1680 (person/h) that was obtained from a census data [12]. Parameters for the analysis is same to that for the simple network in the Sect. 3.1. In this real network case, it is allowed that generated hub node has only one link connecting to the nearest point of the street network.

Figure 8 shows the routes of generated lines and the location of the hub node. In all cases, the optimized hub location is about 2 Km away in southward direction from the Mitaka station. The line number of the best solution is 20. The vehicle number is 51. The total travel time is 190 h, in which moving time is 145 h and waiting time is 45 h.

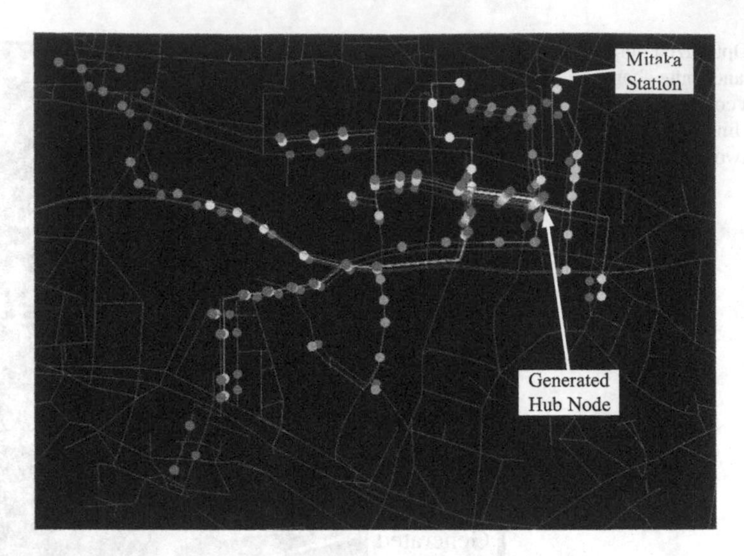

Fig. 8 Optimized hub location and route chart around Mitaka railway station

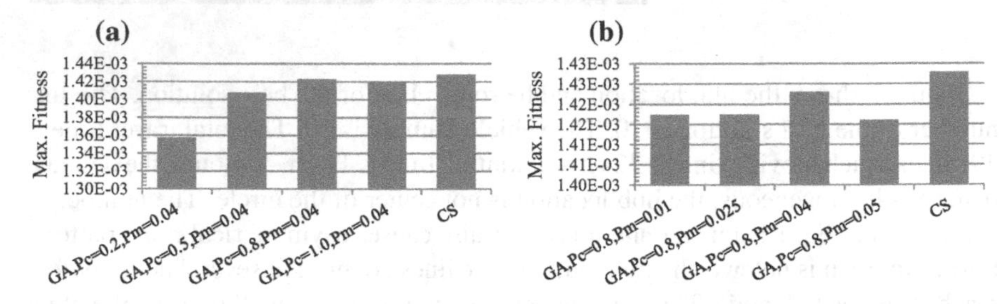

Fig. 9 Fitness of final solution averaged on 10 trials. **a** Pm = 0.04, **b** Pc = 0.8

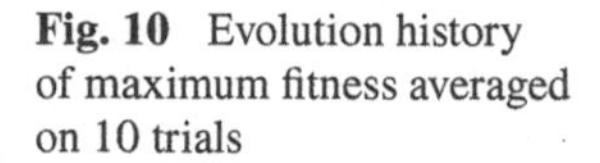

Fig. 10 Evolution history of maximum fitness averaged on 10 trials

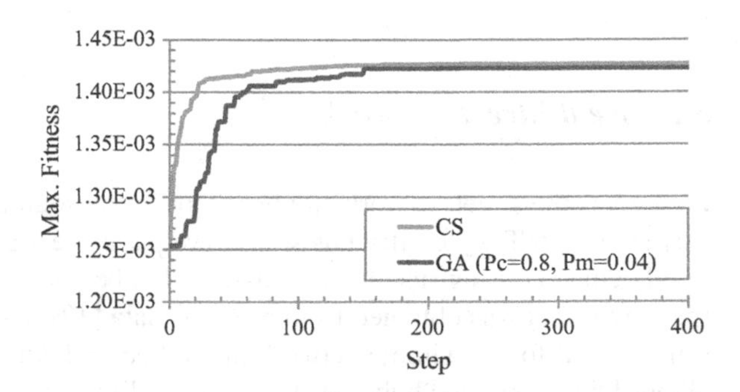

Figure 9 shows the maximum fitness of the final solutions averaged on 10 trials. The result of the CS is better than that of GA. Although the difference between GA(Pc $\geq$ 0.8) and CS is small, Fig. 10 shows that CS reach the best solution at smaller step than that of CS.

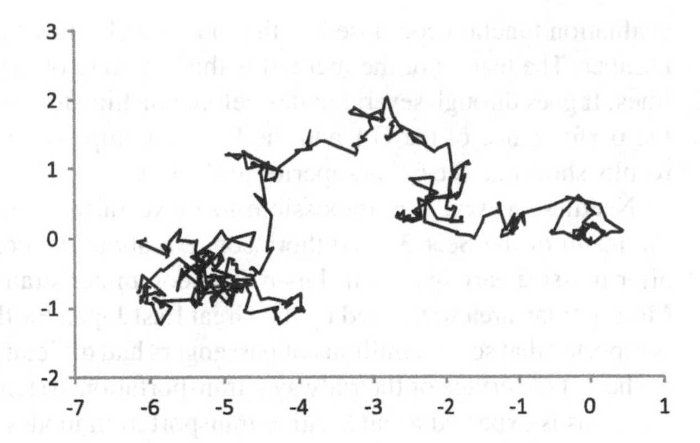

Fig. 11 Trajectory Example of Lévy flight

4 Discussion

It is confirmed that both GA and CS succeed in increasing the fitness of the populations smoothly. However, the numerical experimental results show that the performance of the CS is better than that of the GA. The CS finds out the best solutions earlier than the GA. (The computation time of the two algorithms is almost same. In the simple network problem, the computation time of the GA (Pc = 0.8, Pm = 0.04) is 0.46 s and that of the CS is 0.28 s. In the real street network, the computation time of the GA (Pc = 0.8, Pm = 0.04) is 110 s and that of the CS is 95 s. All numerical experiments were conducted by a laptop PC with Core i7, 3.10 GHz.) This result implies that the CS has capability to save computation resources, which is crucial for large scale optimization problems.

Figure 11 shows the reason why the CS has capability to find better solutions immediately. The position searching process of the CS is based on the Lévy flight. It means that the searching space is focused around a certain position in most case, but sometime it jumps as shown in the Fig. 11 This mechanism is compatible with the position searching process in the real world.

The parameter, w in Eq. 1 controls the balance of the cost between passengers and transportation companies. This value greatly influences the solutions. However, considering the averaged salary per hour of Japanese workers and running cost of the transportation vehicles in Japan, it becomes order of 10. This is the reason that w was fixed to 10 in this paper.

5 Conclusion

The method optimizing hub location for the PTNs is developed by combining the growing network model and two optimizing location algorithms, Genetic Algorithm and Cuckoo Search Algorithm. It can find out optimized hub location and transportation line network simultaneously. Degree of optimization was measured by the evaluation function composed of the total travel time of passengers and the vehicle number. The feature of the method is the topology of the transit route of generated lines. It goes through several nodes before reaching to the hub node. In this problem, the performance of the GA and the CS are compared and numerical experimental results show that the CS is superior to the GA.

Needless to say, it is impossible to move railway stations to a new point like the result of the Sect. 3.2. Authors concern about the commuter stranded problem after massive earthquakes in Japan. The commuter stranded problem in the Tokyo Metropolitan area was posed by the Great East Japan Earthquake in the year 2011. It is reported that several millions of passengers had difficulty to return their homes due to the out of service of the railway's transportation system. In such situation, bus or waterbus is expected as alternative transportation modes and effective hub stations connecting local feeder lines and trunk lines should be generated from scratch. The method in this paper is complementary relationship with the method generating trunk line network [13] to organize whole network for the stranded commuter problem in the future.

Acknowledgments This work was supported by JSPS KAKENHI Grant Numbers 17360424, 25280116, and 16H03157.

References

1. T. Majima, K. Takadama, D. Watanabe, and M. Katuhara: Generating Hub-Spoke Network for Public Transportation, SWARM 2015: The First International Symposium on Swarm Behavior and Bio-Inspired Robotics, pp. 48–51 (2015)
2. D. Dubois, G. Bel and M. Llibre: A set of Methods in Transportation Network Synthesis and Analysis, Journal of the Operational Research Society, Vol.30, No.9, pp.797–808 (1979)
3. M.T. Gastner and M.E.J. Newman, Shape and Efficiency in Spatial Distribution Networks, Journal of Statistical Mechanics, P01015, (2006)
4. W. Lampkin and P.D. Saalmans, "The Design of Routes, Service Frequencies and Schedules for a Municipal Bus Undertaking: A case study", *Operation Research Quarterly*, No. 18, pp. 375–397, 1967.
5. T. Majima, K. Takadama, D. Watanabe, and M. Katuhara: Characteristic and Application of Network Evolution Model for Public Transport Network, Multiagent and Grid Systems, Vol. 12, No. 1, pp. 1–11 (2016)
6. M.E. O'Kelly: The location of Interacting Hub Facilities, Transportation Science, Vol. 20, pp. 92–106 (1986)
7. Z. Drezner, H.W. Hamacher (Eds.) : Facility Location, Applications and Theory, Springer (2002)
8. S. Gelareh and S. Nickel: A Benders Decomposition for Hub Location Problems Arising in Public Transport, Operations Research Proceedings, Vol. 2007, Springer, pp. 129–134 (2008)
9. S. Nickel, A. Schobel and T. Sonneborn: Hub Location Problems in Urban Traffic Networks, Mathematical methods on optimization in transportation systems, Kluwer Academic Publishers, pp. 95-107 (2001)
10. Xin-She Yang: Nature-Inspired Metaheuristic Algorithms, Second Edition Luniver Press (2010)
11. Iztok Fister Jr., Xin-She Yang, Dusan Fister, and Iztok Fister: Cuckoo Search: A Brief Literature Review, Cuckoo Search and Firefly Algorithm: Theory and Applications, Springer (2014)

12. Ministry of Land Infrastructure, Transport and Tourism Website, https://www.mlit.go.jp/sogoseisaku/transport/sosei_transport_tk_000007.html
13. T. Majima, K. Takadama, D. Watanabe, and M. Katuhara: Generation of Public Transportation Network for Commuter Stranded Problem, Proceedings of WEIN'16, pp. 1–8 (2016)

Randomising Block Sizes for BlockCopy-Based Wind Farm Layout Optimisation

Michael Mayo, Maisa Daoud and Chen Zheng

Abstract The BlockCopy stochastic local search algorithm is a state-of-the-art optimiser for the Wind Farm Layout Optimisation problem. Unlike many other metaheuristics-based optimisers, BlockCopy requires the specification of only one key parameter, namely a block size. In this paper, we investigate the effect on different block sizes on the optimisation results. Using standard benchmarks for the Wind Farm Layout Optimisation problem, we show that smaller fixed block sizes (relative to overall layout size) produce better optimised layouts than larger fixed block sizes. More interestingly, we also show that randomising the block size parameter results in optimisation performance at the same or a better level than that produced by the best algorithm with a fixed block size. Effectively, this means that the user can ignore the need to tune the block size parameter and simply randomise it instead. Such a strategy results in what is effectively a parameterless, but none-the-less effective, optimisation algorithm for the Wind Farm Layout Optimisation problem.

Keywords Wind farm layout optimisation · Blockcopy · Local search · Parameter tuning

1 Introduction

The Wind Farm Layout Optimisation problem concerns finding the optimal positions for wind turbines (termed "micro-siting" in the literature) in a planned wind farm [1–4]. In a wider context, the process of micro-siting occurs after (i) the site and boundaries for the wind farm has been chosen, (ii) the characteristics (i.e. typical distributions across wind speeds and directions) for the site have been measured,

---

M. Mayo (✉) · M. Daoud · C. Zheng
Department of Computer Science, University of Waikato, Hamilton, New Zealand
e-mail: mttd1@students.waikato.ac.nz; mmayo@waikato.ac.nz

C. Zheng
e-mail: cz90@students.waikato.ac.nz

G. Leu et al. (eds.), *Intelligent and Evolutionary Systems*,
Proceedings in Adaptation, Learning and Optimization 8,
DOI 10.1007/978-3-319-49049-6_20

and (iii) the make, model and other details of the particular turbines (which are manufacturer-dependent) have been chosen.

Micro-siting is critical in wind farm design for several reasons. The primary reason, which happens to be the one focussed on in this work, is the phenomenon of wake interference: if two or more adjacent turbines are incorrectly sited, then the wake (i.e. the reduced velocity downstream wind) of one turbine can unduly negatively influence the other turbines. This negative influence manifests as reduced downstream power generation or, alternatively, as an increased probability of damage to downstream turbines due to increased turbulence.

The Wind Farm Layout Optimisation problem therefore addresses the problem of turbine micro-siting in order to mitigate these negative effects of wake interference.

In terms of practical relevance, solving this problem is of immense importance to the global wind industry—and therefore the world economy in general. Wind farms worldwide are typically becoming larger and larger (for example, the London Array [5] generates 630 MW of power that could power 490,000 households) and therefore small efficiency gains in design may lead to power for significantly more households.

Conversely, the problem is also of considerable interest to researchers in meta-heuristics and related fields. This is for two reasons. Firstly, the solutions to the problem are fundamentally multi-dimensional rather than one dimensional as is typically the case for many blackbox test functions. In the simplest case, a wind farm can be represented as a two-dimensional (flat) layout with turbine positions specified using two coordinates x and y. In more complex cases, additional attributes such as turbine height further increase the dimensionality of the problem. Since wake effects are non-linear and concern interactions between neighbouring turbines, then correspondingly the optimisation problem is non-linear with considerable epistasis between the variables being optimised.

The second reason why the Wind Farm Layout Optimisation is of interest to researchers is because it represents a problem with a significantly complex evaluation function. Since wake effects between every pair of turbines must be computed individually, the time complexity of the simplest wake models is at least $\mathcal{O}(n^2)$ where n is the number of turbines. The makes the evaluation of a significant quantity of potential layouts infeasible, and therefore many methods fail to achieve satisfactory results.

In this paper, we continue our exploration of a recently proposed stochastic local search method for optimising wind farm layouts. The method, known as BlockCopy [6], divides a wind farm layout into square regions and performs stochastic local search by copying blocks of turbines from one region to another.

Previously, we were able to show that BlockCopy significantly outperforms another state-of-the-art algorithm called the Turbine Displacement Algorithm [7].

In this paper, we address one significant issue that arises immediately when the BlockCopy algorithm is used: how large should the blocks be? Our findings indicate that block size has a significant impact on the optimisation performance of the algorithm when the block size is fixed throughout a single run of the algorithm. However, if the algorithm is modified so that instead of a fixed block size, the search operator instead picks a random block size every time it is applied, then the approach performs as well as (or in some cases, better than) the best approach with a fixed block size. The net effect of the finding reported here is that the user is obviated from the need to select a block size, and the BlockCopy local search algorithm effectively becomes parameterless.

2 Background

2.1 Wind Farm Layout Optimisation Problem

In its simplest form, the Wind Farm Layout Optimisation belongs to the family of two-dimensional circle-packing problems: i.e. the problem is to find optimal positions of a set of points (x, y)—each point representing a turbine in our case—such that no two points lie too closely together and all points lie inside a pre-specified 2D shape.

Unlike typical circle packing problems, however, the objective in the wind farm case is not to maximise the number of packed circles but instead to maximise either the total energy efficiency of the farm or to minimise the cost of energy produced by the farm. The minimum distance constraint between the points represents the minimum inter-turbine distance: if turbines are placed to closely together, then wake effect models generally become inaccurate [4] and layouts cannot therefore be evaluated properly. Following Samorani [3] we set the minimum turbine distance to 120 m, or three times the rotor diameter for turbines with a 40 m diameter rotor.

In this work, we use the Jensen far wake model [8, 9] to evaluate potential layouts. The Jensen wake model is an analytic approach for assessing wake interferences for two-dimensional layouts where the distribution of potential wind speeds and directions at the farm site is assumed a constant.

Although originally proposed in the mid-1980s, the Jensen is still used widely in the community. To illustrate, Samorani [3] describes it precisely in a recent 2013 introductory survey on the Wind Farm Layout Optimisation problem, and Shakoor et al. [4] performed an extensive comparison of several different wake effect models, concluding that "...Jenson's far wake model is a good choice to solve the wind farm layout optimisation problem due to its simplicity and relatively high degree of accuracy."

A graphical illustration of the wake effect according to the Jensen model is given in Fig. 1. To explain briefly, incoming wind travelling at speed u_0 m/s reaches a turbine with rotor radius r_r. As the wind passes through the turbine's blades, its velocity is reduced. At a distance of x meters downstream, the wind speed is u_j m/s, which must be less than u_0. Moreover, the wake radius increases linearly in size with distance. At x meters downstream, its radius becomes $r_1 > r_r$.

The exact mathematical details of the Jenson wake model, including how the reduced wind speeds and the spreading wake radius should be calculated, are adequately explained by Samorani [3]. We follow the exact same approach in our

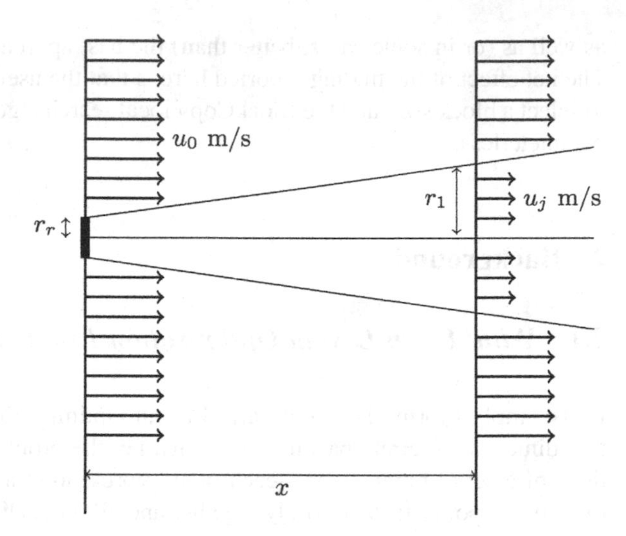

Fig. 1 Depiction of the wake effect (reproduced from [3])

implementation, and therefore the interested reader is referred to Samorani [3] for the specific mathematics behind the model (which we do not present here for reasons of space).

It suffices to say, however, that one key characteristic of the Jensen model is that the wind velocity deficit caused by a turbine is inversely proportion to the distance x squared from the turbine that caused it; therefore, eventually, most wakes become negligible for large enough x. However, if a turbine lies inside multiple wakes (which is highly likely for small, dense layouts) then the wakes aggregate and so even small wake effects must be calculated. Again, the interested reader should refer to Samorani [3] for the method of aggregating wakes.

One important aspect of the Jensen model is how a turbine's power output is computed. Generally speaking, such a calculation depends on the type of wind turbines being installed and is therefore manufacturer-specific. Therefore, in this paper, we used an idealised turbine model described originally in Mosetti et al. [2] and later described again by Samorani [3]. In this model, the power output of a single farm is proportional to the cube of the incoming wind speed between two boundary wind speeds. The boundary wind speeds are known as the cut-in speed and the nominal speed. If the wind speed increases beyond the nominal wind speed, then power generation becomes a constant until a cut-out speed is reached. For turbines with a cut-in speed of 2 m/s, a nominal speed of 12.8 m/s and a cut out of 18 m/s, then the power curve used to model individual turbines is:

$$power(u) = \begin{cases} 0 \text{ kw} & \text{where } u < 2 \text{ m/s} \\ 0.3u^3 \text{ kw} & \text{where } 2 \text{ m/s } \leq u < 12.8 \text{ m/s} \\ 629.1 \text{ kw} & \text{where } 12.8 \text{ m/s } \leq u < 18 \text{ m/s} \\ 0 \text{ kw} & \text{where } u > 18 \text{ m/s} \end{cases} \quad (1)$$

This is precisely the wind turbine model that we use in this research.

Finally, we briefly turn to the objective function for the Wind Farm Layout Optimisation problem. We have already alluded to the fact that it is focussed on minimising the wake effect. More precisely, if $\mathbf{l}$ is a layout, then the objective function can be written down as:

$$F(\mathbf{l}) = \sum_{s \in S} r_s \frac{\sum_{j \in \mathbf{l}} power\left(u_s(1 - v^s_{def}(j))\right)}{\sum_{j \in \mathbf{l}} power(u_s)} \tag{2}$$

where j is a turbine's position in the layout, and $v^s_{def}(j)$ is the total velocity deficit (i.e. total loss in power as a proportion between 0 and 1 due to wake interference) at j. Th total velocity deficit is a quantity aggregated across all other turbines in the layout that may be affecting the turbine at position j. S is a set of wind scenarios, in which each element of S is a potential wind speed/direction pair. This must be measured at the wind farm site and is considered to be a constant across the entire layout. The variable r_s represents the probability of scenario $s \in S$ and u_s is the specific wind speed under scenario s. It should be evident that $\sum_{s \in S} r_s = 1.0$ in order to compute proper expected power values.

The objective value of a layout, therefore, is defined as the total power output of the farm *with* wakes divided by the total, potential, power output of the farm *without* wakes. Clearly, this is a ratio that is maximised at 1.0, and therefore our aim is to find layouts with an objective value as close to 1.0 as is possible.

2.2 BlockCopy-Based Stochastic Local Search

The BlockCopy stochastic local search algorithm, first described by Mayo and Zhen [6], optimises wind farm layouts by defining a search operator that copies entire groups of turbines at a time from one place on the layout to another. It iteratively applies this operator, starting with a randomly-generated layout called the "current" layout, to generate a progression of new layouts. The current layout always represents the fittest layout found so far. If a new layout has improved fitness compared to the current layout (as a result of one block copy operation), then the new layout becomes the new "current" layout of the search. Otherwise, if the new layout has a lesser fitness, it is discarded.

The basic search operator is illustrated in Fig. 2. In our current formulation of the algorithm, the regions that are copied are square "blocks" of a fixed size such that the layout can be divided into blocks that do not overlap and that exhaustively cover the entire layout.

One application of the BlockCopy operator proceeds by randomly selecting a source and destination block. The turbines in the destination block are deleted, and then turbines from the source block are copied to the destination block one at a time. Such an approach basically preserves the local configuration of turbines in the source block.

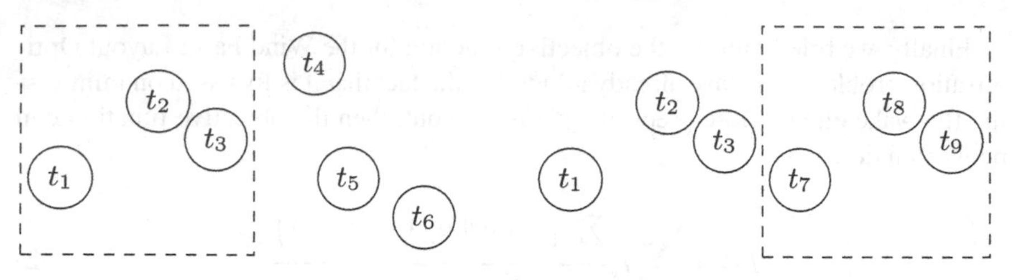

(a) Initial layout with selected source block.

(b) Final layout showing new target block.

Fig. 2 Illustration of the BlockCopy operator. In this example, the *left* block of a small layout is duplicated to the *right* hand side of the layout

After the copy operation is complete, if the total number of turbines in the layout has either increased or decreased (because of differences in the number of turbines present at the source and destination block), then turbines are either randomly added or randomly purged from the layout globally in order to keep the total number of turbines in the layout at a constant value.

All constraints must also be adhered to by the operator so that new layouts are always legal: if the copying of any turbine would result in a constraint violation, then it is simply not copied.

The reasoning behind this approach is that by maintaining the relative configuration of turbines whenever a block is copied, then if a particularly good local configuration of turbines is present in the layout, then this configuration will quickly replicate itself across the layout via successive BlockCopy operations.

In the initial recently published evaluation of the algorithm [6], it was shown that BlockCopy outperforms one state-of-the-art approach called the Turbine Displacement Algorithm [7] on a set of benchmark problems using a cost-based objective function and a different far wake model than Jensen.

Moreover, it was also shown in the previous paper that the local search variant of BlockCopy (with only a single current layout held in memory) outperformed a population-based evolutionary variant also using the BlockCopy operator for mutation and/or crossover. We believe that the primary reason for this is the limited number of evaluations that the problem affords: local search effectively performs significantly more exploitation of the search space than population-based approaches do. Population-based strategies, in contrast, are better balancers of exploitation and exploration. The cost of this, however, is that more evaluations are required to reach the same level of fitness. We believe this makes local search a better option given the time complexity of the evaluation function.

A significant disadvantage of the BlockCopy approach, however, is that it is not clear what the best default size for the blocks should be. We therefore conducted the set of experiments described in the next section to assess the impact about different block size decisions on the optimisation results.

3 Experimental Setup

In order to perform experiments related to Wind Farm Layout Optimisation, one must firstly define and/or obtain some benchmark problems so that any results can, in the future, be compared with other results from the literature.

In this paper, therefore, we assume that the layout is comprised of a fixed quantity of 64 turbines, and that this number can neither increase nor decrease. Furthermore, we also assume that the layout is a square of size 1.5 km × 1.5 km. Turbines cannot be placed outside of the layout boundaries. Fixing the layout to such a small size effectively causes the wake effects to become a non-trivial negative influence on the wind farm's efficiency, which makes the problem interesting.

The optimisation algorithms are all initialised with different random starting layouts. The algorithm for constructing a random starting layout is straightforward: turbines are iteratively added at random locations on the layout, as long as they do not cause any constraint violations (i.e. the new turbine must not be placed too closely to an already-placed turbine). If a constraint would be violated by a placement, then a new random position is chosen for the current turbine. This is repeated until all 64 turbines have been initially placed.

Next, we utilise Samorani's [3] three different problems for benchmarking wind farm layout optimisation algorithms. The benchmark problems are defined by Table 1.

Problem A is the simplest benchmark, and consists of only a single wind scenario in which wind blows with uniform expected speed and in a single direction. The set of scenarios S for the objective function therefore consists of only a single element.

Problem B, alternatively, consists of 36 different wind scenarios. Each scenario differs only in the wind direction. while the expected wind speed is a constant. Unlike Problem A, therefore, this benchmark has no single dominant direction.

Problem C is the most interesting and challenging of the three benchmarks. As is the case with Problem B, there are 36 possible wind directions. In Problem C's case, however, for each different wind direction, there are also three different expected wind speeds. Furthermore, there is a clear dominant wind direction: 310° is the direction with the highest probability of the greatest wind speed, and therefore it is also the direction of the greatest power production. In total, Problem C consists of 108 different wind scenarios.

Samorani [3] describes Problem C graphically by means of a histogram of wind speeds versus directions. In order to implement this benchmark, we therefore reverse-

Table 1 Problems from Samorani [3]

Problem	Direction(s)	Expected speed(s)	#Wind scenarios
A	$\{0°\}$	{12 m/s}	1
B	$\{0°, 10°, \ldots, 350°\}$	{12 m/s}	36
C	$\{0°, 10°, \ldots, 350°\}$	{8 m/s, 12 m/s, 17 m/s}	108

Table 2 Probabilities used for the 108 wind scenarios under Problem C (rounded to three significant figures) derived from a chart in [3]

Direction	$u_s = 8$ m/s	$u_s = 12$ m/s	$u_s = 17$ m/s
0°–260°	0.00404	0.00865	0.0115
270°	0.00404	0.0107	0.0127
280°	0.00404	0.0121	0.0156
290°	0.00404	0.0141	0.0185
300°	0.00404	0.0138	0.0300
310°	0.00404	0.0190	0.0352
320°	0.00404	0.0138	0.0300
330°	0.00404	0.0141	0.0185
340°	0.00404	0.0121	0.0156
350°	0.00404	0.0107	0.0127

engineered the probabilities from his publication. The probabilities we used for Problem C are given in Table 2.

Now that the three benchmark problems have been described, we next describe the variants of the BlockCopy local search algorithm that we tested.

The basic difference between the variants is how the size of the block is chosen. We selected four sensible fixed sizes for the blocks that were chosen because they divide the 1.5 km × 1.5 km layout evenly. The fixed sizes were: 125 m (which divides the layout into 12 × 12 blocks), 250 m (dividing the layout into 6 × 6 blocks), 500 m (making 3 × 3 blocks) and 750 m (which is 2 × 2 blocks). Each fixed size corresponds to one algorithm variant. We also tested a fifth algorithm which selects a block size from the above set of four sizes at random each time it performs a BlockCopy operation. We call this algorithm simply "random".

In order to obtain statistical results, we ran each algorithm 30 times on each benchmark. Each algorithm was run for 20,000 iterations before terminating. Therefore the total number of experimental runs performed was 3 benchmarks × 5 algorithms × 30 repeats, or 450 runs in total.

4 Results

The results of our experiments are given in Figs. 3, 4 and 5 for each of the problems A, B and C, respectively. Each figure has two parts: a subfigure (a) showing the convergence curves for each of the five algorithms (averaged across 30 runs), and a box-and-whiskers plot (b) showing the distribution of final results for each algorithm.

Examining the convergence curves firstly, we can see that the choice of block size has a significant impact on convergence performance. In particular, the very large

Fig. 3 Results for Problem A (see online colour version for best viewing)

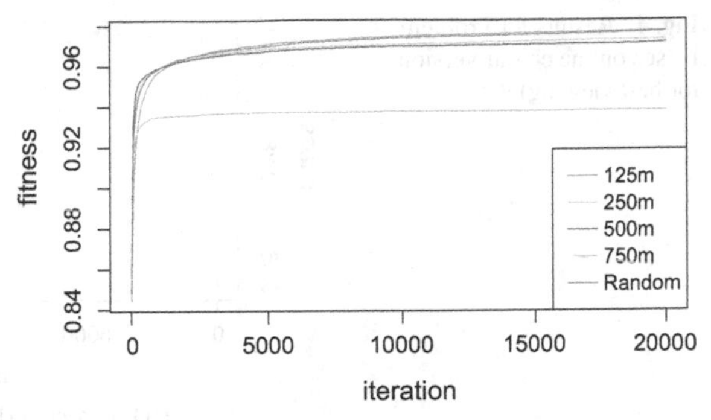

(a) Averaged convergence curves.

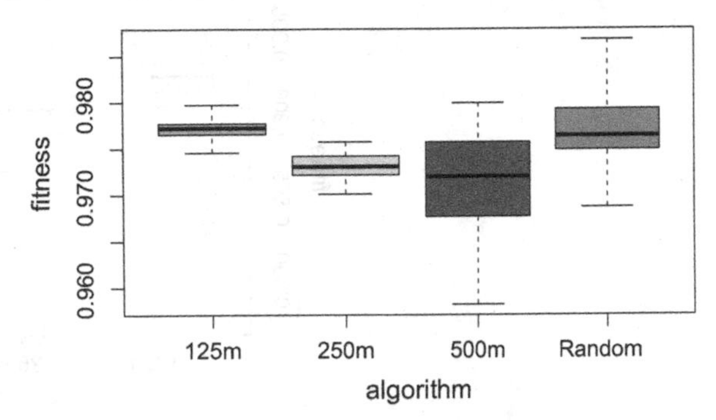

(b) Final fitness distributions.

block sizes (750 m for Problem A and 500 m + for Problems B and C) result in significantly worse convergence curves. This is to be expected since larger block sizes correspond to larger steps in the search space—and such large steps may miss nearby local optima. Conversely, for the small block sizes (125 and 250 m) the convergence curves are all fairly similar.

Figures 3b, 4b and 5b depict the final distribution of fitness values after each of the 30 runs. Generally speaking, larger fixed block sizes clearly shift the median fitness down compared to smaller fixed block sizes. This negative effect is most dramatic for the algorithm with a fixed block size of 750 m, which also consistently has the worst convergence curve. This algorithm's fitness distribution is so low that its particular set of results have been excluded from the plots so that the rest of the results can be read clearly. The worsening effect of larger block sizes is clearly apparent for the 500 m block size when inspecting the box plots.

In terms of answering the question of which is the best overall fixed block size, the answer to that question appears to be 125 m—which is also the smallest fixed block size considered. In all cases, the median and maximum fitnesses achieved by the 125 m variant exceed those of all the other fixed block size algorithms.

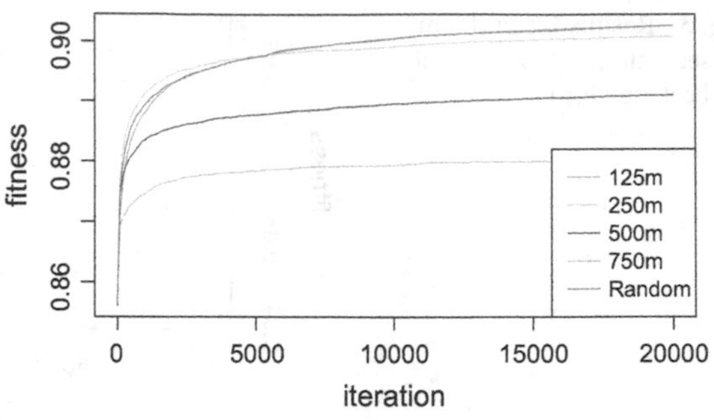

(a) Averaged convergence curves.

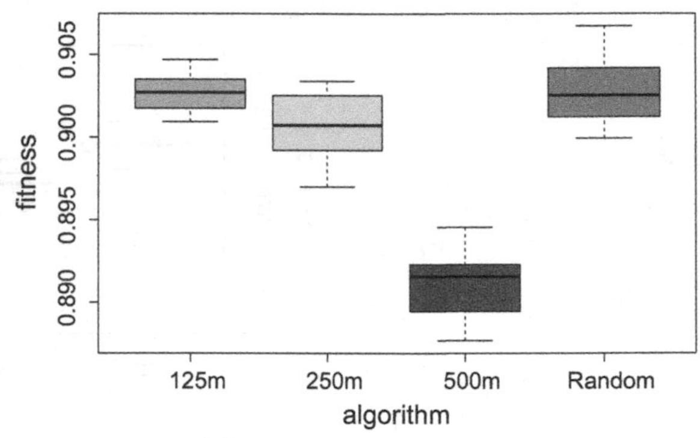

(b) Final fitness distributions.

Fig. 4 Results for Problem B (see online colour version for best viewing)

The exception to the trend, however, is the random block size algorithm. Interestingly, this algorithm has quite a different behaviour to the fixed-size algorithms.

Firstly, it is always competitive with the other algorithms. In terms of median overall performance, it always performs slightly below the 125 m fixed algorithm's median, but this performance is obviously not statistically significant due to the close overlap of the distributions.

Most interestingly, however, the random algorithm's maximum (as opposed to median) achieved fitness over all runs turns out to be the greatest for each of the benchmark problems. The explanation for this appears to be largely due the random algorithm having a greater variance in final fitness results. This is increased variance is especially evident in the box-and-whisker plots for Problems A and B. It is also somewhat evident visually in the Problem C plots. As a consequence of this higher variance, over thirty runs, the random algorithm always finds the best single layout compared to the 125 m algorithm. This difference is quite small for Problem C, but an inspection of the numeric results used to generate the plot indicate that the random algorithm does indeed produce the overall best layout.

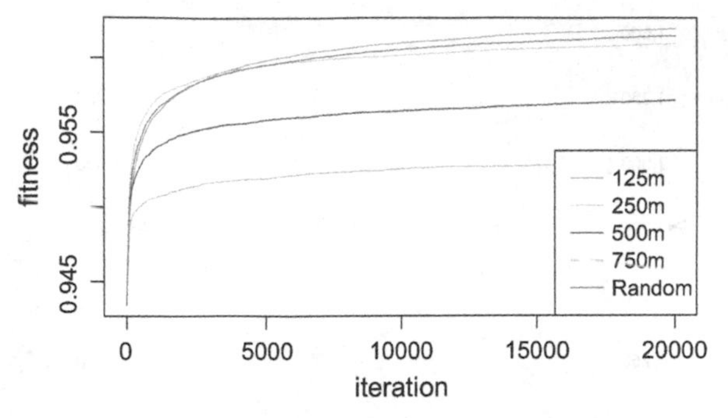

(a) Averaged convergence curves.

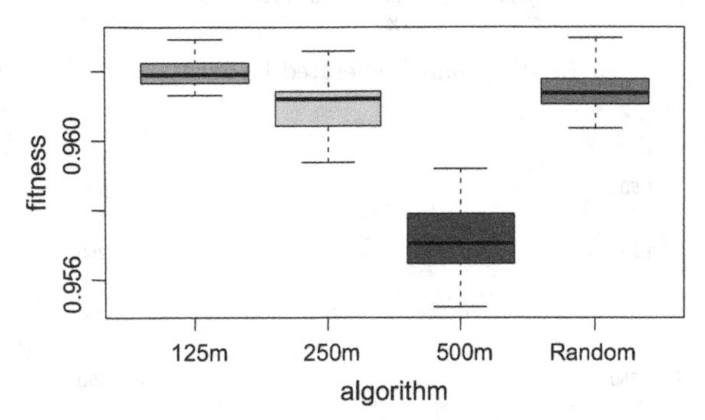

(b) Final fitness distributions.

Fig. 5 Results for Problem C (see online colour version for best viewing)

To conclude this section, some of the best optimised layouts found by the random algorithm are presented in Fig. 6. By way of contrast, the layouts are shown alongside one example of a random unoptimised layout.

5 Conclusion

Overall, the results show that the randomised block size algorithm is a better choice than the algorithms that have a fixed block size, as long as the goal is to find the single best layout over multiple runs of the BlockCopy local search algorithm. Such a multiple-restart approach is in fact a sensible approach to take for most difficult optimisation problems.

Moreover, the fact that the random algorithm is competitive with the best algorithm using a fixed block size means effectively that the user can bypass the problem of parameter tuning. Instead, the algorithm needs simply to create a set of appropriate block sizes for random block size selection. The tuning and selection of parameters

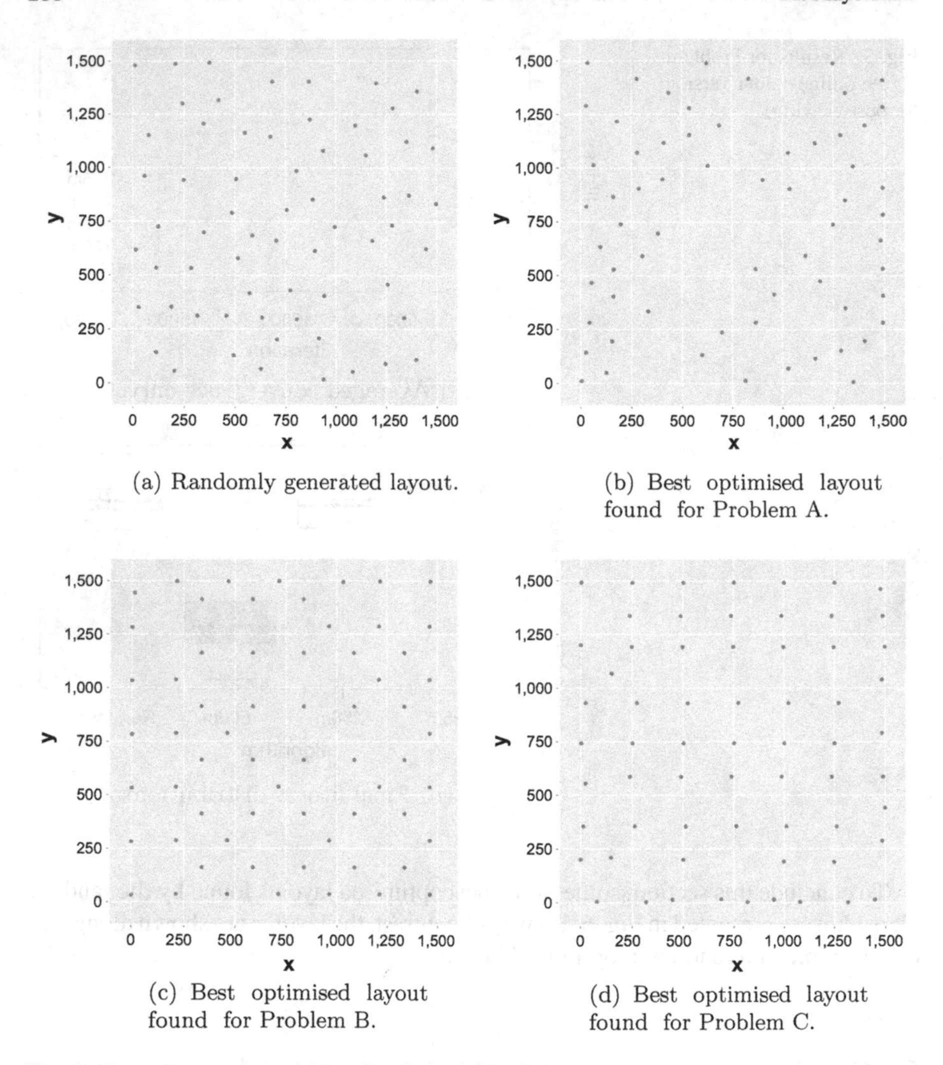

(a) Randomly generated layout.

(b) Best optimised layout found for Problem A.

(c) Best optimised layout found for Problem B.

(d) Best optimised layout found for Problem C.

Fig. 6 Example unoptimised (**a**) and optimised (**b**)–(**d**) layouts

is known to have a significant impact on the performance of metaheuristic optimisation algorithms [10]. By-passing this problem by simply randomising the main key parameter of the algorithm has been shown to be effective.

To conclude, future work will continue address the problem of improving the behaviour of the BlockCopy local search algorithm for the Wind Farm Layout Optimisation problem.

References

1. Mayo, M., Daoud, M.: Informed mutation of wind farm layouts to maximise energy harvest. Renewable Energy 89, 437–448 (2016)
2. Mosetti, G., Poloni, C., Diviacco, B.: Optimization of wind turbine positioning in large wind farms by means of a genetic algorithm. Journal of Wind Engineering and Industrial Aerodynamics 51(1), 105–116 (1994)
3. Samorani, M.: The wind farm layout optimization problem. In: Pardolas, P. (ed.) Handbook of Wind Power Systems, pp. 21–38. Springer-Verlag (2013)
4. Shakoor, R., Hassan, M., Raheem, A., Wu, Y.: Wake effect modelling: A review of wind farm layout optimization using Jensen's model. Renewable and Sustainable Energy Reviews 58, 1048–1059 (2016)
5. London Array brochure. Online PDF brochure, retrieved 9 Nov 2015, http://www.londonarray.com/wp-content/uploads/London-Array-Brochure.pdf
6. Mayo, M., Zhen, C.: Blockcopy-based operators for evolving efficient wind farm layouts. In: Proc. IEEE World Conference on Computational Intelligence, WCCI. p. to appear (2016)
7. Wagner, M., Day, J., Neumann, F.: A fast and effective local search algorithm for optimizing the placement of wind turbines. Renewable Energy 51, 64–70 (2013)
8. Jensen, N.: A note on wind generator interaction. Tech. rep., Risø DTU National Laboratory for Sustainable Energy (1983)
9. Katic, I., Høstrup, J., Jensen, N.: A simple model for cluster efficiency. In: Proc. Europe and Wind Energy Association Conference and Exhibition (1986)
10. Karafotias, G., Hoogendoorn, M., Eiben, A.E.: Parameter control in evolutionary algorithms: Trends and challenges. IEEE Transactions on Evolutionary Computation 19(2), 167–187 (2015)

Optimization of Aircraft Landing Route and Order Based on Novelty Search

Akinori Murata, Hiroyuki Sato and Keiki Takadama

Abstract This paper focuses on the Aircraft Landing Problem (ALP) and proposes the efficient aircraft landing route and order optimization method compared to the conventional method. As a difficulty in solving ALP, both landing route and order of *all* aircrafts should be optimized together, meaning that they cannot be optimized independently. To tackle this problem, our method employs *novelty search* to generate variety candidates of aircraft landing routes, which are indispensable to generate the feasible landing order of all aircraft. Through the experiment on a benchmark problem, it has revealed that the proposed method can reduce the occupancy time of aircrafts in an airport.

Keywords Aircraft scheduling · Landing route optimization · Landing order optimization · Evolutionary computation · Novelty search

1 Introduction

In aircraft landing, air traffic controllers should determine both the aircraft landing routes and their landing order to minimize an occupancy time of aircrafts in an airport as air transportation service [1]. Such landing route and landing order are important issue because of directly affecting the occupancy time. This problem is called as the aircraft landing problem (ALP) [2]. For this issue, the conventional research tackled this problem by dividing into the following two problems: (1) the landing route optimization problem and (2) the landing order optimization problem.

A. Murata (✉) · H. Sato · K. Takadama
The University of Electro-Communications,
1-5-1 Chofugaoka, Chofu, Tokyo 182-8585, Japan
e-mail: kouho.aki@cas.hcuec.ac.jp

H. Sato
e-mail: sato@hcuec.ac.jp

K. Takadama
e-mail: keiki@infuec.ac.jp

G. Leu et al. (eds.), *Intelligent and Evolutionary Systems*,
Proceedings in Adaptation, Learning and Optimization 8,
DOI 10.1007/978-3-319-49049-6_21

Regarding the first problem, Tajima et al. proposed the real-time path planning method that can evolve the landing routes of the aircraft by evolutionary computation [3]. Regarding the second problem, on the other hand, Xiao proposed the binary-representation-based genetic algorithm method that can evolve the landing orders of the aircraft by evolutionary computation as the aircraft arrival sequencing and scheduling problem [4]. Many previous works have dealt with either one of those specific problems [5]. What should be noted here is that there is no method that can cope with the aircraft landing problem, meaning to solve both the landing route and landing order optimization together.

To tackle this problem, we previously proposed the new mechanism for ALP, which can address both the landing route and landing order optimization together. In detail, aircrafts individually generate candidates of their own routes toward the destination and appropriate combinations of routes of all aircrafts are explored by multi-objective evolutionary computation. This mechanism successfully optimized the landing route and order of aircrafts [6], but it has some limitations on finding the *robust* landing route and order of aircrafts, which can cope with a change of their landing routes and order due to delay of other aircrafts or weather change such as typhoon. Since our previous method aims at minimizing the distance of individual landing route, it is difficult to find the robust landing routes. To overcome this issue, this paper proposes the new method that generates the robust landing route and order of aircrafts by employing the *novelty search* [7] while minimizing the distance of individual landing routes. We employ the novelty search because this method searches the not-explored solution space (i.e., the different landing routes and order in ALP) in high priority, which contributes to finding the good routes and orders needed for delay of other aircrafts or weather change.

This paper is organized as follows. Section 2 gives a brief description of related works on the aircraft landing problem and novelty search. Section 3 proposes the aircraft landing route and order optimization method which employs novelty search. Section 4 conducts the experiments and Sect. 5 discusses its result. Finally, we summarize the contribution of this paper and show future works in Sect. 6.

2 Related Works

2.1 Hierarchical Evolutionary Computation

In my previous work, we proposed optimization method based on hierarchical evolutionary computation. our approach aims at optimizing both landing routes and the landing order together to minimize the occupancy time of airport. Specifically, different from the Xiao's method [4], our method optimizes not only the landing order but also the landing routes that follow the optimized landing order; unlike the Tajima's method [3], our method optimizes the landing routes for aircrafts.

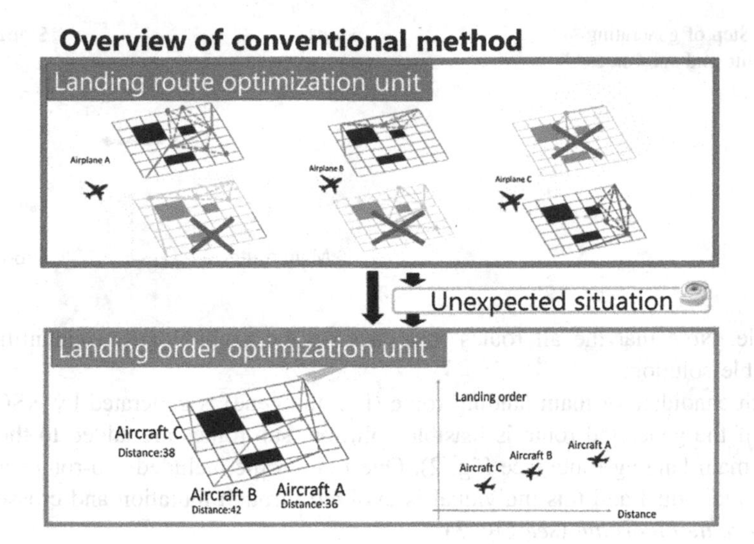

Fig. 1 Architecture

As shown in Fig. 1, our method is that produces candidates of the robust landing routes that can be flexibly customized depending on the changeable situations beforehand (i.e., wake turbulence and a congestion of aircrafts). Our method is based on NSGA-II as a multi-objective optimization technique [8] and composed of the following two units;

Landing route optimization unit: which makes candidates of landing route from the current position to the destination airport. Each landing route is represented by a gene as in Tajima's method; the gene represents a set of waypoints. Then, some sub-routes are added to each generated main route as options that enable the main route to be flexibly customized depending on the situations.

Arrival sequence optimization unit: which selects the candidates of main route or customizes the main landing route by replacing with the sub-routes depending on the situations. Then, it optimizes a combination of landing routes of multiple aircrafts in order to minimize the occupancy time of destination's runway.

Landing Route Optimization Unit

As shown in Fig. 2, this unit first generates candidates of main route and then, adds sub-routes to each of the generated main landing route. Then the candidates are evaluated in terms of two aspects; (1) the total distance of main route from the current position to the destination airport and (2) the robustness of the route can be quantified as the number of possible sub-routes included in the main route. Thus, this unit eventually produces the landing routes which indicates short distance to the destination airport and can be customized for adapting as many situations as

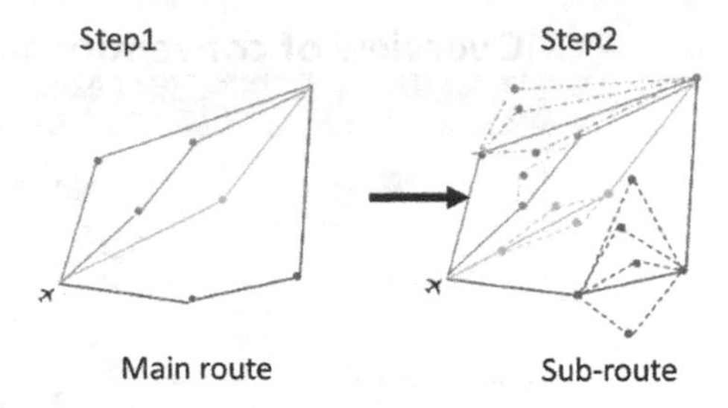

Fig. 2 Step of generating main route and sub-routes

possible. Note that the all routes that pass on the prohibit area is identified as infeasible solution.

Each candidate of main landing route (i.e., the gene) is generated by NSGA-II. Then, if the generated route is feasible solution, sub-routes are added to the generated main landing route (see Fig. 2). One main route included sub-routes are set of one individual and this individual is evolved through mutation and crossover;

Sub-route algorithm (*see* Fig. 3)

Step 1: For one of main route, calculating each the distance between the waypoints.

Step 2: Creating a center point which is a midpoint between the waypoints and drawing a line on the vertical bisector through this center point. Setting M points on the line at regular intervals. This points become new waypoints. Maximum angle θ is the parameter.

Step 3: New route added the new waypoint created in Step 2 is calculated the distance.

Step 4: This route is saved as one of sub-route if the limitations of this route are met.

Step 5: Repeating from Step 2 to Step 4 until the termination condition is satisfied.

After adding the sub-routes to each landing route, they are evaluated with the fitness function and applied to genetic operators (i.e., the crossover and mutation) to find better solutions with a high fitness which have small distances to the destination and many sub-routes. Overall procedure can be described as follows;

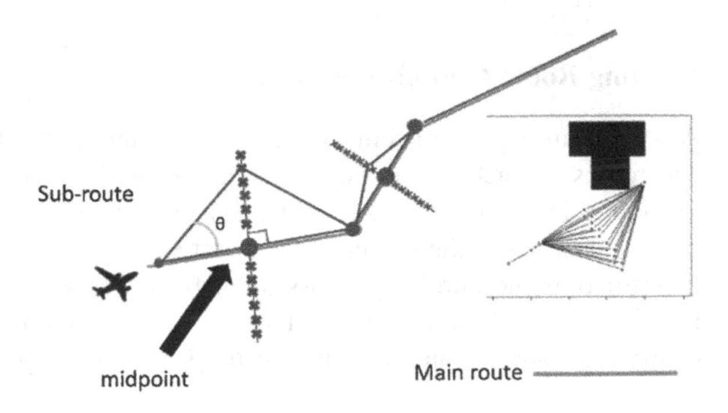

Fig. 3 The main landing route with the sub-routes

Calculating fitness of each route in terms of the distance to the destination airport and the number of sub-routes. Selecting routes as parents and copying them as the offspring. Then, the crossover and the mutation are applied to generated offspring. Producing a next generation population by the non-dominant sorting and degree of congestion tournament selection.

Evaluation Function

Each route is evaluated based on two factors. One is a distance to the destination airport. The other is robustness for changing environment. An evaluation function for the distance is as following equation.

$$\text{Distance evaluation} = \frac{1}{distance} \tag{1}$$

where distance indicates total length of the route. From the equation, the shortest distance to the destination airport is the maximum evaluation value of the candidates. Next, an evaluation function of the robustness is calculating as follow.

$$\text{Robustness} = \frac{\textit{the number of feasible detour routes}}{\textit{generating total detour routes}} \tag{2}$$

where generating total detour routes means all detour routes or, to put it in another way, the number of feasible and infeasible detour routes. And the number of feasible detour routes indicates that the routes are not on the prohibit area.

Arrival Sequence Generating Unit

The generating arrival sequence unit picks out the top N candidates by the non-dominant sorting and degree of congestion tournament selection and searches the best combination of landing routes for some aircrafts by GA.

Each locus of gene is represented in the integer number and the selected route from the candidates of landing route for the aircraft. This unit takes the following steps;

Step 1: All aircraft choose one route from the candidates of landing route
Step 2: Checking on the interval of aircrafts whether meet constraint condition or not.
Step 3: Calculating fitness that meets limitation
Step 4: Selecting parents from among solutions that meet limitation by using tournament selection
Step 5: Crossing between parents and mutation
Step 6: Repeating from Step 2 to Step 5 until the termination condition is satisfied

After this step, main route of each aircraft has been determined. If new aircrafts appear in next step, the aircraft which determined the main route takes an

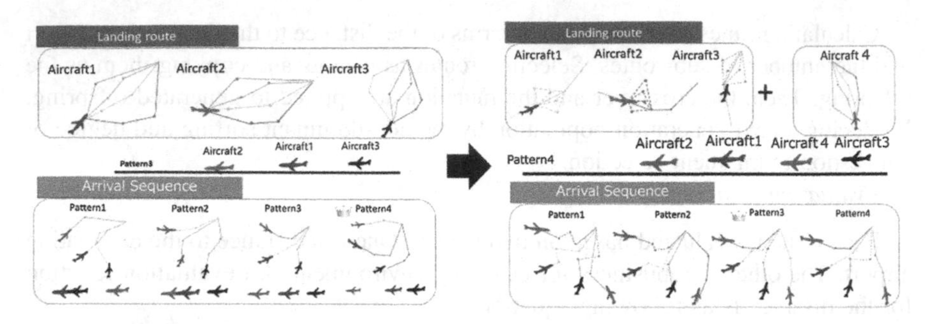

Fig. 4 Overview of arrival sequence generating unit

opportunity to switch sub-route (Fig. 4-left). In case of this, arrival sequence generating unit replace sub-routes for main routes and research the best combination of landing routes by GA (Fig. 4-right).

The following are described the limitation condition that all intervals of the arrival sequence keep the distance appropriately, how to calculate evaluation value of the solution (Step 2 and Step 3) and how to mutate and crossover (Step 5) in detail.

Constraints and evaluation value after generating of the combination of routes for some aircrafts, the arrival sequence generating unit determines the landing order of aircrafts. Here, when the landing order indicates that an interval of landing distance is small, its landing order can be identified as infeasible. Since an aircraft lands on the airport in safety, an enough interval is required. Thus, to identify the generated landing order either one of feasible solution or infeasible one, we add the following constraint condition for landing order;

$$d_n - d_p > r \tag{3}$$

Symbol d_n and d_p represent the distances to the destination of two aircrafts n and p; thus $d_n - d_p$ represents the interval between the two aircrafts. Then, if the interval is larger than a threshold r for all possible route combination of two aircrafts, its landing order is identified as a feasible solution.

Additionally, to evaluate the generated landing order we introduce the following equation as calculation of fitness;

$$\text{Fitness} = \sum_{i=1}^{n} d_i \tag{4}$$

The fitness of landing order is simply calculated as the summation of distance of each aircraft. It is the distance of main route. Thus, the small value of fitness means that its landing order can reduce the occupancy time of the destination airport.

2.2 Novelty-Search

Novelty-search is radical evaluation method of individual, introduced by Lehman [8]. What should be noted here is that novelty-search does not use fitness function and simple evaluation design. The fitness is depended on a problem. So, it is hard to design fitness function. Especially, deceptive problem, i.e. maze problem, is inferior to other evaluation method. On the other hand, novelty-search is high-performance of this problem because it attaches importance to a uniqueness. The uniqueness means difference in other individuals in that a relative position. Measuring distance between target individual and other individuals by Euclid or Mahalanobis, or other metric [9].

2.2.1 Novelty Metric

Novelty-search measures how unique this is in all solutions and novelty metric is calculated as following an equation.

$$\rho(x) = \frac{1}{K}\sum_{i=1}^{K} dist(x, \mu_i) \tag{5}$$

where μ_i indicates the individual which is ith-nearest neighbor in the population and x means the evaluated individual. The "dist" function is used Euclid ($\|x - \mu_i\|$) as usual. For scoring novelty, a target group as the neighbor is used an archive or the population which is current generation. The number of K is vary from the population and problems, and algorithms. K value is most used 15 [10].

3 Proposed Method

This paper introduces a new novelty metric, novelty scaling. The new metric is expected to keep important candidates. This metric is calculated by following law.

$$\rho(x) = \frac{1}{K}\sum_{i=1}^{K} dist(x, \mu_i)\frac{1}{a^{\left(\frac{distance}{d\_min} - 1\right)}} \tag{6}$$

where distance is and d_min indicates the minimum distance in the candidates at the current generation, and a is the parameter. This function makes weighted factor as the center of the minimum distance. The more a is increased, the narrower scale of distance is. Though the conventional method use a robustness of the situation change as the objective function, the proposed method replaces this objective function with this function.

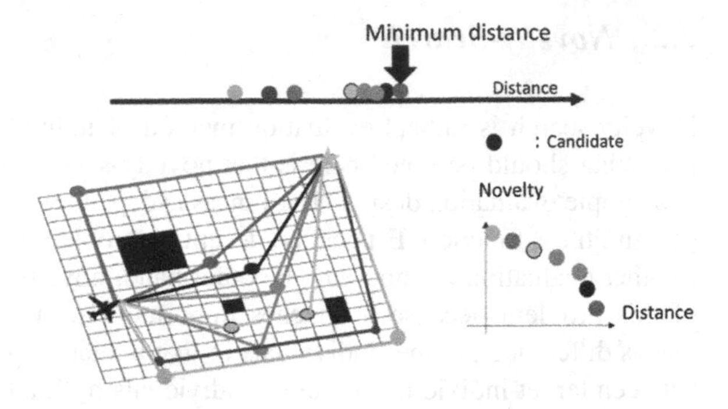

Fig. 5 Two objective functions based on novelty and distance

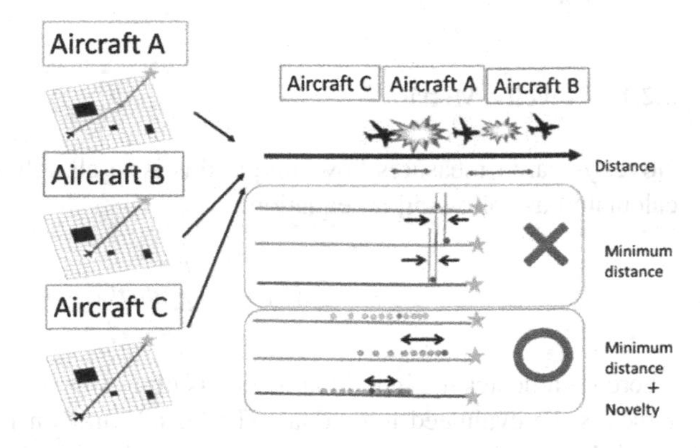

Fig. 6 In comparison to conventional and proposed method

To keep the appropriate distance of the all intervals, all aircrafts are required generating a diversity of distance. We introduce novelty metric. As described earlier, an individual which is different from a population are high evaluation value by novelty search. For example, far distance are high evaluation value based on novelty search as shown in Fig. 5. But the shortest distance cannot be found only using novelty metric.

In Fig. 6, generating method in a case where only distance estimate can search shortest distance. If all aircraft only have shortest distance, this case cannot be met the constraint condition. Here it can be seen that we introduce multi-objective approach based on NSGA-II. We use two objective functions which are novelty search and distance.

4 Experiment

4.1 Experiment Setting

We conduct experiments on the grid map as introduced in [4]. As shown in Fig. 7 is a grid-map where the aircraft was flying. Noted that the black squares represent

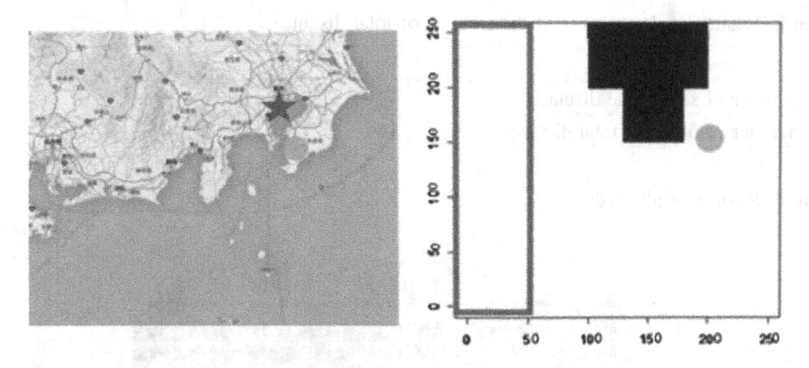

Fig. 7 Field map

obstacles which indicate a prohibit area. Around Haneda airport, are set up outbound aircraft from this airport and for military purpose. Two dimensions map and this map we use here is a 250 km × 250 km.

In this experiment, we use Haneda airport as a reference. This airport is highest air traffic in Japan. Aircrafts which arrives at Haneda airport come from east or west. There are four runways so from east and west side aircraft can own runway separately. Given this situation, we take scheduling only west side.

We define nine aircrafts (denoted ID1–ID9 respectively) are staying at the starting area denoted by the blue-rectangle in the figure. To investigate whether our method successfully optimizes its order that reduces the occupancy time of the destination airport, we conduct the following one experimental case; This system optimizes each aircraft route and arrival sequence at 5 min intervals The maximum number of appearance aircrafts is three at one time. About thirty aircrafts land on Haneda airport per an hour, therefore average appearance aircraft is considered three. Each aircraft chooses an appropriate route which is considered interval of length of another aircrafts routes. Once aircraft routes had been determined, the aircraft must cruse along the course of chosen route. We conduct three steps per one seed and 25 trials.

We used the following parameter setting for NSGA-II and simple GA; the population size is set to 100, the maximal generation is 500, the crossover rate is 1.0 and mutation rate is 0.5. Additionally, r is set to 9.26 km. Each aircraft generates twenty sub-routes.

4.2 Results

Table 1 shows the number of different total distance and value out of trials and Table 2 indicates the result of total distance in all trials. A red label means that the total distance of all aircrafts is shorter than that one and a blue one means longer.

Table 1 Results of the number of differences of total distance

	Step 1	Step 2	Step 3
The number of same total distance	11	6	1
The number of different total distance	14	19	24

Table 2 Results of all seeds

	Step1		Step2		Step3	
	proposed	conventional	proposed	conventional	proposed	conventional
1	669.40	681.55	1116.43	1183.88	1428.59	1521.474
2	611.01	609.42	1032.00	1024.81	1326.049	1328.031
3	624.12	625.99	1094.67	1096.89	1394.068	1405.213
4	682.99	682.95	1173.23	1220.37	1458.314	1551.554
5	655.91	655.91	1125.08	1148.40	1426.832	1466.707
6	561.02	573.66	1005.12	1074.24	1300.339	1436.878
7	595.16	598.71	1044.66	1046.50	1320.439	1343.845
8	634.51	634.51	1141.02	1148.72	1444.829	1474.628
9	636.65	639.94	1152.80	1169.73	1465.756	1493.244
10	598.71	598.71	1052.59	1053.82	1353.618	1364.017
11	604.07	603.76	1049.57	1057.13	1318.352	1337.106
12	603.85	608.27	1047.55	1052.17	1314.428	1319.041
13	545.79	545.79	997.52	997.52	1278.422	1278.422
14	589.76	589.76	1059.07	1059.07	1358.534	1363.829
15	568.34	568.34	1029.77	1037.11	1355.681	1370.815
16*	655.17	675.66	1112.09	1146.75	1382.789	1458.704
17	579.51	579.51	1018.98	1030.27	1399.732	1412.438
18	540.69	540.69	993.10	993.10	1349.673	1349.144
19	592.98	592.98	1103.40	1103.40	1406.542	1409.294
20	597.30	597.30	1098.00	1098.00	1484.789	1508.992
21	590.65	590.07	1082.36	1097.76	1479.839	1496.058
22	553.70	565.70	933.86	942.13	1252.052	1260.486
23*	543.78	543.78	1009.83	1009.83	1382.936	1385.548
24	720.99	721.53	1178.49	1179.04	1529.12	1529.671
25	609.18	610.98	1046.84	1097.06	1399.137	1448.891

Figures 8 and 9 shows an example of the routes and initial positions of each aircraft in each step. In Step 1, ID1–ID3 appeared in this map. As is the case with Step 1, three aircraft came out in Step 2 and Step 3.

Figures 10 and 11 (the upper side) indicates intervals between a leading aircraft and a following aircraft. Vertical axis indicates the difference in distance between

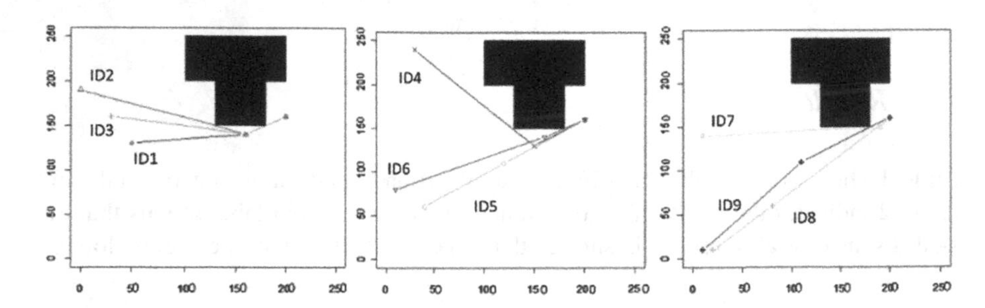

Fig. 8 Routes of proposed method (seed 23)

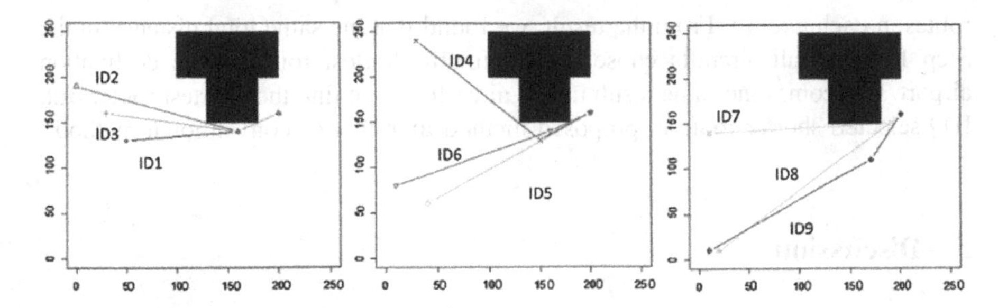

Fig. 9 Routes of conventional method (seed 23)

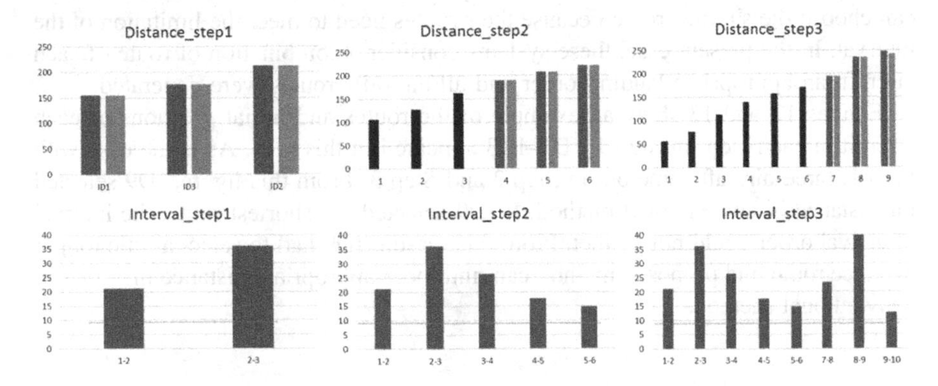

Fig. 10 The result of distance and interval of proposed method (seed 23)

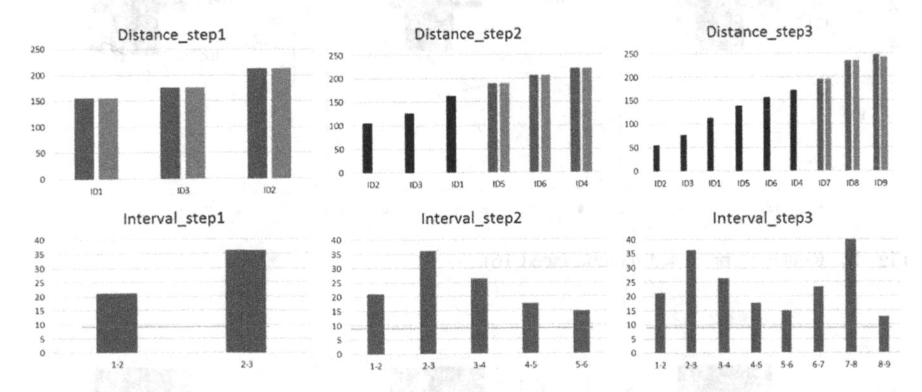

Fig. 11 The result of distance and interval of conventional method (seed 23)

them. Horizontal axis means the arrival order of the leading and following aircraft. Red line indicates minimum distance landing to the destination airport (9.26 km).

Figures 10 and 11 (the lower side) shows route distance of each aircraft. This graph is sorted in ascending order of distance. A blue bar indicates distance of selected route and orange bar indicates the shortest routes in the candidates of routes in each aircraft. From the result, we found that the same total distance in the Step 1 means all aircrafts chose the minimum shortest route to the destination airport. It becomes the same result unless aircraft cannot find the shortest route. But, ID9 selected shorter route in proposed method than that of conventional method.

5 Discussion

Next, the results show it is different from the total distance between proposed method and conventional method in 1 step (seed 16). In this case, not all aircrafts can choose the shortest route because their routes need to meet the limitation of the interval. In the present case, these systems consider a combination of route of each aircraft, an appropriate landing order and all aircrafts routes were generated.

Figures 12 and 13 show an example of the routes and initial positions of each aircraft in each step. In Step 1, ID1–ID3 appeared in this map. As is the case with Step 1, three aircraft came out in Step 2 and Step 3. From this figure, ID9 selected far distance in conventional method. If ID9 selected the shortest route, the interval of arrival order could not be met. From this reason, ID9 had to select a little longer distance route and proposed method can find more appropriate distance than that of conventional method.

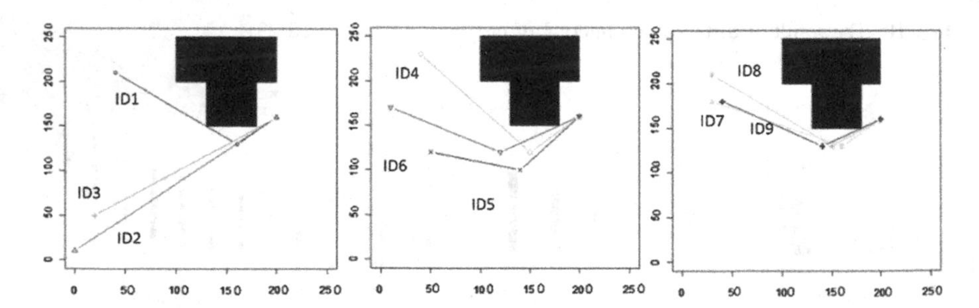

Fig. 12 Routes of proposed method (seed 16)

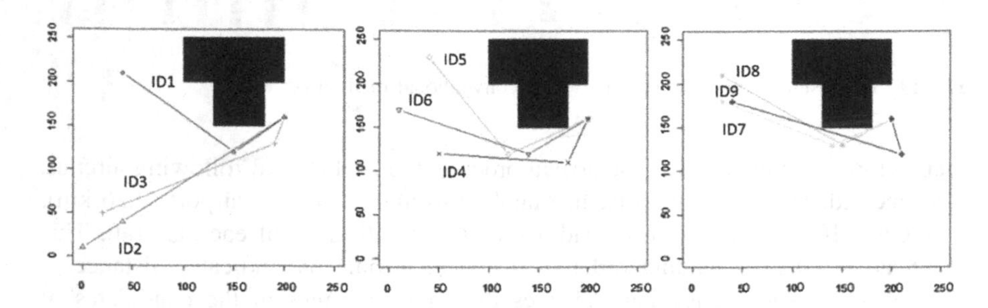

Fig. 13 Routes of conventional method (seed 16)

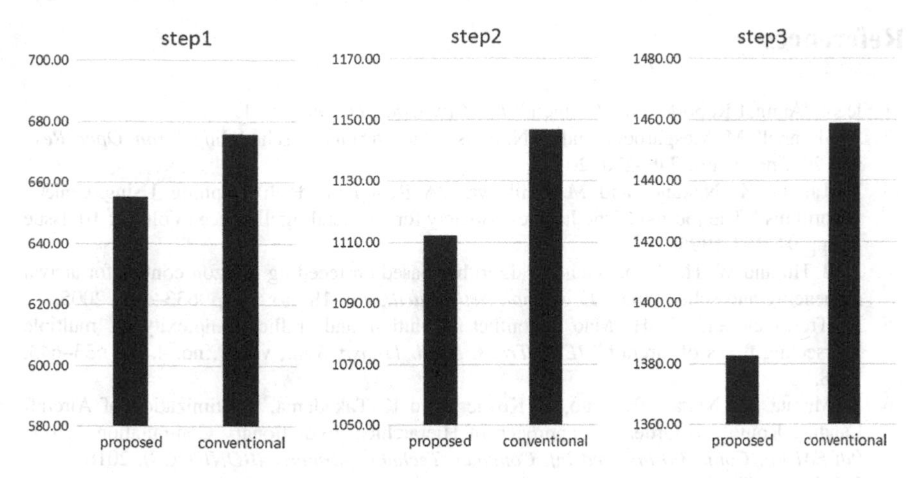

Fig. 14 Difference of total distance

Figure 14 indicates the difference of total distance in each step. As can be seen from this figure, the total distances are gradually increased step by step. For this reason, proposed method generates candidates of importance of length lead to good results.

6 Conclusions

This paper focused on the aircraft landing problem (ALP) which historically addressed the landing routes and order of aircrafts separately, and proposed the new method which optimized both aircraft landing route and order together to minimize an occupancy time of aircrafts in an airport. For this purpose, our proposed method employed *novelty search* to generate variety candidates of aircraft landing routes, which are indispensable for the feasible landing order of all aircrafts especially in the case of delay of other aircrafts or weather change such as typhoon. To investigate effectiveness of the proposed method, we tested it on the grid map which simulated the flying area of the aircrafts. The experimental results revealed that our proposed method can reduce the occupancy time of aircrafts in an airport by optimizing the landing routes and order of the aircrafts in comparison with the conventional method. Since novelty search contributed to generating the various and desirable distance among aircrafts for arrival order, our proposed method successfully found the appropriate landing routes and order of the aircrafts.

What should be noted here is that the obtained results have only been obtained simple grid map. Therefore, further careful qualification and justifications such as considering a velocity of aircraft are needed to generalize our results. A challenge for future research remains in terms of conducting simulation which considers real model (e.g., physics model, 3 dimension).

References

1. Icao, "Annual Report of the Council," *Int. Civ. Aviat. Organ.*, 2011.
2. J. a. Bennell, M. Mesgarpour, and C. N. Potts, "Airport runway scheduling," *Ann. Oper. Res.*, vol. 204, no. 1, pp. 249–270, 2013.
3. T. Tajima, K. Nakano, and M. Ichikawa, "A Real-Time Path Planning Using Genetic Algorithms." The journal of the Japanese Society for Artificial Intelligence (Volume: 10, Issue 14) pp 94–104 1995.
4. X. B. Hu and W. H. Chen, "Genetic algorithm based on receding horizon control for arrival sequencing and scheduling," *Eng. Appl. Artif. Intell.*, vol. 18, no. 5, pp. 633–642, 2005.
5. K. Treleaven and Z. H. Mao, "Conflict resolution and traffic complexity of multiple intersecting flows of aircraft," *IEEE Trans. Intell. Transp. Syst.*, vol. 9, no. 4, pp. 633–643, 2008.
6. A. Murata, M. Nakata, H. Sato, T. Kovacs, and K. Takadama, "Optimization of Aircraft Landing Route and Order: An approach of Hierarchical Evolutionary Computation," *Proc. 9th EAI Int. Conf. Bio-inspired Inf. Commun. Technol. (formerly BIONETICS)*, 2016.
7. J. Lehman, "Evolution Through the Search for Novelty," p. 223, 2007.
8. K. Deb, A. Pratap, S. Agarwal, and T. Meyarivan, "A fast and elitist multiobjective genetic algorithm: NSGA-II," *IEEE Trans. Evol. Comput.*, vol. 6, no. 2, pp. 182–197, 2002.
9. E. Naredo and L. Trujillo, "Searching for Novel Clustering Programs," *Proc. 15th Annu. Conf. Genet. Evol. Comput. - GECCO 2013*, pp. 1093–1100, 2013.
10. J. B. Mouret, "Novelty-Based Multiobjectivization," *Stud. Comput. Intell.*, vol. 341, pp. 139–154, 2011.

Design Strategy Generation for a Sounding Hybrid Rocket via Evolutionary Rule-Based Data Mining System

Masaya Nakata and Kazuhisa Chiba

Abstract This paper aims to reveal design strategies of a sounding hybrid rocket to efficiently perform extinction-reignition. On previous studies, we have derived design candidates which each is a solution optimized on a design optimization problem for the sounding hybrid rocket. To extract design strategies from the design candidates, this paper applies an evolutionary mining system to design candidates, considered as a knowledge discovery task. Specifically, we introduce an evolutionary rule-based data mining system for analyzing a set of Pareto optimal solutions. Our system acquires rules indicating local design information and also we intuitively discover anomalous design strategies.

Keywords Evolutionary machine learning · Learning classifier system · Knowledge discovery · Hybrid rockets

1 Introduction

Hybrid rocket engines (HREs) have been expected as an innovative technology for advanced rocket design. Compared with the solid rocket engine which is currently operated in many countries, HREs have several advantages: lower cost, higher safety, and pollution free flight due to no gunpowder use. HREs are actively researched especially in the E.U. [9] and the U.S. [14]; Virgin Galactic "SpaceShipOne" [15] which uses HREs, is practically operated for manned private spaceflights. In Japan, Institute of Space and Astronautical Science (ISAS)/Japan Aerospace Exploration

M. Nakata (✉)
Faculty of Enginering, Yokohama National University,
79-5, Tokiwadai, Hodogaya, Yokohama 240-8501, Japan
e-mail: nakata-masaya-tb@ynu.ac.jp

K. Chiba
Graduate School of Informatics and Engineering,
The University of Electro-Communications,
1-5-1, Chofugaoka, Chofu, Tokyo 182-8585, Japan

G. Leu et al. (eds.), *Intelligent and Evolutionary Systems*,
Proceedings in Adaptation, Learning and Optimization 8,
DOI 10.1007/978-3-319-49049-6_22

Agency (JAXA) has also begun to develop a next-generation space transportation with a hybrid rockets using HREs.

As a part of the ISAS/JAXA's sounding hybrid rocket project [11], our challenge is, with artificial intelligence techniques such as evolutionary computation and machine learning, to clarify the advantage of hybrid rocket and to extract design knowledge of the hybrid rocket. For instance, we defined an optimization problem on single-time ignition, which is the identical condition of the current solid rocket, to understand the performance difference between solid and hybrid rockets [2]. Then, we revealed the implication of solid fuels in the sounding hybrid rocket performance and identified that a regression rate is a key element for sounding hybrid rocket performance [3].

In this study, we next attempt to understand design knowledge that the hybrid rocket efficiently performs extinction-reignition. This is an important issue of HREs since HREs still have a technical problem to improve the thrust of HREs with extinction-reignition. Specifically, HREs are designed with different phases between fuel and oxidizer which a solid fuel and liquid/gas oxidizer is generally used. Since HREs have low regression rate of solid fuel due to turbulent boundary layer combustion, the thrust of HREs is less than that of pure-solid/liquid engines to implement premixed combustion [10].

For our purpose, we take the following steps. Firstly, we define the design optimization problem of the hybrid rocket which aims at optimizing the efficiency of extinction-reignition, and next we apply an evolutionary multi-objective optimization method to its problem [4]. Consequently, we have derived design candidates which each is a solution optimized on its optimization problem. While each candidate can be a specific example of the design, it still remains unclear that, how design variables affect extinction-reignition depending on other variables, and thus, how the design variable should be set to. Hence, the further next step is to extract design information indicating general or unique design strategies, to feed back the hybrid rocket design.

Accordingly, this paper investigates those design candidates to reveal design strategies to efficiently perform extinction-reignition. Here, we introduce an evolutionary rule-based mining system as a kind of Learning Classifier Systems (LCSs) [8]. Since LCSs are capable to evolve a minimal rule set resulting in a human-readable solution [13], LCSs are actively applied to a wide range of data mining task. Different from the typical use of LCSs as a mining system, we here deal with design candidates forming a set of Pareto optimal solutions with three objective functions (see Sect. 2). Hence, our LCS is customized for evolving rules considered for those objective functions. In brief, since there could be a trade-off between objective functions, that is, there may not be rules having good values for all three objective functions, our system evolves rules considered for a different number of objective functions; however, it still attempts to evolve rules having good values for as many objective functions as possible.

In next section, we explain the design optimization problem of the hybrid rocket and show design candidates derived from its problem. Section 3 introduces our LCS for analyzing the Pareto optimal solutions. Section 4 shows acquired rules and we interpret the rules to understand design strategies. Finally, our conclusion is given in Sect. 5.

2 Sounding Hybrid Rocket Problem and Design Candidates

2.1 Sounding Hybrid Rocket Problem

We consider a conceptual design for a single-stage sounding hybrid rocket, simply composed of a payload chamber, an oxidizer tank, a combustion chamber, and a nozzle [11]. A launch vehicle for aurora scientific observation will be focused because more efficient sounding rockets are desired due to successful obtaining new scientific knowledge on the aurora observation by ISAS/JAXA in 2009. In addition, a single-stage hybrid rocket problem fits for resolving fundamental physics regarding HREs because of its simplicity.

The problem definition is identical with the previous study [6] except for the design variables. Since the acquired hypothesis indicates that 2nd combustion should be feeble to merely sustain vehicle gross weight, we anew prepare design variables regarding oxidizer mass flow for each combustion.

1. Objective Functions. Three objective functions are defined. First objective is maximizing the downrange in the lower thermosphere (altitude from 90 to 150 [km]) R_d [km] (OBJ_1). Second is maximizing the duration in the lower thermosphere T_d [s] (OBJ_2). It recently turns out that atmosphere has furious and intricate motion in the lower thermosphere due to energy injection, from which derives aurora, from high altitude. The view of these objective functions is to secure the horizontal distance and time for competently observing atmospheric temperature and the wind so that the thermal energy balance is elucidated on atmospheric dynamics. Third objective is minimizing the initial gross weight of launch vehicle $M_{tot}(0)$ [kg] (OBJ_3), which is generally the primary proposition for space transportation. Current missions of scientific observations are merely implemented on ballistic trajectories. If major expansion of R_d or T_d can be fulfilled due to efficient extinction-reignition operations, we obtain a distinct paradigm shift of scientific observation missions.

2. Design Variables. We use 10 design variables: oxidizer mass flow on 1st combustion $\dot{m}_{ox}^{(1st)}$ [kg/s] (dv_1), oxidizer mass flow on 2nd combustion $\dot{m}_{ox}^{(2nd)}$ [kg/s] (dv_2), fuel length L_{fuel} [m] (dv_3), initial radius of port $r_{port}(0)$ [m] (dv_4), total combustion time $t_{burn}^{(total)}$ [s] (dv_5), first combustion time $t_{burn}^{(1st)}$ [s] (dv_6), extinction time from the end of first combustion to the beginning of second combustion t_{ext} [s] (dv_7), initial pressure in combustion chamber $P_{cc}(0)$ [MPa] (dv_8), aperture ratio of nozzle ϵ [-] (dv_9), and elevation at launch time $\phi(0)$ [deg] (dv_{10}). We set two combustion periods as follows:

Table 1 Upper/lower limits of each design variable

Serial number	Design variable	Unit	Design space
dv_1	Oxidizer mass flow on 1st combustion	[kg/s]	$1.0 \leq \dot{m}_{ox}^{(1st)} \leq 30.0$
dv_2	Oxidizer mass flow on 2nd combustion	[kg/s]	$1.0 \leq \dot{m}_{ox}^{(2nd)} \leq 30.0$
dv_3	Fuel length	[m]	$1.0 \leq L_{fuel} \leq 10.0$
dv_4	Initial radius of port	[m]	$0.01 \leq r_{port}(0) \leq 0.30$
dv_5	Total combustion time	[s]	$20.0 \leq t_{burn}^{(total)} \leq 60.0$
dv_6	1st combustion time	[s]	$10.0 \leq t_{burn}^{(1st)} \leq 40.0$
dv_7	Extinction time	[s]	$1.0 \leq t_{ext} \leq 300.0$
dv_8	Initial pressure in combustion chamber	[MPa]	$3.0 \leq P_{cc}(0) \leq 6.0$
dv_9	Aperture ratio of nozzle	[–]	$5.0 \leq \epsilon \leq 8.0$
dv_{10}	Elevation at launch time	[deg]	$60.0 \leq \phi(0) \leq 90.0$

$$
\begin{aligned}
t_{burn}^{(1st)} &= \begin{cases} t_{burn}^{(total)} & (t_{burn}^{(total)} < t_{burn}^{(1st)}) \\ t_{burn}^{(1st)} & (t_{burn}^{(total)} \geq t_{burn}^{(1st)}) \end{cases}, \\
t_{burn}^{(2nd)} &= \begin{cases} 0 & (t_{burn}^{(total)} < t_{burn}^{(1st)}) \\ t_{burn}^{(total)} - t_{burn}^{(1st)} & (t_{burn}^{(total)} \geq t_{burn}^{(1st)}) \end{cases}.
\end{aligned} \tag{1}
$$

Under $t_{burn}^{(total)} < t_{burn}^{(1st)}$ condition, it is defined that $t_{burn}^{(1st)}$ is set to be $t_{burn}^{(total)}$ and second-time combustion is not performed. Since $t_{burn}^{(2nd)}$ is significant factor to understand extinction-reignition behavior, we include it in dataset for data mining be named "dv^*". There is no constraint except upper/lower limits of each design variable summarized in Table 1. These upper/lower values are exhaustively covering the region of the design space which is physically admitted. When there is a sweet spot (the region that all objective functions proceed optimum directions) in the objective-function space, the exploration space would intentionally become narrow due to range adaptation on the evolutionary computation.

2.2 Design Candidate Derivation

We used a hybrid evolutionary computation (EC) between the differential evolution and the genetic algorithm [6] to efficiently generate a hypothetical dataset using exact solutions. The detail of this result is summarized in [4]. The population size is set to be 18 in one generation; the hybrid EC is performed until 4,500 generations. The generation number is decided by evolution convergence. The hybrid EC consequently generates two discontinuous connecting and convex nondominated surfaces

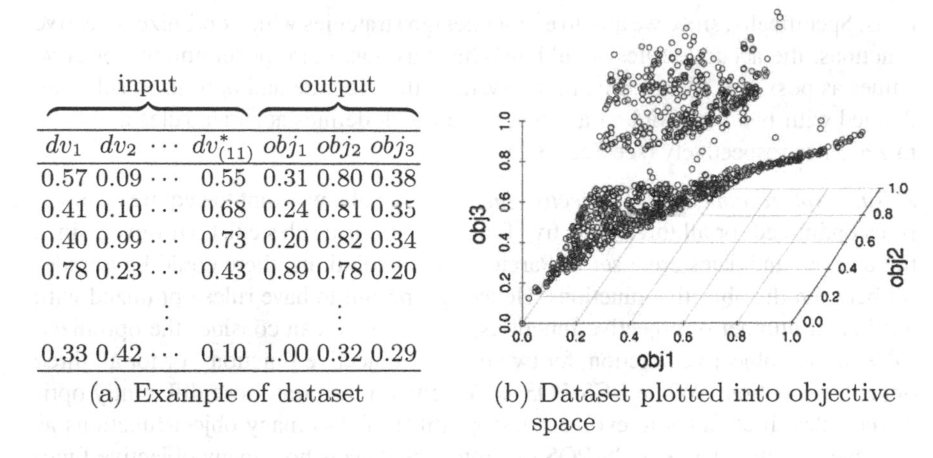

input				output		
dv_1	dv_2	$\cdots$	$dv^*_{(11)}$	obj_1	obj_2	obj_3
0.57	0.09	$\cdots$	0.55	0.31	0.80	0.38
0.41	0.10	$\cdots$	0.68	0.24	0.81	0.35
0.40	0.99	$\cdots$	0.73	0.20	0.82	0.34
0.78	0.23	$\cdots$	0.43	0.89	0.78	0.20
		$\vdots$			$\vdots$	
0.33	0.42	$\cdots$	0.10	1.00	0.32	0.29

(a) Example of dataset (b) Dataset plotted into objective space

Fig. 1 The dataset consisting of pareto optimal solutions

except several isolated individuals. A dataset consists of 1,000 Pareto optimal solutions derived from a multi-objective optimization technique on the hybrid rocket design problem. Figure 1b confirms that the pareto optimal solutions in the dataset have a tradeoff between three object values. As shown in Fig. 1a, each solution is composed of 11 optimized values of design variables dv_1, dv_2, ..., ..., dv_{10}, $dv^*_{(11)}$ (only $dv^*_{(11)}$ is implicitly optimized) and 3 objective values obj_1–obj_3; all values are converted to be normalized values.

3 Evolutionary Rule-Based Data Mining System

Our system is inspired from Learning Classifier System called LCS [8] which is an approach of evolutionary rule-based machine learning. We introduce an LCS as a mining tool for Pareto optimal solutions, called LCS-POS. The rest of section describes a brief concept of LCS-POS and then a specific mechanism.

3.1 Concept

1. Learning accurate and optimized rules. We use a condition-prediction rule to represent a design strategy. The condition part indicates a combination of design variable values dv_i, and the prediction part indicates expected objective values obj_j when the rule is applied. Here, we want to acquire rules which accurately predict the objective values. While many modern LCSs are designed to learn such accurate rules, e.g., XCS [16], we further want to learn rules optimized for the objective functions. Specifically, since we aim to extract design strategies which optimize objective functions, the accurate rules should also have as maximum or minimum objective values as possible. Technically, in the system, the accurate and optimized rules are defined with two threshold parameters which both defines accurate rules and optimized rules respectively (see Sect. 3.2).

2. *Rule exploration on multi-objective function.* Ideally, we want to acquire accurate rules optimized for all three objective functions if those rules exist. However, since the design candidates are a set of Pareto optimal solutions, there could be a trade-off between the objective functions. Hence, we permit to have rules optimized with a different number of objective functions, and then we can consider the optimized rules for one objective function, for two of three objective functions, or for all three objective functions. While LCS-POS is designed to acquire those differently optimized rules, it attempts to evolve rules optimized for as many object functions as possible. In other words, LCS-POS evolutionary detects how many objective functions can be optimized together. For knowledge discovery, this can be an advantage of LCS-POS, because LCS-POS would extract various design strategies for different combinations of objective functions.

3.2 Rule Format

A rule consists of a condition and the following five parameters; a prediction p_j, a prediction error ϵ_j, a niche size *ns*, a fitness F and a number of optimized objective functions n_{obj}. The condition C specifies inputs (or design candidates in the dataset) which its rule can be applied. The condition is a set of sub-conditions, denoted by $C = \{c_0, c_1, \dots, c_{11}\}$, where c_i is a sub-condition applied for a design variable dv_i. Each c_i consists of a lower value l_i and an upper value u_i such that $0 \leq l_i \leq u_i \leq 1$, both specifying a numerical range of design variable. The prediction $p = \{p_1, p_2, p_3\}$ and the prediction error $\epsilon = \{\epsilon_1, \epsilon_2, \epsilon_3\}$ are used for calculating F; each prediction p_j and each prediction error ϵ_j indicate an average of objective value obj_j and its error respectively. F is used for identifying an accurate and optimized rule. *ns* indicates a number of overlapping rules. Finally, n_{obj} indicates that how many objective functions its rule has been optimized for.

To define the accurate and optimized rules, we introduce the following two threshold parameters; the maximum permitted error ϵ_θ and the maximum permitted prediction p_θ. We define that accurate rules for the objective function OBJ_j have ϵ_j less than ϵ_θ. Similarly, we define that the optimized rules for OBJ_j have p_j less than p_θ. Note that, while OBJ_1, OBJ_2 are maximizing functions, LCS-POS internally considers them as minimizing functions to simplify the mechanism; the objective values obj_1 and obj_2 are simply converted to $1 - obj_1$ and $1 - obj_2$ when identifying the optimized rules.

3.3 Mechanism

Our system consists of following four components: *rule-initialization*, *rule-evaluation*, *rule-subsumption* and *rule-evolution*. The rule-initialization is called once at the start, and then, other three components are repeated.

1. Rule initialization. To build the initial rule set, a *covering* operator takes place [1]; for each design candidate in the dataset, the system generates a rule which can be applied to its input. Specifically, the condition of generated rule for a input is determined as; for the design variable value dv_i of the input, l_i and u_i of each sub-condition c_i are set to a random value but should be satisfied with $l_i \leq dv_i \leq u_i$. Note that, if the rule set has already included rules applicable to a input, the covering operator does not take place.

2. Rule evaluation. After the rule initialization or the rule evolution is executed, the parameters of each rule r are updated with the following order; p_j, ϵ_j, ns, n_{obj}, and F. p_j for obj_j, as in Eq. (2a), is simply calculated as the average of objective values of inputs which its rule can be applied. Note that, in Eq. (2a), $s \in \boldsymbol{s}_r$ indicates an input s belonging to a subset of inputs $\boldsymbol{s}_r$ which the rule r can be applied; obj_j^s is the j-th objective value of s; and $|\boldsymbol{s}_r|$ is a number of input in the subset $\boldsymbol{s}_r$; we call $|\boldsymbol{s}_r|$ as a correspondent number, which indicates a generality of rule. Then, the prediction error is calculated as an absolute error of p_j against the objective value (see Eq. (2b)).

$$p_j = \sum_{s \in \boldsymbol{s}_r} \frac{obj_j^s}{|\boldsymbol{s}_r|}, \tag{2a}$$

$$\epsilon_j = \sum_{s \in \boldsymbol{s}_r} \frac{|p_j - obj_j^s|}{|\boldsymbol{s}_r|}. \tag{2b}$$

To calculate F, we firstly calculate a rule's *validity* $\eta = \{\eta_1, \eta_2, \eta_3\}$ which is used for identifying the accurate and optimized rule. η_j for OBJ_j is calculated from an *accuracy* E_j and an *optimality* P_j.

$$E_j = \begin{cases} 1 & \text{if} \quad \epsilon_j < \epsilon_\theta, \\ \epsilon_\theta / \epsilon_j & \text{otherwise.} \end{cases} \tag{3a}$$

$$P_j = \begin{cases} 1 & \text{if} \quad p'_j < p_\theta, \\ p_\theta / p'_j & \text{otherwise.} \end{cases} \tag{3b}$$

E_j identifies an accurate rule, and as in Eq. (3a), E_j is set to the maximum value 1 when ϵ_j is less than ϵ_θ, indicating an accurate rule; otherwise, a value less than 1 depending on ϵ_j. Similar to E_j, P_j is maximally set to 1 when p'_j is less than p_θ; otherwise a value less than 1 (See Eq. (3b)). Note that, to consider the minimizing functions, p'_j is set to $1 - p_1$ and $1 - p_2$ for $j = 1, 2$; for $j = 3$, p'_j is set to p_3. Then, each

η_j is calculated as $\eta_j = E_j \cdot P_j$. Hence, we can identify a rule with $\eta_j = 1$ as an accurate and optimized rule for the j-th object function OBJ_j. ns of rule is set to a number of overlapping rules which can be applied to the same inputs matched with its rule as in [1]. As a result, n_{obj} is set to a count value of $\eta_j = 1$ in η. Finally, F is calculated from η. If n_{obj} is equal to or larger than 1, that is, the rule has been identified as the accurate and optimized rule for at least one objective function, its fitness is set to 1; otherwise, F is calculated from the maximum validity with a parameter ν;

$$F = \begin{cases} 1 & \text{if} \quad n_{obj} \geq 1, \\ \max\limits_j \eta_j{}^{\nu} & \text{otherwise.} \end{cases} \tag{4}$$

3. Rule subsumption. The rule-subsumption performs to promote the generalization of rule [1]; a rule is subsumed to a more general rule. Specifically, for two rules r_k with F=1 and r_l with any value of F, when all lower values l_i^k and upper values u_i^k of r_k are smaller and larger than that of r_l respectively (i.e., $l_i^k < l_i^l$, $u_i^k > u_i^l$,), r_k is defined as having a more general condition than r_l. Then, the system decides whether the rule r_l can be deleted as; if the r_l is not accurate or optimized rule with $F < 1$, r_l is deleted to remove redundant rules; in addition, if the r_l has been identified as accurate and optimized rule (having $F = 1$) for the same objective functions of the r_k, r_l is also deleted.

4. Rule evolution. Finally, the system evolves rules with the genetic algorithm [7]; the system replaces 50 % of the existing rules with generated offspring. It selects two rules from the rule set with the roulette wheel selection based on the rule's fitness [1]; it copies them and performs crossover and mutation on the copies' conditions with the probability χ and μ respectively. The resulting offspring are inserted into the rule set and the existing rules are deleted if the number of rule in the rule set is larger than a maximum rule set size N_{rule} to keep the rule set size constant. The deleted rules are selected with a probability P_{del} which is calculated as $(ns/F)^{-n_{obj}}$. Hence, the rules with a low fitness value and a large niche size would be deleted; in addition, the accurate and optimized rules for a smaller number of objective functions are also selected for deletion.

4 Result

Nondominated solutions to construct dataset are not uniformly in design space; search bias on the hybrid EC may affect correspondent number. However, since the convergence of the hybrid EC was confirmed by hypervolume, nondominated solutions are thoroughly distributed in the feasible region of objective-function space shown in Fig. 1b.

The advantages to generate rules using EC are summarized in the following three point. Firstly, automatically intelligent discretization of design-variable space (in this study, 11 dimensional space) can be performed. Second, the values of all objective functions can be used as an indicator to extract significant rules from enormous those, i.e., important design-variable space can be extracted via objective-function space. Finally, the difference of its important design strategy can be intuitively compared with other design strategies due to heatmap visualization. The rough set theory [12] is a representative rule mining method; correspondent number is employed as a rule selection indicator [5]. However, since an evolutionary rule-based data mining system can avoid using rough indicator, it is considerable to be an advanced and beneficial tool for rule mining.

4.1 Preliminary Experiments

Analysis with different ϵ_θ settings LCS-POS optimizes the expected values of objective functions by generated rules. But we can deal with the allowable error ϵ_θ between prediction and exact values. If ϵ_θ becomes large, an obscure rule may be permitted. Figure 2 indicates that the number of generated rules are increased as ϵ_θ becomes large. Although rules for design variables are similar, objective values are different; the number of rules consequently swells. The range of predicted objective values is expanded due to ϵ_θ gain, LCS-POS severally generates rules which would be merged if ϵ_θ were small. If we will obtain rough rules, ϵ_θ might set to be large. However, since large ϵ_θ causes low reliability of predicted rules, ϵ_θ should set to be low at the constant as $\epsilon_\theta = 0.05$ in this study.

Influence of allowable range of p_θ to generate rules The optimization for all objective functions may recognize sweet-spots existence and generate the rule to be in sweet spots. It is natural for the equality among objective functions that LCS-POS does not have the arbitrariness and it generates the rules to optimize all of them. However, if target range p_θ for optimizing objective functions is too small, there may be no correspondent individual; no rule is generated. We should set an appropriate value on p_θ, but there is no indicator due to problem dependency; a parameter study should be accordingly performed.

The result of the parameter study regarding p_θ in this study is shown in Fig. 3. When $p_\theta = 0.25$, no rule is generated; if $p_\theta = 0.30$, just two rules are consequently generated. If $p_\theta \geq 0.35$, we can obtain sufficient number of rules. We acquire the design information regarding the minimum degree to compromise. Furthermore, this result reveals that there is a sweet spot in the design space if we compromise p_θ of 0.30 for each objective function. We will use the result of $p_\theta = 0.35$ below to observe many candidates around the sweet spot region.

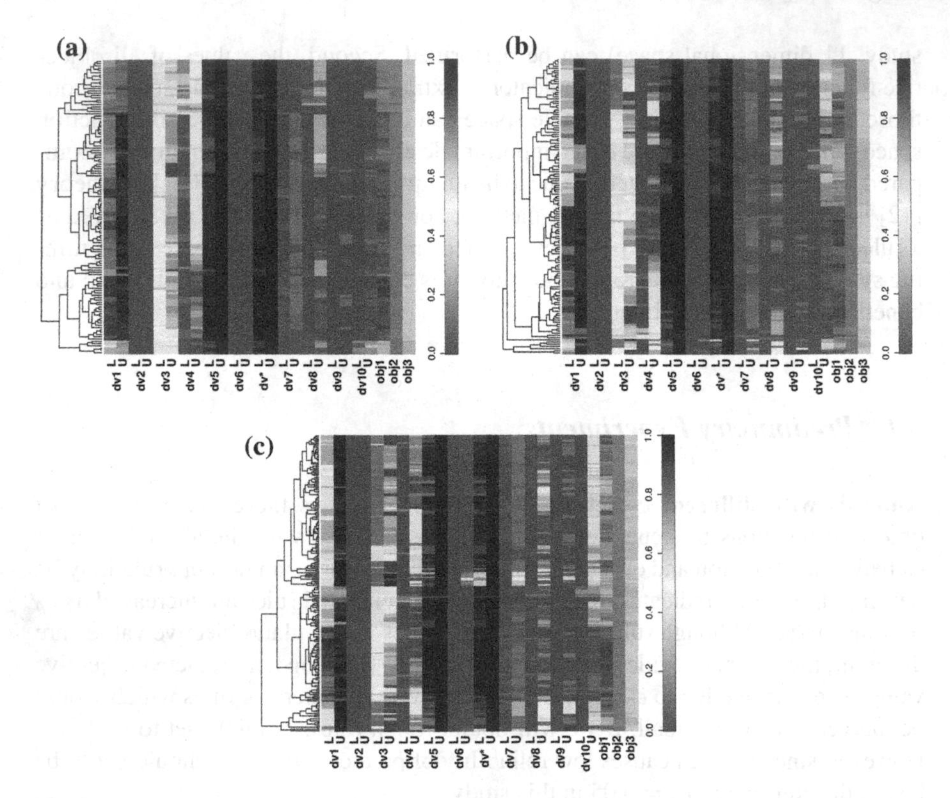

Fig. 2 Visualization of predicted rules, which three objective functions are optimized with $p_\theta = 0.35$. There are six constraints for values of three objective functions and their prediction torelance ϵ_θ; **a** $\epsilon_\theta = 0.05$, **b** $\epsilon_\theta = 0.10$, and c) $\epsilon_\theta = 0.15$. Note that L and U respectively denote lower value and upper one of each design variable

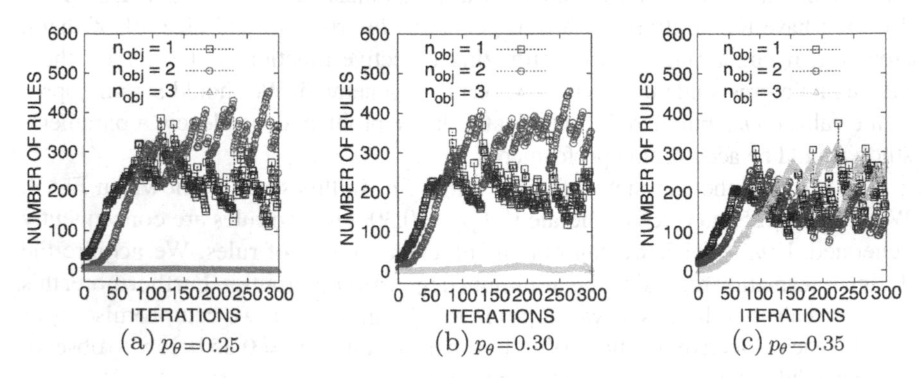

Fig. 3 Number of the accurate and optimized rules for one objective function $n_{obj} = 1$, for two objective functions $n_{obj} = 2$, and for all functions $n_{obj} = 3$

4.2 Comparison of Compromise Results Between p_θ and n_{obj}

The present design problem has a sweet spot in the design space. However, there is no flexibility because the range of all three objective functions has no diverse value in Fig. 2a. Although the behaviors of dv_1 ($\dot{m}_{ox}^{(1st)}$), dv_4 ($r_{port}(0)$), and dv_8 ($P_{cc}(0)$) are with perturbations, there are no correlations between $r_{port}(0)$ and the objective functions as well as between $P_{cc}(0)$ and the objective functions. As the upper bound of $\dot{m}_{ox}^{(1st)}$ is large, obj_1 (R_d) roughly enlarges; $\dot{m}_{ox}^{(1st)}$ is an essential design variable in this sweet spot.

When considering all three objective functions, there was especially no flexibility of obj_2 (T_d) and obj_3 ($M_{tot}(0)$). Hence, the number of optimized objective functions is set to be two. Figure 4 displays obtained rules on a heatmap; it indicates that obj_1 (R_d) is released from optimization and constraints. The phylogenetic tree in Fig. 4a obviously reveals unique rules. Figure 4b shows the heatmap with the order sorted by obj_1 (R_d) values; we recognize the influence of unique rules for the objective functions and also the difference of design-variable behaviors in the vicinity. Figure 4b indicates that obj_3 ($M_{tot}(0)$) becomes low as obj_1 (R_d) is low, that is, there is a severe tradeoff between R_d and $M_{tot}(0)$. If we drastically extend R_d, we should choose a compromise strategy to absolutely sacrifice $M_{tot}(0)$. In contrast, obj_2 (T_d) does not act in concert with R_d behavior. This fact indicates that other strategies need for optimizing R_d and T_d. Figure 4 also indicates the design variable behaviors. Since the variations of dv_1, dv_3, and dv_{10} correspond to that of obj_1, $\dot{m}_{ox}^{(1st)}$, L_{fuel}, and $\phi(0)$ give an effect on extending R_d. $\dot{m}_{ox}(0)$ rise gives an effect on $\dot{r}_{port}(t)$ growth and thrust gain. L_{fuel} also gives an effect on R_d swell. If L_{fuel} increases, fuel mass rises; long-time combustion can be performed. $\phi(0)$ decrease gives an effect on R_d augmentation. If $\phi(0)$ is low, flight paths become horizontal; R_d increases. These correlations are physically rea-

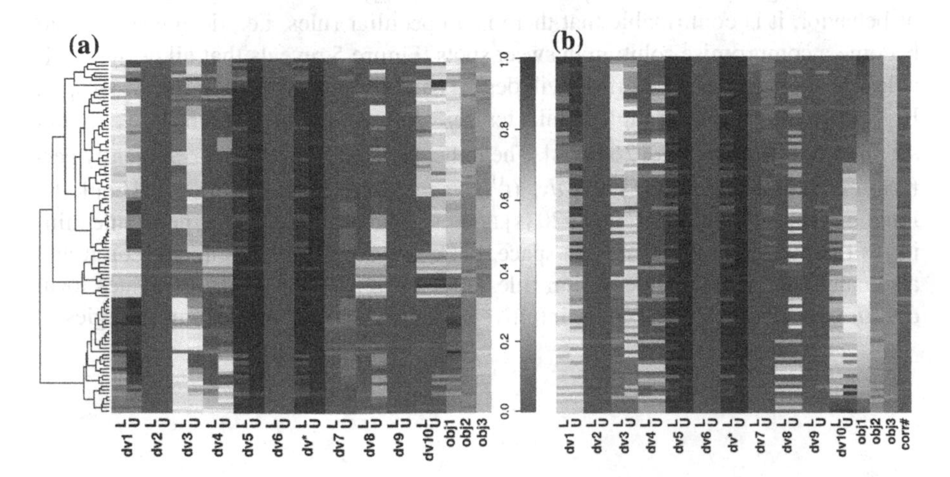

Fig. 4 Visualization of predicted rules, $p_\theta = 0.30$, under optimizing two objective function condition, **a** the order by phylogenetic tree and **b** the order sorted by obj1

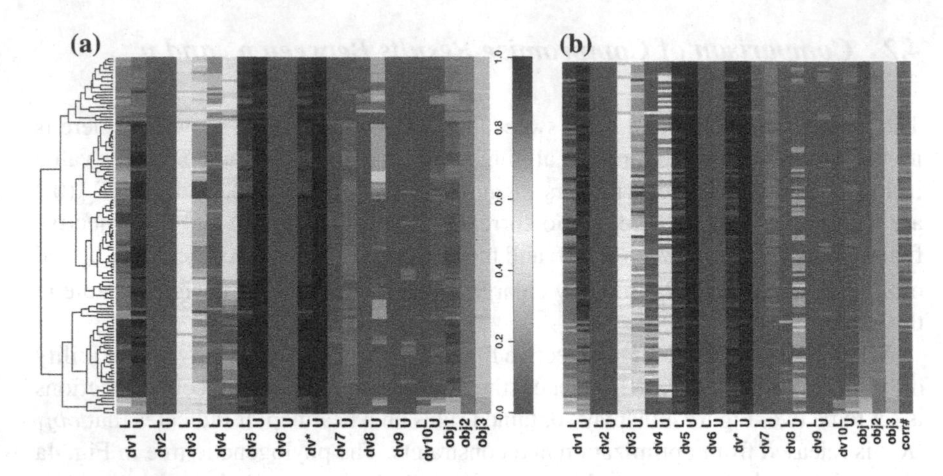

Fig. 5 Visualization of predicted rules, $p_\theta = 0.35$, under optimizing three objective function condition, **a** the order by phylogenetic tree and **b** the order sorted by obj1

sonable. Figure 5 shows the generated rules on heatmap under $p_\theta = 0.35$ condition. We can observe that Fig. 5 is similar to the clipping heatmap from the upper region of Fig. 4b. If severe tradeoff relationships exist among objective functions, no rule is generated even when p_θ rises. Thereupon, we should unavoidably decrease the number of optimized objective functions on the LCS-POS, but it swerves from the purpose of optimizing all objective functions for real-world problems and of searching compromise solutions. Indeed, the lower region on Fig. 4b is unnecessary design information. Rules should be generally produced for all objective functions set in problem definition.

Since the generated heatmaps do not have design-variable rules which get out of behavior, it is confirmable that there is no peculiar rules, i.e., design strategy to become a compromise solution in sweet spots. Figure 5 reveals that all design variables do not have flexibility if we will design an individual existed in the sweet spot. Especially, dv_1 ($\dot{m}_{\rm ox}^{(1{\rm st})}$) strongly dominates obj_1 (R_d); if we achieve high R_d, we should set $\dot{m}_{\rm ox}^{(1{\rm st})}$ to be greater than 26 [kg/s]. The indicated design strategy to efficiently perform extinction-reignition is that dv_6 ($t_{\rm burn}^{(1{\rm st})}$) should be short; $dv*$ ($t_{\rm burn}^{(2{\rm nd})}$) should be long; dv_7 ($t_{\rm ext}$) should be around 120 [s] at most. The bottleneck of general rule mining is the discretization of design space, but our system does not depend on it and also can generate adaptable design rules. Our system evolutionary generates local design information as well as we intuitively identify anomalous design strategies.

5 Conclusions

This study have been investigated design strategies to efficiently perform extinction and reignition on a sounding hybrid rocket design problem by introducing an evolutionary rule-based mining system be named Learning Classifier System for Pareto optimal solutions. The algorithm is not depend on discretization of design space and is designed to evolve flexible rules considered for those objective functions. Consequently, we obtain the design rule to efficiently implement extinction-reignition: 1st combustion should be short; 2nd combustion should be long; extinction time might be on the periphery of 120 [s] at most. We revealed the local design information and also we intuitively discover anomalous design strategies; the acquired rules show how design variables affect extinction-reignition depending on other variables.

References

1. Butz, M.V., Wilson, S.W.: An algorithmic description of XCS. Soft Computing 6(3–4), 144–153 (2002)
2. Chiba, K., Kanazaki, M., Nakamiya, M., Kitagawa, K., Shimada, T.: Conceptual design of single-stage launch vehicle with hybrid rocket engine for scientific observation using design informatics. Journal of Space Engineering 6(1), 15–27 (2013)
3. Chiba, K., Kanazaki, M., Nakamiya, M., Kitagawa, K., Shimada, T.: Diversity of design knowledge for launch vehicle in view of fuels on hybrid rocket engine. Journal of Advanced Mechanical Design, Systems, and Manufacturing 8(3), JAMDSM0023, 1–14 (2014)
4. Chiba, K., Kanazaki, M., Shimada, T.: Effective operations of extinction-reignition with simple control of oxidizer flux on a single-stage sounding hybrid rocket. In: AIAA Paper 2017-xxxx on the 55th AIAA Aerospace Science Meeting. AIAA (2017)
5. Chiba, K., Obayashi, S.: Knowledge discovery in aerodynamic design space for flyback-booster wing using data mining. Journal of Spacecraft and Rockets 45(5), 975–987 (2008)
6. Chiba, K., Yoda, H., Ito, S., Kanazaki, M.: Visualization of design-space constitution for single-stage hybrid rocket with rigid body in view of extinction-reignition. In: Proceedings on IEEE Symposium Series on Computational Intelligence. pp. 933–940. IEEE (2015)
7. Goldberg, D.E.: Genetic Algorithms in Search, Optimization, and Machine Learning. Addison Wesley (1989)
8. Holland, J.H.: Escaping Brittleness: The Possibilities of General Purpose Learning Algorithms Applied to Parallel Rule-based system. Machine learning: An artificial intelligence approach 2, 593–623 (1986)
9. Karabeyoglu, M.A.: Advanced hybrid rockets for future space launch. In: Proceedings on 5th European Conference for Aeronautics and Space Sciences. EUCASS (2013)
10. Karabeyoglu, M.A., Altman, D., Cantwell, B.J.: Combustion of liquefying hybrid propellants: Part 1, general theory. Journal of Propulsion and Power 18(3), 610–620 (2002)
11. Kosugi, Y., Oyama, A., Fujii, K., Kanazaki, M.: Multidisciplinary and multi-objective design exploration methodology for conceptual design of a hybrid rocket. In: AIAA Paper 2011-1634. AIAA (2011)
12. Kusiak, A.: Rough set theory: A data mining tool for semiconductor manufacturing. IEEE Transactions on Electronics Packaging Manufacturing 24(1), 44–50 (2001)
13. Lanzi, P.L.: Learning classifier systems: Then and now. Evolutionary Intelligence 1, 63–82 (2008)
14. Simurda, L., Zilliac, G., Zaseck, C.: High performance hybrid propulsion system for small satellites. In: AIAA Paper 2013-3635. AIAA (2013)

15. Thicksten, Z.: Handling considerations of nitrous oxide in hybrid rocket motor testing. In: AIAA Paper 2008-4830. AIAA (2008), 44th AIAA/ASME/SAE/ASEE Joint Propulsion Conference and Exhibit
16. Wilson, S.W.: Classifier Fitness Based on Accuracy. Evolutionary Computation 3(2), 149–175 (Jun 1995)

A Novel Binary Particle Swarm Optimization Algorithm and Its Applications on Knapsack and Feature Selection Problems

Bach Hoai Nguyen, Bing Xue and Peter Andreae

Abstract Particle swarm optimisation (PSO) is a well-known evolutionary computation technique, which has been applied to solve many optimisation problems. There are two main types of PSO, which are continuous PSO (CPSO) and binary PSO (BPSO). Since PSO is originally proposed to address continuous problems, CPSO has been studied extensively while there are only a few studies about BPSO. In a standard PSO algorithm, momentum is an important component, which preserves the swarm's diversity. However, since movements in binary search spaces and continuous search spaces are different, it is not appropriate to apply directly the momentum concept of CPSO to BPSO. This paper introduces a new momentum concept to BPSO, which leads to a novel BPSO algorithm, named SBPSO. SBPSO is compared with a recent BPSO algorithm, named PBPSO, in two well-known binary problems: knapsack and feature selection. The experimental results on knapsack datasets show that SBPSO can find better solutions than PBPSO. In feature selection problems, SBPSO can select a smaller number of features and still achieve similar or better accuracies than PBPSO and using all the original features in a comparative computation time.

Keywords Particle swarm optimisation · Feature selection · Knapsack · Binary optimisation

B.H. Nguyen (✉) · B. Xue · P. Andreae
School of Engineering and Computer Science, Victoria University of Wellington, 600, Wellington 6140, New Zealand
e-mail: Hoai.Bach.Nguyen@ecs.vuw.ac.nz

B. Xue
e-mail: Bing.Xue@ecs.vuw.ac.nz

P. Andreae
e-mail: Peter.Andreae@ecs.vuw.ac.nz

G. Leu et al. (eds.), *Intelligent and Evolutionary Systems*, Proceedings in Adaptation, Learning and Optimization 8,
DOI 10.1007/978-3-319-49049-6_23

1 Introduction

Optimisation tasks play an important role in many real-world problems. In order to solve such problems, there are two main approaches [8], which are classical optimisation [5] and heuristic optimisation. The classical approaches, including linear or non-linear programming, have limitations when there are too many variables. Heuristic approaches [17] are proposed when the search space is complicated or the classical methods' assumptions such as continuity and differentiability are not satisfied [8]. Instead of providing an exact solution, the heuristic search usually aims to find out an approximate solution, which is good enough in an acceptable computation time. Evolutionary computation (EC) is a kind of heuristic search inspired by biological principles. Some well-known EC techniques are genetic algorithms [23], genetic programming [13] and particle swarm optimisation [18].

Particle swarm optimisation (PSO) is proposed in [9], which mimics the social behaviour of bird flocking. In PSO, a particle represents a candidate solution and has its own position (x) and velocity (v). The particle also records its own best position (*pbest*) and its neighbours' best position (*gbest*) so far, which guide it to follow promising trajectories. PSO has been successfully applied to solve many real-world optimisation problems such as financial forecasting [2], steel frame designing [7] and unmanned combat air vehicle path planning [26].

There are two main types of PSO, which are continuous PSO [24] (CPSO) and binary PSO [10] (BPSO). BPSO has been widely applied to many combinatorial problems, for instance knapsack problems [12] and feature selection [22]. In the original BPSO [10], each position entry has two possible values 0 or 1. The velocity is used to determine how likely the corresponding position's element is set to 1, as in Eqs. (1) and (2).

$$x_d^{t+1} = \begin{cases} 1, \text{ if } rand() < \frac{1}{1+e^{-v_d^{t+1}}} \\ 0, \text{ otherwise} \end{cases} \tag{1}$$

$$v_d^{t+1} = w * v_d^t + c_1 * r_1 * (pbest_d - x_d^t) + c_2 * r_2 * (gbest_d - x_d^t) \tag{2}$$

where t is the tth iteration, d is the dth dimension in the search space, w is an inertia weight, c_1 and c_2 are two acceleration constants, r_1, r_2 are two random variables uniformly distributed in [0, 1].

There are some issues in the original BPSO [9]. Firstly, the new position element is determined without considering its previous position. In addition, the velocity formula of BPSO is exactly same as the formula of CPSO, which might not be suitable since in a binary space, particles do not move as smoothly as in a continuous space. Particularly, in CPSO, the velocity defines the length and the direction that particles follow in the next step. However, in BPSO, each position's element has only two possible values and particles move by flipping its bits. Therefore, in a binary search space, the idea of moving in a direction, as in the *velocity* term, is not applicable. In PSO, there are three most important components contributing to the search ability of PSO via the velocity, which are *momentum*, the *cognitive*-component (defined as *pbest*) and the *social*-component (defined as *gbest*). Once the velocity does not describe accurately the movement of binary particles, the three components, especially the best experiences are not effectively used to guide the particles. For example in BPSO the parameter setting in the velocity has an opposite effect in comparison with CPSO [11, 12]. These might be the reasons for the limited performance of BPSO in comparison with CPSO, even in combinatorial problems [3, 21]. In order to make PSO adapt with binary search spaces, it is necessary to define new *velocity* and *momentum* concepts, which can capture the movements of binary particles in a more appropriate way.

Pampara et al. [16] utilise an angle modulation to map from a high-dimensional binary search space into a 4-dimensional continuous search space. The experimental results show that the proposed algorithm is better than the traditional BPSO. However, in the benchmark functions, there are a small number of variables, which are totally independent. In [25], a BPSO algorithm is proposed to avoid premature convergence, which is a common problem in BPSO. In the proposed BPSO, a mutation operator is used to improve the diversity of the swarm. The mutation rate is set to $\frac{1}{N}$ with an expectation that at least one bit in a particle is mutated, where N is the length of the particle position. The proposed BPSO (MBPSO) is used to select features in spam detection problems. The experiment shows that MBPSO can evolve better feature subsets than other EC algorithms such as ACO, PSO or GA. However, it is not clear how mutation operators are applied to the PSO algorithm. In comparison with BPSO, MBPSO are sometimes more expensive. In addition, adding only a mutation operator may not be able to overcome other limitations of the original BPSO, such as velocity and previous position ignorance. Ganesh et al. [6] uses BPSO to narrow down the search space size for ear detection problems, where the task is to select pixels with high entropy values and discard low entropy pixels to find the most promising areas of the original search space. The only difference between BPSO in [6] and the original PSO is that a random number is multiplied to the previous velocity, which aims to prevent the premature convergence in BPSO. The experimental results show that the proposed algorithm has promising results on four benchmark face datasets. However, the effect of the random number is not analysed i.e. there is no comparison between the proposed BPSO and the original BPSO. In addition, the velocity concept from CPSO is still used in this work. Xue et al. [22] proposes a new BPSO algorithm (PBPSO), in which a probability vector is used to replace the velocity vector, in which each entry shows the flipping probability of the corresponding position's bit. The experimental results show that PBPSO outperforms the standard BPSO in both effectiveness and the efficiency in feature selection problems. Although the probability vector reflects the binary particle's movement more accurately, the *momentum* concept, one of three most important components, is not defined. Since *momentum* and *movement* are two intimate factors, a new *movement* concept should be followed by a corresponding new *momentum* definition. Therefore, in this work we propose a new *momentum* concept for BPSO based on the probability vector.

Goals: The overall goal of this paper is to develop a new BPSO algorithm for binary optimisation problems. In order to achieve this goal, we introduce a new *momentum* concept and updating mechanism, which can help particles move in a more meaningful way than the original BPSO. The proposed BPSO algorithm, Sticky BPSO (SBPSO), is compared with a recently developed BPSO algorithm (PBPSO) [22]. The comparison is performed on two well-known combinatorial problems, which are knapsack and feature selection. Knapsack is a traditional binary problem, which has been used widely to evaluate searching abilities of different binary optimisation algorithms [12, 19]. Feature selection is even more difficult than Knapsack because of the complex interactions between features. Specifically, we will investigate:

- whether introducing momentum can enhance the BPSO's ability to find out solutions in knapsack problems with higher profits than PBPSO,
- whether the proposed algorithm, SBPSO can evolve smaller feature sets with similar or better accuracies than PBPSO in the same computational time.

2 Proposed Approach

The key idea of a standard PSO algorithm is to guide the particle move to promising areas by utilising three factors including *pbest*, *gbest* and *momentum*. The *momentum* corresponds to a tendency to keep moving in the same direction. It has an important effect, which ensures the exploration of particles when both *gbest* and *pbest* are trapped in local optima. In other words, *momentum* preserves the diversity of the swarm. Therefore, we aim to develop a new BPSO algorithm, which can cope with binary search spaces and maintain three important properties of a PSO algorithm: *momentum*, *cognitive* and *social* components.

Since the original momentum in CPSO is not meaningful in BPSO, we define a new momentum concept for BPSO. In a binary search space, a particle moves by flipping its position entries; and therefore can not "keep moving" in the same direction. But we can view *momentum* as a tendency to stay at the current position, which is known as *stickiness* property. The *stickiness* should decay with time, so that when the bit is just flipped, it should initially has a high probability of remaining at the new value, which will reduce over successive iterations. While a bit is "stuck" at a new value, the particle will search around this value by mutating other bits. We represent the *stickiness* of the dth bit using a $currentLife_d$ variable, which records the number of iterations since a bit was just flipped. If the bit has just flipped then its *stickiness* in the immediate following iteration should be 1. As long as the bit stays continuously at the current value, the $currentLife_d$ increases linearly with respect to the iteration numbers. Particularly, after an iteration, if the bit does not change its value, the bit's $currentLife_d$ increases by 1, which also reduces the bit's *stickiness* property. The $currentLife_d$ is bounded by an upper value, called *maxLife*. When calculating the probability of switching a bit, we calculate its stickiness as belows:

$$stickiness_d = 1 - \frac{currentLife_d}{maxLife} \tag{3}$$

As can be seen from Eq. (3), $stickiness_d$ linearly decreases from 1 to 0 after *maxLife* iterations. The flipping probability of the dth bit, called p_d, is given in Eq. (4), which contains a momentum component.

$$p_d = i_m * (1 - stickiness_d) + i_p * |pbest_d - x_d| + i_g * |gbest_d - x_d| \tag{4}$$

In the above equation, i_m, i_p and i_g represent the proportions that *momentum*, *cognitive* (*pbest*) and *social* (*gbest*) factors contribute to the flipping probability, respectively. It can be seen that if a particle is different from *gbest* and *pbest* in at least *maxLife* iterations, the flipping probability is $i_m + i_g + i_p$. In this case, to ensure that the particle moves back to *gbest*, $i_m + i_p + i_g$ is set to 1. Given the *pbest* and *gbest* are not changed, the smaller the stickiness, the more likely the corresponding bit is mutated. Based on the new flipping probability vector, the new position of a particle is calculated by Eq. (5).

$$x_d^{t+1} = \begin{cases} 1 - x_d^t, \text{ if } random() \leq p_d \\ x_d^t, \text{ otherwise} \end{cases} \tag{5}$$

3 Experimental Studies

In this section, the effect of the new momentum concept is evaluated via a comparison between our proposed BPSO algorithm, called SBPSO, and a recently developed BPSO algorithm, PBPSO [22], in two binary problems: knapsack and feature selection. SBPSO and PBPSO have the same parameter settings: $i_m = 0.25, i_p = 0.25, i_g = 0.5$, which ensures that *gbest* plays a more important role than *gbest*. To test their performances, a statistical significance test, Wilcoxon test, is performed to compare the two methods. The confidence interval is 95 %.

3.1 Knapsack Problem

In a knapsack problem, there are n items, each of which has a positive profit $p_j > 0$. Besides the profit, each item also has m positive resource consumptions, $\{r_{j1}, r_{j2}, \dots, r_{jm}\}$, which respectively correspond to m types of resources. A knapsack has m capacities for each resource type, $\{C_1, C_2, \dots, C_m\}$. The task is to select a subset of items so that the total profit is maximised meanwhile the total resource consumption does not exceed the corresponding capacity of the knapsack. The problem can be shown in the following formulae:

$$max \sum_{j=1}^{n} p_j * x_j, \tag{6}$$

$$s.t.: \sum_{j=1}^{n} r_{jk} * x_j \leq C_k, \forall k \in \{1, 2, \ldots, m\}, \tag{7}$$

$$x_j \in \{0, 1\}, \forall j \in \{1, 2, \ldots, n\} \tag{8}$$

where $x_j = 1$ or 0 represents that the jth item is selected or not selected, respectively.

A single-dimensional knapsack problem (SKP) has only one resource, which is m =1. When there are more than one resources, the problem is called a multi-dimensional knapsack problem (MKP). In this work, 30 MKP datasets with at least 50 items are selected from [4] to compare the two algorithms, SBPSO and PBPSO.

3.1.1 BPSO for Knapsack Problem

In terms of representation, each particle's position, x_i is a binary vector, whose length is equal to the number of items, n. Particularly the ith particle's position is denoted as $x_i = (x_{i1}, x_{i2}, \ldots, x_{in})$ where $x_{ij} \in \{0, 1\}, \forall j \in \{1, 2, \ldots, n\}$, where $x_{ij} = 0$ shows that in the ith particle, the jth item is not selected.

Since the knapsack problem is a constrained problem, a penalty function strategy is applied in the fitness function to convert this problem to an unconstrained problem. Particularly, BPSO needs to maximise the following fitness function:

$$fitness_{Knap} = \sum_{j=1}^{n} p_j * x_j + \alpha * \sum_{k=1}^{m} \min(C_k - \sum_{j=1}^{n} r_{jk} * x_j, 0) \tag{9}$$

where α is set to 1000 to ensure that feasible solutions are evolved by BPSO algorithms.

Both SBPSO and PBPSO are run 50 independent times using 50 different random seeds. Each run contains 3000 iterations and the population size is equal to the number of items. *maxLife* in SBPSO is set to 50.

3.1.2 Experimental Results

Table 1 shows the experimental results of SBPSO and PBPSO on 14 out of 30 Knapsack datasets and the same pattern was observed on the other 16 datasets, but the detailed results are not presented due to the page limit. In the tables, "#Is" and "#Cs" represent the number of items and costs in each dataset, which correspond to n and m in Eq. (9), respectively. "Optimal" shows either the theoretical optimal or best known profit. "Hits" means the number of runs in which a BPSO algorithm evolves a solution with an "optimal" profit. "AveProfit" and "StD" stand for the average and the

Table 1 Experimental results on knapsack datasets

Dataset	Method	#Is	#Cs	Optimal	Hits	AveProfit ± StD	Test	Time(ms)
Pet7	PBPSO	50	5	16537.0	0	16391.94 ± 45.09		507.86
	SBPSO	50	5	16537.0	5	16477.2 ± 40.39	+	564.0
Sento1	PBPSO	60	30	7772.0	0	7414.32 ± 101.74		966.14
	SBPSO	60	30	7772.0	5	7704.76 ± 47.14	+	1417.24
Sento2	PBPSO	60	30	8722.0	0	8549.48 ± 36.31		1184.02
	SBPSO	60	30	8722.0	1	8676.44 ± 23.29	+	1230.78
Weish10	PBPSO	50	5	6339.0	0	6184.22 ± 80.49		507.04
	SBPSO	50	5	6339.0	28	6326.98 ± 22.49	+	558.32
Weish15	PBPSO	60	5	7486.0	0	7100.64 ± 110.22		677.86
	SBPSO	60	5	7486.0	4	7438.84 ± 29.55	+	957.52
Weish20	PBPSO	70	5	9450.0	0	8926.7 ± 116.68		902.42
	SBPSO	70	5	9450.0	0	9348.16 ± 48.2	+	1045.78
Weish25	PBPSO	80	5	9939.0	0	9246.76 ± 96.21		1202.18
	SBPSO	80	5	9939.0	0	9760.58 ± 61.14	+	1244.32
Weish30	PBPSO	90	5	11191.0	0	10426.06 ± 113.37		1477.02
	SBPSO	90	5	11191.0	0	10947.28 ± 57.29	+	1638.74
Gk01	PBPSO	100	15	3766.0	0	3657.0 ± 10.32		2111.88
	SBPSO	100	15	3766.0	0	3679.42 ± 9.51	+	2274.7
Gk02	PBPSO	100	25	3958.0	0	3846.06 ± 8.09		2454.6
	SBPSO	100	25	3958.0	0	3867.78 ± 10.75	+	2682.28
Gk03	PBPSO	150	25	5656.0	0	5490.08 ± 8.88		5052.5
	SBPSO	150	25	5656.0	0	5512.46 ± 12.79	+	5749.02
Gk04	PBPSO	150	50	5652.0	0	5608.56 ± 9.25		6723.28
	SBPSO	150	50	5652.0	1	5625.88 ± 9.75	+	7441.88
Weing7	PBPSO	105	2	1095445.0	0	1052973.24 ± 5801.6		1592.22
	SBPSO	105	2	1095445.0	0	1080452.0 ± 3632.63	+	1806.74
Weing8	PBPSO	105	2	624319.0	0	467870.78 ± 20783.97		1506.44
	SBPSO	105	2	624319.0	0	562477.06 ± 15752.94	+	1759.8

standard derivation of the 50 profits obtained from the 50 independent runs. "Test" shows the results of the Wilcoxon significance tests between SBPSO and PBPSO, where "+" ("−") means that SBPSO is significantly better (worse) than PBPSO, and "=" means they are similar. "Time" shows the average computation time of the algorithms in milliseconds.

As can be seen from Table 1, in terms of the average profit, SBPSO achieves better profit than PBPSO. For example, in Weish25, the average profit of SBPSO is about 6 % better than PBPSO. In addition, on 27 out of the 30 datasets, the standard deviations of SBPSO are smaller than PBPSO, which demonstrates that SBPSO is more stable. The other two datasets are large datsets with more than 100 items, in which SBPSO's standard deviation is a little bit larger than PBPSO. Probably due to the difficult search spaces, SBPSO mostly finds the same solutions as PBPSO and sometimes, SBPSO can discover a significantly better solution, which makes SBPSO's standard deviation larger than PBPSO. This can be seen in Gk04, in which SBPSO hits the optimal point only 1 time. In terms of the hit rate, SBPSO is able to discover the optimal solutions in half of the datsets, while PBPSO can not reach any optimal point. For example, in Weish10, SBPSO achieves the optimal solution in 28 out of the 50 times, while the hit rate of PBPSO is 0. The significance test results show that, on all datasets, SBPSO is significantly better than PBPSO. However, in terms of efficiency, SBPSO is more expensive than PBPSO, which can be seen in the "Time" column. The main reason is that SBPSO has extra steps to maintain the *currentLife* vector and calculate the momentum.

In general, the experimental results show that because of using momentum, SBPSO is more stable and explores the search space better than PBPSO. However, SBPSO's computation time is a little bit longer than PBPSO's computation time.

3.2 Feature Selection Problems

With advancements in technology, "big dimensionality" becomes a big problem due to the "curse of dimensionality". Classification, one of the most important tasks in machine learning, also suffers the deterioration in accuracies and a long learning time. The main reason is that a large feature set usually contains a number of irrelevant or redundant features, which may hide useful information from the relevant features [27]. In order to deal with this problem, feature selection is proposed mostly as a pre-processing step to select a small relevant feature subset by removing irrelevant and redundant features. It is expected that feature selection can shorten the training time and improve the classification performance over using all features.

However, feature selection is a difficult combinatorial problem, firstly because of its large search space. Suppose there are N original features, the total number of possible feature subsets is 2^N. In comparison with knapsack problems, feature selection is more challenging because there are interactions between features while items in the knapsack problem are independent. For example, two relevant features might provide the same information, which makes one of them become a redundant feature. Additionally two weakly relevant features might be significantly relevant when working together [20]. Since PSO has been widely applied to achieve feature selection [14, 15], feature selection is chosen as a benchmark problem to compare the searching abilities of the two PSO algorithms, SBPSO and PBPSO.

3.2.1 BPSO for Feature Selection Problems

In BPSO-based feature selection approaches, each position entry corresponds to one original feature. The value of one entry is either 1 or 0, which respectively indicates whether the corresponding feature is selected or not.

Feature selection has two main objectives, which are to maximise the classification accuracy and minimise the number of selected features. Therefore, the following minimisation fitness function is used:

$$fitness_{FS} = \alpha * ErrorRate + (1 - \alpha) * \frac{\#selected}{\#all} \tag{10}$$

where *ErrorRate* means the classification error rate of the selected features, *#selected* represents the number of selected features and *#all* is the total number of original features. α is used to control the contributions of the classification performance and the number of selected features. Since in feature selection problems, the classification performance is preferable, α is set to 0.9.

14 datasets from the UCI machine learning repository [1] are chosen to test the performance of SBPSO and PBPSO. These datasets can be seen in Table 2. Each dataset is randomly divided into two sets with different proportions: 70 % for training set and 30 % for test set. During the evolutionary feature selection process, a 10-fold cross-validation is applied to training set to calculate the accuracy to evaluate the fitness of the selected features. The 5-nearest neighbour classification algorithm is used in the experiments. For each dataset, each algorithm is run 40 independent times. Each run contains 100 iterations. The total number of particles in the swarm is 30.

Table 2 Datasets

Dataset	#Features	#Classes	#Instances
Wine	13	3	178
Australian	14	2	178
Zoo	17	7	101
Vehicle	18	4	846
German	24	2	1000
WBCD	30	2	569
Ionosphere	34	2	351
Sonar	60	2	208
Movementlibras	90	15	360
Hillvalley	100	2	606
Musk1	166	2	476
Arrhythmia	279	16	452
Madelon	500	2	4400
Multiple features	649	10	2000

Table 3 Experimental results on feature selection

Dataset	Method	AveSize	Training accuracy		Testing accuracy		Time(s)
			AveAcc ± Std	T	AveAcc ± Std	T	
Wine	Full	13.0	80.11 ± 0.00	–	76.54 ± 0.00	–	
	PBPSO	4.0	95.59 ± 0.42	=	96.45 ± 0.73	=	5.3
	SBPSO	3.8	95.48 ± 0.76		96.52 ± 2.02		5.3
Australian	Full	14.0	67.08 ± 0.00	–	70.05 ± 0.00	–	
	PBPSO	2.0	84.27 ± 0.00	=	85.02 ± 0.00	=	55.3
	SBPSO	2.1	84.15 ± 0.57		85.00 ± 0.09		55.6
Zoo	Full	17.0	83.90 ± 0.00	–	80.00 ± 0.00	–	
	PBPSO	8.1	94.06 ± 2.00	=	93.75 ± 1.45	=	2.5
	SBPSO	8.0	94.17 ± 1.91		93.79 ± 1.07		2.4
Vehicle	Full	18.0	81.33 ± 0.00	–	84.06 ± 0.00	+	
	PBPSO	5.1	85.01 ± 0.41	=	83.25 ± 0.74	=	86.2
	SBPSO	5.1	84.99 ± 0.39		83.35 ± 0.79		88.0
German	Full	24.0	69.86 ± 0.00	–	68.00 ± 0.00	–	
	PBPSO	7.1	72.86 ± 2.24	=	69.75 ± 2.30	=	116.4
	SBPSO	7.1	72.31 ± 2.23		68.99 ± 2.41		118.4
WBCD	Full	30.0	93.22 ± 0.00	–	92.98 ± 0.00	–	
	PBPSO	2.7	94.71 ± 0.55	–	93.17 ± 2.44	–	42.4
	SBPSO	2.1	94.96 ± 0.22		94.33 ± 1.23		40.9
Ionosphere	Full	34.0	81.30 ± 0.00	–	83.81 ± 0.00	–	
	PBPSO	7.5	90.20 ± 0.97	–	89.14 ± 2.03	=	20.1
	SBPSO	6.5	91.47 ± 0.83		89.45 ± 1.56		18.6
Sonar	Full	60.0	74.48 ± 0.00	–	76.19 ± 0.00	–	
	PBPSO	20.3	86.03 ± 1.82	–	79.78 ± 3.70	=	9.4
	SBPSO	19.3	87.70 ± 1.67		79.62 ± 3.13		9.9
Movementlibras	Full	90.0	96.56 ± 0.00	=	94.69 ± 0.00	+	
	PBPSO	20.4	96.53 ± 0.20	=	94.26 ± 0.39	=	27.0
	SBPSO	16.5	96.58 ± 0.21		94.29 ± 0.41		27.1
Hillvalley	Full	100.0	55.19 ± 0.00	–	56.59 ± 0.00	–	
	PBPSO	34.4	57.38 ± 0.83	–	57.55 ± 1.52	=	261.9
	SBPSO	32.3	58.09 ± 0.79		57.80 ± 1.78		263.7
Musk1	Full	166.0	81.68 ± 0.00	–	83.92 ± 0.00	–	
	PBPSO	72.6	89.60 ± 0.96	–	85.06 ± 2.34	=	67.4
	SBPSO	71.4	90.06 ± 1.19		85.51 ± 2.10		68.3
Arrhythmia	Full	278.0	93.52 ± 0.00	–	93.78 ± 0.00	–	
	PBPSO	97.9	94.12 ± 0.18	–	94.28 ± 0.33	–	82.8
	SBPSO	89.8	94.36 ± 0.24		94.50 ± 0.33		79.2
Madelon	Full	500.0	71.26 ± 0.00	–	70.90 ± 0.00	–	
	PBPSO	233.7	78.63 ± 0.65	–	77.44 ± 1.01	–	3228.5
	SBPSO	225.4	79.42 ± 0.66		78.19 ± 1.25		3328.4
Multiple features	Full	649.0	99.01 ± 0.00	–	98.53 ± 0.00	–	
	PBPSO	250.2	99.26 ± 0.05	=	98.92 ± 0.11	=	2360.2
	SBPSO	231.7	99.28 ± 0.06		98.93 ± 0.10		2225.3

3.2.2 Experimental Results

Table 3 shows the experimental results of SBPSO and PBPSO on the 14 datasets. In this table, "Full" means that all features are used. "AveSize" is the average number of selected features over 40 independent runs. "AveAcc" and "Std" stand for the average and the standard deviation of 40 accuracies in the 40 runs. "T" shows the significance test results, which compare between either using all features or PBPSO and SBPSO. "–"/ "+"/ "=" mean that the other methods are significantly worse/ better/ similar to SBPSO. The average computation time of two methods is shown in the "Time" column.

From Table 3, it can be seen that in most cases SBPSO can select a smaller number of features and achieve similar or better testing accuracy than PBPSO. For example, on Arrhythmia, SBPSO's accuracy is slightly better than PBPSO while SBPSO selects about 8 features fewer than PBPSO. In terms of training accuracy, on all datasets SBPSO always achieves similar or better average accuracy than PBPSO while selecting a smaller number of features.

As can be seen in the "Time" column, SBPSO is faster than PBPSO on 6 out of the 14 datasets. On the remaining datasets, SBPSO's computation time is only around 3 % higher than PBPSO, which is still in an acceptable range. Table 3 shows that SBPSO can use approximately the same computation time to evolve smaller feature subsets while maintaining or improving the classification performance over PBSO.

The evolutionary processes of the two BPSO algorithms are shown in Fig. 1. Three datasets, Zoo, WBCD and Multiple Features are selected in the order of increasing the number of features, which corresponds to the difficulty levels of the search spaces in different datasets. Other datasets show a similar pattern. In Fig. 1, the horizontal axis represents the number of iterations and the vertical axis shows the average fitness values of the *gbest* in the 40 independent runs. Since the task is to minimise the fitness value i.e. the error, the closer to the horizontal axis, the better the algorithm. As can be seen in Fig. 1, when the number of features is small like the Zoo dataset, SBPSO and PBPSO's evolutionary processes are quite similar. However, when the number of features increases, SBPSO seems to be superior to PBPSO. Despite of starting from the same positions, SBPSO just stays with PBPSO

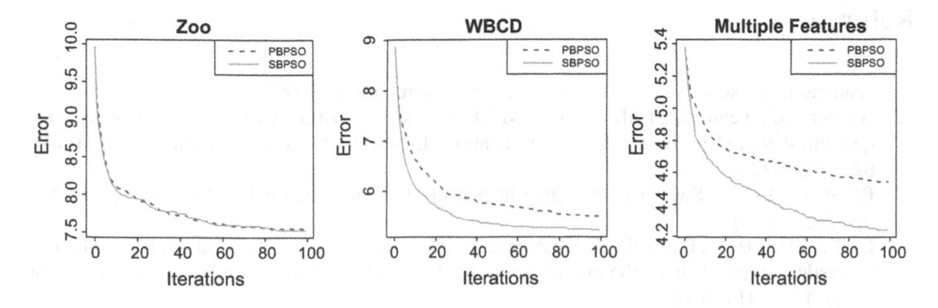

Fig. 1 Evolutionary Process of SBPSO and PBPSO in feature selection problems

for a few iterations (around 5 iterations) before quickly leaving PBPSO behind in the following iterations by using the momentum. In WBCD, the distance between SBPSO and PBPSO is closer at the later iterations, which means that given more iterations, PBPSO might catch up SBPSO. However, when the dataset's search space is even more complicated, for instance in the Multiple Features dataset, the gap between SBPSO and PBPSO is even larger with respect to the number of iterations. So the evolutionary processes clearly show that by introducing the *momentum* term in the updating mechanisms, SBPSO can better explore the search space, especially when the search space is large and complicated.

4 Conclusions and Future Work

The paper introduced a new *momentum* concept to BPSO based on which a new updating mechanism was developed. The proposed BPSO, called SBPSO is compared with a recently developed BPSO (named PBPSO) [22] in two well-known binary problems: knapsack and feature selection. The experimental results show that in knapsack problems, SBPSO has a superior performance on all datasets. In feature selection, SBPSO usually selects a smaller number of features and achieves similar or better testing accuracies. The training accuracies and evolutionary processes have shown that the introduced *momentum* can help SBPSO to improve the search ability over PBPSO, especially when the number of features is large. In terms of computation cost, SBPSO is slightly more expensive than PBPSO since SBPSO needs to maintain the *currentLife* vector.

This work focuses mainly on introducing the *momentum* concept to BPSO, more works need to be conducted to optimise its parameters such as i_m, i_p, i_g and *maxLife*. In addition, the proposed PSO is for binary problems only, which can be extended to general discrete problems in future. We will also extend SBPSO to a multi-objective BPSO algorithm to consider the trade-off and optimise multiple conflicting objectives simultaneously.

References

1. Asuncion, A., Newman, D.: UCI machine learning repository (2007)
2. Bagheri, A., Peyhani, H.M., Akbari, M.: Financial forecasting using anfis networks with quantum-behaved particle swarm optimization. Expert Systems with Applications 41(14), 6235–6250 (2014)
3. Blum, C., Li, X.: Swarm intelligence in optimization. In: Swarm Intelligence, pp. 43–85. Springer (2008)
4. Drake, J.H., Özcan, E., Burke, E.K.: A case study of controlling crossover in a selection hyper-heuristic framework using the multidimensional knapsack problem. Evolutionary computation 24(1), 113–141 (2016)
5. Foulds, L.R.: Optimization techniques: an introduction. Springer Science & Business Media (2012)

6. Ganesh, M.R., Krishna, R., Manikantan, K., Ramachandran, S.: Entropy based binary particle swarm optimization and classification for ear detection. Engineering Applications of Artificial Intelligence 27, 115–128 (2014)
7. Gholizadeh, S., Moghadas, R.: Performance-based optimum design of steel frames by an improved quantum particle swarm optimization. Advances in Structural Engineering 17(2), 143–156 (2014)
8. Jordehi, A.R., Jasni, J.: Particle swarm optimisation for discrete optimisation problems: a review. Artificial Intelligence Review 43(2), 243–258 (2015)
9. Kennedy, J., Eberhart, R., et al.: Particle swarm optimization. In: Proceedings of IEEE international conference on neural networks. vol. 4, pp. 1942–1948. Perth, Australia (1995)
10. Kennedy, J., Eberhart, R.C.: A discrete binary version of the particle swarm algorithm. In: Systems, Man, and Cybernetics. Computational Cybernetics and Simulation. IEEE International Conference on. vol. 5, pp. 4104–4108 (1997)
11. Khanesar, M.A., Teshnehlab, M., Shoorehdeli, M.A.: A novel binary particle swarm optimization. In: Control Automation, 2007. MED '07. Mediterranean Conference on. pp. 1–6 (June 2007)
12. Liu, J., Mei, Y., Li, X.: An analysis of the inertia weight parameter for binary particle swarm optimization. IEEE Transactions on Evolutionary Computation PP(99), 1–1. doi:10.1109/TEVC.2015.2503422 (2015)
13. Neshatian, K., Zhang, M.: Genetic programming for feature subset ranking in binary classification problems. In: Genetic programming, pp. 121–132. Springer (2009)
14. Nguyen, H., Xue, B., Liu, I., Zhang, M.: Filter based backward elimination in wrapper based pso for feature selection in classification. In: Evolutionary Computation (CEC), 2014 IEEE Congress on. pp. 3111–3118 (July 2014)
15. Nguyen, H., Xue, B., Liu, I., Zhang, M.: Pso and statistical clustering for feature selection: A new representation. In: Simulated Evolution and Learning, Lecture Notes in Computer Science, vol. 8886, pp. 569–581. Springer International Publishing (2014)
16. Pampara, G., Franken, N., Engelbrecht, A.P.: Combining particle swarm optimisation with angle modulation to solve binary problems. In: 2005 IEEE Congress on Evolutionary Computation. vol. 1, pp. 89–96 Vol. 1 (Sept 2005)
17. Pearl, J.: Heuristics: intelligent search strategies for computer problem solving (1984)
18. Unler, A., Murat, A.: A discrete particle swarm optimization method for feature selection in binary classification problems. European Journal of Operational Research 206(3), 528–539 (2010)
19. Wang, L., long Zheng, X., yao Wang, S.: A novel binary fruit fly optimization algorithm for solving the multidimensional knapsack problem. Knowledge-Based Systems 48, 17–23 (2013)
20. Xue, B., Zhang, M., Browne, W.N., Yao, X.: A survey on evolutionary computation approaches to feature selection. IEEE Transactions on Evolutionary Computation 20(4), 606–626 (Aug 2016)
21. Xue, B., Cervante, L., Shang, L., Browne, W.N., Zhang, M.: A multi-objective particle swarm optimisation for filter-based feature selection in classification problems. Connection Science (2-3), 91–116 (2012)
22. Xue, B., Nguyen, S., Zhang, M.: A new binary particle swarm optimisation algorithm for feature selection. In: European Conference on the Applications of Evolutionary Computation. pp. 501–513. Springer (2014)
23. Yuan, H., Tseng, S.S., Gangshan, W., Fuyan, Z.: A two-phase feature selection method using both filter and wrapper. In: Systems, Man, and Cybernetics. IEEE International Conference on. vol. 2, pp. 132–136. IEEE (1999)
24. Zambrano-Bigiarini, M., Clerc, M., Rojas, R.: Standard particle swarm optimisation 2011: A baseline for future pso improvements. In: 2013 IEEE Congress on Evolutionary Computation. pp. 2337–2344 (2013)
25. Zhang, Y., Wang, S., Phillips, P., Ji, G.: Binary pso with mutation operator for feature selection using decision tree applied to spam detection. Knowledge-Based Systems 64, 22–31 (2014)

26. Zhang, Y., Wu, L., Wang, S.: Ucav path planning by fitness-scaling adaptive chaotic particle swarm optimization. Mathematical Problems in Engineering 2013, 1–8 (2013)
27. Zhao, H., Sinha, A.P., Ge, W.: Effects of feature construction on classification performance: An empirical study in bank failure prediction. Expert Systems with Applications 36(2), 2633–2644 (2009)

Particle Swarm Optimization for Yard Truck Scheduling in Container Terminal with a Cooperative Strategy

Ben Niu, Fangfang Zhang, Li Li and Lang Wu

Abstract Trucks play a significant role in transporting containers between the seaside and storage yard at a container terminal. This paper exhibits a cooperative strategy for scheduling trucks, which allows trucks working or acting together toward a common purpose that can reduce truck-unload rate and cut back the make span. The objective is to minimize the total time cost of the sum of the delay of requests and the travel time of yard trucks. Particle swarm optimization (PSO) algorithm and three of its variants are applied to deal with the scheduling problem. The effectiveness of PSOs are analyzed by four typical different level-scale test problems. The results demonstrate that social learning PSO (SLPSO) can obtain better results than other algorithms for different scale cases.

Keywords Yard truck scheduling · Cooperative scheduling · Particle swarm optimization

1 Introduction

Due to the world trade expansion, container traffic has been growing steadily and this trend is expected to continue. This calls for efficient container terminal operations. Therefore, the optimal management for container terminals is desperately needed.

There are three fundamental equipment in typical container terminals: quay cranes (QCs), yard trucks (YTs) and yard cranes (YCs) [1]. When a vessel arrives at a port, containers are discharged by QCs. And then, YTs are utilized to transport the containers to the storage yard. The storage yard refers to the area where containers handling, transport, storage and transfer are occurred. YTs play a significant role in the process of transportation between the seaside and storage yard.

B. Niu (✉) · F. Zhang · L. Li · L. Wu
College of Management, Shenzhen University, Shenzhen, China
e-mail: Drniuben@163.com

G. Leu et al. (eds.), *Intelligent and Evolutionary Systems*,
Proceedings in Adaptation, Learning and Optimization 8,
DOI 10.1007/978-3-319-49049-6_24

The yard truck scheduling problem has characteristics of backhaul due to the repeated pickup and delivery processes involved. In traditional trucks scheduling in container terminals, the itinerary of a single-truck consists of three major steps. Firstly, a truck goes to a quay crane in discharging (called as CD) to pick up a container. Secondly, it delivers the container to an assigned storage area (called as AD). Finally, it turns back to the original CD. This is a typically static scheduling between the seaside and storage yard [1]. For more dynamic scenario, for example, a truck may go to another CD to discharge containers or a storage area for export containers (called as AL) to load a container after the completion of previous delivery operation.

There are many studies on the yard truck scheduling problem. In [2], a mixed-integer programming model was raised to decide the yard truck fleet size and allocate delivery jobs to YTs. In [3], the time-space network technique was utilized to describe the potential movements of yard trucks, thus to decrease the pollution of yard trucks operations. Taking the yard truck scheduling and storage allocation as a whole, Wang et al. [4] studied the influence of yard truck configurations on the truck employment strategy. In [5], a dynamic truck scheduling model with strong applicability was designed to reduce truck-load rate, and shorten the time of handling task. Wang et al. [6] gave weight to both internal truck scheduling and storage allocation and put forward a model that determines the strategy of owning and renting trucks in container terminals. Emphasizing on sequence-dependent processing time and different preparation time, the problem of scheduling a fleet of trucks to perform a set of transportation jobs was investigated in [7].

Unfortunately, most of the research above is not directly applicable to container terminal operations while disregarding their dynamic nature. Yet, as we mentioned before, on most dynamic scenario, container terminal operations need greater flexibility. And the development of models should take into account the characteristics and constraints associated with container terminals.

Taking dynamic characteristics into consideration, this paper addresses the truck scheduling problem in the container terminal using cooperative scheduling strategy, where trucks are normally considered to load a container to the assigned quay crane in AL for export containers in loading operation (called as CL) after the delivery to AD. Figure 1 is a truck-map which shows the cooperative scheduling of trucks in the terminal.

As can be seen from Fig. 1, truck 1 turns to the AL (33) for export container to load a container after delivering its container in AD (23) while truck 2 goes to the AL (25) for export container to load a container after handing over its container in AD (15). Obviously, this cooperative scheduling strategy will greatly reduce the unload time compared with traditional static scheduling strategy.

Scheduling problem is NP-hard [8]. Heuristic algorithms have been widely applied to deal with this kind of problem and achieve more effective solutions. Lee et al. [9] used the preparation time for jobs as the representation of the chromosome, instead of using job sequence which is generally employed in the typical genetic algorithm. Chung et al. [10] presented a new hybrid genetic algorithm with exhaustive searching in order to achieve fine local searching to determine the

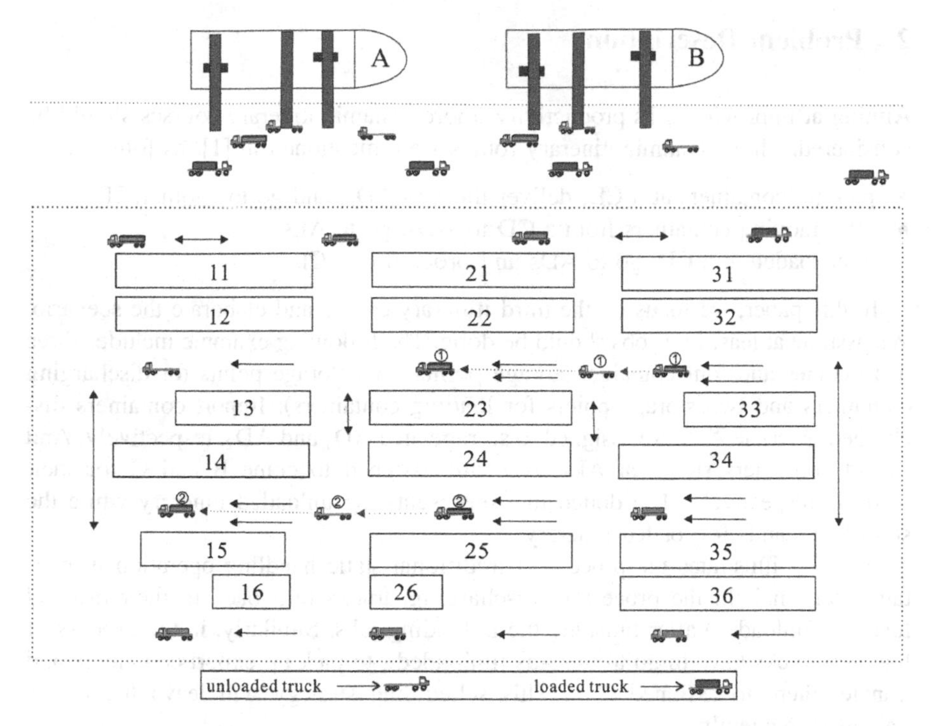

Fig. 1 An overview of cooperative scheduling for trucks

production schedule in the factories. Niu et al. [11] designed a mapping schema using bacterial foraging optimization to deal with integrated yard truck scheduling and storage allocation.

Motivated by the foraging behavior of birds, PSO algorithm was developed by Kennedy et al. in 1995 [12]. Then many researchers have studied the mechanism of PSO algorithm and proposed variety of its variants, such as [13–15]. Particle swarm optimization algorithm, as a significant branch of swarm intelligence, is also widely used in many application fields. In order to dig deeper in PSO algorithm and evaluate its performance in scheduling problems, this paper adopts PSOs to verify the model we proposed.

This study optimizes the yard truck scheduling in container terminal with the cooperative strategy. In the next section we describe the problem in detail. The scheduling problem is formulated in Sect. 3. The PSOs for solving the scheduling problem are discussed in Sect. 4. The results of computational experiments are presented in Sect. 5. The final section concludes the paper.

2 Problem Description

Aiming at improving YTs productivity, more dynamic itinerary courses should be considered. Three dynamic itinerary routes were mentioned in [1], as follows.

- pick up containers at a CD, deliver them to ADs, and go to another CD
- after moving containers from a CD to ADs, go to ALs
- after loaded at a CD, go to ADs and proceed to a CL

In this paper, we focus on the third itinerary course and elaborate the scenario. We assume at least four jobs should be done. The following example includes three quay cranes and four container storage points (two storage points for discharging containers and two storage points for loading containers). Import containers discharged by crane A, B are assigned to storage area AD_1 and AD_2, respectively. And export containers stored at AL_1, AL_2 are assigned to crane B and C for their loading, respectively. The dotted line represents the unloaded itinerary while the solid line stands for loaded itinerary.

Figure 2 illustrates the process of traditional static handling operation in container terminal. In the process of discharging, trucks turn back to their original location (unloaded) after finishing the unloading jobs. Similarly, in the process of loading, trucks have to go to the Ads (unloaded), to pick up export containers and transfer them to assigned QCs. This scheduling strategy will leave the trucks unloaded frequently.

The cooperative scheduling addressed in this paper is defined with a given set of jobs, as shown in Fig. 3. When a vessel arrives at the terminal, containers are discharged on trucks by the QCs. For example, a truck transfers the container to the assigned storage area AD_1. Then the truck moves to AL_1 to pick a container for export. So, the container picked up from AL_1 will be transported to the assigned QC incidentally. However, in a busy terminal, many containers are waiting to be transported nearby the quay crane B, so the truck can undertake one job (for example, transfers a container to AD_2). The same rule is true for the following scheduling. The truck moves to AL_2 to pick up a container for export. And finally, the truck stops nearby quay crane C and waits for a new job. Obviously, this cooperative scheduling strategy can decrease the rate of unloaded itinerary to a large extent.

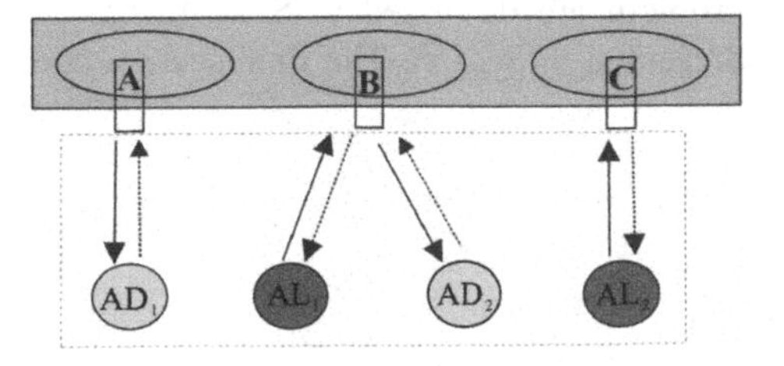

Fig. 2 Before cooperative scheduling

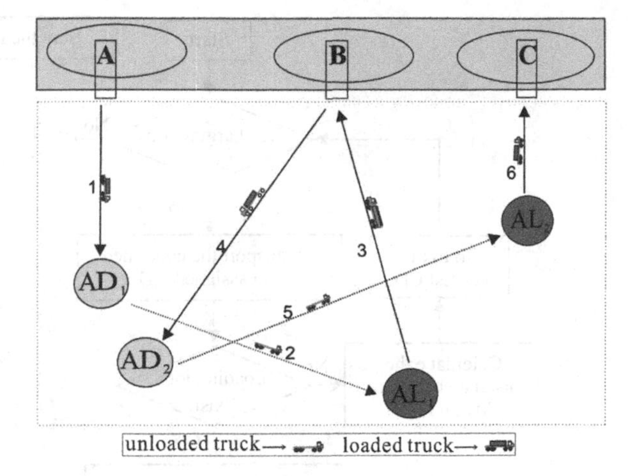

Fig. 3 After cooperative scheduling

As described above, the cooperative scheduling strategy we proposed is more advantageous than the traditional static one.

3 Model Development

In this section, we build a model to minimize the total time cost, that is, the sum of the delay of requests and the travel time of yard trucks. And the flowchart of the cooperative scheduling strategy is shown in Fig. 4.

3.1 Modeling Assumptions

According to the previous research on yard truck scheduling [8, 16], the following assumptions are made in this study.

- The number of trucks is limited.
- The number of storage locations is no less than the number of discharging containers.
- The quay crane and yard crane are always available. That is to say, once the yard truck arrives at the quay crane or yard crane, it can be served immediately.
- Congestion among yard trucks is not considered.
- The pick-up and drop-off locations of each job are known and uniquely identified by their (x, y) coordinates.
- The truck travel speed is the same for both loaded and empty trips.

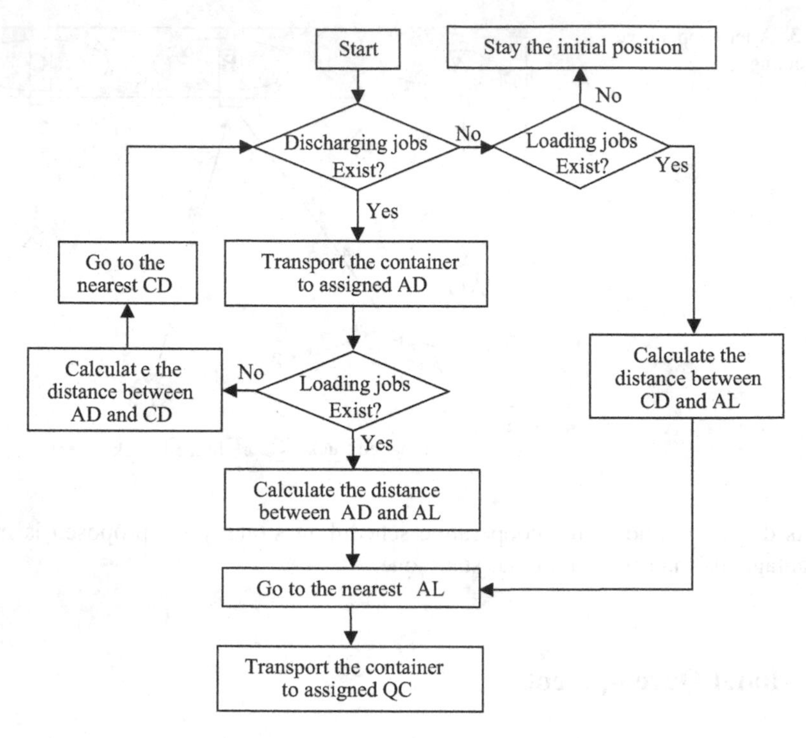

Fig. 4 The flowchart for the cooperative scheduling strategy

3.2 Notations

The following notations are used in this study.

i, j	Index of jobs, $i \neq j$
p, q	Index of location
N	The number of containers to be transported
M	The number of trucks to be used
J^-	Set of discharging jobs
J^+	Set of loading jobs
J'	Union set of all jobs and initial status, $J' = J \cup \{l_r\}$
J''	Union set of all jobs and final status, $J'' = J \cup \{k_r\}$
w_i	Starting time of job i
$[a_i, b_i)$	A soft time window for each job. Ii is a period of time involving the earliest possible a_i and the due time b_i
c_i	Completion time of request i.
d_i	Delay of request i. $d_i = max\{0, c_i - b_i\}$.
$\tau_{p,q}$	The travel time between location p and location q
t_i	The processing time of job i

	$\tau_{oi,ei}$, if job i is a loading job.
	$\tau_{oi,\zeta k}$, if job i is a discharging job and allocated to storage location k.
L_m	The initial location for a truck.
r_m	The initial departure time for a truck
P_i	The pick-up location of job i
Q_i	The drop-off location of job i.
$S_{i,j}$	Setup time of trucks from the destination of job i to the origin of job j.
	τ_{ei}, o_j, if job i and job j are discharging jobs.
	$\tau_{ei,\zeta j}$, if job i is a discharging job and job j is a loading job.
x_{ik}	1, if container i is allocated to storage location k.
	0, otherwise.
X_{ijm}	1, if truck m ($\forall m \in M$) processes job j after job i.
	0, otherwise.
Y_{im}	1, if truck m ($\forall m \in M$) processes job i ($\forall i \in N$).
	0, otherwise.

3.3 Model Formulation

In the process of cooperative scheduling, we aim at decreasing the unloaded itinerary rate to minimize the total make span of the transportation jobs. The processing time of job i has two components: the travel time for the empty trip to P_i (if there is any) and the complete time of job i. The problem formulation is modified based on the model provided by Ng et al. [17] and Lee et al. [8]. However, Ng et al. [17] proposed to schedule a fleet of trucks to perform a set of discharging jobs, ignoring the loading jobs. And Lee et al. [8] only considered the typical static operation pattern in the process of scheduling. In this paper, we take the loading jobs and dynamic itinerary routes into consideration. The revised model is given as follows:

$$\textit{Minimize}: Z = \alpha_1 \sum_{i \in N} d_i + \alpha_2 \left(\sum_{i \in N} t_i + \sum_{i,j \in J} s_{ij} X_{ijm} \right) \tag{1}$$

$$\sum_{m=1}^{M} Y_{im} = 1 \quad \forall i \in N \tag{2}$$

$$\sum_{j=1 \text{ and } i \neq j}^{N+1} X_{ijm} \leqslant Y_{im} \quad \forall i \in N \tag{3}$$

$$\sum_{j=1 \text{ and } i \neq j}^{N} X_{ijm} \leqslant Y_{im} \quad \forall i \in N \tag{4}$$

$$1 \geqslant X_{ijm} + X_{ijm} \geqslant Y_{im} + Y_{jm} - 1 \quad \forall i,j \in N \quad and \quad i \neq j; \quad \forall m \in M \tag{5}$$

$$w_i + S_{ij} + t_i \leqslant K(1 - X_{ijm}) + w_j \quad \forall i \in J^{'} \quad and \quad \forall i \in J^{''} \tag{6}$$

$$a_i + d_i \leqslant c_i \quad \forall i \in N \tag{7}$$

$$r_m + t_{L_m, P_i} + d_i \leqslant c_i \quad \forall i \in N; \quad \forall m \in M \tag{8}$$

$$s_{ij} = \tau_{e_i, o_j} \quad \forall i, j \in J^{-} \tag{9}$$

$$s_{ij} = \sum_{k \in K} \tau_{e_i, \zeta_j} x_{jk} \quad \forall i \in J^{-} \quad and \quad \forall j \in J^{+} \tag{10}$$

$$t_i = \tau_{o_i, e_i} \quad \forall \in J^{+} \tag{11}$$

$$t_i = \sum_{k \in K} \tau_{o_i, \zeta_k} x_{ik} \quad \forall i \in J^{-} \tag{12}$$

$$x_{ik}, X_{ijm}, Y_{im} \in \{0, 1\}, \forall i \in J^{'}, \forall i \in J^{''} \quad and \quad \forall k \in K \tag{13}$$

$$X_{ijm}, Y_{im} \in \{0, 1\}, \quad \forall i \in N - 1; \forall i \in N \quad \forall m \in M \tag{14}$$

$$w_i \in R \quad \forall i \in J^{'} \cup J^{''} \tag{15}$$

$$t_i \in R \quad \forall i \in J \tag{16}$$

$$S_{ij} \in R \quad \forall i \in J \quad and \quad \forall j \in J \tag{17}$$

$$d_i \geqslant 0 \quad \forall i \in J^{'} \cup J^{'} \tag{18}$$

The movement of a container from its origin to destination is defined as a job, denoted by i and j. Two types of jobs are considered in this paper, loading jobs and discharging jobs. Let J^{+} and J^{-} represent the set of loading jobs and the set of discharging jobs, respectively. A soft time window $[a_i, b_i)$ for each job is given as a constant.

Constraint (2) states that each job is processed by only the same truck. Constraints (3)–(5) give the relationship between X and Y for jobs handled by the same truck. Constraint (6) gives the connection of the starting time of a job and that of its successor. Constraint (7) defines the relationship between the completion time, preparation time and duration of a job. Constraint (8) gives the relationship between the duration and completion time, the truck preparation time and the travel time of a truck from its initial location to the pick-up location. Constraints (9) and (10) define the setup time of trucks from the destination of job i to the origin of job j. Constraints (11) and (12) define the processing time of job i. Constraints (13)–(18) are simple constraints which define the range of values of some variables.

4 Solution Approach

This paper aims at employing PSOs in solving the scheduling problem. Tasgetiren et al. [18] presented the completion time for n-job m-machine problem by equations. Similarly, in the yard truck scheduling problem, given the processing times t_{im} for job i on truck m $(k = 1, 2, \ldots, m)$, and a job permutation $\pi_i = [\pi_{i1}, \pi_{i2}, \ldots, \pi_{in}]$, n jobs $(J = 1, 2, \ldots, n)$ will be sequenced through m trucks. Let $T(\pi_J, m)$ denotes the completion time of job π_J on truck m. The calculation of completion time for n-job m-truck problem is given as follows:

$$T(\pi_1, 1) = t_{\pi 1,1} \tag{19}$$

$$T(\pi_J, 1) = T(\pi_{J-1}, 1) + t_{\pi J,1} \quad J = 2, 3, \ldots n \tag{20}$$

$$T(\pi_J, k) = T(\pi_{J-1}, k) + t_{\pi J,k} \quad J = 2, 3, \ldots n, \quad k = 2, 3, \ldots, m \tag{21}$$

Then the yard truck scheduling is to find a permutation $\pi*$ in the set of all permutations $\prod$. As demonstrated by the following equation.

$$T(\pi *) \leqslant T(\pi_n, m) \quad \forall \pi \in \prod \tag{22}$$

4.1 Particle Swarm Optimization Algorithms (PSOs)

After Kennedy proposed particle swarm optimization in 1995 [12], inertia weight was introduced into PSO algorithm (called SPSO) to provide a balance between global and local exploration abilities by Shi et al. in 1998 [19]. The key optimization mechanism of SPSO algorithm is described as follows.

$$V_{id} = w * V_{id} + c_1 * rand() * (P_{id} - X_{id}) + c_2 * rand()(P_{gd} - X_{id}) \tag{23}$$

$$X_{id} = X_{id} + V_{id} \tag{24}$$

The vector $P_{id} - X_{id}$ represents the distance from individual's current positions (X_{id}) to the individual's previous best position (P_{id}). The $G_{gd} - X_{id}$ indicates the distance between current positions (X_{id}) and the best position (P_{gd}) that has been found by any member of the neighborhood. c_1 and c_2 are two positive constants, *rand*() is a uniformly distributed random function in the range [0, 1], and w is the inertia weight.

Three improved PSOs, including CLPSO, LPSO and SLPSO can be referred to literature [13–15], respectively. The main updating equations of the three PSOs are described in Table 1.

Table 1 The chosen PSOs for comparison

Algorithm	Updating equations
CLPSO	$V_{id} = w * V_{id} + c * rand() * (P_{id} - X_{id})$
LPSO	$V_{id} = w * V_{id} + c_1 * rand() * (P_{id} - X_{id}) + c_2 * rand()(P_{gd} - X_{id})$ *with topology structure of Square*
SLPSO	$\Delta X_{ij}(t+1) = r_1 * \Delta X_{ij}(t) + r_2(t) * I_{ij}(t) + r_3 * \epsilon * C_{ij}(t)$ *with* $I_{ij}(t) = X_{kj}(t) - X_{ij}(t), C_{ij}(t) = X_j(t) - \overline{X}_{ij}(t)$

PSOs are used to solve difficult continuous optimization problems. However, the scheduling problem in this paper is a discrete one. So, we have to enable the PSOs to be applicable to the continuous problem.

4.2 Solution Representation

In our former research, solution representation was discussed in [20]. The position vector of each particle $\pi_{i(n^+ + n^- + l + m)}$ with $(n^+ + n^- + l + m)$ dimensions is divided into three parts. The $(n^+ + n^-)$ dimensions $\pi_i^J = [\pi_{i1}, \pi_{i2}, \ldots, \pi_{in}(n^+ + n^-)]$ denote scheduling permutation of jobs. The l dimension $\pi_i^l = [\pi_{i(n^+ + n^- + 1)}, \pi_{i(n^+ + n^- + 2)}, \ldots, \pi_{i(n^+ + n^- + l)}]$ denotes potential locations available to the discharging containers. We distinguish them from job permutation part with negative numbers. The m dimension $\pi_i^m = [\pi_{i(n^+ + n^- + l + 1)}, \pi_{i(n^+ + n^- + l + 2)}, \ldots, \pi_{i(n^+ + n^- + l + m)}]$ denotes workload assignment, namely, the number of jobs assigned to each truck.

In order to design a corresponding relationship between the scheduling problem and the particles, a suitable mapping to convert continuous position of particles $X_i^J = [X_{i1}, X_{i2}, \ldots, X_{i(n^+ + n^-)}]$ into job sequence $\pi_i^J = [\pi_{i1}, \pi_{i2}, \ldots, \pi_{in(n^+ + n^-)}]$ in PSOs is needed. The smallest position value (SPV) rule [18] is employed in this study.

Table 2 exhibits the solution representation of particle X_i^J for PSOs with its corresponding sequence. According to the SPV rule, the smallest position value is -1.57, so the dimension $j = 5$ is assigned to be the first job in the processing sequence; the second smallest position value is 0.03, so the dimension $j = 2$ is assigned to be the second job in the processing sequence, and so on.

Table 2 Solution representation of particle X_i

j	1	2	3	4	5	6
X_i^J	2.34	0.03	3.13	0.78	−1.57	1.87
π_i^J	5	2	6	3	1	4

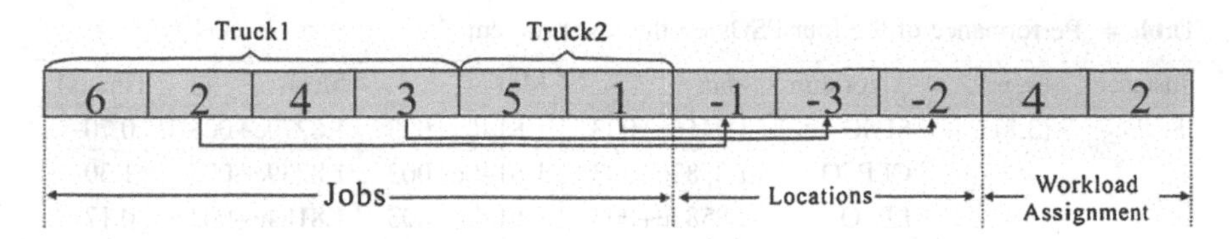

Fig. 5 An example for encoding scheme

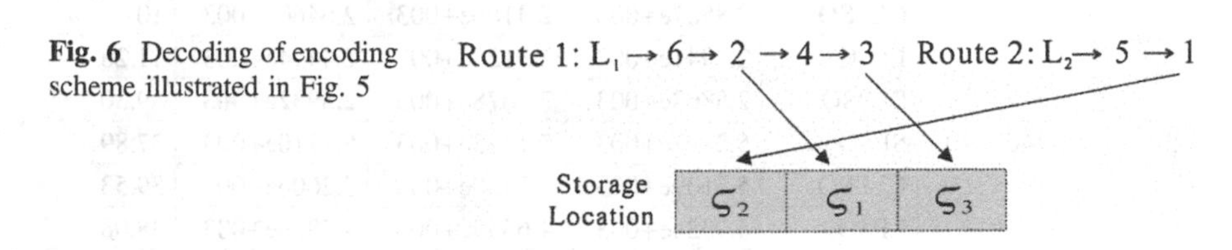

Fig. 6 Decoding of encoding scheme illustrated in Fig. 5

The l dimension $\pi_i^l = [\pi_{i(n^+ + n^- + 1)}, \pi_{i(n^+ + n^- + 2)}, \ldots, \pi_{i(n^+ + n^- + l)}]$ is the permutation of the integers from -1 to L. The m dimension $\pi_i^m = [\pi_{i(n^+ + n^- + l + 1)}, \pi_{i(n^+ + n^- + l + 2)}, \ldots, \pi_{i(n^+ + n^- + l + m)}]$ are workload assignment for trucks. As can be seen in Fig. 5, two trucks are arranged to work on six scheduling jobs. And Truck 1 is responsible for four jobs while Truck 2 is assigned two jobs. We assume that the first three jobs are discharging jobs and the second three jobs are loading jobs. The scheduling solution is that Truck 1 will handle jobs 6, 2, 4, 3, sequentially, while Truck 2 is assigned to handle jobs 5, 1, sequentially. According to the location solution, the first discharging job is located in ζ_2, the second discharging job is located in ζ_1 and the last discharging job is located in ζ_3, as shown in Fig. 6.

5 Computational Experiments

The computational experiments used to evaluate the performance of PSOs are discussed in this section. Four test problems of scheduling trucks are solved by Matlab R2001b running on a PC with Intel Core i5 2.20 GHz and 4 GB RAM.

The number of jobs (n) ranges from 8 to 300 while the number of trucks (m) ranges from 3 to 50 [17]. The four typical different scales of n and m are listed in Table 3. The initial location of trucks and pick-up/drop-off location are created following a uniform distribution in the two-dimension square from 0 * 0 m^2 to 1500 * 1500 m^2. And the earliest possible time of the jobs is randomly generated

Table 3 Four representative combinations of n and m

Instance	n	m
1	8	3
2	40	15
3	160	40
4	300	50

Table 4 Performance of the four PSOs on the test problem

Instance	(n, m)	Algorithm	Max	Min	Mean	Time(s)
1	(3,8)	SPSO	1.9466e+003	1.8140e+003	1.8279e+003	0.70
		CLPSO	1.8582e+003	1.8140e+003	1.8239e+003	1.30
		LPSO	1.8583e+003	1.8140e+003	1.8184e+003	0.17
		SLPSO	1.8147e+003	1.8140e+003	**1.8140e+003**	0.30
2	(15,40)	SPSO	2.7748e+003	2.3547e+003	2.5558e+003	13.26
		CLPSO	2.8587e+003	2.4160e+003	2.6460e+003	10.44
		LPSO	2.7941e+003	2.3125e+003	2.4494e+003	11.28
		SLPSO	2.5863e+003	2.2678e+003	**2.3932e+003**	9.30
3	(40,160)	SPSO	5.2197e+003	5.1623e+003	5.1710e+003	37.89
		CLPSO	5.3109e+003	5.2450e+003	5.2606e+003	39.53
		LPSO	4.9324e+003	4.6359e+003	4.7829e+003	38.08
		SLPSO	4.7277e+003	4.5305e+003	**4.3096e+003**	36.08
4	(50,300)	SPSO	6.9758e+003	6.7766e+003	6.8016e+003	49.30
		CLPSO	7.5080e+003	7.2925e+003	7.3626e+003	61.73
		LPSO	7.3979e+003	7.1155e+003	7.1690e+003	59.65
		SLPSO	6.3277e+003	6.1905e+003	**6.2306e+003**	48.08

following a uniform distribution of $\cup(0, 1500)$ (unit: second) and the due time of jobs is generated following a uniform distribution of $\cup(200, 500)$ (unit: second) [16]. The travel speed of trucks is 11.11 m/s and the two weight α_1 and α_2 are set to 0.6 and 0.4 as described in literature [8]. Each experiment is run twenty times.

It can be seen from the results presented in Table 4 that, with the increase of the number of jobs and trucks, the computational time grows rapidly. For the small-scale instance (i.e. $n = 3, m = 8$), all of the PSOs can find the minimum schedules for the problem. Figure 7 shows the average convergence rate on small-scale instance.

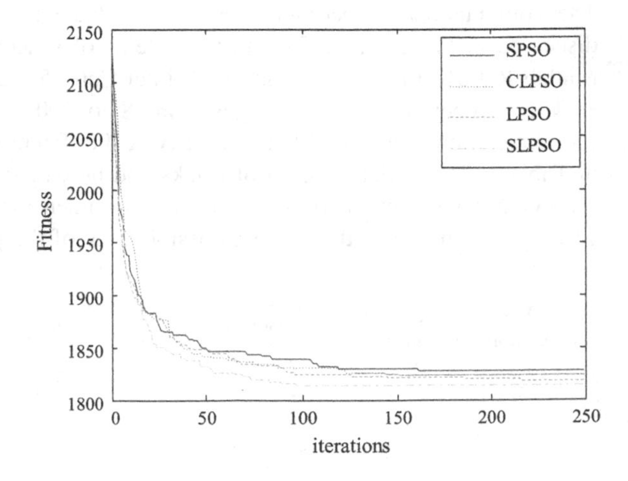

Fig. 7 The average evolution curve for small-scale instance ($n = 3, m = 8$)

Experimental study shows that the SLPSO outperforms the other three PSOs on all test problems. Our comparative results show that SLPSO performs well on small-scale problems and is promising for solving large-scale problems as well. It may be attributed to the fact that social learning mechanisms have the advantage of allowing individuals to learn behaviors from others without incurring the costs of individual trial-and-errors.

6 Conclusions

In this study, we addressed the cooperative scheduling strategy to reduce the unloaded rate and thus to cut back the make span. And PSO and its variants are applied to find optimal schedule strategy for the problem. A comprehensive set of test problems are used to compare the performance of the PSOs. The computational results demonstrated that the SLPSO performs better than all three PSOs on small-scale problems and large-scale problems as well. In future research, the development of more practical application will still be an emphasis on the study of integrated optimization model in container terminal operations.

Acknowledgments This work is partially supported by the National Natural Science Foundation of China (Grants nos. 71571120, 71271140, 71461027, 71471158 and 71501132), the Natural Science Foundation of Guangdong Province (Grant nos. 2016A030310074, 2016A030310067) and the Innovation and Development Fund Project of Shenzhen University (Grant no. 16XSCX04).

References

1. Nishimura, E., Imai, A., Papadimitriou, S.: Yard Trailer Routing at a Maritime Container Terminal. Transportation Research Part E Logistics & Transportation Review. 41(1):53–76 (2005)
2. Shang, J.: A heuristic Algorithm for the Integrated Yard Truck Scheduling in Container Terminal with Twin 40-Foot Quay Crane. Computer, Mechatronics, Control and Electronic Engineering (CMCE). 386–389 (2010)
3. Tsai, F.M., Chang, Y.M.: Optimization of Yard Truck Scheduling and Routing Model with Time-Space Network. Sustainable Development in Shipping and Transport Logistics. (2014)
4. Wang, Z.X., Chan, F.T.S., Chung, S.H., Niu, B.: Yard Truck Configurations for Efficient Operation of Container Terminals in Hong Kong. International Conference on Information Science, Electronics and Electrical Engineering. (2014)
5. Zheng, X., Shu, F., Mi, W.: Research on Genetic Algorithm-Based Simulation of Dynamic Container Truck Scheduling. IEEE International Conference on Computer Science and Automation Engineering. 131–135 (2012)
6. Wang, Z.X., Chan, F.T.S., Chung, S.H., Niu, B.: A decision support method for internal truck employment. Industrial Management & Data Systems. 114(9):1378–1395 (2014)
7. Ng, W.C., Mak, K.L., Zhang, Y.X.: Scheduling Trucks in Container Terminals Using a Genetic Algorithm. Engineering Optimization. 39(1):33–47 (2007)

8. Lee, D.H., Cao, J.X., Shi, Q., Chen, J.H.: A Heuristic Algorithm for Yard Truck Scheduling and Storage Allocation Problems. Transportation Research Part E Logistics & Transportation Review. 45(5): 810–820 (2009)
9. Lee, L.H., Chew, E.P., Tan, K.C., Wang, Y.: Vehicle Dispatching Algorithms for Container Transshipment Hubs. Operations Research-Spektrum. 32(3): 663–685 (2010)
10. Chung, S.H., Chan, F.T.S., Ip, W.H.: Minimization of Order Tardiness Through Collaboration Strategy in Multifactory Production System. IEEE Systems Journal. 5(1): 40–49 (2011)
11. Niu, B., Xie, T., Bi, Y., Liu, J.: Bacterial Colony Optimization for Integrated Yard Truck Scheduling and Storage Allocation Problem. IEEE Congress on Evolutionary Computation. 431–437 (2014)
12. Kennedy, J., Eberhart, R.: Particle swarm optimization. IEEE International Conference on Neural Networks, 1995. Proceedings. 1942–1948 (1995)
13. Liang, J.J., Qin, A.K., Suganthan, P.N., Baskar, S.: Comprehensive Learning Particle Swarm Optimizer for Global Optimization of Multimodal Functions. Evolutionary Computation IEEE Transactions on. 10(3):281–295 (2006)
14. Kennedy, J., Mendes, R.: Population Structure and Particle Swarm Performance. Evolutionary Computation, 2002. CEC '02. Proceedings of the 2002 Congress on. 1671–1676 (2002)
15. Cheng, R., Jin, Y.: A Social Learning Particle Swarm Optimization Algorithm for Scalable Optimization. Information Sciences. 291(6):43–60 (2015)
16. Wang, Z.X., Chan, F.T.S., Chung, S.H., Niu, B.: Minimization of Delay and Travel Time of Yard Trucks in Container Terminals Using an Improved GA with Guidance Search. Mathematical Problems in Engineering. 2015:1–12 (2015)
17. Ng, W.C., Mak, K.L., Zhang, Y.X.: Scheduling Trucks in Container Terminals Using a Genetic Algorithm. Engineering Optimization. 39(1):33–47 (2007)
18. Tasgetiren, M.F., Sevkli, M., Liang, Y.C., Gencyilmaz, G.: Particle Swarm Optimization Algorithm for Permutation Flowshop Sequencing Problem. Ant Colony Optimization and Swarm Intelligence, International Workshop. 382–389 (2004)
19. Shi, Y., Eberhart, R.C.: Parameter Selection in Particle Swarm Optimization. International Conference on Evolutionary Programming Vii. Springer-Verlag. 591–600 (1998)
20. Niu, B., Xie, T., Tan, L.J., Bi, Y., Wang, Z.X.: Swarm Intelligence Algorithms for Yard Truck Scheduling and Storage Allocation Problems. Neurocomputing. 188: 284–293 (2015)

A Method to Reduce the Amount of Inventoried Stock in Thai Supply Chain

Tomohito Okada, Akira Namatame, Hiroshi Sato and Saori Iwanaga

Abstract A global industrial enterprise is a complex network of different distributed production plants that produce, inventory, and distribute products. An agent-based model can be used to solve complex network problems that involve independent actors. The global economy and the increase in both demand fluctuation and pressure to lower costs while satisfying customers have put a premium on smart supply chain management. It is important to undertake a risk benefit analysis of supply chain design alternatives before making decisions. Simulation is an effective approach to comparative analysis and evaluation of such alternatives. In this paper, we describe an agent-based simulation tool for the design of smart supply chain networks and logistics networks. In the agent-based approach, supply chain models comprise supply chain agents. The agent-based simulation tool is useful to predict the effects of local and system-level activities on multi-plant performance and to improve the tactical and strategic decision-making at the enterprise level. Specifically, this model can reveal the optimal transport method under demand fluctuation and network disruption conditions. We found that selecting transport methods according to maximum stock is effective and can reduce cut the amount of stock in the whole supply chain in Thailand.

Keywords Agent-based model • Multi-agents • Supply chain management • Logistics • Multi-echelon • Manufacturing • Risk management • Demand fluctuation • Network disruption • Supply chain cost

---

T. Okada · A. Namatame · H. Sato
Department of Computer Science, National Defense Academy of Japan, Yokosuka, Japan

S. Iwanaga (✉)
Department of Maritime Safety Technology, Japan Coast Guard Academy, Kure Hiroshima, Japan
e-mail: s-iwanaga@jcga.ac.jp

G. Leu et al. (eds.), *Intelligent and Evolutionary Systems*,
Proceedings in Adaptation, Learning and Optimization 8,
DOI 10.1007/978-3-319-49049-6_25

1 Introduction

Supply chain and logistics networks have become larger and more complex due to globalization. This tendency makes it difficult to manage the supply chain and meet market demand. Collaboration among many independent contractors and suppliers is necessary to meet market demand. Companies should have an alternative strategic solution for their supply chain to ensure continued existence of their company. One solutions is to become a multi-plant enterprise. This has several advantages such as low raw material costs, flexibility to change product, and the ability to change the product flow network [1]. However, with a multi-plant enterprise, the supply chain network involves many actors and it is operated collectively. The logistics network, i.e., the physical transportation network, also becomes an interconnected complex system that affects the behavior of the supply chain network.

Many analytical methods [2–8] for modeling and optimizing different scenarios in a multi-plant enterprise have been proposed. These methods involve mathematical formulations for operations management in multi-plant industrial networks. These studies have proposed methods to solve the combined production and distribution scheduling problem in multi-plant environments using mathematical programming approaches.

2 Agent-Based Supply Chain Architecture

Lee and Billington [2] provided an insightful survey of common pitfalls in supply chain management practices. Some studies demonstrate that market and supplier relationships depend on quality, delivery time, and contract flexibility rather than cost. Many studies have analyzed inventory problems in a multi-echelon supply chain. Svoronos and Zipkin [3] investigated a multi-echelon system with multiple tiers in the supply chain. In a multi-echelon system, it is assumed that the company manages the supply chain and exercises centralized control.

Towill et al. [4] applied simulation technology to evaluate the effect of different supply chain strategies when demand increases. Swaminathan et al. [5] presented a modeling and simulation framework for developing decision support tools for supply chain management. They developed a framework with two basic elements, i.e., object modeling of supply chain flows and agent modeling of supply chain entities.

The Just-in-Time philosophy affects conventional supply chain management styles. The supply chain affected by this philosophy becomes globalization, use of third parties, and reducing the lead-time. These trends reduce supply chain costs and increase the company's competitiveness relative to other companies in the same market.

Supply chain distribution can be modeled as a network problem. Player in the supply chain and relationships between players can be represented as nodes and links. Considering a supply chain as a network problem, a disruption means that a node is shut down or a link is served.

Supply chains are defined as a collection of business centers through which products pass at various stages of completion from the provision of raw materials to final sales. Individual companies in supply chain are only aware of limited part of the supply chain. Thus, it is difficult for an individual company to estimate demand. The players in a supply chain depend on their own information, which may differ from the information by headquarter management. The amount of order may be larger than the headquarter one. The more amplified, the amount of order. These problems can occur with a dynamic supply chain. Consequently, each company makes incorrect demand estimations and inventory is larger than that required by the actual demand. This is known as the bullwhip effect [5].

Information sharing among companies in the supply chain has been proposed as a solution to the bullwhip effect. Information systems that support supply chain management have unique characteristics. First, they should support distributed collaboration among companies. Second, a single company cannot manage multiple players in the supply chain directly, but there are need to coordinate each company. Third, high intelligence for strategy, planning, and flexibility adaptation is required. For these reasons, agent modeling is suitable to support supply chain management [6].

Multi-agent technology has many beneficial features for autonomous, collaborative, and intelligent systems in distributed environments, which makes it one of the best candidates for complex supply chain management [7]. Agent-based modeling is a suitable approach to analyze a system that is influenced by autonomous agents [8]. Such a system is affected by the behavior of the players in the system and their interactions. The agent-based model comprises a network of decentralized agents. Generally, agents can make decisions autonomously. They can all decisions without centralized management. In addition, agents can change decisions when conditions change.

A multi-plant enterprise can be modeled as a modular, decentralized, changeable agent network. Such networks have many agents of different types. In addition, all agents set their own goals in the supply chain.

The agent-based model can be used to analyze different stages of the supply chain in order to determine what could happen under different scenarios, e.g., in an aid supply chain where aid was not sufficient to meet demand. Analyzing different stages of supply chain can provide information about possible side effects or delayed consequences, such as the bullwhip effect. This is important because distorted information in the supply chain can lead to significant inefficiencies. In the case of an aid supply chain, inventory moves up the chain and fluctuates as more donations are recollected in distribution centers, which distorts demand information. The model represents the flow of emergency goods and how it is affected by information feedback. Thus, the model can explain the existence of bullwhip effects at different times depending on the initial stock of stored goods in different parts of the chain and on exogenous variables.

3 Model and Simulation

In Southeast Asia, floods have caused extensive damage. Despite this, there have been little anti-disaster operations that deal with floods. We should consider network disruptions caused by such disasters. Disasters can disrupt a supply chain network, e.g., factories can shut down and products cannot be transported to markets.

This study simulates a supply chain model in the automobile industry in Thailand that supports supply chain management. In former study, we dealt with Southeast Asia [9]. Then, we can get data of open industrial data in Thailand. In Thailand, there are 17 distributed industrial areas [10]. We clarify which routes are effective to supply products produced in these industrial areas to the market in Thailand.

3.1 Agents and Networks

The agent model is constructed from two types of agents, i.e., factory agents and market agents. In this simulation, there are 17 factory agents and 30 market agents (47 agents in total). Each agent is independent and can make autonomous decisions. The function, location and the network are described as follows.

- Factory agent
  Factory agents produce auto mobiles. In the automobile supply chain network, a factory agent is an upstream supplier in the supply chain network. Each factory agent produces products based on various demands and delivers products based on orders from downstream retailers.
- Market agent
  Market agents are consumers. Market agents are downstream in the supply chain network. It is impossible to estimate consumption precisely; however, it is possible to estimate consumption to some extent based on centralized sales promotion campaigns. If consumption is high, the available stock will be depleted. Lack of available stock results in lost sales opportunities. Consequently, the company's reputation will suffer, which affects the future management. The market controls inventory to avoid lack of stock due to unpredictable consumption and to keep stock less within the inventory which is not running short.

- Agent locations

We selected 17 factory agents from the industrial area [10] and 30 market agents, including tourist spots, like Phuket and so on (Table 1).

- Agent networks
  The transport path between tradable agents in this model is a bipartite graph. A bipartite graph is a layered graph with two layers, i.e., the factory layer and the market layer. All factory agents can interact with all market agents. In this network, agents of same type cannot interact, i.e., transactions between factory agents are prohibited, as are transaction between market agents. We obtained lead time data about in the transport route "Google map" as the land transportation line (Table 2). We used Google Maps to identify transportation routes and determined lead times based on the route. Here, lead time represents the time from the start to end of a given process. For a market agent, lead time represents the time between placing an order and product delivery.

Agent Rules

In this study, transportation routes are land transportation lines that connect factories (factory agents) and markets (market agents) in Thailand. We analyzed which route is selected frequently by simulating the supply chain network.

Table 1 Agent names, types and IDs

ID	Market agent	ID	Market agent	ID	Factory agent
0	Chiang Rai	17	Cha-am	30	Lamphun
1	Mae Hong Son	18	Rayong	31	Pichit
2	Chiang Mai	19	Huahin	32	Singburi
3	Lampang	20	Ban Pae Pier	33	Khon Kaen
4	Phrae	21	Chang	34	Nakhon Ratchasima
5	Nong Khai	22	Tao	35	Prachinburi
6	Sukhothai	23	Surat Thani	36	Ratchaburi
7	Phitanulok	24	Phangnga	37	Ayudthaya
8	Khon Kaen	25	Kaolack	38	Saraburi
9	Surin	26	Phuket	39	Chachoengsao
10	Khorat	27	Krabi	40	Rayong
11	Phra Nakhon Si Ayutthaya	28	Tambon Khlong Khanan	41	Chonburi
12	Prachinburi	29	Lanta Yai	42	Pathum Thani
13	Kanchanaburi			43	Bangkok
14	Nakhon Pathom			44	Samut Prakarn
15	Bangkok			45	Samut Sakhon
16	Bang Lamung, Phattaya			46	Songkhla

Table 2 Lead time between agents

		Factory agent (ID)																
		30	31	32	33	34	35	36	37	38	39	40	41	42	43	44	45	46
Market agent (ID)	0	4.0	8.0	9.0	11.0	11.0	12.0	11.0	9.0	10.0	12.0	14.0	12.0	11.0	11.0	11.0	11.0	22.0
	1	5.3	9.8	11.5	14.2	15.3	15.0	15.1	12.2	13.1	14.6	15.5	13.8	17.8	17.5	14.3	14.4	28.0
	2	0.8	5.5	6.7	9.7	9.5	12.0	10.0	7.5	8.0	10.5	10.6	11.5	9.5	8.8	9.0	9.2	21.0
	3	1.0	4.8	6.1	9.1	8.8	8.9	9.2	6.2	7.1	10.9	12.2	11.5	8.3	7.9	8.5	8.4	19.8
	4	2.4	3.8	5.2	8.4	8.5	8.9	8.2	5.9	6.9	6.8	9.8	8.6	7.8	7.5	8.0	7.5	19.1
	5	9.7	7.4	7.7	2.4	6.0	7.9	10.8	8.0	7.0	8.8	9.9	8.8	9.2	9.5	9.8	9.9	20.3
	6	3.7	2.1	4.2	7.0	6.5	6.7	6.7	4.5	5.5	6.8	10.2	6.1	5.5	5.5	6.2	6.5	17.6
	7	4.5	1.5	3.5	4.5	5.6	6.5	6.0	3.8	4.5	6.2	9.5	6.5	5.5	5.1	5.7	5.5	17.0
	8	8.0	5.2	6.1	0.0	3.5	6.0	8.8	7.0	5.0	6.5	7.6	6.5	8.1	7.0	8.0	7.5	18.0
	9	11.4	7.0	6.5	3.6	2.9	5.5	7.1	6.7	6.0	5.2	6.9	6.5	6.5	6.5	6.5	6.5	17.9
	10	10.7	6.5	5.0	3.0	2.2	4.8	6.5	6.3	4.5	5.5	6.0	5.8	5.9	5.7	5.5	6.5	17.2
	11	7.7	3.9	1.2	5.9	4.0	2.7	2.8	0.0	1.1	2.0	3.4	2.1	1.2	1.3	1.6	1.9	13.2
	12	12.2	5.5	3.5	6.2	2.2	0.0	4.4	2.5	2.4	1.8	3.0	3.0	2.8	2.7	2.8	3.2	14.8
	13	8.9	5.0	2.5	7.4	5.5	4.2	1.5	2.1	2.8	3.5	4.7	3.2	2.5	2.1	2.9	2.7	12.8
	14	8.5	4.9	2.7	7.6	4.0	3.8	1.0	1.9	2.2	2.1	3.5	2.1	1.4	0.9	1.5	1.0	12.2
	15	8.1	4.5	2.1	7.5	4.6	2.8	1.9	1.4	1.6	1.4	2.5	1.2	0.5	0.0	0.4	1.0	12.5
	16	10.1	6.5	4.3	7.4	4.8	3.1	3.6	2.9	3.5	2.1	1.2	1.3	2.1	2.4	2.0	2.6	14.1
	17	9.9	6.2	3.9	8.0	5.8	4.5	1.2	3.1	3.5	3.1	4.4	3.0	2.3	2.3	2.2	1.6	10.5
	18	13.4	8.9	4.5	7.9	5.1	3.4	3.9	3.1	3.6	2.1	0.0	1.9	2.3	2.2	2.2	2.8	14.4
	19	10.1	6.6	4.2	8.4	6.1	5.0	1.6	3.5	3.9	3.4	4.9	3.5	2.7	2.6	2.7	2.0	10.2
	20	13.5	7.1	4.0	8.0	5.5	3.9	4.4	4.2	4.0	2.3	0.3	2.0	2.8	3.0	2.9	3.5	14.8

(continued)

Table 2 (continued)

	Factory agent (ID)																
	30	31	32	33	34	35	36	37	38	39	40	41	42	43	44	45	46
	15.5	10.5	9.3	10.4	7.5	5.5	7.3	6.3	6.9	5.9	4.9	5.3	6.9	6.8	6.8	6.8	17.8
22	17.6	14.3	11.7	15.8	13.8	12.5	8.9	10.9	11.2	10.9	12.2	10.8	10.0	10.0	9.9	9.3	7.5
23	19.1	15.9	13.2	17.3	15.3	14.0	10.3	12.3	12.7	12.4	13.8	12.2	11.5	11.6	11.5	10.8	7.3
24	17.6	15.0	11.6	15.8	13.8	12.5	8.9	10.9	11.2	10.9	12.2	10.7	10.0	10.0	10.0	9.3	5.5
25	18.1	14.8	12.2	16.3	14.2	13.1	9.5	11.5	11.9	11.5	12.9	11.4	10.5	10.6	10.5	9.9	6.7
26	18.7	15.3	12.7	16.9	14.8	13.5	9.9	11.9	12.3	11.9	13.3	11.8	10.9	11.0	11.0	10.3	6.5
27	17.4	14.1	11.5	15.6	13.6	12.4	8.7	10.7	11.1	10.7	12.1	10.5	9.8	9.9	9.8	9.1	4.5
28	18.0	14.6	12.2	16.1	14.1	13.0	9.2	11.2	11.8	11.3	12.8	11.1	10.4	10.5	10.3	9.7	4.5
29	19.3	15.9	13.4	17.5	15.4	14.3	10.7	12.6	13.1	12.7	14.0	12.5	11.7	11.8	11.7	11.1	5.9

- Market agents

Market agents consume products. We set the consumption pattern as a lognormality distribution, which the standard consumption pattern follows normal distribution. However, actually, there can be a remarkable rise or fall of sudden demands; thus, it is acceptable to appropriate to other distributions. Thus, we employ a lognormality distribution as the consumption pattern. This is the distribution of a random variable for which the logarithm of the variable has normal distribution, which is given as follows.

$$D_{t+1} = \mu + \rho \times D_t \tag{1}$$

Here, D_t is demand at step t, μ is the basement value of consumption at step t (5 in this simulation, ρ is a value in the range −1.0 to 1.0 at normal distribution, and the average is 0.1.

The safety stock is the threshold. And it is out of stock if it less than the threshold. The safety stock is expressed as follows.

$$\mathrm{SS} = \mathrm{AvgC} \times \mathrm{MaxLT} \tag{2}$$

Here, SS represents the safety stock, AvgC represents the average stock consumption and MaxLT is the longest lead time among factory agents for the given market agent. AvgC is the averaged stock consumption over the previous 10 steps. This value is updated for each time step after the tenth time step. For the first nine time steps, we calculate is the averaged stock consumption as the total past stock consumption divided by t at time step t.

The market agent calculates the safety stock, the order point, and the order amount, and manages the inventory. When the current stock is less than an order point, the market agent orders products. The order point (OP) is expressed by the following function.

$$\mathrm{OP} = \mathrm{SS} + \mathrm{AvgC} \times \mathrm{LT} \tag{3}$$

Here, LT is the lead time for a particular factory agent for a given market agent.

AO represents the order amount and depends on the current stock at the order point. Typically, the market agent orders the required products for the lead time.

$$\mathrm{AO} = \mathrm{AvgC} \times \mathrm{LT} + \mathrm{OP} - \mathrm{CS} \tag{4}$$

Here, CS represents current stock.

Sometimes, products cannot be delivered on in time because the factory agent is out of stock. If the order amount is greater than the order point, a market agent orders same as ordering point.

$$\mathrm{AO} = \mathrm{AvgC} \times \mathrm{LT} \tag{5}$$

We employ three methods to select a factory agent that can transact with a market agent, “random selection”, “minimum lead time” and “maximum stock”.

For “random selection”, each market agent chooses a factory agent randomly (Case 1). For “minimum lead time”, each market agent chooses a factory whose lead time is the least among the factory agents (Case 2). For “maximum stock”, each market agent chooses a factory agent whose stock is greatest among all factory agents (Case 3).

We set the interval of ordering to once in 10 time steps for each agent. The upper limit for the product order amount is 100. If this order amount exceeds 100, the order amount substitutes for 100.

- Factory agents

Based on an order from a market agent, the factory agent sends out products. The factory agent calculates the consumption of products and sends out the products and manages inventory. If the inventory amount is less than a certain production point, it produces products based on the amount of production.

The factory agent calculates the safety stock, the ordering point, the amount of the order, and manages the inventory. These functions are similar to that of the market agent. The safety stock SS is the threshold. And it is out of stock if it less than the threshold. It is given by the following function.

$$SS = AvgC \times MinLT \times N \quad (6)$$

Here, AvgC is the average stock consumption and MinLT is the shortest lead time of the market agents for the factory agent. The average stock consumption is the averaged stock consumption over the past 10 step. This value is updated for each time step after the tenth time step. Until nine time steps, we calculate the average stock as the total past stock consumption divided by t at time step t. N is the number of market agents that transact with the factory agent. When current stock is less than a given production point, the factory agent produces products based on the amount of production. The amount of production depends on the stock at the given production point. Normally, the factory agent produces the required amount of products for the lead time. We denote current stock as CS.

$$PP = SS + AvgC \times LT \times N \quad (7)$$

$$AP = PP - SS + PP - CS \quad (8)$$

However, products are sometimes out of stock. In this case, the factory agent produces products as same as ordering point if the amount of produce is greater than production point.

$$AP = PP - CS \quad (9)$$

3.2 Simulation Settings

We define a single step as one hour. Each simulation has 8760 steps, i.e., approximately one year. The factory agent produces products when the amount of inventory is under the order point. Here, the production cost is $90. The market agent stocks and orders inventory form a factory agent. When the inventory is less than the order point, the market agent orders the inventory shortage. In the market, the inventory is consumed by the customer. The profit for a single product is $120. When the inventory is less than the order point, the market agent orders inventory. The delivery cost is $1 per inventory for a single time step. The store cost is $0.1 per inventory for a single time step. This simulation runs three times. We use the average data from three simulations.

3.3 Simulation Results

The simulation results are shown in Table 3. The average amount of stock and the standard deviation are described by the average stock per single time step in the simulation. We found that the average amount of stock of market agent is the most in the three cases if market agents select a factory agent randomly. If the market agents select a factory agent whose lead time is the least in the factory agents, the averaged stock is the least in the three cases. In other words, selecting by "minimum lead time" is the best way to minimize the amount of stock for market agents.

On the other hand, for factory agents, selecting by "maximum stock" is the best way to minimize the amount of stock and the averaged stock is the averaged stock is the least in the three cases. Because the amount of stock for factory agents are larger than that of market agents, as for entire supply chain, selecting by "maximum stock" is the best way to minimize the amount of stock for all agents and the averaged stock is the averaged stock is the least in the three cases.

As shown in Fig. 1, the transportation routes differ relative to the manner in which the factory agent is selected.

- Case 1: Random selection

When each market agent selects a factory agent randomly, the agents transport products by all possible transportation routes as shown in Fig. 1a.

Table 3 Averaged stock in the simulation

		Case 1	Case 2	Case 3
Supply of market agent	Average	31.63	13.57	14.39
	Standard deviation	42.49	10.56	13.60
Supply of factory agent	Average	909.14	1252.58	100.99
	Standard deviation	822.89	2731.21	16.04
Entire supply	Average	349.03	461.72	45.717
	Standard deviation	651.06	1748.27	44.08

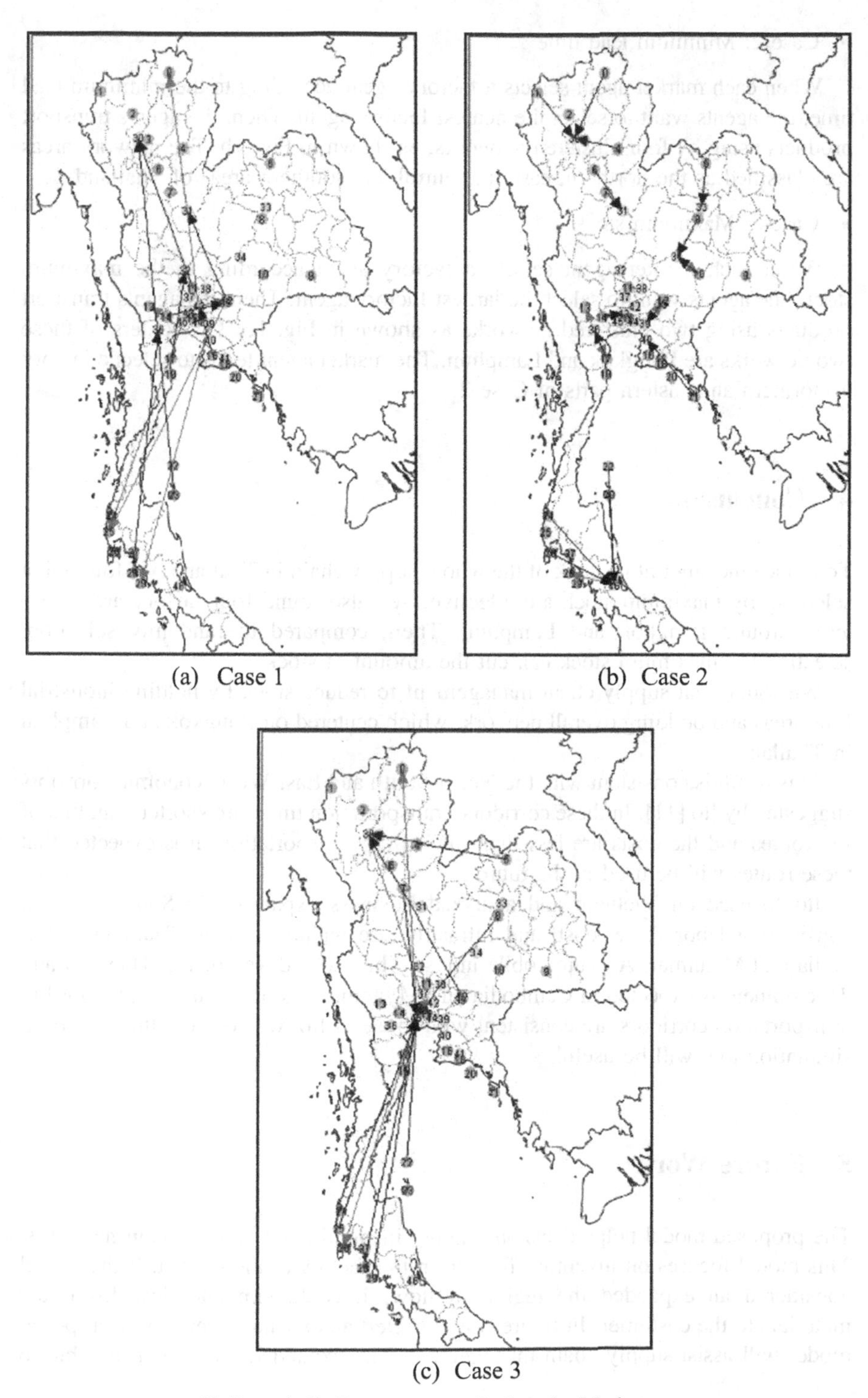

(a) Case 1

(b) Case 2

(c) Case 3

Yellow circle: Factory agent, Red circle: Market agent

Fig. 1 Supply chain networks

- Case 2: Minimum lead time

When each market agent selects a factory agent according to the minimum lead time, the agents want to select the nearest factory agent. Then, the agents transport products using by four clustered networks, as shown in Fig. 1b. The network areas are classified as the northern, eastern, central and southern areas of Thailand.

- Case 3: Maximum stock

When each market agent selects a factory agent according to the maximum stock, the agents want to select the largest factory agent. Then, the agents transport products using two clustered networks as shown in Fig. 1c. The centers of these two networks are Bangkok and Lamphun. The market agent tends to select a factory in northern and eastern parts of Case 2.

4 Conclusion

To reduce the amount in stock of the whole supply chain in Thailand, we found that selecting by maximum stock are effective. We also found four factory areas centered around Bangkok and Lamphun. Then, compared to randomly selection, selecting the maximum stock can cut the amount of stock.

We found that supply chain management to reduce stock by holding industrial four areas and building overall network, which centered on Bangkok and Lamphun in Thailand.

This result is consistent with the North-South and East-West economic corridors suggested by Ito [11]. In these corridors, transportation times are shorter than that of sea routes and the costs are less than that of air transportation. It is expected that these routes will be used in the future.

Ito focused on Vietnam and analyzed business expansion in Southeast Asia considering labor force, cost, and infrastructure maintenance in Thailand, Cambodia and Myanmar. An automobile industry has been developed in Thailand and development is expected in Cambodia and Myanmar. As our results with regard to transportation corridors are consistent with those of Ito, we consider that proposed simulation tool will be useful.

5 Future Work

The proposed model helps decision making in multi-plant supply chain networks. This model focuses on inventory flow from the factory to the market. If the model considered an expanded the region of flow, it could simulate flow from raw materials to the customer. In future, we expected an expanded version of proposed model will assist supply chain management. The expanded model will be able to determine optimal delivery routes under various scenarios and the demand from the manager. We expect that the expanded model will have additional functionality and make use of expanded data to enable modeling of demand fluctuations, market disruptions.

This study has demonstrated the potential of using agent-based modeling for the supply chain management. Decision making in supply chain network is affected by various types of agents and the relationships among agents. The agent-based modeling approach can help decision making to manage the supply chain relative dynamic networks and environmental conditions. Considering unpredictable market changes and the need to adapt to such charges, it is expected that agent-based modeling will reveal optimal decisions.

References

1. Behdani, B., Zofia, L., Arief, A., Rajagopalan, S.: Agent-based Modeling to Support Operations Management in a Multi-plant Enterprise. In: International Conference on Networking, Sensing and Control (ICNSC '09) pp. 323–328 (2010)
2. Lee, H.L, Billington, C.: Managing supply chain inventory: pitfalls and opportunities. Slone Management Review **33**(3), 65–73 (1992)
3. Svoronos, A., Zipkin P.: Evaluation of one-for-one replenishment policies for multiechelon inventory systems. Management Science **37**(1), 68–83 (1991)
4. Towill, D.R., Naim, M.M., Wikner, J.: Industrial dynamics simulation models in the design of supply chains, International Journal of Physical Distribution & Logistics Management, **22** (5), 3 – 13 (1992)
5. Swaminathan, J.M., Smith, S.F., Sadeh, N.M.: Modeling Supply Chain Dynamics, Decision Sciences **29**(3), 607–632 (1998)
6. IMF-World Economic Outlook Databases, March (2015)
7. Naoaki, H.: Introduction of Supply Chain Management. Nikkei Bunko (1999)
8. Fernando, D, M., Gonzalo, G., Antonio, E., Luis Puigjaner.: An agent-based approach for supply chain retrofitting under uncertainly (2007)
9. Tomohito, O., Akira, N., Hiroshi, S.: An Agent-Based Model of Smart Supply Chain Networks. Intelligent and Evolutionary Systems Vol.5 of the series Proceedings in Adaptation, Learning and Optimization pp 373–384(2015)
10. Tokyo Development Consultants.: Tokyo Development Consultants, Available: http://www.tdc-thai.com/factory/area/
11. Sayaka, I.: Rising Like a Dragon. Available: http://www.sumitomocorp.co.jp/english/business/kouhou-person/article/id=26806.

Increasing Stability of Human Interaction Against Time Delay on Perceptual Crossing Experiment

Sohtaroh Saitoh, Hiroyuki Iizuka and Masahito Yamamoto

Abstract Perceptual Crossing experiment is the method for analyzing human interaction with low degrees of freedom of perception and motion. In this experiment, participants need to establish a cooperative interaction to achieve a task. In this study, we introduce time delays of participant's perception in this experiment, and evaluate the stability of human social interaction against the time delay. Our result shows that human social interactions have the stability against the time delays by changing their behaviors adaptively. We also investigate if the stability can be enhanced by modulating the participant's behavior conversely. It is shown that it is possible to increase the stability of social interaction by slowing down behaviors of participants.

1 Introduction

Human social interaction has been studied intensely in order to develop a robot that can interact with human naturally and smoothly. Those robots are expected to be used for various kinds of purposes such as entertainment, business, and rehabilitation. For entertainment and business purposes, human can enjoy their interaction with a robot as a friend and it may explain and recommend new products instead of sales staffs [1]. For the rehabilitation purposes, the robot is applied to improve mental disabilities through interactions [2]. In order to understand human social interaction for such an application, the analyses tend to be complicated because of several interaction modalities such as verbal communication, facial expressions including gaze directions, and gestures by hands [3]. Auvray et al. have investigated the social dynamic interaction in a minimal shared virtual environment, which is called perceptual crossing experiment where two participants interact with each other with restricted and simple motion and sensation [4]. The studies using the experimental framework revealed that exact interaction timing is important to sustain the human interactions on such a environment [5].

S. Saitoh (✉) · H. Iizuka · M. Yamamoto
Hokkaido University, North 14, West 9, Sapporo, Hokkaido 060-0814, Japan
e-mail: sohtaroh@complex.ist.hokudai.ac.jp

G. Leu et al. (eds.), *Intelligent and Evolutionary Systems*,
Proceedings in Adaptation, Learning and Optimization 8,
DOI 10.1007/978-3-319-49049-6_26

Up to present, a lot of social robots and software agents that interact with human have been developed and some of them are deployed in a public area to interact with human customers at shops or restaurants [6]. However, it is difficult for a robot to sustain the interaction with a human for a long time. Human feels uncomfortable and easily get bored due to delayed and poor responses. In order to avoid such uncomfortable situations, it is required to clarify the conditions of interaction timings to sustain the interaction.

The aim of this paper is to clarify the stability conditions of the social interaction by introducing time delay on perceptual crossing experiment and to investigate how we can increase the stability of human interaction on such delay environment. By introducing time delay, it is shown how the social interaction is established and how we can modulate the stability of interaction. Time delay effects of individual perception are well studied as cognitive science [7, 8] however this paper investigates how the delay is suppressed or enhanced in the social interaction.

2 Perceptual Crossing Experiment

A schematic view of our experimental setup is shown in Fig. 1. It is built based on the original perceptual crossing experiment [4]. Two participants are embodied as avatars in a 1D virtual environment in which they can move left or right by using a trackball mouse. The translational left/right movements are transformed to the avatar's motion and up/down movements are simply ignored. The both ends of the 1D virtual space are connected, i.e., the space becomes circular. The participants hold a voice coil motor on the left hand and receive tactile stimuli as vibrations from the motor for as long as their avatar overlaps with another virtual objects. In the virtual environment, each participant can encounter two different virtual objects: an object of their partner's avatar, and an obstacle object called "shadow" that keeps a constant distance to the partner's avatar, which means that the shadow moves in an exactly same manner with the avatar (In addition to these, a static object exists in the original Auvray's experiments, which is eliminated here). There are neither size nor

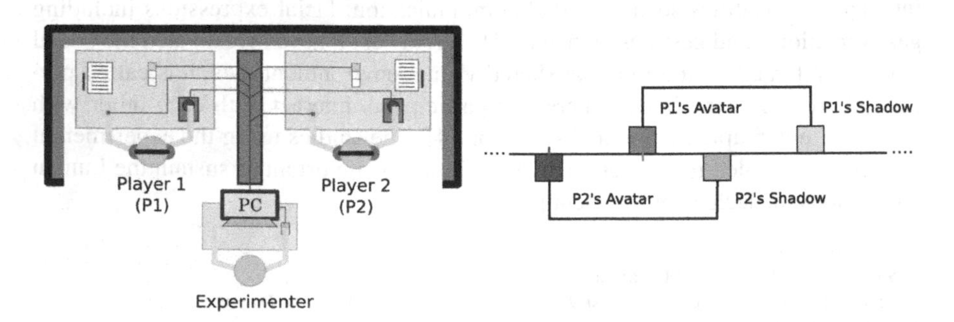

Fig. 1 Experimental setup and virtual environment

feedback differences between avatar and shadow objects. It means they are identical except for the position. The task is to find the partner's avatar without being attracted by the shadow object and participants are instructed to click in order to report that they have found the partner's avatar during a trial. No feedback about the success or failure for the task is provided during the experiment. The discrimination of the partner's avatar with the shadow object can be achieved only by the social cooperated and coordinated interaction.

A trial of experiment consists of 60 s. The participants are allowed to click once during a trial when they think that the object that they are interacting with (i.e., the object that causes vibrations) is the partner's avatar. When they have not been confident during a trial, they do not have to click. They are instructed to get a point in each trial according to the timing of the click: +1 point if clicking when encountering the avatar, −1 point if clicking when not encountering the avatar, 0 point without clicking in the trial. Their task through the experiment is to maximize the total points of paired participants as a cooperation game. The participants are not informed about the partner's click, which means that they have to keep interactions even after their own click. It is necessary for both participants to cooperate with each other and click at the appropriate time.

3 Time Delay

The previous studies show that the participant can discriminate the partner's avatar with the shadow object even their movements are objectively same and can correctly click when interact with the partner [5]. In order to see how the successful coordinated interaction is sustained, we introduce into the perceptual crossing experiment the time delay between the time when avatar touches another objects in the virtual environment and the time when the participants are given the actual feedback caused by the contacts. In the experiment, there are basically six different time delays, i.e., 0 ms (no time delay), 100, 200,... 500 ms delays. There are a few additional time delay conditions, which is described later. For example, under 100 ms time delay condition the participants perceive a tactile stimulus 100 ms after the actual avatar's contacts with objects in the virtual environment. In this paper, both participants have the same time delays.

Six participants (three pairs) joined this experiment. 20 trials were performed for each delay condition. There was rest between different conditions.

4 Result

The example of participants' behaviors (Pair 1) without time delay is shown in Fig. 2. In the experiment, the length of the 1D space is 600 units long (from −300 to 300 in the figure). Because they have no feedback at the beginning, they just search for

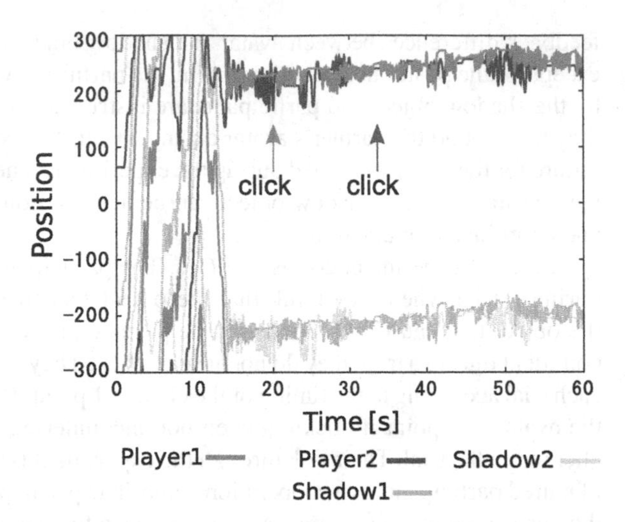

Fig. 2 Example behaviors of pair 1 without the time delay

objects by moving toward one direction to touch anything. When passing the shadow object and getting feedbacks, the participants tried to start interaction. However, the shadow object did not respond and the coordinated interaction could not be initialized. When the participants touched each other, their behavior were organized and they sticked together. Their behaviors formed turn-taking behaviors in this case where one participant oscillates around while anther stays there and their roles are exchanged alternately. The turn-taking behaviors are often observed in the perceptual crossing experiments [9] however it is not absolutely necessary for this task. After establishing turn-taking behavior, both participants could click to report that they were interacting with the partner's avatar. Figure 3 shows the example behaviors when the 500 ms time delay was introduced. It should be noted that the time delay is so small that it is not observable in the scale of the graph. The task was not achieved in this example. When the participant touched something, they tried to form a coordinated behavior in the same manner of the above. However, even when touching the partner's avatar, they could not keep touching. They touched each other for a while and they split away. The coordination could not be achieved by the end of the trial. The time delay destabilized the interaction and prohibited to form a coordination. In the successful trials even under the delayed environment (not shown), they somehow formed the coordination successfully and clicked during the coordination.

Figure 4 shows the successful rates of clicks under different time delay conditions. The results of each participant are separately shown. If both participants can click correctly, it means that they can establish a cooperative interaction to achieve the task. When the cooperative interaction can be maintained even under the time delay conditions, we could say that the cooperative interaction is stable. In that sense, the successful rates of the interactions under the time delay conditions are regarded as the stability. Pair 1 and 2 could achieve the task without time delay because the percentage of the correct click rate with 0 ms time delay is over 90 %. To see the stability against the time delay, the performances are compared in different amounts

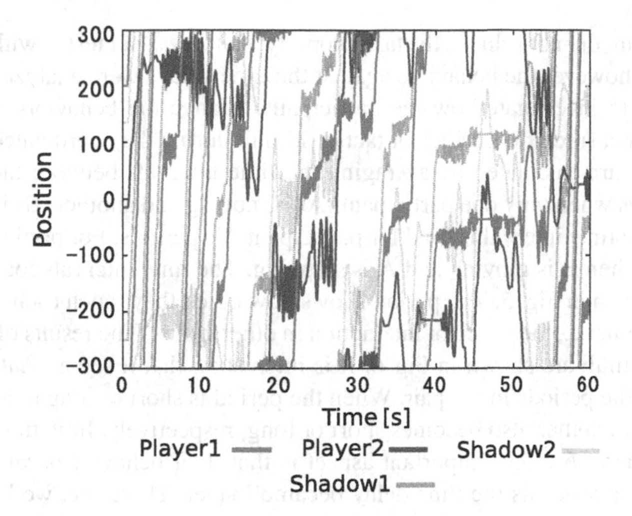

Fig. 3 Example behaviors of pair 1 with 500 ms time delay

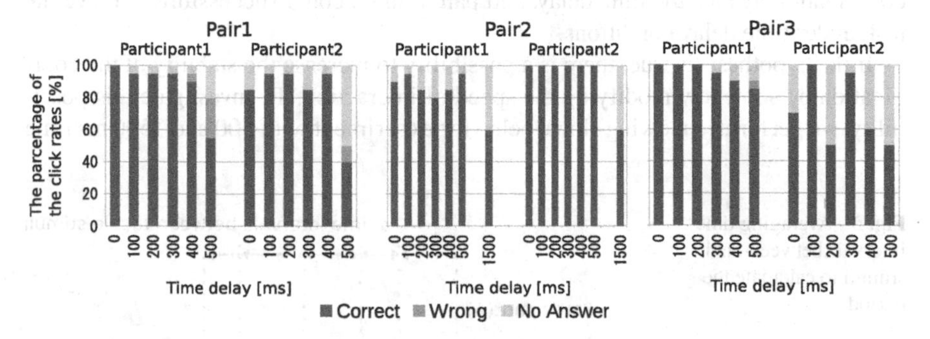

Fig. 4 Click rates of all participants under different time delay conditions

of time delay. Pair 1 was stable until 400 ms time delay. For pair 2, because their performances were not affected by 500 ms time delays, we performed additional experiments with 700, 900, 1200, and 1500 ms delay conditions. The performances decreased at 1500 ms delay condition, which is shown on the right-most of pair 2 graphs in Fig. 4. Pair 2 was more stable than pair 1 in terms of time delays. The difference of the stabilities of pair 1 and 2 has not been analyzed yet but it must depend on the shape of the coordination. On the other hand, the successful rate of the participant 2 of pair 3 was very low even without the time delay while the click result of the partner was perfect. The successful rates of the participant 2 in other time delay conditions were not consistent. The reason for that is not obvious however the participant did not make wrong click and just did not click when the performances were low. The participant's confidence to do clicks might not be consistent. It is also shown that a few pairs could not achieve the task in the previous studies [5]. Because the participant did not achieve the task properly, pair 3 was not tested with the further experiment however the behaviors against the time delays were analyzed.

In order to investigate how the participants change the behaviors against time delays, we measured the periods of tactile stimuli during the coordinated behaviors. The periods are calculated by averaging the time intervals between tactile stimuli which occurs when only one participant (A) is moving and another (B) is staying. In this case, the time interval is used for participant A's periods. For participant B, it is calculated when B is moving and A is stopping. The time intervals for the calculations are shown in Fig. 5. The period shows how quick they touch each other during a coordination, i.e., the speed of interaction in other words. The results of the periods of tactile stimuli are shown in Fig. 6. It is interesting that it seems that the participants share the periods in the pair. When the period is short or long in a participant, the period of another also becomes short or long, respectively. Imitation might have happened there. Another important aspect is that their behavior became slower in terms of the periods as the time delay became longer. Therefore, we hypothesized that the participants change the speed of the interaction to establish or sustain the coordination against the time delay, and pair 1 and 2 could successfully achieve the task under time delay conditions.

If the hypothesis is true, there is a possibility to increase the stability of the coordination by somehow modifying the speed of interaction. To investigate this possibility, we performed slowing-down behavior experiment with 500 and 1500 ms time

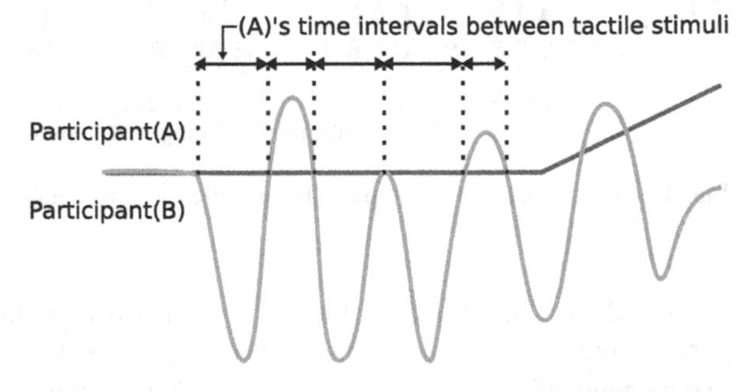

Fig. 5 Averaging time intervals between tactile stimuli to calculate the period

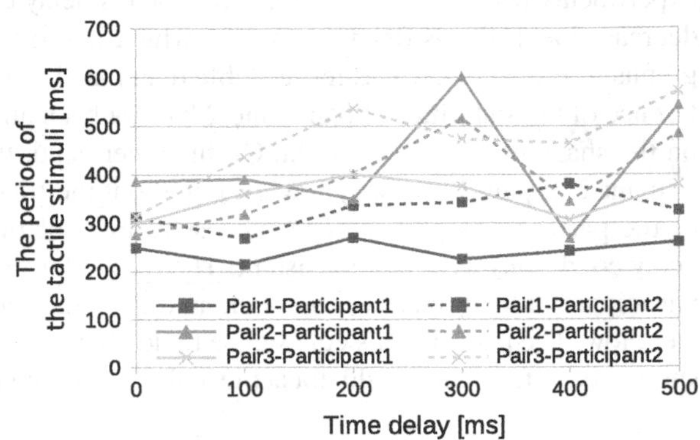

Fig. 6 Time intervals during coordination in different time delays

delay conditions for pair 1 and 2, respectively, where their performances of the task were low. If it can increase the stability, the performance must rise. In the experiments, the participants' movements are slowed down. The slowing-down was simply implemented by changing the gain of transformation from the trackball mouse movements to the movements in the virtual environment. It dropped as much as 50 % of the previous experiment in the slowing-down behavior experiment. It should be noted that the slowing-down does not necessarily cause the decrease of the interaction speed, which is defined as the period of the tactile stimuli. If the participants move the trackball mouse during the coordination in the same manner as the previous experiment, the period does not change. What changes in the slowing-down experiment is the time duration of the vibration when passing through the partner. We expected that such a change affects the participants' behavior and causes the decrease of the interaction speed. The fact that their movements were slowed down was not informed to the participants.

Figure 7 shows that the successful rates of clicks in the previous time-delay experiment and slowing-down experiment with the same amount of the time delay. The performances of the slowing-down experiment became better than the previous experiment for both pair 1 and 2. We also measured the interaction speed by the time period of the tactile stimuli in the same way as Fig. 6. The results are shown in Fig. 8. Despite the fact that slowing-down modulation does not mean decrease of the interaction speed, the periods actually became slower. Figure 9 shows the example of the participants' movements with 500 ms time delay in the slowing-down experiment. The participants could form the coordination and achieved the task. This result shows that modifying the speed of the avatar makes it possible to indirectly control the speed of the social interaction, and thus to increase the stability of the social interaction.

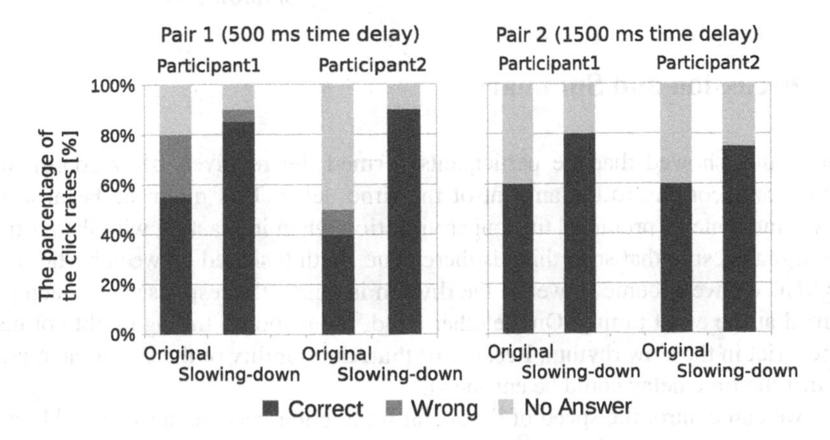

Fig. 7 Comparison of the click rates in the original and slowing-down conditions

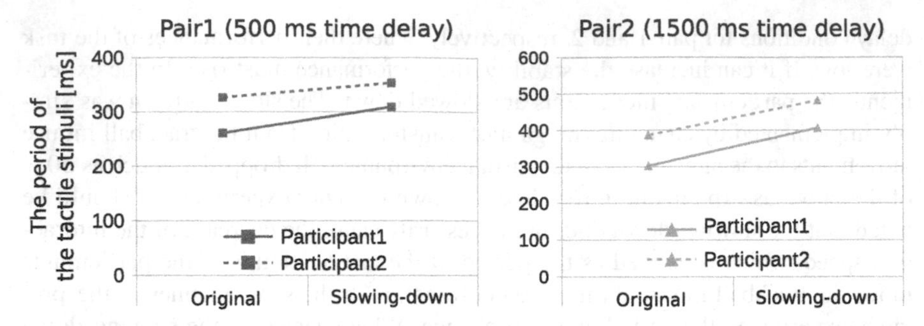

Fig. 8 Comparison of the time intervals of the tactile stimuli in the original and slowing-down conditions

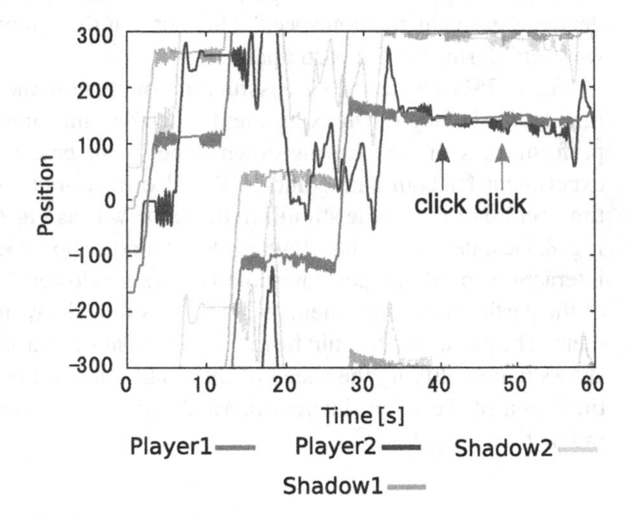

Fig. 9 Example behaviors of pair 1 with 500 ms time delay in the slowing-down condition

5 Discussion and Summary

Our results showed that the participants formed the relatively slow coordinated behaviors according to the amount of the time delay. This might be because the slower movements produced the longer vibration when interacting with the partner, which makes sure that something is there. The rhythm shared between participants might also have become slower. If the rhythm is quick, the response have been performed at the exact timing. On the other hand, the response timing might not need to be strict in the slow rhythm. Because of that, the stability of the social interaction against the time delay could be enhanced.

If we can control the speed of the social interaction between a robot and human somehow by an interface device as our experiment did, we might be able to establish more fruitful interaction between human and a robot. For example, slowing the speaking speed by an interface device can produce more stable interaction between human and robot. The modulation of the gesture speed in the interaction might have the same effect. These could be investigated as future works.

References

1. Shimizu, Y., Yoshida, S., Shimazaki, J., Kubota, N.: An interactive support system for activating shopping streets using robot partners in informationally structured space. In: 2013 IEEE Workshop on Advanced Robotics and its Social Impacts. (2013) 70–75
2. Billard, A., Robins, B., Nadel, J., Dautenhahn, K.: Building robota, a mini-humanoid robot for the rehabilitation of children with autism. Assistive Technology **19**(1) (2007) 37–49
3. Okada, S., Bono, M., Takanashi, K., Sumi, Y., Nitta, K.: Context-based conversational hand gesture classification in narrative interaction. In: Proceedings of the 15th ACM on International Conference on Multimodal Interaction. ICMI '13, New York, NY, USA, ACM (2013) 303–310
4. Auvray, M., Lenay, C., Stewart, J.: The attribution of intentionality in a simulated environment : The case of minimalist devices. Tenth Meeting of the Association for the Scientific Study of Consciousness, Oxford, UK, 2006 (2006) 23–26
5. Froese, T., Iizuka, H., Ikegami, T.: Using minimal human-computer interfaces for studying the interactive development of social awareness. Frontiers in Psychology **5(1061)** (2014)
6. Aldebaran: Who is pepper? http://www.aldebaran.com/en/cool-robots/pepper (Accessed 1 September 2016)
7. Blakemore, S.J., Wolpert, D., Frith, C.: Why can't you tickle yourself? Neuroreport **11**(11) (2000) 11–16
8. Shimada, S., Qi, Y., Hiraki, K.: Detection of visual feedback delay in active and passive self-body movements. Experimental Brain Research **201**(2) (2010) 359–364
9. Iizuka, H., Ando, H., Maeda, T.: Emergence of communication and turn-taking behavior in nonverbal interaction (in japanese). IEICE Transactions on Fundamentals of Electronics, Communication and Computer Science **J95-A** (2012) 165–174

Differential Evolution with Landscape-Based Operator Selection for Solving Numerical Optimization Problems

Karam M. Sallam, Saber M. Elsayed, Ruhul A. Sarker and Daryl L. Essam

Abstract In this paper, a new differential evolution framework is proposed. In it, the best-performing differential evolution mutation strategy, from a given set, is dynamically determined based on a problem's landscape, as well as the performance history of each operator. The performance of the proposed algorithm has been tested on a set of 30 unconstrained single objective real-parameter optimization problems. The experimental results show that the proposed algorithm is capable of producing good solutions that are clearly better than those obtained from a set of considered state-of-the-art algorithms.

1 Introduction

Optimization is an important decision making tool in many fields, including, but not limited to, operations research, engineering design and data mining. Without loss of generality, a global unconstrained single objective optimization problem, as considered in this paper, can be stated as finding the values of a decision vector $\vec{x} = (x_1, x_2, \ldots, x_D) \in \mathbb{R}^D$, which satisfies the variable bounds, $x^{min} \leq x \leq x^{max}$ and minimizes or maximizes an objective function $f(\vec{x})$, where x^{min} and x^{max} are the lower and upper boundaries, respectively. In these problems, the decision variables may be integer, real, discrete, or mixed [10] and the objective function can be linear or

---

K.M. Sallam (✉) · S.M. Elsayed · R.A. Sarker · D.L. Essam
School of Engineering and Information Technology, University of New South Wales, Canberra, Australia
e-mail: karam.sallam@student.adfa.edu.au

S.M. Elsayed
e-mail: s.elsayed@adfa.edu.au

R.A. Sarker
e-mail: r.sarker@adfa.edu.au

D.L. Essam
e-mail: d.essam@adfa.edu.au

G. Leu et al. (eds.), *Intelligent and Evolutionary Systems*,
Proceedings in Adaptation, Learning and Optimization 8,
DOI 10.1007/978-3-319-49049-6_27

nonlinear, convex or non-convex, continuous or not continuous, and uni-modal or multi-modal [9].

As gradient based methods usually encounter many difficulties when solving such complex problems [16], evolutionary algorithms (EAs) have received much interest over the last few decades. EAs are population-based search strategies that have demonstrated promising results in solving complex optimization problems [29]. The reasons for this popularity are (1) they do not require the satisfaction of specific mathematical properties; (2) they are flexible to dynamic changes; and (3) they have the capability for self-organization [12]. However, as EAs are stochastic algorithms, there is no guarantee that they will reach an optimal solution in every run. To add to this, the performance of EAs depends on parameter settings.

The family of EAs contains various algorithms, such as differential evolution (DE) [30], genetic algorithm (GA) [13] and evolution strategy (ES) [27]. The major difference between these algorithms, is in the way they produce new solutions. Among those algorithms, DE has gained popularity in solving continuous optimization problems [7, 28]. However, there is no guarantee that a DE algorithm, which performs well for one problem, or a certain class of problems, will work well for another, or on a range of problems. One reason for this is the variability of the underlying mathematical properties of optimization problems.

As a consequence, researchers have proposed multi-operator and multi-method based algorithms to solve complex optimization problems [9, 11]. However, the way of combining these operators and/or methods in the best way is still a challenging task. In the evolutionary algorithms, the selection of operators for use in a search process is made based on different criteria, such as the improvement in the quality of solutions, and/or constraint violations and/or the feasibility rate [9], re-enforcement learning mechanisms [1, 17], convergence differences and progress ratios [14]. However, the use of landscape information in the selection process is rare, even though it may boost the performance of an algorithm if it is carefully incorporated [2, 6]. However, for these methods that do exist, they have some limitations: (1) the landscape analysis was performed using an off-line mode, i.e., initial experiments were conducted to calculate landscape statistics values independently of the evolutionary process used for solving the problem [22, 23]; (2) the calculation of the landscape measures was computationally expensive [23]; and (3) a training and testing mechanism is used, which may mean the algorithm is biased towards the considered test problems, and hence its performance can deteriorate when solving another set of problems.

In this paper, a new DE framework is proposed, in which a function's landscape information is considered, in addition to the usual performance history of the operators in selecting the best-performing DE operator during the evolutionary process. We also consider linear population size reduction, in which population size is reduced continuously with a linear function. In linear population size reduction, the worst individual is deleted to resize the population. In this paper, before deleting the worst-ranking individuals, a modified technique is used, the 2 worst solutions and the centroid of the entire population are used to generate a new individual. If the new one is better than the second worst one, it replaces it. To speed up the convergence of the proposed algorithm, the sequential quadratic programming (SQP) technique is periodically applied, once every predefined number of generations. This DE algorithm with landscape based operator selection is named DE-LOS.

To judge the performance of the proposed framework, a total of 30 test functions were solved from the CEC2014 competition [18]. These benchmark sets have different mathematical properties, and are of 10, 30, 50 and 100 dimensions. The computational results show that the performance of DE-LOS is much better than the top two algorithms from the CEC2014 competition.

The rest of this paper is organized as follows: in Sect. 2, a review of DE algorithms and operators are reviewed, along with some landscape measures. Section 3 presents the proposed framework. The simulation results on benchmark problems, and the value of parameters are provided in Sect. 4. Finally, Sect. 5 provides conclusions and possible future research directions.

2 Related Work

In this section, a literature review of DE and the concept of landscape analysis are discussed.

2.1 Differential Evolution Algorithm

DE was proposed by Storn and Price [30]. It is a popular EA because it usually converges fast, is simple in implementation, and the same settings can be used for many different optimization problems. As of the literature, DE showed good performance in comparison to several other EAs on a wide variety of problems [8]. The DE algorithm uses three operators (mutation, crossover and selection) to evolve a population of individuals during the search process.

2.2 Improved DE Algorithms

In this section, some of the improved variants of DE are discussed.

2.2.1 Single Operator de Variants

An adaptive DE algorithm with an optional external memory (JADE) was proposed by Zhang et al. [35], in which the CR_i of each individual x_i at each generation was independently generated according to a normal distribution of mean μCr and standard deviation 0.1, where when the value of CR_i falls outside [0,1], it is repaired to a value in [0,1]. Also, the value of, F_i, of each individual, x_i, was independently generated according to a Cauchy distribution with parameter μF and scale parameter 0.1. If its value is greater than 1, then it is truncated to 1, or regenerated if $F_i < 0$.

Success-history based parameter adaptation for differential evolution (SHADE), which is an improved version of JADE, uses a history based parameter adaptation method. In SHADE, instead of using a single pair (μCR, μF) to guide parameter adaptation, the mean values of *SCR* and *SF* for each generation, were stored in memory as *MCR* and *MF*.

The L-SHADE [31] algorithm is a SHADE algorithm that uses linear population size reduction (LPSR) to dynamically re-size its population during a run. LPSR reduces the population linearly as the number of fitness evaluations increases. LSHADE showed good performance, in comparison with other algorithms over a set of unconstrained optimization problems.

Sallam et al. [28] proposed a neurodynamic differential evolution algorithm for solving the CEC2015 single objective optimization problems. An adaptive mechanism was proposed for the appropriate use of LSHADE and neuro-dynamic during the search process.

2.2.2 Multi-operator DE Variants

In this section, a brief review of multi-operator based DE and self-adaptive DE is provided.

Self adaptive multi-operator differential evolution (SAMO-DE) was proposed by Elsayed et al. [9] for solving constrained optimization problems. In their proposed algorithm, each operator has its own sub-population which are evolved by different DE operators. Based on an improvement measure, in which the solution quality, constraint violation and feasibility ratio were used to calculate the success of each operator, the number of individuals in each sub-population was adaptively updated, and more emphasis was given to the operator with the highest success. The results showed that SAMO-DE performed better than other-state-of-the-art algorithms.

Composite DE (CoDE) was proposed by Wang et al. [33] for solving optimization problems. In CoDE, three mutation strategies were randomly combined with three fixed control parameter settings for generating a new trial vector at each generation. To generate a new solution, three vectors were generated, then the best one among them was selected to enter the next generation. From the experimental results, it was concluded that CoDE is a promising DE algorithm for solving optimization problems.

A self-adaptive DE (SaDE) was proposed by Qin et al. [26] for solving unconstrained real-parameter optimization. In SaDE, both the trial vector generation strategy and its associated control parameter values, were gradually self-adapted according to a success rate, that was calculated based on previous learning experience. At the beginning, all mutation strategies had equal probability to generate a new solution, and the probability was updated after an initial *LP* generations, accordingly as follows: at the end of each generation, after evaluating all the generated trial vectors, the number of trial vectors generated by each strategy that successfully entered the next generation was recorded in its success memory and the number of trial vectors generated by each strategy that failed to enter the next generation was recorded in its failure memory. This algorithm performed much better than both the traditional DE algorithm and several state-of-the-art adaptive parameter DE variants.

All of the above mentioned methods did not incorporate any landscape information in the selection phase.

2.3 Landscape Analysis

Generally, a fitness landscape consists of: (1) a set of solutions (populations of individuals), (2) fitness values (objective function values) of individuals, and (3) a neighborhood operator which can be used as a distance measure [19, 22]. Measuring the fitness landscape of a problem aids researchers to classify a problem as easy or hard to solve [25]. Many landscape measures have been proposed to understand and analyze different characteristics of a problem [19, 24], and this section reviews some of them.

Auto-correlation is often used to measure the ruggedness of a fitness landscape [5, 24]. Fitness distance correlation (FDC), proposed by Jones and Forrest [15], is another method used to measure problem difficulty [32]. It measures the correlation between the objective value and the distance to the nearest optimum in the search domain. Among landscape measures is also the searchability of a problem. To measure the searchability of a problem, which is the ability of the search operator to move to a region of a search space of better fitness value, an information landscape metric exists, which is computed based on the difference between the information landscape vector of the problem to be solved and a reference landscape vector. The reference landscape is the landscape of a function that is easy to be optimized by any optimization algorithm in any dimension [3].

An information matrix $M = [a_{i,j}]$ for a minimization problem, is constructed using Eq. 1

$$a_{i,j} = \begin{cases} 1 & \text{if } f(x_i) < f(x_j) \\ 0.5 & \text{if } f(x_i) = f(x_j) \\ 0 & \text{otherwise} \end{cases} \quad (1)$$

Not all of the entries in the information landscape are necessary for defining the information landscape [3, 4]. There is duplication in the entries due to symmetry (so the lower triangle should be omitted), the entries on the diagonal are always 0.5 (and also should be omitted), and the row and column of the optimum solution should also be omitted. So, the information matrix can be reduced to a vector $LS = (ls1, ls_2, ..., ls_{|LS|})$, where the number of elements in LS, $|LS| = \frac{(NP-1)\times(NP-2)}{2}$.
Continuing from this:

$$LD = \frac{1}{|LS|} \times \sum_{i=1}^{|LS|} |(ls_i)_f - (ls_i)_p| \quad (2)$$

where $(ls_i)_p$ is the information landscape vector of the problem to be solved, and $(ls_i)_f$ is the information landscape vector of the reference function. When LD is near 0, the problem is considered easy, while $LD = 1$, means the problem is difficult.

In the recent past, researchers and practitioners have used fitness landscape to determine and select an appropriate algorithm or operator for solving optimization problems. In [20], a prediction model was developed to predict when a particle swarm optimization (PSO) algorithm would fail to solve a particular optimization problem. Decision trees were employed to predict the failure of seven different PSO algorithms, by using a number of different fitness landscape metrics. In [6], an adaptive operator selection mechanism, based on a set of four fitness landscape analysis techniques, was used to train an online regression learning model (dynamic weighted majority), which was used to predict the weight of each operator in each generation. Their proposed mechanism was used to determine the most suitable crossover operator, among four crossover operators, to solve a set of Capacitated Arc Routing Problem (CARP) instances. The authors used instantaneous reward, in which the reward was considered as the value computed at the last evaluation. In comparison with some of the-state-of-the-art algorithms, the algorithm did not show significant benefit.

3 Landscape-Based Adaptive Operator Selection DE

In this section, our novel DE-LOS algorithm is presented.

3.1 DE-LOS

The existing multi-operator algorithms use an adaptive operator selection mechanism, which is usually based on the success of generating new offspring. In this section, a DE-LOS algorithm is proposed, which uses problem landscape information, as well as the performance of operators, to adaptively place emphasis on the most suitable DE operator. The general steps in DE-LOS are given in Algorithm 1.

To begin with, three mutation strategies (DE/φbest/1, DE/current-to-φbest/1/archive and DE/current-to-φbest/1/without archive) are used. Initially, NP random individuals are generated within the variable bounds using a Latin Hypercube design. Then, each operator is randomly assigned to the same number of individuals. Next, a new solution is generated using its assigned mutation strategy. At the same time, the information landscape negative searchability metric and performance history, using Eqs. 4 and 2, respectively, are calculated for each single operator. This process con-

Algorithm 1 Proposed algorithm

1: $C \leftarrow 0$; Generate an initial population (X) of size NP using Latin Hypercube Design;
2: Calculate the fitness values of X;
3: $FES \leftarrow FES + NP$;
4: **while** $FES \leq MAX_{FES}$ **do**
5: $\quad C \leftarrow C + 1$;
6: $\quad$ **if** $FES \leq limit$ **then**
7: $\quad\quad$ **if** $C < CS$ **then**
8: $\quad\quad\quad$ Evolve the population using m DE operators;
9: $\quad\quad$ **else if** $C \geq CS$ and $C < 3CS$ **then**
10: $\quad\quad\quad$ **if** mod(C, CS==0) **then**
11: $\quad\quad\quad\quad$ Calculate the average normalized value ANV for each DE operator using Eq. 7;
12: $\quad\quad\quad\quad$ $m \leftarrow m - 1$ - i.e., the best $m - 1$ DE operators;
13: $\quad\quad\quad\quad$ Evolve the population using the best m DE operators;
14: $\quad\quad\quad$ **else if** mod($C, 2CS$==0) **then**
15: $\quad\quad\quad\quad$ Calculate average normalized value ANV of each DE operator using Eq. 7;
16: $\quad\quad\quad\quad$ $m = m - 1$, i.e., discard the worst DE operator;
17: $\quad\quad\quad$ **end if**
18: $\quad\quad\quad$ Evolve the population using m operators;
19: $\quad\quad$ **else if** mod($C, 3CS$==0) **then**
20: $\quad\quad\quad$ $C \leftarrow 0$, reset m to 3, and go to step 5;
21: $\quad\quad$ **end if**
22: $\quad$ **else**
23: $\quad\quad$ **if** mod($iter$,100==0) **then**
24: $\quad\quad\quad$ apply SQP as a local search up to a fixed number of fitness evaluations.
25: $\quad\quad$ **end if**
26: $\quad\quad$ Evolve the population using the best DE operator;
27: $\quad$ **end if**
28: $\quad$ $FES \leftarrow FES + NP$
29: $\quad$ Update population using Eq. 8
30: **end while**

tinues for a certain number of generations, say CS generations. After CS generations, the average value of the landscape metric and performance history are computed for every operator, using Eqs. 5 and 6, respectively. Subsequently, the normalized value of both measures is computed using Eq. 7. Based on this value, the best two operators are selected to be used in the subsequent cycle. Throughout the next cycle, at each generation, offspring are generated using one of those two operators, while the performance measure and landscape value are calculated for each operator. Then, the normalized values are calculated for the two mutation strategies. Based on the overall mean normalized performance measure (Eq. 7), the worst operator (the one with the minimum value) is discarded. Subsequently, the remaining best operator is used to evolve the entire population, for the subsequent CS generations. Note that after every CS generations, the success and landscape metrics are reset to zero. The above process is repeated every $3CS$ generations, however, after a predefined number of fitness evaluations is reached, the best-performing operator so far, is used to evolve the population until a stopping criterion is reached. Furthermore, during this stage, SQP is periodically applied to the best individual from the whole population.

Algorithm 2 Algorithm for computing the Information landscape negative searchability.

1: Input: population of individuals (X) updated by operator op;
2: Determine the location of the best individual in the sample, x^*.
3: Construct the pairwise comparison matrix M using Eq. 1;
4: Construct vector LS_f that represents the information matrix of the problem.
5: Construct the reference function, f_{ref}, by using Eq. 3.
6: Construct the vector LS_{ref} that represents the information landscape of the reference function.
7: Compute the value of the Information Landscape negative searchability index using Eq. 2.

3.2 *The Selection Phase*

3.2.1 Information Landscape Negative Searchability Measure

The information landscape negative searchability measure, which is based on the difference between the information landscape vector of the problem to be solved and a well-known spherical function as a reference landscape, is considered in this research, due to its simplicity and scalability [21].

The reference function $f_{ref}(\vec{x})$ is constructed using Eq. 3.

$$f_{ref}(\vec{x}) = \sum_{j=1}^{D}(x_j - x_j^*)^2 \tag{3}$$

where $\vec{x}_i^*$ is the best individual in the sample.

In this paper, Latin Hypercube Design is used to generate an initial population [34] that properly covers the search space of the problem. After constructing the vector landscape of the problem to be optimized (LS_f) and the vector landscape of the reference function (LS_{ref}), the information landscape negative searchability measure is computed using Eq. 2, this is done as part of Algorithm 2.

3.2.2 Average Normalized Value (ANV)

After the information landscape negative searchability value for each operator was computed, the success rate (SR) of each operator is computed. The success rate of each operator (SR_{op}) is defined as the number of successful offspring generated by a search operator (op), divided by the number of individuals assigned to op, as shown in Eq. 4:

$$SR_{op} = \frac{\text{Number of improved offsprings}}{\text{Number of all individuals evolved by operator}} \tag{4}$$

The normalized value for the *SR* and landscape metrics are calculated using Eqs. 5 and 6, respectively.

$$NM_{SR} = \frac{M_{SR_{OP}}}{\sum_{OP=1}^{m} M_{SR_{OP}}} \tag{5}$$

$$NM_{LD} = \frac{(1 - M_{LD_{OP}})}{\sum_{OP=1}^{m}(1 - M_{LD_{OP}})} \tag{6}$$

where M_{SR} and M_{LD} are the mean value of the success rate and landscape value, respectively.

Subsequently, the normalized performance of each operator is computed using Eq. 7:

$$ANV_{OP} = (NM_{SR_{OP}} + NM_{LD})/2 \tag{7}$$

3.3 Population Updating Method

A linear population size reduction scheme is used to adaptively re-size *NP* during the evolutionary process [31], as follows:

$$NP_{iter} = round[(\frac{NP^{min} - NP^{max}}{FES_{max}}) \times cfe + NP^{max}] \tag{8}$$

where NP^{min} is the smallest number of individuals that the proposed algorithm can use. *cfe* is the current number of fitness evaluations, FES_{max} is the maximum number of fitness evaluations. The default value of NP^{max} is set as $18D$, NP^{min} is set as 7.

To get some benefit from the worst individuals before deleting them, a new solution is generated using information from the worst two individuals and the centroid of the population ($X_{cent} = \frac{\sum_{j=1}^{D}\sum_{i=1}^{NP} x_{i,j}}{NP}$), as

$$X_{new} = X_{cent} + rand \times (X_{NP} - X_{NP-1}) \tag{9}$$

Then the worst individual is deleted, and a decision is made to decide if X_{NP-1} is replaced by X_{new} or not, based on the objective value.

4 Experimental Results

In this section, the performance of the proposed algorithm is tested by solving a set of problems taken from the CEC2014 competition on learning-based real-parameter single objective optimization [18]. The CEC2014 benchmark test set contains 30 test problems. The search space for all the problems is $[-100, 100]^D$. The proposed algorithm was run following the guidelines of the competition. That required 51 independent runs for each test problem with up to $FES_{MAX} = 10,000D$ fitness evaluations. In the experimentation, if the deviation of the best fitness value from the optimal solution is less than or equal to $1.0e-8$, it was considered as zero. The algorithm was coded using Matlab R2014a, and was run on a PC with a 3.4 GHz Core I7 processor with 16 GB RAM, and windows 7.

4.1 Algorithm Parameters and Operators

The default values of NP^{init}, and NP^{min} were set based on our experimental analysis, $NP^{init} = 18D$ and $NP^{min} = 7$. φ was set at a value of 0.6 for DE/φbest/1 to maintain diversity, while its value was 0.1, for the other two variants, to speed up the convergence rate. A is the archive rate, and it was set at a value of 1.4. H, the memory size, was set at the value of 5. *limit* the maximum limit to run the multi-operator phase, where as after it the best performing operator evolves the population until the end of the run, was set at the value of $\frac{2}{3} \times FES_{MAX}$, and CS was100. The scaling factor F and the crossover probability CR were set as in [31].

4.2 Detailed Results for 10, 30, and 50D

The computational results of DE-LOS for 10, 30, and 50D are shown in Table 1. For 10D, from the results obtained, the proposed algorithm provided the optimal solutions for all unimodal functions ($F01 - F03$). For the multimodal functions ($F04 - F16$), DE-LOS was able to obtain the optimal solutions on six problems, while it was very close for the rest. For hybrid functions ($F17 - F22$), DE-LOS was able to obtain the optimal solution for only $F17$, and was very close for the rest of the test problems. However it became stuck in local solutions for all the composition test problems, $F23 - F30$.

For 30D, from the results, DE-LOS was able to obtain the optimal solution on all the unimodal problems. For multimodal problems, DE-LOS was able to obtain the optimal solution for $F04$, $F06$, $F07$, $F08$ and $F10$, while it was very close to the optimal solution for the rest. For hybrid functions, the best solutions obtained were close to the optimal. Again, for the composition problems, DE-LOS got stuck in local solutions.

Table 1 Detailed Results for 10D

	10D			30D			50D		
	Best	Mean	Std.	Best	Mean	Std.	Best	Mean	Std.
F01	**0.0000E+00**	**0.0000E+00**	**0.0000E+00**	**0.0000E+00**	**0.0000E+00**	**0.0000E+00**	1.4941E-02	9.7381E+00	1.6581E+01
F02	**0.0000E+00**	**0.0000E+00**	**0.0000E+00**	**0.0000E+00**	**0.0000E+00**	**0.0000E+00**	**0.0000E+00**	**0.0000E+00**	**0.0000E+00**
F03	**0.0000E+00**	**0.0000E+00**	**0.0000E+00**	**0.0000E+00**	**0.0000E+00**	**0.0000E+00**	**0.0000E+00**	**0.0000E+00**	**0.0000E+00**
F04	**0.0000E+00**	2.5318E+01	1.5547E+01	**0.0000E+00**	**0.0000E+00**	**0.0000E+00**	**0.0000E+00**	3.2701E+01	4.6706E+01
F05	2.1219E-05	1.4533E+01	8.2010E+00	1.9999E+01	2.0000E+01	8.4719E-05	2.0000E+01	2.0000E+01	3.0879E-04
F06	**0.0000E+00**	**0.0000E+00**	**0.0000E+00**	**0.0000E+00**	2.2540E-02	9.4797E-02	2.1569E-04	6.6749E-01	8.4004E-01
F07	**0.0000E+00**	8.7003E-04	2.6924E-03	**0.0000E+00**	**0.0000E+00**	**0.0000E+00**	**0.0000E+00**	1.9342E-04	1.3813E-03
F08	**0.0000E+00**	**0.0000E+00**	**0.0000E+00**	**0.0000E+00**	**0.0000E+00**	**0.0000E+00**	**0.0000E+00**	**0.0000E+00**	**0.0000E+00**
F09	**0.0000E+00**	2.0289E+00	2.1428E+00	1.9899E+00	1.1179E+01	7.5265E+00	1.0945E+01	2.2215E+01	8.0015E+00
F10	**0.0000E+00**	1.2246E-03	8.7454E-03	0.0000E+00	1.6329E-03	7.0209E-03	0.0000E+00	5.4217E-02	6.8031E-02
F11	1.8736E-01	4.5646E+01	7.0302E+01	7.2417E+02	1.2808E+03	2.1279E+02	2.0309E+03	3.2116E+03	4.4592E+02
F12	0.0000E+00	4.5128E-02	3.5151E-02	2.1114E-02	6.5626E-02	3.0115E-02	1.7739E-02	6.0733E-02	2.5329E-02
F13	1.9079E-02	4.4725E-02	1.2332E-02	8.9616E-02	1.2119E-01	1.8950E-02	1.3417E-01	1.7983E-01	2.1522E-02
F14	2.0373E-02	8.6169E-02	2.8643E-02	1.4846E-01	2.2851E-01	2.5230E-02	2.7022E-01	2.9956E-01	1.8229E-02
F15	2.4961E-01	5.1461E-01	1.7170E-01	1.3835E+00	1.9349E+00	2.9087E-01	2.9296E+00	4.5685E+00	8.5283E-01
F16	2.8912E-01	1.1718E+00	3.6937E-01	7.5888E+00	8.8845E+00	5.2729E-01	1.6029E+01	1.7271E+01	6.9797E-01
F17	**0.0000E+00**	1.0520E+00	1.7762E+00	4.4058E+01	1.4910E+02	8.3895E+01	4.2814E+02	1.0661E+03	3.7011E+02
F18	6.8422E-04	8.7598E-02	1.3293E-01	1.2766E+00	5.4211E+00	2.4883E+00	4.9062E+01	8.6760E+01	1.2591E+01
F19	2.5697E-02	9.0563E-02	7.8481E-02	1.1901E+00	2.9193E+00	6.6671E-01	6.1461E+00	8.9485E+00	1.8277E+00
F20	8.8738E-04	1.7272E-01	1.5260E-01	9.0747E-01	3.4603E+00	1.3381E+00	5.7622E+00	1.2356E+01	3.8644E+00

(continued)

Table 1 (continued)

	10D			30D			50D		
	Best	Mean	Std.	Best	Mean	Std.	Best	Mean	Std.
F21	1.2221E-04	4.0420E-01	3.2446E-01	1.5087E+00	7.9776E+01	7.0129E+01	2.4379E+02	4.2392E+02	9.4926E+01
F22	1.7450E-02	6.0789E-02	3.8817E-02	1.1833E+01	2.4106E+01	5.9125E+00	2.7924E+01	1.4015E+02	9.1661E+01
F23	2.0000E+02	3.2438E+02	2.5379E+01	2.0000E+02	3.0688E+02	2.9265E+01	2.7186E+02	3.4259E+02	1.0102E+01
F24	1.0000E+02	1.0976E+02	2.2786E+00	2.0000E+02	2.1692E+02	1.0554E+01	2.0000E+02	2.5604E+02	3.1495E+01
F25	1.0000E+02	1.1584E+02	1.4053E+01	2.0000E+02	2.0240E+02	7.0748E-01	2.0000E+02	2.0497E+02	1.2932E+00
F26	1.0001E+02	1.0004E+02	1.6358E-02	1.0008E+02	1.0012E+02	1.5014E-02	1.0011E+02	1.0016E+02	1.8533E-02
F27	8.1331E-01	2.1034E+01	7.9706E+01	2.0000E+02	3.0000E+02	2.0000E+01	3.0023E+02	3.3037E+02	3.1344E+01
F28	2.0000E+02	3.6736E+02	4.5559E+01	2.0000E+02	8.1699E+02	1.1047E+02	2.0000E+02	1.0982E+03	1.3089E+02
F29	1.3175E+02	2.2014E+02	1.2629E+01	7.1325E+02	7.1643E+02	3.7186E+00	6.9851E+02	8.0150E+02	3.9877E+01
F30	4.5429E+02	4.6307E+02	5.7407E+00	4.0223E+02	6.6248E+02	2.4821E+02	8.0234E+03	8.8013E+03	5.6733E+02

For 50*D*, DE-LOS was able to obtain the optimal solutions in *F*02 and F03, while for *F*01 it obtained very close solutions to the optimal one. For multimodal problems, DE-LOS was robust in solving *F*08, efficient in solving *F*04, *F*07, *F*08 and *F*10, while it got stuck in local solutions for the rest of the test problems. This was also the situation for the hybrid and composition problems, although its performance in solving the hybrid problems was a little bit better than its performance in solving the composition problems.

4.3 DE-LOS Versus State-of-the-art Algorithms on CEC2014

DE-LOS was compared with the top two algorithms in the literature LSHADE [31] and UMOEAs [11]. The matlab source codes for LSHADE and UMOEAs were downloaded online. We ran these algorithms using the same parameters suggested by the authors in their papers and the other conditions were the same as the competition guidelines. To make a fair comparison, all the algorithms were run using the same seeds.

Table 2 shows a comparison summary of the results obtained from DE-LOS and the other two algorithms for 10*D*, 30*D*, and 50*D* problems. A non-parametric test, Wilcoxon rank-sum test, was chosen, to judge the difference between any paired algorithms. The results regarding the best and average fitness functions are presented in Table 2. The significance level was set at a value of 10 %. Based on the test results/rankings, one of three signs (+, −, and ≈) was assigned for the comparison of any two algorithms (shown in the last column), where the "+" sign means that the first algorithm is significantly better than the second, the "−" sign means that the first algorithm is significantly worse, and the "≈" sign means that there is no significant difference between the two algorithms. Considering the quality of solutions, and from the results in Table 2, it is clear that DE-LOS is always better than the other algorithms, based on the best and average results obtained, and this is obvious for 30*D* and 50*D*.

Based on the statistical test, DE-LOS is better than UMOEAs in 10*D*, 30*D*, and 50*D* in regard to best and average results, except for the best results in 10*D* and 50*D*, where there is no significant difference between DE-LOS and UMOEAs. Considering the comparison between DE-LOS and LSHADE, DE-LOS is significantly better than LSHADE in 10*D*, 30*D*, and 50*D*.

In addition, based on the average results obtained, the average ranking of DE-LOS, LSHADE and UMOEAs, as produced by the Friedman test, is summarized in Table 3. The results in Table 3 are consistent with the results in Table 2, in which DE-LOS had the best rank.

Table 2 A comparison summary between DE-LOS and other state-of-the-art algorithms

	10D					30D				50D			
DE-LOS		Better	Equal	Worse	Dec.	Better	Equal	Worse	Dec.	Better	Equal	Worse	Dec.
Versus LSHADE	Best	14	10	6	+	17	9	4	+	18	4	8	+
	Mean	18	6	6	+	14	8	8	+	16	10	4	+
Versus UMOEAs	Best	10	13	7	≈	15	8	7	+	12	6	12	≈
	Mean	18	4	8	+	22	4	4	+	24	1	5	+

Table 3 Friedman's test results

Algorithm	10D	30D	50D
	Rank	Rank	Rank
DE-LOS	**1.63**	**1.60**	**1.58**
LSHADE	2.08	1.85	1.85
UMOEAs	2.28	2.55	2.57

5 Conclusion and Future Work

During the last few decades, DE algorithms have shown superior performance to many other-state-of-the-art algorithms in solving both unconstrained and constrained optimization problems. It is known that no single algorithm or operator is able to solve all kinds of optimization problems. Even though for a single run, an algorithm or operator may perform well in the earlier generations, its performance often decreases during later generations. So the selection of an appropriate algorithm or operator is not an easy task. In this paper, the DE-LOS algorithm has been presented. It used landscape and normalized performance measures to dynamically place more emphasis of the best-performing DE mutation.

The algorithm has been tested on 30 bound constrained numerical optimization problems from the CEC2014 competition. The results obtained were better than those obtained from the best two algorithms in the literature.

In future work, we will investigate the use of more than one landscape measure, and will incorporate some of them with multi-method-based algorithms.

References

1. Auer, P., Cesa-Bianchi, N., Fischer, P.: Finite-time analysis of the multiarmed bandit problem. Machine learning 47(2-3), 235–256 (2002)
2. Bischl, B., Mersmann, O., Trautmann, H., Preuß, M.: Algorithm selection based on exploratory landscape analysis and cost-sensitive learning. In: Proceedings of the 14th annual conference on Genetic and evolutionary computation. pp. 313–320. ACM (2012)
3. Borenstein, Y., Poli, R.: Information landscapes. In: Proceedings of the 7th annual conference on Genetic and evolutionary computation. pp. 1515–1522. ACM (2005)
4. Borenstein, Y., Poli, R.: Decomposition of fitness functions in random heuristic search. In: Foundations of Genetic Algorithms, pp. 123–137. Springer (2007)
5. Chicano, F., Luque, G., Alba, E.: Autocorrelation measures for the quadratic assignment problem. Applied Mathematics Letters 25(4), 698–705 (2012)
6. Consoli, P.A., Minku, L.L., Yao, X.: Dynamic selection of evolutionary algorithm operators based on online learning and fitness landscape metrics. In: Simulated Evolution and Learning, pp. 359–370. Springer (2014)
7. Das, S., Suganthan, P.N.: Differential evolution: a survey of the state-of-the-art. Evolutionary Computation, IEEE Transactions on 15(1), 4–31 (2011)

8. Elsayed, S.M., Sarker, R.A., Essam, D.L.: Differential evolution with multiple strategies for solving cec2011 real-world numerical optimization problems. In: Evolutionary Computation (CEC), 2011 IEEE Congress on. pp. 1041–1048. IEEE (2011)
9. Elsayed, S.M., Sarker, R.A., Essam, D.L.: Multi-operator based evolutionary algorithms for solving constrained optimization problems. Computers & operations research 38(12), 1877–1896 (2011)
10. Elsayed, S.M., Sarker, R.A., Essam, D.L.: Memetic multi-topology particle swarm optimizer for constrained optimization. In: Evolutionary Computation (CEC), 2012 IEEE Congress on. pp. 1–8. IEEE (2012)
11. Elsayed, S.M., Sarker, R.A., Essam, D.L., Hamza, N.M.: Testing united multi-operator evolutionary algorithms on the cec2014 real-parameter numerical optimization. In: Evolutionary Computation (CEC), 2014 IEEE Congress on. pp. 1650–1657. IEEE (2014)
12. Fogel, L.J., Owens, A.J., Walsh, M.J.: Artificial intelligence through simulated evolution (1966)
13. Goldberg, D.E., Holland, J.H.: Genetic algorithms and machine learning. Machine learning 3(2), 95–99 (1988)
14. Gordián-Rivera, L.A., Mezura-Montes, E.: A combination of specialized differential evolution variants for constrained optimization. In: Advances in Artificial Intelligence–IBERAMIA 2012, pp. 261–270. Springer (2012)
15. Jones, T., Forrest, S., et al.: Fitness distance correlation as a measure of problem difficulty for genetic algorithms. In: ICGA. vol. 95, pp. 184–192 (1995)
16. K. Deb: Optimization for engineering design: Algorithms and examples. PHI Learning Pvt. Ltd. (2012)
17. Li, K., Fialho, A., Kwong, S., Zhang, Q.: Adaptive operator selection with bandits for a multiobjective evolutionary algorithm based on decomposition. Evolutionary Computation, IEEE Transactions on 18(1), 114–130 (2014)
18. Liang, J., Qu, B., Suganthan, P.: Problem definitions and evaluation criteria for the cec 2014 special session and competition on single objective real-parameter numerical optimization. Computational Intelligence Laboratory, Zhengzhou University, Zhengzhou China and Technical Report, Nanyang Technological University, Singapore (2013)
19. Malan, K.M., Engelbrecht, A.P.: A survey of techniques for characterising fitness landscapes and some possible ways forward. Information Sciences 241, 148–163 (2013)
20. Malan, K.M., Engelbrecht, A.P.: Particle swarm optimisation failure prediction based on fitness landscape characteristics. In: Swarm Intelligence (SIS), 2014 IEEE Symposium on. pp. 1–9. IEEE (2014)
21. Malan, K., Engelbrecht, A.: Characterising the searchability of continuous optimisation problems for pso. Swarm Intelligence 8(4), 275–302 (2014)
22. Mersmann, O., Bischl, B., Trautmann, H., Preuss, M., Weihs, C., Rudolph, G.: Exploratory landscape analysis. In: Proceedings of the 13th annual conference on Genetic and evolutionary computation. pp. 829–836. ACM (2011)
23. Muñoz, M.A., Kirley, M., Halgamuge, S.K.: A meta-learning prediction model of algorithm performance for continuous optimization problems. In: Parallel Problem Solving from Nature-PPSN XII, pp. 226–235. Springer (2012)
24. Pitzer, E., Affenzeller, M.: A comprehensive survey on fitness landscape analysis. In: Recent Advances in Intelligent Engineering Systems, pp. 161–191. Springer (2012)
25. Poursoltan, S., Neumann, F.: Ruggedness quantifying for constrained continuous fitness landscapes. In: Evolutionary Constrained Optimization, pp. 29–50. Springer (2015)
26. Qin, A.K., Huang, V.L., Suganthan, P.N.: Differential evolution algorithm with strategy adaptation for global numerical optimization. Evolutionary Computation, IEEE Transactions on 13(2), 398–417 (2009)
27. Rechenberg, I.: Evolution strategy. Computational Intelligence: Imitating Life 1 (1994)
28. Sallam, K.M., Sarker, R.A., Essam, D.L., Elsayed, S.M.: Neurodynamic differential evolution algorithm and solving cec2015 competition problems. In: Evolutionary Computation (CEC), 2015 IEEE Congress on. pp. 1033–1040. IEEE (2015)

29. Sarker, R., Kamruzzaman, J., Newton, C.: Evolutionary optimization (evopt): a brief review and analysis. International Journal of Computational Intelligence and Applications 3(04), 311–330 (2003)
30. Storn, R., Price, K.: Differential evolution a simple and efficient adaptive scheme for global optimization over continuous spaces, international computer science institute, berkeley. Berkeley, CA (1995)
31. Tanabe, R., Fukunaga, A.S.: Improving the search performance of shade using linear population size reduction. In: Evolutionary Computation (CEC), 2014 IEEE Congress on. pp. 1658–1665. IEEE (2014)
32. Tomassini, M., Vanneschi, L., Collard, P., Clergue, M.: A study of fitness distance correlation as a difficulty measure in genetic programming. Evolutionary Computation 13(2), 213–239 (2005)
33. Wang, Y., Cai, Z., Zhang, Q.: Differential evolution with composite trial vector generation strategies and control parameters. Evolutionary Computation, IEEE Transactions on 15(1), 55–66 (2011)
34. Ye, K.Q.: Orthogonal column latin hypercubes and their application in computer experiments. Journal of the American Statistical Association 93(444), 1430–1439 (1998)
35. Zhang, J., Sanderson, A.C.: Jade: adaptive differential evolution with optional external archive. Evolutionary Computation, IEEE Transactions on 13(5), 945–958 (2009)

The Effect of Word-of-Mouth in U-Mart Artificial Futures Market

Hiroshi Sato, Tomohiro Shirakawa and Daisuke Nakagawa

Abstract In many stock markets, it is believed that word-of-mouth effect makes the market unstable. The market nowadays is becoming vulnerable because the trading style has shifted to online trading. The signal or noise can be spread very fast in the network of online traders. By U-Mart artificial market simulator, we show that this assumption is correct in most cases, but there is a special situation in which word-of-mouth effect can make the market stable.

Keywords Agent-based simulation · Artificial market · U-Mart · Word-of-mouth · Stylized fact

1 Introduction

As with all developed country, there are growing demands on online trading in Japan. Introduction of new products like mini-stock spur this trend [1]. In Japan, there are about 22 million accounts used for online trading and the number of the accounts continues to rise about 5 % in a year.

It is often observed that the human traders in the real stock market make their decision based on the information in the Internet. There are many online communities such as bulletin boards or SNS which can show and predict price movement or traders actions. For example, at the beginning of 2017, SNS site "Stocks in Everyone" gathered more than 360,000 people and they exchange their predictions about almost all of individual stocks in the SNS.

H. Sato (✉) · T. Shirakawa · D. Nakagawa
Department of Computer Science, National Defense Academy,
1-10-20 Hashirimizu, Yokosuka, Kanagawa 239-8686, Japan
e-mail: hsato@nda.ac.jp

T. Shirakawa
e-mail: sirakawa@nda.ac.jp

D. Nakagawa
e-mail: syo1maxmaro@gmail.com

G. Leu et al. (eds.), *Intelligent and Evolutionary Systems*,
Proceedings in Adaptation, Learning and Optimization 8,
DOI 10.1007/978-3-319-49049-6_28

These communications generate the word-of-mouth effects. Shimokawa shows that a small market or minor brands are easily affected by the word-of-mouth effects [12, 14]. However, it had been outside the scope of traditional economics. Unlike the traditional economics, traders in the real world are not so rational.

Recently, agent based simulation is used for analysis of this type of phenomenon. This approach is called artificial market. There are two ways to implement artificial market: one is simple simulation based on KISS principle [8]. KISS is the abbreviation of "Keep It Simple, Stupid." KISS principle advocates that simplicity is the key in design. The other is high fidelity simulation. This approach aims to create the realistic model for the simulation as far as possible [11].

In this paper, we place ourselves at realistic simulator point of view because the increase of computing power and obtaining detailed information about trading behaviors enable us to conduct realistic simulation. We adopt U-Mart artificial market simulator as an realistic simulator [13]. U-Mart simulates virtual futures market of real stock index in order to connect virtual to real. It provides a variety of trading agents called standard agent set [6].

The purpose of this study is to analyze the effect of the word-of-mouth. It is conducted by introducing the function of word-of-mouth in the standard agent set of U-Mart simulator and comparing the results changing the rate of word-of-mouth type agents in the whole population of traders.

2 Word-of-Mouth in Market

2.1 Literatures

The study of psychological effect in economics and finance is gaining attention. The following two phenomena are important:

- **Asymmetry of Information** It is impossible to get all information about the investment. Each trader has different information. Many traders usually adopt the strategies which is the majority in the population.
- **Consistency of Traders** Traders don't want to change their decision. This means that they rely on a group of agents and trust them even when they are no longer reliable. The traders' decision will be changed finally if they find more reliable group.

These phenomena generate herding behaviors and herding usually remains in the population. The word-of-mouth is a typical example of herding behaviors.

There are several studies which considers word-of-mouth effect in market. The main type of study is regression using real data to the model [9, 10]. This type of study can be said as top-down type research. As written in previous section, however, we would like to use agent-based approach to treat this problem. This type of study can be said as bottom-up type research.

The regression can only explain existing situation of the market. On the other hand, the agent-based simulation can evaluate the numerous situations through numerous settings.

2.2 *Model of Word-of-Mouth*

In order to analyze of word-of-mouth effects, we use agent-based modeling and conduct simulation. We adopt the model which is proposed by Misawa [5] and modify the model to implement word-of-mouth phenomenon. The following is the description of the model.

Communication In order to implement the communication between agents, we introduce the two-dimensional lattice type landscape where agents exist. Each agent moves on the landscape randomly in every time step. If it sees other agent in its range of vision, it decides whether it imitates the strategy or not. This landscape will be real (e.g. geographical landscape) or virtual (e.g. in Internet geographic landscape). Figure 1 is the structure of the landscape. A cluster of the agents in the landscape is called community.

Decision Making The trading agent in this paper can imitate the majority decision in his neighborhood. Every agent i has its own decision $d_i(t)$ at time t. It also has reliability threshold a_i. In every time step, the agent makes its decision (e.g. buy, sell or do nothing). It also gather the information of decisions of its neighbors in the landscape and calculate the majority decision d_i^{maj}. It compares the majority decision with its own decision. If the number of majority decision in its neighbors exceed its reliability threshold, it adopts majority's decision.

$$d_i(t) = \begin{cases} d_i^{maj}(t) \ (number\ of\ d_i^{maj}(t) > a_i(t)) \\ d_i(t) \quad (otherwise) \end{cases} \tag{1}$$

3 Virtual Market

We adopt U-Mart virtual features market simulator as virtual market simulator. This is an high-fidelity simulator made by U-Mart project [13]. The advantages of this simulator compared to other market simulators are:

- **Price Formation** U-Mart simulator decide the contract price from the method that is used in real stock market. "Itayose" is used in this study.
- **Variety of Agents** U-Mart simulator provides the variety of trading agents. Many representative technical analysis based methods are implemented as agents.
- **Connection to the Real** In U-Mart simulator, agents trade the virtual futures of the real stock index. Both prices are correlated and this connects virtual and real.

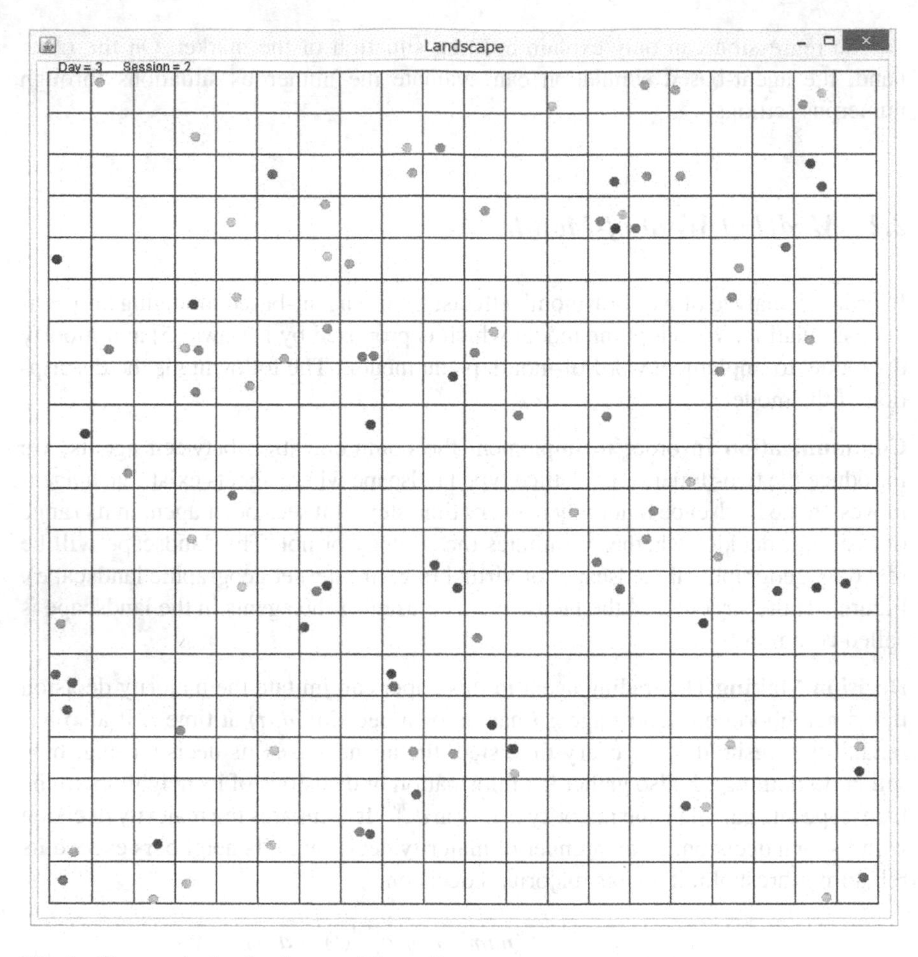

Fig. 1 Communication landscape of the trading agents

- **Open simulator** It is available from the book and internet. If we use common tools, we can share and compare the results.

Figure 2 shows the screenshot of the ongoing simulation by U-Mart system.

4 Computer Simulation

In this section, we simulate the word-of-mouth effects using U-Mart simulator. U-Mart simulator version 2 is used here.

Fig. 2 A screenshot of U-Mart virtual futures market simulator

4.1 Simulation Settings

Parameter Setting In order to analyze the word-of-mouth effect, we focus on the following three parameters:

- **The range of communication** (r) The range of the vision in the landscape. (range: 0, 1, 2, 3, 4, 5, 6, 7, 8, 9)
- **The ratio of word-of-mouth agent** (s) How many agents use word-of-mouth and how many don't. (range: 0.1, 0.2, 0.3, 0.4, 0.5, 0.6, 0.7, 0.8, 0.9, 1.0)
- **The reliability threshold** (a) The threshold whether the agent adopts the majority decision of its neighbor (range: 0.4, 0.5, 0.6, 0.7, 0.8)

The number of possible combination of these parameters is 500 (=$10 \times 10 \times 5$).

Market Setting We prepare four types of price movements which are used as the price of real stock index (spot price) for virtual futures in the market—ascent, descent, reverse, and oscillation. Figure 3 shows these spot prices.

Agent Setting The following nine types of agents are used in this experiment: Trend strategy, Anti-trend strategy, Random strategy, S-Random strategy, RSI strategy, SRSI strategy, Moving average strategy, S-Moving average strategy, Day trade strategy. These agents are the member of the standard agent set in the U-Mart simulator.

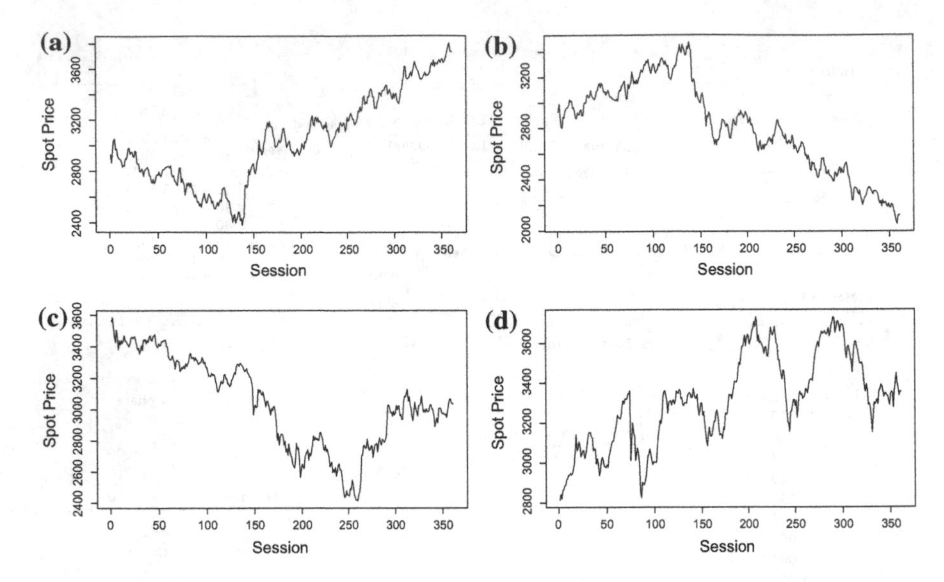

Fig. 3 The spot price sequence for virtual futures used in the experiment: **a** ascent, **b** descent, **c** reverse, **d** oscillation

In fact, each agent has its own parameters. The default parameter value are used in this paper. The number of agent in the market is 100—Ten agents in each type, and ten more RandomStrategy agents.

4.2 Experiment 1: Baseline Results

We sweep the three parameters described in Sect. 4.1 for word-of-mouth effect. All 500 combinations of the parameters are simulated. In each parameter setting, four spot price movements are used. In each price movement, 100 times simulation are done.

4.2.1 Ratio of "Nothing Done"

We focus on the parameter s and see how the market changes. Figure 4 shows the failure rate in transaction, in other words, the ratio of "nothing done" in the market.

From Fig. 4, we can see that if the parameter s—the number of agents who use the word-of-mouth—increases, the failure rate also increases. The same trend can be observed for the other parameters—r and a.

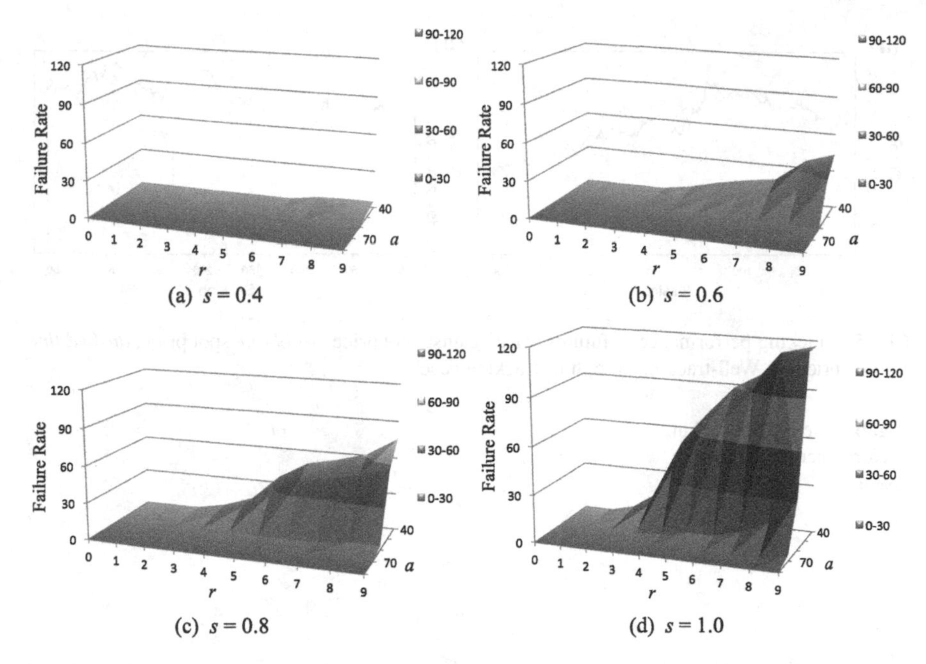

Fig. 4 Failure rate of trading The screenshot of U-Mart virtual futures market simulator. x-axis: parameter r, y-axis: parameter a, z-axis: failure rate of trading. **a** $s = 0.4$, **b** $s = 0.6$, **c** $s = 0.8$, **d** $s = 1.0$

4.2.2 Tracking Performance of Futures to Spot

Then, we analyze the tracking performance of future prices which are produced by the agents to the spot prices. In theory, the future prices must track the spot prices. However, if there are many irrational agents, future price does not follow the spot price. Figure 5 depicts the two cases of tracking. We measure the tracking performance by the disassociation degree D defined in Eq. (2).

$$D = \sum^{t} |p_s(t) - p_f(t)| \tag{2}$$

where, $p_s(t)$ is the spot price at time t and $p_f(t)$ is the futures price at time t.

Figures 6 and 7 show the top 50 and the bottom 50 tracking cases in the parameter space. In both cases, large value of s lead to these extreme cases. Well tracking cases are spread near the a-s plane. On the other hand, there are two clusters in parameter space in Fig. 7.

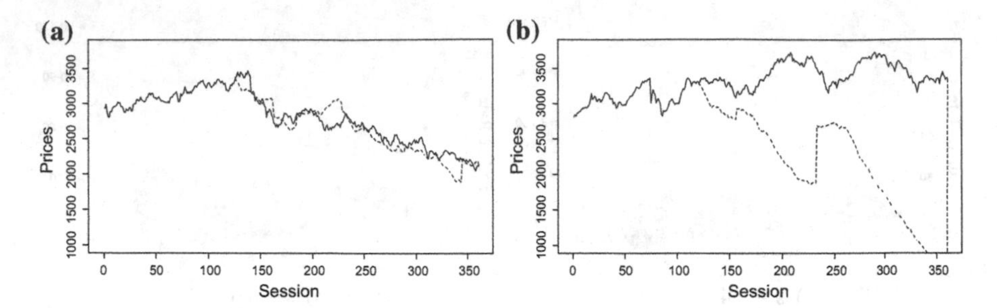

Fig. 5 Tracking performance of futures price against spot price (*solid line* spot price, *dashed line* futures price). **a** Well-tracking case, **b** ill-tracking case

Fig. 6 The top 50 tracking performance in parameter space

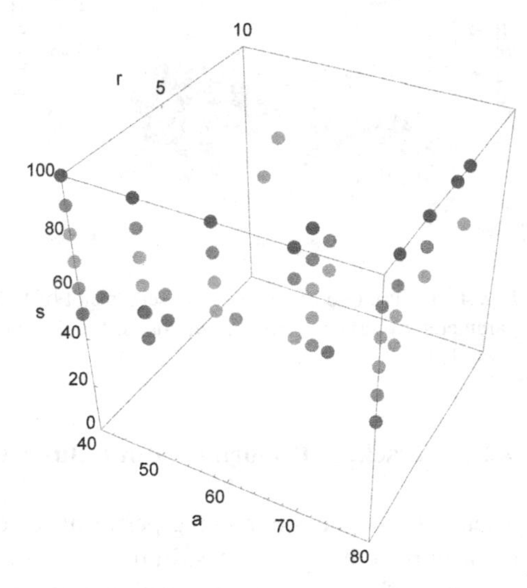

Fig. 7 The bottom 50 tracking performance in parameter space

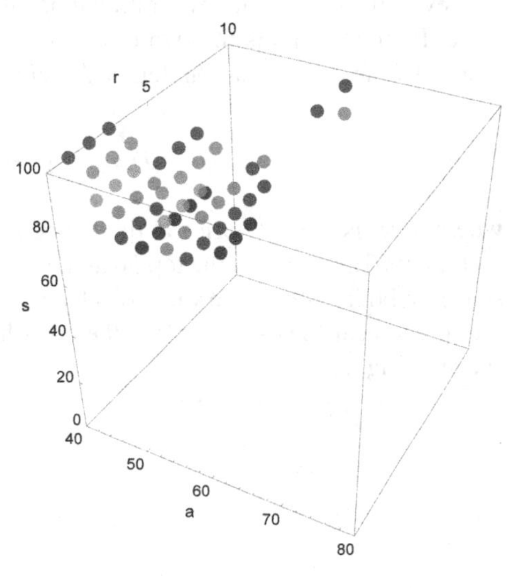

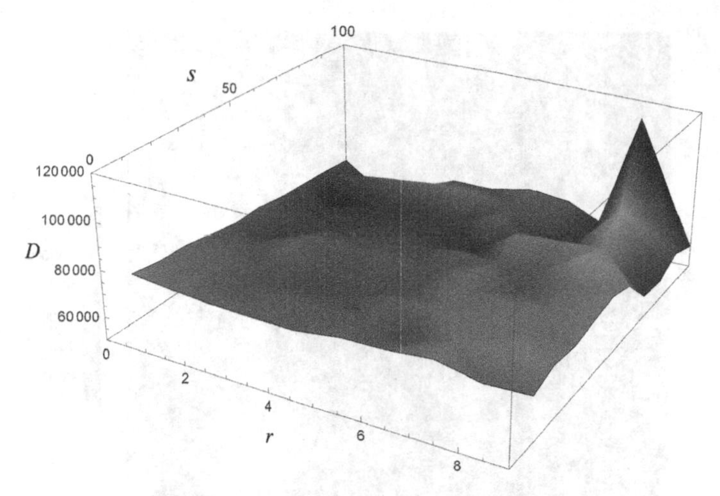

Fig. 8 Tracking performance in randomly distributed reliability threshold (3D plot). x-axis: r, y-axis: s, z-axis: the disassociation degree defined in Eq. 2

4.3 *Experiment 2: Heterogeneous Reliability Threshold*

In previous experiment, all parameters about word-of-mouth are the same. But it is more natural to think there is the variety in the parameter values. In this experiment, we set the random value of a—reliability threshold—to each agent. The distribution of a is the uniform.

Figure 8 shows the disassociation degree with 3D-plot. Figure 9 shows the same situation in contour plot. From these figures, we can find that larger value of r occurs high disassociation between spot price and future price. There is a peak in the region of the large r and large s, but it is not in the edge of the parameter space. It is surprising that there are very few disassociation in the region of largest r and the largest s.

5 Conclusion

In many stock markets, it is believed that word-of-mouth effect makes the market unstable. The market nowadays is becoming vulnerable because the trading style has shifted to online trading. The signal or noise can be spread very fast in the network of online traders. By U-Mart artificial market simulator, we show that this assumption is correct in the most cases, but there is a special situation in which word-of-mouth effect can make the market stable.

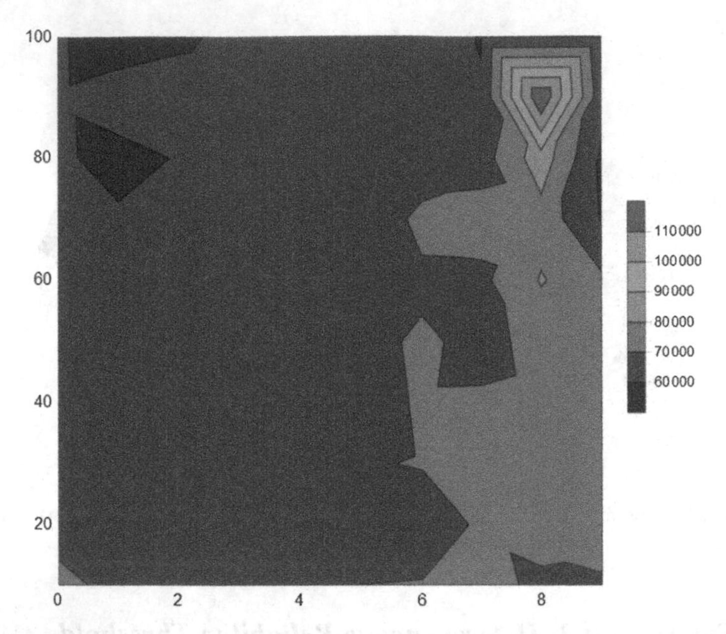

Fig. 9 Tracking performance in randomly distributed reliability threshold (contour plot). x-axis: r, y-axis: s, *color* the disassociation degree defined in Eq. 2

Although we use the relatively realistic artificial market simulator, there is room to introduce more realistic regulations in the simulation, for example, circuit-breaker, special quotes and so on. There is also room to introduce more realistic properties in the decision of the trading agents, especially in communication between traders.

References

1. Japan Securities Dealers Association: Working paper on Internet Transaction, http://www.jsda.or.jp/shiryo/chousa/interan.html (2016) (in Japanese)
2. Kusada, K., Mizuta, T., Hayakawa, S., Izumi, K., Yoshimura, S.: Analysis of the Market Makers Spread's Impact to Markets Volume Shares using an Artificial Market, Proc. of the 28th Annual Conference of the Japanese Society for Artificial Intelligence, 3L3-OS-26a-2, pp. 1–4 (2014) (in Japanese)
3. Luenberger, D. G.: Investment Science, Oxford University Press (1997)
4. Matsumoto, M., Fukui, K., Moriyama, K., Kurihara, S., Numao, M.: A design and evaluation of Q-learning agents on U-Mart, Proc. of 20th Annual Conference of the Japanese Society for Artificial Intelligence, 1B2-2, pp. 1–4 (2006) (in Japanese)
5. Misawa, T., Suzuki, K., Okano, Y., Shimokawa, T.: Emergence of Opinion Leaders Based on Agent Model and Its Impact to Stock Prices, The transactions of the Institute of Electrical Engineers of Japan. C, A publication of Electronics, Information and System Society 129(2), pp. 336–343 (2009) (in Japanese)
6. Ono, I., Sato, H.: Building Artificial Markets for Evaluating Market Institutions and Trading Strategies, Realistic Simulation of Financial Market, pp. 59–85, Springer Japan, Tokyo (2016)

7. Shiozawa, Y.: Significance of Artificial Markets for Economics (Special Issue: "Artificial Markets"), Japanese Society for Artificial Intelligence, Vol. 15, No. 6, pp. 951–957 (2000) (in Japanese)
8. Epstein, J. M., Axtell, R. L.: Growing Artificial Societies: Social Science from the Bottom Up, A Bradford Book (1996)
9. Hong, H., Kubik, J. D., Stein, J. C.: Thy Neighbor's Portfolio: Word-of-Mouth Effects in the Holdings and Trades of Money Managers, Journal of Fiance, Vol. 60, No. 6, pp. 2801–2824 (2005)
10. Liu, Y-J., Meng, You, W., Zhao, L.: Word-of-Mouth Communication, Observational Learning, and Stock Market Participation, Social Science Research Network, http://dx.doi.org/10.2139/ssrn.2251570 (2014)
11. Ono, I, Sato, H., Mori, N., Nakajima, Y., Matsui, H., Koyama, Y., Kita, H.: U-Mart System: A Market Simulator for Analyzing and Designing Institutions, Evolutionary and Institutional Economic Review, Vol. 5, No. 1, pp. 63–79 (2008)
12. Shimokawa, T., Misawa, T., Watanabe, K.: Word of Mouth: An Agent-based Approach to Predictability of Stock Prices, Transactions of the Japanese Society for Artificial Intelligence, Vol. 21, No. 4, pp. 340–349 (2006) (in Japanese)
13. Shiozawa, Y., Nakajima, Y., Matsui, H., Koyama, H., Taniguchi, K., Hashimoto, F.: Artificial Market Experiments with U-Mart System, Springer Japan, Tokyo (2008)
14. Tumarkin, R. Whitelaw, R. F.: News or Noise? Internet Posting and Stock Prices, Financial Analysts Journal, Vol. 57, Issue 3, pp. 41–51 (2001)

Multiple Imputation and Ensemble Learning for Classification with Incomplete Data

Cao Truong Tran, Mengjie Zhang, Peter Andreae, Bing Xue and Lam Thu Bui

Abstract Missing values are a common issue in many real-world datasets, and therefore coping with such datasets is an essential requirement of classification since inadequate treatment of missing values often leads to large classification errors. One of the most popular ways to address incomplete data is to use imputation methods to fill missing fields with plausible values. Multiple imputation, which fills each missing field with a set of plausible values, is a powerful approach to dealing with incomplete data, but is mainly used for statistical analysis. Ensemble learning which constructs a set of classifiers instead of one classifier has proven capable of improving classification accuracy, but has been mainly applied to complete data. This paper proposes a combination of multiple imputation and ensemble learning to build an ensemble of classifiers for incomplete data classification tasks. A multiple imputation method is used to generate a set of diverse imputed datasets which is then used to build a set of diverse classifiers. Experiments on ten benchmark datasets use a decision tree as classification algorithm and compare the proposed approach with two other popular approaches to dealing with incomplete data. The results show that, in almost all cases, the proposed method achieves significantly better classification accuracy than the other methods.

C.T. Tran (✉) · M. Zhang · P. Andreae · B. Xue
School of Engineering and Computer Science, Victoria University of Wellington,
PO Box 600, 6140 Wellington, New Zealand
e-mail: cao.truong.tran@ecs.vuw.ac.nz

M. Zhang
e-mail: mengjie.zhang@ecs.vuw.ac.nz

P. Andreae
e-mail: peter.andreae@ecs.vuw.ac.nz

B. Xue
e-mail: bing.xue@ecs.vuw.ac.nz

C.T. Tran · L.T. Bui
Faculty of Information Technology, Le Qui Don Technical University,
Hanoi, Vietnam
e-mail: lam.bui07@gmail.com

G. Leu et al. (eds.), *Intelligent and Evolutionary Systems*,
Proceedings in Adaptation, Learning and Optimization 8,
DOI 10.1007/978-3-319-49049-6_29

Keywords Incomplete data · Multiple imputation · Ensemble learning · Classification

1 Introduction

Classification is one of the main tasks in data mining and machine learning. Classification has been successfully applied to many scientific areas such as computer science, engineering, statistic, medicine, biology, etc [4]. In spite of receiving great attention over many decades, there are still open issues in classification; one of these issues is incomplete data [10].

An incomplete dataset is a dataset containing some fields which are missing values. Missing values are a unavoidable problem in many real-world datasets [15, 18]. For instance, 45 % of the datasets in the UCI repository [1], which is one of the most popular data repositories for machine leaning, have the issue of missing values [10]. The reasons for missing values are various. For example, in a social survey, respondents often ignore to answer some questions; some results collected from industrial experiments may be missing values due to mechanical failures while collecting data; medical datasets are often incomplete because not all tests can be run on every patient [9].

Missing values lead to severe issues for classification. One of the most severe issues is non-applicability of many classification algorithms. Although some classification algorithms are able to deal with incomplete data, many others require complete data. Therefore, these classification algorithms cannot directly work with incomplete data. Even for algorithms that can cope with incomplete data, missing values often result in large classification errors [10, 21].

One approach to handling classification with incomplete data is to use imputation methods to replace missing fields with plausible values before using classification algorithms. For example, mean imputation replaces each missing field with the average of the complete values of the same feature. Imputation methods provide complete data that can be then used by any classifier. Consequently, imputation methods are one of the most popular approaches to addressing classification with incomplete data [10].

Multiple imputation is an approach to tackling incomplete data by creating multiple imputed datasets to reflect better the uncertainty in incomplete data. In statistical fields, multiple imputation has become increasingly popular because of its convenience and flexibility [15, 18, 20]. Multiple imputation also has been a powerful technique for addressing classification with incomplete data [9, 19, 23]. However, when multiple imputation is used for classification with incomplete data, multiple imputed datasets are simply averaged to generate a single imputed dataset which is then used by classification algorithms [9, 23]. The disadvantage of this approach is that it ignores the ability of multiple imputation to reflect the uncertainty of incomplete data. How to exploit this ability of multiple imputation in classification with incomplete data is still an open issue.

Ensemble learning algorithms can build a set of classifiers for classification task instead of a single classifier. After that, a new instance is classified by taking a vote of their predictions. Both theoretical development and empirical research have showed that an ensemble can help to improve classification accuracy [8, 16]. However, ensemble methods are mainly applied to complete data. Therefore, how to use ensemble methods for improving classification with incomplete data should be further investigated.

1.1 Research Goals

The goal of this paper is to propose a combination of multiple imputation with ensemble learning for improving classification with incomplete data. The proposed method is compared with two other popular approaches to dealing with missing values. One approach is to use single imputation to generate a single imputed dataset. Another approach is to use multiple imputation to generate a single imputed dataset by averaging multiple imputed datasets. Results from experiments are used to address the following objectives:

1. Whether the combination of multiple imputation with ensemble learning can achieve better classification than using single imputation; and
2. Whether the combination of multiple imputation with ensemble learning can achieve better classification than using multiple imputation to generate a single imputed dataset by averaging multiple imputed datasets.

1.2 Organisation

The rest of the paper is organised as follows. Section 2 discusses related work. Section 3 outlines the proposed method. Section 4 presents experiment design. Section 5 shows results and analysis. Section 6 draws conclusions and presents future work.

2 Related Work

This section discusses related work including classification with missing data, imputation methods and ensemble learning.

2.1 Classification with Missing Data

There are four major approaches to addressing classification with incomplete data including the removal approach, the imputation approach, the model-based approach and the machine learning approach [10].

The removal approach eliminates all instances containing missing values before using classifiers. The main benefit of this approach is to provide complete data that can be then classified by any classifiers. Nevertheless, incomplete instances are not classified by the classifier. Therefore, this approach is only able to be applied to the training process and when a dataset includes a small number of incomplete instances [9].

The imputation approach uses imputation methods to replace missing values with suitable values before using classifiers. For instance, mean imputation fills all missing fields in a feature with the average of complete values in the feature. The main benefit of this approach is to provide complete data which can be used by any classification algorithm. By using imputation methods, both complete and incomplete instances are attended in the classification process. Furthermore, most imputation methods can enhance classification accuracy compared to the corresponding methods without using imputation. Therefore, the imputation approach is a main way to address classification with incomplete datasets [9].

The model-based approach generates a data distribution model from input data. Thereafter, a combination of the data distribution model and Bayesian decision theory [3] is used to classify both complete and incomplete instances. Although this approach can classify both complete and incomplete instances, it requires to make assumptions about the joint distribution of all features in the model [10].

The machine learning approach makes classifiers that are able to directly classify incomplete datasets without using nay imputation methods. For instance, C4.5 [17] can tack with missing values in both training data and test data by using a probabilistic approach.

2.2 Imputation Methods

The goal of imputation methods is to fill missing fields with plausible values [15]. Imputation methods can be categorized into single imputation and multiple imputation [9]. While single imputation methods search one value for each missing value, multiple imputation methods search multiple values for each missing value.

2.2.1 Single Imputation

Each missing field is filled by one value in single imputation methods. This paper uses three single imputation methods: mean imputation, hot deck imputation and K nearest neighbours-based imputation.

Mean imputation replaces all missing fields in each feature with the average of the complete values in the feature. The advantage of this method is that it maintains the mean of each feature, but it under-represents the variability in the data since all missing fields in each feature have the same value [10].

In hot deck imputation, for each incomplete instance, the most similar instance with the incomplete instance is found, and missing fields are replaced with complete values from the most similar instance. The main merit of hot deck imputation is that it fills missing fields by real values from the data. Nonetheless, this method only utilises the information of one instance; thus, it ignores all global properties of the data [15].

KNN-based imputation is based on K-nearest neighbors algorithm for classification. For each incomplete instance, firstly, it finds the K most similar instances with the incomplete instance, and then fills missing fields of the incomplete instance with the average of values in the K most similar instances. KNN-based imputation often performs better than mean imputation and hot deck imputation [2]. However, this method is often computationally intensive owing to having to search through all instances to find the K most similar instances for each incomplete instance [10].

2.2.2 Multiple Imputation

Multiple imputation has three main steps. Firstly, incomplete data is put N times (N > 1) into an imputation model incorporating random variation to build N different imputed datasets. After that, each imputed dataset is separately analysed by standard procedures for complete data. The second step provides N analysis results. Finally, the N analysis results are combined to provide a final result [15, 18].

Multiple imputation has become more and more popular because of several reasons. Firstly, multiple imputation often reflects better uncertainty related to a particular model used for imputation, though it is computationally more expensive than single imputation [9]. Moreover, many recent software developments have based on the multiple imputation framework [12].

One of the most convenient and powerful multiple imputation methods is multivariate imputation by chained equations (MICE) [22]. The first step to generate multiple imputed datasets in MICE is multiple imputation by chained equations. MICE utilises a set of regression methods such as classification and regression trees (CART) [5] and Random forest [14]. Initially, each missing field is replaced by a complete value randomly chosen from the same feature. Afterwards, each incomplete feature is regressed on all other features to compute a better estimate for the feature. The process is repeated several times for all incomplete features to generate a single imputed dataset. The whole procedure is repeated N times to generate

N imputed datasets which are then used to calculate the final imputed dataset [22]. MICE software [6] makes it easy to use this method.

2.3 Ensemble Learning

Ensemble learning is the process that builds a set of classifiers for classification. Thereafter, a new instance is classified by voting the decision of the individual classifiers. Ensemble learning has been proved capable of achieving better classification accuracy than any single classifier [8, 16].

An ensemble of classifiers is good if the individual classifiers in the ensemble is accurate and diverse. Bagging and Boosting are two popular approaches to building accurate ensembles [16]. Both Bagging and Boosting use "resampling" techniques to manipulate the training data. Bagging manipulates the original training dataset of N instances by randomly drawing with replacement instances. Therefore, in the resulting training dataset, some of the original instances may appear multiple times while others might disappear. Bagging is often effective on "unstable" learning algorithms such as neural networks and decision trees where small changes in the training dataset lead to major changes in predictions. Experimental results show that Bagging ensemble almost always performs better than a single classifier. Boosting manipulates the original dataset for each individual classifier by using the performance of the previous classifier(s). In Boosting, instances which are incorrectly classified by previous classifiers are selected more often than instances which are correctly classified. Therefore, Boosting tries to build new classifiers that are better to classify instances for which the current ensemble's performance is poor. Empirical results show that with little or no classification noise, Boosting ensemble also almost always performs better than a single classifier, and it is sometimes more accurate than Bagging ensemble. However, in situations with substantial classification noise, Boosting ensemble is often less accurate than a single classifier because Boosting often overfits noisy datasets [16].

An ensemble of classifiers trained with random subsets of features is presented in [13] to classify with incomplete data. In this approach, each base classifier is trained with a randomly selected subset of features. In [7], a combination of data analysis and ensemble learning is proposed to deal with classification with incomplete data. Firstly, the incomplete data is analysed and grouped into complete data subsets, and then each data subset is used to train one classifier. In the both approaches, when an incomplete instance needs be classified, only those classifiers trained with those features that are available in the instance are used to classify the instance. Although, the two methods are able to cope with incomplete data in some degree, they cannot guarantee to classify all incomplete instances, especially when data contains many missing values. Moreover, combining ensemble learning and multiple imputation has not been investigated. Therefore, using ensemble learning for classification with incomplete data should be more investigated.

3 Multiple Imputation and Ensemble Learning for Classification with Missing Data

The proposed algorithm has two phases: the training process and the application process. The training process uses a multiple imputation method combined with ensemble learning to build a set of classifiers. After that, the application process uses the multiple imputation method and the set of classifiers to classify a new incomplete instance (Fig. 1).

In the training process, a training incomplete dataset is put into a multiple imputation method to build a set of imputed datasets. Afterwards, each imputed dataset is used as a training data by a classification algorithm to train a classifier. As a result, a set of classifiers are generated from the set of imputed datasets.

In the application process, if an instance which needs to be classified is incomplete, the incomplete instance is put into the multiple imputation method (along with the training data) to generate a set of imputed instances. After that, each classifier is applied to each imputed instance to generate a large set of predicted classes. The final predicted class will be the most frequent class of all the predictions. If an instance

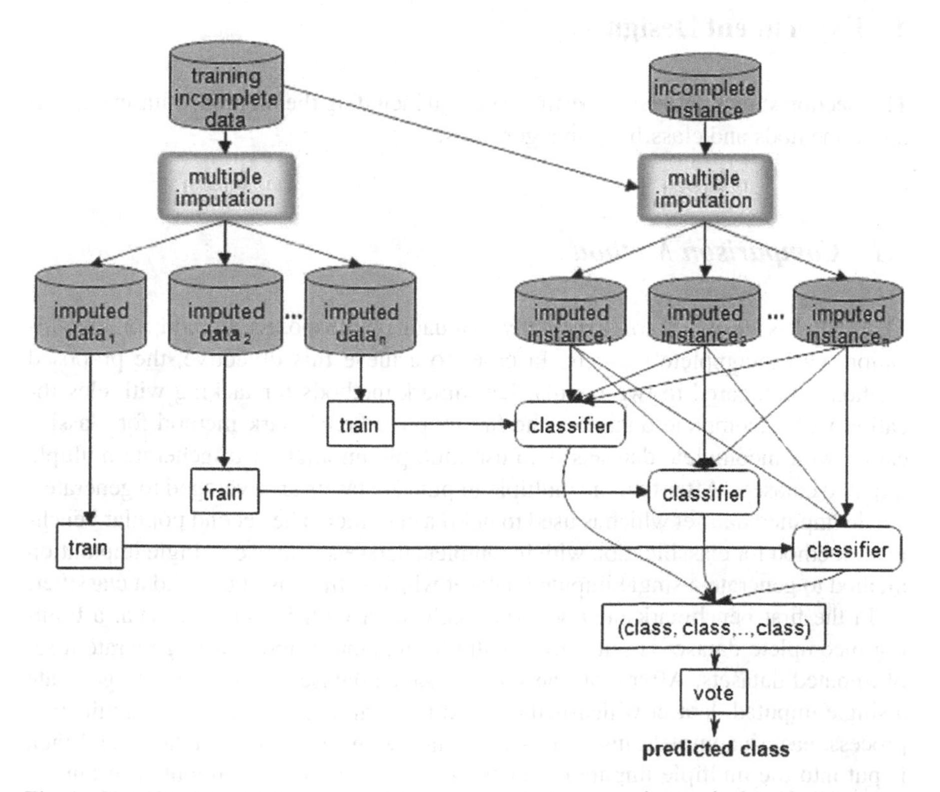

Fig. 1 Classification with incomplete data using a multiple imputation method and building a set of classifiers

which needs to be classified is complete, the complete instance does not need the imputation method. Rather, they are classified directly by each of the classifiers and the most frequent class is chosen.

A key requirement of ensemble methods is that the set of classifiers should be diverse. The key idea of the proposed algorithm is that it exploits the ability of the multiple imputation method to build a set of diverse imputed datasets from which diverse classifiers can be constructed. This is in contrast to the usual use of multiple imputation for classification which averages the imputed datasets into a single dataset. From one incomplete dataset, multiple imputation is able to generate a set of diverse imputed datasets because the initial step of the multiple imputation is to fill each missing field with a randomly chosen complete value. Therefore, the initial step generates different temporary imputed datasets. Although the same regression method is then used to improve the temporary imputed datasets, the multiple imputation method is able to generate a set of diverse imputed datasets, especially when the training dataset contains many missing fields. As a result, a classifier ensemble which is then built by using the set of imputed datasets is hopefully diverse.

4 Experiment Design

This section shows detailed experiment design including the method, datasets, imputation methods and classification algorithms.

4.1 Comparison Method

This study is designed to empirically evaluate the proposed method for classification with incomplete datasets. In order to achieve this objective, the proposed method is compared to two popular benchmark methods for tacking with classification with incomplete datasets. The first popular benchmark method for classification with incomplete datasets is to use multiple imputation to generate multiple imputed datasets. After that the multiple imputed datasets are averaged to generate a single imputed dataset which is used to build a classifier. The second popular benchmark method for classification with incomplete datasets is to use a single imputation method to generate a single imputed dataset which is then used to build a classifier.

In the first benchmark method for classification with incomplete data, a training incomplete dataset is put into a multiple imputation method to generate a set of imputed datasets. After that, the set of imputed datasets is averaged to generate a single imputed dataset which is then used to learn a classifier. In the application process, each incomplete instance is combined with the training dataset, and then is put into the multiple imputation method to generate a set of imputed instances. Subsequently, the set of imputed instances is averaged to generate a single imputed instance which is then classified by the classifier.

In the second benchmark method for classification with incomplete data, a training incomplete dataset is put into a single imputation method to generate a single imputed dataset. Thereafter, the imputed dataset is used to learn a classifier. In the application process, each incomplete instance is combined with the training dataset, and then is put into the single imputation method to generate a single imputed instance. Afterwards, the single imputed instance is classified by the classier.

4.2 Datasets

Ten datasets, summarised in Table 1, are used in the experiments. These are taken from the UCI Repository of Machine Learning Databases [1]. Each dataset is presented in one row in Table 1 including the number of instances, the number of features, the number of classes, the proportion of instances containing at least one missing field and the proportion of missing values.

The first five datasets suffer from missing values in a "natural" way. In the datasets, we do not know any information related to the randomness of missing values, so we make assumption that missing values in the datasets are distributed in a *missing at random* (MAR) way [15].

In order to test the performance of the proposed feature selection method with datasets containing different levels of missing values, the *missing completely at random* (MCAR) mechanism [15] was utilised to introduce missing values into the last five complete datasets. Three different levels of missing values: 10 %, 30 % and 50 % were used to introduce missing values into the datasets. With each dataset in the last five datasets and each level of missing values in the three levels, repeat 30 times: introduce randomly the level of missing values in all features. Hence, from one dataset and one level of missing values, 30 artificial datasets containing missing values were generated. Therefore, from one complete dataset, 90 (=30 × 3) artifi-

Table 1 The datasets used in the experiments

Dataset	#instances	#features	#classes	Incomplete inst (%)	Missing values (%)
Bands	539	19	2	32.28	5.38
Hepatitis	155	19	2	48.39	5.67
Horse-colic	368	23	2	98.1	22.76
Housevotes	435	16	2	46.67	5.63
Mammographic	961	5	2	13.63	3.37
Heartstatlog	270	13	2	0	0
Iris	150	4	3	0	0
Liver	345	7	2	0	0
Parkinsons	197	23	2	0	0
Seedst	210	7	3	0	0

cial datasets containing missing values were generated and a total of 450 (=90 × 5) artificial datasets containing missing values were used in the experiments.

None of the datasets in the experiments comes with a specific test set. Moreover, in some datasets, the number of instances is relatively small. Therefore, the ten-fold cross-validation method was used to measure the performance of the learned classifiers. With the first five incomplete datasets, the ten-fold cross-validation method was performed 30 times. With the last five complete datasets, with each dataset and each level of missing values, the ten-fold cross-validation method was performed on the 30 incomplete datasets. Consequently, for each incomplete dataset in the first five datasets and each level of missing values on one dataset in the last five datasets, 300 pairs of training and testing sets were generated.

4.3 Imputation Algorithms

The experiments used multiple imputation MICE [6] with the random forest as a regression method. In the multiple imputation method, each incomplete feature was repeatedly regressed on other features 10 times. With each incomplete dataset, the multiple imputation method was performed 20 times to procedure 20 imputed datasets.

Three single imputation methods including mean imputation, hot deck imputation and KNN-based imputation were used in the experiment. The three single imputations were in-house implementations. With KNN-based imputation, the number of neighbors K were set five.

4.4 Classification Algorithms

The experiment used C4.5 [17] to classify data. For the classifiers, WEKA's implementation [11] was used and all parameters were set to WEKA's defaults. The number of classifiers in an ensemble is equal to the number of imputed datasets generated by multiple imputation; therefore, the number of classifier in an ensemble is set 20.

5 Results and Analysis

This section presents the comparison between the proposed method with other methods on classification accuracy, and further analysis.

5.1 Results

Table 2 shows the average of classification accuracy and standard deviation using C4.5. In the tables, and in the following ones, MIEL column presents results by using the proposed method, AvgMI column presents results by using the first benchmark method; Mean, HDI and KNNI columns present results from the second benchmark method by using mean imputation, hot deck-based imputation and KNN-based imputation, respectively. With each dataset in the first five datasets, the classification accuracy is the average of accuracies of the 30 times performing ten-fold cross-validation ($30 \times 10 = 300$ experiments).

Table 3 shows the average of classification accuracy and standard deviation using C4.5 with three levels of missing values. With each dataset and each missing level in the last five datasets, the classification accuracy is the average of accuracies of the 30 generated incomplete datasets at each missing level and ten-fold cross-validation ($30 \times 10 = 300$ experiments).

To compare the performance of MIEL with the other methods, the Wilcoxon signed-ranks tests at 95 % confidence interval is used to compare the classification accuracy achieved by MIEL with the other methods. "T" columns in Tables 2 and 3 show significant test of the columns before them against MIEL, where "+", "=" and "−" mean MIEL is significantly more accurate, not significantly different and significantly less accurate, respectively.

Table 2 shows that MIEL can achieve significantly better classification accuracy than the other methods in almost all cases with the datasets containing natural missing values. MIEL achieves similar classification accuracy to the other methods on Housevotes dataset and significantly better classification accuracy than the other methods on the other four datasets.

Table 3 shows that MIEL also can achieve significantly better classification accuracy than the other methods in almost all cases with the datasets containing artificial missing values. MIEL achieves significantly better classification accuracy than the other methods on all fifteen cases.

It is clear from the results that AvgMI is generally better than single imputation methods showing that multiple imputation generates a more reliable imputed dataset. Furthermore, a combination of multiple imputation and ensemble learning is significantly better than using multiple imputation to generate a single imputed data by averaging imputed datasets.

In summary, the proposed method combining multiple imputation combined with ensemble learning is able to enhance classification accuracy of a classifier not only with natural incomplete datasets, but also with artificial incomplete datasets.

Table 2 The average of accuracy comparison between MIEL and the other methods using C4.5 for datasets containing natural missing values

Dataset	C4.5								
	MIEL	AvgMI	T	Mean	T	HDI	T	KNNI	T
Bands	**71.96** ± 1.74	68.22 ± 1.94	+	68.39 ± 2.05	+	63.53 ± 2.50	+	65.04 ± 2.39	+
Hepatitis	**79.82** ± 2.12	78.24 ± 1.80	+	77.42 ± 1.90	+	77.30 ± 3.94	+	78.37 ± 2.42	+
Horse-colic	**84.83** ± 0.77	83.69 ± 1.31	+	83.48 ± 1.1	+	83.48 ± 1.16	+	83.48 ± 1.16	+
Housevotes	**95.80** ± 0.52	**95.96** ± 0.80	=	**95.82** ± 0.97	=	**95.97** ± 0.69	=	**95.88** ± 0.96	=
Mammographic	**82.53** ± 0.55	82.10 ± 0.74	+	82.19 ± 0.71	+	81.87 ± 0.75	+	82.07 ± 0.59	+

Table 3 The average of accuracy comparison between MIEL and the other methods using C4.5 for datasets with several missing rates

Dataset	Missing rate (%)	C4.5								
		MIEL	AvgMI	T	Mean	T	HDI	T	KNNI	T
Heartstatlog	10	**79.65** ± 1.92	75.70 ± 2.13	+	75.25 ± 2.36	+	75.12 ± 2.93	+	74.93 ± 2.15	+
	30	**79.74** ± 2.16	74.01 ± 2.99	+	71.46 ± 3.10	+	68.17 ± 3.69	+	71.17 ± 2.76	+
	50	**76.91** ± 1.99	69.65 ± 3.06	+	64.95 ± 4.62	+	63.53 ± 4.31	+	67.16 ± 3.79	+
Iris	10	**93.55** ± 1.01	92.71 ± 1.34	+	92.28 ± 2.18	+	88.66 ± 3.15	+	91.34 ± 2.84	+
	30	**91.44** ± 1.90	89.04 ± 2.15	+	87.27 ± 3.80	+	73.75 ± 7.41	+	85.16 ± 4.89	+
	50	**84.50** ± 2.68	80.71 ± 3.32	+	77.37 ± 6.11	+	61.15 ± 6.38	+	76.22 ± 6.41	+
Liver	10	**64.29** ± 2.33	61.88 ± 3.38	+	61.58 ± 2.69	+	61.49 ± 2.76	+	61.31 ± 2.97	+
	30	**60.88** ± 2.26	59.06 ± 2.46	+	58.27 ± 2.89	+	57.88 ± 2.67	+	58.85 ± 3.05	+
	50	**58.44** ± 0.81	57.20 ± 1.62	+	56.65 ± 2.00	+	55.85 ± 2.53	+	56.28 ± 2.67	+
Parkinsons	10	**87.70** ± 2.31	83.55 ± 3.07	+	81.96 ± 2.69	+	81.41 ± 2.49	+	81.82 ± 2.47	+
	30	**86.85** ± 1.32	83.09 ± 2.66	+	79.40 ± 2.72	+	76.18 ± 3.25	+	78.61 ± 2.54	+
	50	**83.43** ± 1.77	80.08 ± 2.21	+	76.47 ± 3.75	+	75.57 ± 2.37	+	76.44 ± 3.76	+
Seedst	10	**90.64** ± 1.52	89.41 ± 2.20	+	86.92 ± 2.03	+	83.35 ± 2.97	+	85.47 ± 2.96	+
	30	**89.16** ± 1.26	86.87 ± 2.35	+	82.04 ± 2.87	+	69.80 ± 7.68	+	79.14 ± 5.15	+
	50	**85.92** ± 1.69	82.34 ± 1.80	+	76.44 ± 3.87	+	63.84 ± 8.86	+	74.63 ± 3.63	+

6 Conclusions and Future Work

This paper proposed a new combination of multiple imputation and ensemble learning for classification with incomplete data. Firstly, multiple imputation is used to generate a set of imputed datasets from one incomplete dataset. After that, the set of imputed datasets is used to build an ensemble classifier. The proposed approach was compared with two other popular approaches to dealing with incomplete data: one using multiple imputation to generate one single imputed dataset and the other using single imputation to generate a single imputed dataset. The experiments on ten datasets used C4.5 as classification algorithms. The experimental results showed that the proposed method can achieve better classification accuracy than the two other methods. The experimental results also showed that it is advantageous to exploit the natural diversity generated by multiple imputation, rather than averaging the diverse imputed datasets. Even if the averaged imputed datasets is reliable, using the diversity of imputed datasets in an ensemble method leads to a more effective classifier.

The experiments in the paper used random forest as a regression method in MICE. There are some other regression methods in MICE such as linear regression and CART [5]. Further work could perform this investigation with linear regression and CART. Furthermore, the proposed method uses the majority vote. Therefore, another future work could develop a more powerful vote method to improve the proposed method.

References

1. Asuncion, A., Newman, D.: UCI machine learning repository (2007)
2. Batista, G.E., Monard, M.C.: A study of k-nearest neighbour as an imputation method. In: Hybrid Intelligent Systems - HIS. pp. 251–260 (2002)
3. Berger, J.O.: Statistical decision theory and Bayesian analysis. Springer Science & Business Media (2013)
4. Bishop, C.M.: Pattern Recognition and Machine Learning. Springer-Verlag New York, Inc. (2006)
5. Breiman, L., Friedman, J., Stone, C.J., Olshen, R.A.: Classification and regression trees. CRC Press (1984)
6. Buuren, S., Groothuis-Oudshoorn, K.: MICE: Multivariate imputation by chained equations in R. Journal of statistical software 45, 1–67 (2011)
7. Chen, H., Du, Y., Jiang, K.: Classification of incomplete data using classifier ensembles. In: Systems and Informatics (ICSAI), 2012 International Conference on. pp. 2229–2232 (2012)
8. Dietterich, T.G.: Ensemble methods in machine learning. In: International workshop on multiple classifier systems. pp. 1–15 (2000)
9. Farhangfar, A., Kurgan, L.A., Pedrycz, W.: A novel framework for imputation of missing values in databases. Systems, Man and Cybernetics, Part A: Systems and Humans, IEEE Transactions on 37, 692–709 (2007)
10. García-Laencina, P.J., Sancho-Gómez, J.L., Figueiras-Vidal, A.R.: Pattern classification with missing data: a review. Neural Computing and Applications 19, 263–282 (2010)
11. Hall, M., Frank, E., Holmes, G., Pfahringer, B., Reutemann, P., Witten, I.H.: The WEKA data mining software: an update. ACM SIGKDD explorations newsletter 11, 10–18 (2009)

12. Harel, O., Zhou, X.H.: Multiple imputation: review of theory, implementation and software. Statistics in medicine 26, 3057–3077 (2007)
13. Krause, S., Polikar, R.: An ensemble of classifiers approach for the missing feature problem. In: Neural Networks, 2003. Proceedings of the International Joint Conference on. vol. 1, pp. 553–558 (2003)
14. Liaw, A., Wiener, M.: Classification and regression by randomforest. R news 2, 18–22 (2002)
15. Little, R.J., Rubin, D.B.: Statistical analysis with missing data. John Wiley & Sons (2014)
16. Opitz, D., Maclin, R.: Popular ensemble methods: An empirical study. Journal of Artificial Intelligence Research 11, 169–198 (1999)
17. Quinlan, J.R.: C4. 5: programs for machine learning. Elsevier (2014)
18. Schafer, J.L., Graham, J.W.: Missing data: our view of the state of the art. Psychological methods 7, 147 (2002)
19. Tran, C.T., Andreae, P., Zhang, M.: Impact of imputation of missing values on genetic programming based multiple feature construction for classification. In: 2015 IEEE Congress on Evolutionary Computation (CEC). pp. 2398–2405 (2015)
20. Tran, C.T., Zhang, M., Andreae, P.: Multiple imputation for missing data using genetic programming. In: Proceedings of the 2015 Annual Conference on Genetic and Evolutionary Computation. pp. 583–590 (2015)
21. Tran, C.T., Zhang, M., Andreae, P.: A genetic programming-based imputation method for classification with missing data. In: European Conference on Genetic Programming. pp. 149–163 (2016)
22. White, I.R., Royston, P., Wood, A.M.: Multiple imputation using chained equations: issues and guidance for practice. Statistics in medicine 30, 377–399 (2011)
23. Williams, D., Liao, X., Xue, Y., Carin, L., Krishnapuram, B.: On classification with incomplete data. IEEE Transactions on Pattern Analysis and Machine Intelligence 29(3), 427–436 (2007)

CMA-ES with Surrogate Model Adapting to Fitness Landscape

Kento Tsukada, Taku Hasegawa, Naoki Mori and Keinosuke Matsumoto

Abstract One of the most important issues for evolutionary computation (EC) is to consider the number of fitness evaluations. In order to reduce the number of fitness evaluations, we have proposed the novel surrogate model called Rank Space Estimation (RSE) model and the surrogate-assisted EC with RSE model called the Fitness Landscape Learning Evolutionary Computation (FLLEC). This paper presents a novel CMA-ES with RSE model for continuous optimization problems and a scaling method for input data to surrogate model.

Keywords Evolutionary computation · Support vector machine · Continuous optimization

1 Introduction

Evolutionary Computation (EC) has been applied to various kinds of optimization problems, and several advantages of using EC have been reported. On the other hand, a large number of fitness evaluations is required to obtain adequate results with EC. This high cost of these evaluations is one of the most important issues for EC to obtain adequate results for real-world applications.

Surrogate-assisted ECs have been proposed in order to solve this problem, and several methods have reported success [1]. Moreover, almost every surrogate model

K. Tsukada (✉) · T. Hasegawa · N. Mori · K. Matsumoto
Graduate School of Engineering, Osaka Prefecture University,
1-1 Gakuencho, Sakai, Osaka 599-8531, Japan
e-mail: tsukada@ss.cs.osakafu-u.ac.jp

T. Hasegawa
e-mail: hasegawa@ss.cs.osakafu-u.ac.jp

N. Mori
e-mail: mori@cs.osakafu-u.ac.jp

K. Matsumoto
e-mail: matsu@cs.osakafu-u.ac.jp

G. Leu et al. (eds.), *Intelligent and Evolutionary Systems*,
Proceedings in Adaptation, Learning and Optimization 8,
DOI 10.1007/978-3-319-49049-6_30

attempts to estimate the fitness value itself. If a suitable fitness function model can be generated, this is decisive answer for EC with surrogate models. However, it is difficult to estimate the fitness function completely, and doing so often requires substantial effort. To address this challenge, we have proposed a novel surrogate model named the *Rank Space Estimation* (*RSE*) model [2]. The research of a surrogate model based on rank has been reported [3]. However, there is no research of a surrogate model utilizing the ranks determined by relative evaluation between two individuals. An estimation cost of a fitness landscape can be reduced without critical estimation error by means of limiting the prediction in this model to the ranks of two individuals.

We have also proposed a framework for EC using our proposed RSE model. This proposal is called the *Fitness Landscape Learning Evolutionary Computation* (*FLLEC*) framework. Our methods demonstrated their effectiveness using computational experiments such as taking several combinational problems. On the other hand, in the application of ECs to the field of industry, there are many continuous optimization problems and reducing the number of fitness evaluations in those problems have become important.

In this paper, Evolution Strategy with Covariance Matrix Adaptation (CMA-ES) [4], which is a stochastic method for continuous optimization, is introduced to FLLEC. The number of fitness evaluations is especially reduced by utilizing SVM predictions for evaluating individuals instead of computing their fitness.

2 CMA-ES

The CMA-ES is a EC for various continuous optimization problems. The important processes of the CMA-ES are called *Recombination*, *Step Size Adaptation* (*SSA*) and *Covariance Matrix Adaptation* (*CMA*). Various types of updating methods have been proposed. In this study, "(μw, λ)-CMA-ES" is used. It utilizes the following three procedures.

- weighted recombination [4]
- cumulative step size adaptation(CSA) [5]
- hybrid covariance matrix adaptation(Hybrid-CMA) [6]

2.1 Algorithm of CMA-ES

The following is the algorithm of the CMA-ES used in this study:

1. Initialization
   Mean vector $\boldsymbol{m}$, global step size σ and population size λ are set depending on the problems. Then, we initialize a covariance matrix $\boldsymbol{C}$, evolution path $\boldsymbol{p}_c$ and

p_σ ($C = I, p_c = p_\sigma = 0$). Moreover, parents number μ, recombination weight w and other parameters are set according to the previous work [7]. After that, the following procedure is repeated until the termination criterion is satisfied.

2. Sampling
   Individuals x_i are generated according to the following equations

$$z_i \sim \mathcal{N}(\mathbf{0}, \boldsymbol{I}_n), \tag{1}$$

$$y_i = BDz_i, \tag{2}$$

$$x_i^{(g+1)} = m^{(g)} + \sigma^{(g)} y_i \ (i = 1, \ldots, \lambda), \tag{3}$$

where

B is an orthogonal matrix, $B^T B = BB^T = I$. Columns of B form an orthonormal basis of eigenvectors.
D is a diagonal matrix with square roots of eigenvalues of C as diagonal elements.
$\sim$ denotes the same distribution on the left and right side.
$\mathcal{N}$ represents a multivariate normal distribution.

Moreover, the following equation, for generation number $g = 0, 1, 2, \ldots$, reads

$$x_i^{(g+1)} \sim \mathcal{N}(m^{(g)}, (\sigma^{(g)})^2 C^{(g)}). \tag{4}$$

If individuals $x_i^{(g+1)}$ are generated in infeasible domain, they are re-sampled until they are in feasible domain.

3. Selection and Recombination
   Fitness function $f(x_i^{(g+1)})$ for all individuals is computed. Then, a mean vector is updated according to the following equation,

$$m^{(g+1)} = \sum_{i=1}^{\mu} w_i x_{i:\lambda}^{(g+1)}, \tag{5}$$

where $i : \lambda$ represents an index of the individual which has ith best fitness. This update method is called weighted recombination.

4. Step Size Adaptation
   In the CSA algorithm, a global step size is updated by calculating the following evolution path,

$$p_\sigma^{(g+1)} = (1 - c_\sigma) p_\sigma^{(g)} + \sqrt{c_\sigma(2 - c_\sigma)\mu_{\text{eff}}}\, C^{(g)-\frac{1}{2}} \frac{m^{(g+1)} - m^{(g)}}{\sigma^{(g)}}, \tag{6}$$

where $\mu_{\text{eff}} = (\sum_{i=1}^{\mu} w_i^2)^{-1}$. Then, the global step size σ is updated according to the following equation,

$$\sigma^{(g+1)} = \sigma^{(g)} \cdot \exp\left(\frac{c_\sigma}{d_\sigma}\left(\frac{\|\boldsymbol{p}_\sigma^{(g+1)}\|}{\mathrm{E}\|\mathcal{N}(\boldsymbol{0}, \boldsymbol{I}_n)\|} - 1\right)\right), \tag{7}$$

where $\mathrm{E}(\|\mathcal{N}(\boldsymbol{0}, \boldsymbol{I}_n)\|)$ represents an average norm of normally distributed n-dimensional random numbers norm.

5. Covariance Matrix Adaptation
For the update of a covariance Matrix of a multivariate normal distribution, two update methods are combined in the Hybrid-CMA. They are "rank-one update" and "rank-μ update". In the rank-one update, an evolution path, which indicates a transition of the mutation distribution mean, is calculated for the update.

$$\begin{aligned}\boldsymbol{p}_c^{(g+1)} &= (1 - c_c)\boldsymbol{p}_c^{(g)} \\ &+ h_\sigma^{(g+1)}\sqrt{c_c(2 - c_c)\mu_{\text{eff}}}\frac{\boldsymbol{m}^{(g+1)} - \boldsymbol{m}^{(g)}}{\sigma^{(g)}},\end{aligned} \tag{8}$$

On the other hand, in the rank-μ update, the information of μ individuals is utilized for the update. Then, a covariance matrix is updated as follows.

$$\begin{aligned}\boldsymbol{C}^{(g+1)} &= (1 - c_{\text{cov}})\boldsymbol{C}^{(g)} \\ &+ \frac{c_{\text{cov}}}{\mu_{\text{cov}}}(\boldsymbol{p}_c^{(g+1)}\boldsymbol{p}_c^{(g+1)\mathrm{T}} + \delta(h_\sigma^{(g+1)})\boldsymbol{C}^{(g)}). \\ &+ c_{\text{cov}}(1 - \frac{1}{\mu_{\text{cov}}})\sum_{i=1}^{\mu} w_i \mathrm{OP}(\frac{\boldsymbol{x}_{i:\lambda}^{(g+1)} - \boldsymbol{m}^{(g)}}{\sigma^{(g)}}).\end{aligned} \tag{9}$$

Here, $h_\sigma^{(g+1)}$ is as follows.

$$h_\sigma^{(g+1)} = \begin{cases} 1 & \mathit{if}\ \frac{\|\boldsymbol{p}_\sigma^{(g+1)}\|}{\sqrt{1-(1-c_\sigma)^{2(g+1)}}} \\ & \quad < (1.5 + \frac{1}{n-0.5})\mathrm{E}(\|\mathcal{N}(\boldsymbol{0}, \boldsymbol{I}_n)\|) \\ 0 & \mathit{otherwise}. \end{cases} \tag{10}$$

Moreover,

$$\delta(h_\sigma^{(g+1)}) = \begin{cases} 1 & \mathit{if}\ h_\sigma^{(g+1)} = 0 \\ 0 & \mathit{otherwise}. \end{cases} \tag{11}$$

The optimization progresses repeating 2. ~ 5. [4] and [7] are used as reference.

3 Surrogate-Assisted Evolutionary Computation

This section describes the surrogate-assisted ECs for reducing the number of fitness evaluations.

3.1 *Surrogate Model*

Surrogate models generally estimate the fitness function and are directly used in fitness evaluations in order to reduce the number of fitness evaluations. Several surrogate models have been proposed. From previous studies [1, 8], it is important for effective search to select the most suitable models for each fitness functions.

3.2 *Rank Space Estimation (RSE) Model*

We proposed a novel surrogate model based on the ranks of two individuals called the RSE model. In our model, since the ranks is calculated by relative evaluation between two individuals, estimation quality of proposed method is superior to that of existing models by evaluating only one individual. Besides, most surrogate models predict a fitness function directly, whereas our model only focuses on the order of two individuals. Complete prediction of a fitness function is sometimes more difficult than finding an optimum solution. By contrast, ranking of two individuals results in only three types: high, low or equal. If we ignore the equal ranks of different genotypes, it is sufficient to consider only two results. The famous selection operator called tournament selection shows good performance even though this operator only utilizes rank information. This fact guarantees the suitability of RSE model.

The RSE model utilizes SVM as the training method. The input to SVM is the genotype information for two individuals and the output concerns their class labels and their ranks. If an individual in the latter part has a higher fitness than the other one, the class label of the vector is set to 1. On the other hand, if the individual in the first part has a higher fitness than the other one, the class label of the vector is set to -1. Figure 1 shows an outline of our SVM. If the input positions of two individuals are swapped, the output is expected to give the opposite result. Therefore, two items of training data are obtained from one pair of individuals because there are two positions in the SVM input.

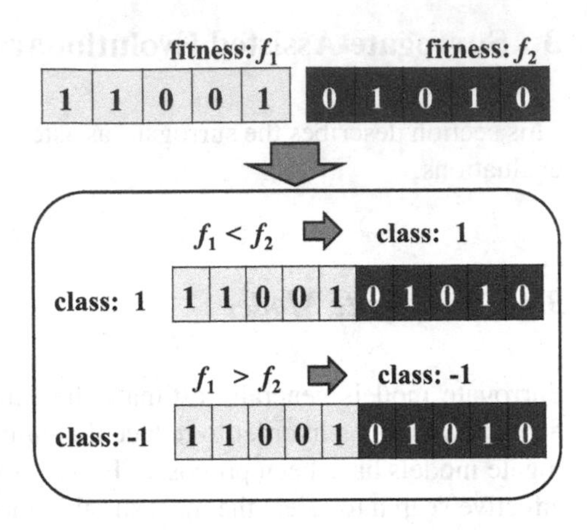

Fig. 1 Outline of SVM

3.3 FLLEC

We proposed a novel evolutionary computational framework which introduces the RSE model. It is called FLLEC. The population can evolve reducing the number of fitness evaluations using predictions from the RSE model in FLLEC.

In search process of ECs, both a surrogate model and an original fitness function should be utilized [9]. In the previous work [1], this was considered as the issue of evolution control and fixed evolution control have often been used. Fixed evolution control has two main approaches:individual-based evolution control and generation-based evolution control. Previous works have reported that generally the surrogate model needs to be trained repeatedly. However, the best interval for relearning depends on respective optimization problems and algorithms. Search processes differ especially depending on control mechanisms for convergence. Therefore, the relearning interval should be set considering those differences.

4 CMA-ES with RSE Model

We reported that FLLEC shows the good performance with binary coding problems [2]. However, the analysis of FLLEC for continuous problems is also very important. This paper presents the FLLEC for continuous problems and analyzes the features of FLLEC in solving continuous problems. Specifically, the CMA-ES is adopted as EC parts of FLLEC and the RSE model is introduced to the CMA-ES according to Sect. 3.3.

4.1 How to Apply RSE Model

In the CMA-ES, fitness of all individuals in a generation must be computed so as to calculate Eq. (5). In this study, the number of those computations is reduced by means of replacing computations of fitness with SVM predictions. However, if the individuals $\boldsymbol{x}_{i:\lambda}(i = 1, 2, \ldots, \mu)$ which has ith best fitness are selected only based on SVM predictions, an incorrect prediction is harmful for the effective search. Therefore, after all individuals in a generation are ranked by SVM, fitness of only N_{e} $(\geq \mu)$ individuals out of population are computed. Here, N_{e} represents the number of fitness evaluations in each generation. N_{e} individuals have the ranks in the range between 1 and N_{e} computed by SVM. This method enables the reduction of the number of fitness evaluations because fitness of all individuals in a generation is not necessarily computed. Besides the harmful influence of incorrect predictions from SVM is mitigated because Eq. (5) is calculated using the re-sorted ranks based on N_{e} individuals' fitness.

Specifically, we introduce a prediction from SVM into the "Selection and Recombination" step mentioned in Sect. 2.1. The detail is as follows.

- Selection and Recombination with SVM
  Firstly, the number of fitness evaluations in each generation (N_{e}) is set. Secondly, N_{e} individuals are selected based on the ranks computed by SVM. The round robin tournament is used to rank individuals by SVM. Thirdly, only N_{e} individuals' fitness is computed. Finally, Eq. (5) is calculated using the re-sorted ranks based on N_{e} individuals' fitness.

A SVM is relearned at a relearning interval (N_{r}) and it is used until next relearning. The fitness information of individuals sampled after the last relearning is used for the training data. Therefore, the more a N_{r} is set at large, the more a fitness information of individuals in the past generation is utilized. These methods are categorized into "individual-based evolution control" mentioned in Sect. 3.3. Figure 2 shows the evolutionary process by it.

4.2 Scaling Method

For our proposed surrogate model using SVM, training and test data should be scaled in order to improve an accuracy rate of the model. In ECs for continuous optimization problems, population density is biased. This bias is harmful to learn a fitness landscape effectively. Hence, in this study, the biased population density is smoothed by means of scaling input data for SVM. In CMA-ES, the population density is represented as a normal distribution. This biased population density should be converted to a uniform distribution for the effective learning of a fitness landscape. In order to reduce a computation cost, we approximate a normal distribution in CMA-ES sampling as a logistic distribution. A sigmoid function, which is the cumulative distrib-

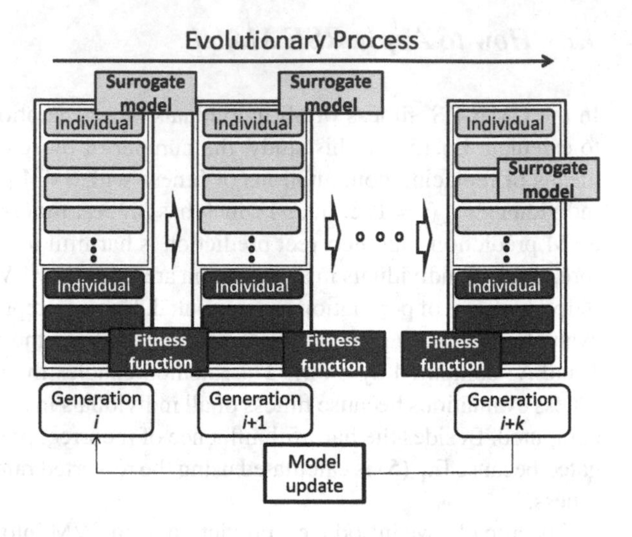

Fig. 2 Individual-based evolution control

ution function of a logistic distribution, is utilized for that scaling. The elements of input data for SVM are mapped onto a sigmoid function. Figure 3 shows the validity of the logistic distribution approximation. The upper figure shows the relative frequency of standardized normally distributed random numbers and the lower figure shows the relative frequency of these numbers mapped onto the sigmoid function. Figure 3 indicates that the mapping onto the sigmoid function smooths the biased population density. The sigmoid function follows

$$f_i(x) = \frac{1}{1 + \exp\{-\frac{x-m_i}{s_i}\}}, \quad i = 1, 2, \ldots, n \tag{12}$$

where m_i is the ith element of the mutation distribution mean vector and s_i is the parameter of the sigmoid function. s_i is calculated from global step size σ and $\boldsymbol{D}$ in the generations where training data are sampled. The variance of a logistic distribution is given as $\pi^2 s_i^2/3$ and that of a normal distribution in CMA-ES sampling is given as $(\sigma d_i)^2$. Therefore, s_i is calculated as

$$s_i = \frac{\sqrt{3}\sigma d_i}{\pi}. \tag{13}$$

We described our proposed scaling method considering the biased population density as follows.

$$\boldsymbol{x}' = \boldsymbol{B}^{\mathrm{T}}(\boldsymbol{x} - \boldsymbol{m}). \tag{14}$$

$\boldsymbol{B}^{\mathrm{T}}$ rotates $\boldsymbol{x} - \boldsymbol{m}$ into the coordinate axes. That is, the principal axes of the distribution $\mathcal{N}(\boldsymbol{0}, \boldsymbol{C})$ are rotated into the coordinate axes.

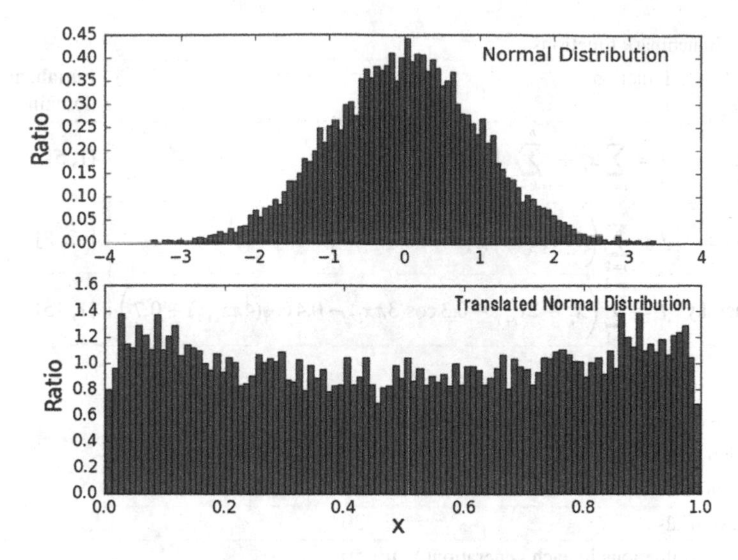

Fig. 3 Frequency distribution of standardized normally distributed random number

$$x'' = (f_1(x'_1), f_2(x'_2), \ldots, f_i(x'_i))^{\mathrm{T}}. \quad (15)$$

$f_i(x'_i)$ represents the elements mapping from an SVM input vector. It is considered that the above scaling method enables surrogate models to learn a fitness landscape efficiently.

5 Experiments

In this section, the effectiveness of the CMA-ES with RSE model and scaling algorithm mentioned in Sect. 4.2 are demonstrated. For the comparison experiment, the following methods are prepared.

- CMA-ES:original, mentioned in Sect. 2.1.
- CMA-ES with RSE model:using a linear scaling
- CMA-ES with RSE model:using our proposed scaling

A linear scaling is used to confirm the difference between scaling methods. The linear scaling maps input data for SVM into $[-1, 1]$.

Table 1 shows the benchmark functions utilized in our experiment and Table 2 shows the experimental conditions. The upper part shows the common parameters for a CMA-ES and the lower part shows the parameters for the RSE model. These parameters are set through the preliminary experiments and the strategy parameters for CMA-ES are set as mentioned in Sect. 2.1. In each trial, the termination criterion

Table 1 Benchmark functions

Name	Function	Initialization domain
k-tablet ($k = n/4$)	$f = \sum_{i=1}^{k} x_i^2 + \sum_{i=k+1}^{n} (100x_i)^2$	[1, 5]
Rosenbrock	$f = \sum_{i=1}^{n-1} \left(100(x_i^2 - x_{i+1})^2 + (x_i - 1)^2 \right)$	[−2, 2]
Bohachevsky	$f = \sum_{i=1}^{n-1} \left(x_i^2 + 2x_{i+1}^2 - 0.3\cos(3\pi x_i) - 0.4\cos(4\pi x_{i+1}) + 0.7 \right)$	[1, 15]

Table 2 Experimental conditions (experiment 1)

Population size	100
Dimension size	10
Number of trials	30
Number of evaluations in each generation(N_e)	50
Training data size(N_{ts})	1000
Interval of relearning(N_r)	1
Kernel function of the SVM	Polynomial (cubic)

is that the objective function value of the best individual reaches under 1.0×10^{-10}. If the number of fitness evaluations is over $n \times \lambda \times 10^3$ before the termination criterion is satisfied, the search is regarded as failure.

5.1 Results and Discussion

Figures 4, 5 and 6 show the average objective function value with the number of fitness evaluations. The abscissa shows the number of fitness evaluations and the ordinate is a logarithmic scale axes and shows the objective function value. In the experiment, the original CMA-ES and the method using our proposed scaling achieved the global optima in each trial. On the other hand, the method using the linear scaling did not achieve that in each trial.

These figures indicate that better solutions are obtained at an early stage of the search in linear scaling compared with other methods. On the other hand, when the objective function value reaches around 1.0×10^{-1}, searching better solution began to go wrong. It is assumed that the trend resulted from the error prediction from SVM. In the stagnant stage of the search, the variance of the distribution in Eq. (4) remained large. That indicates the population did not successfully converge on the

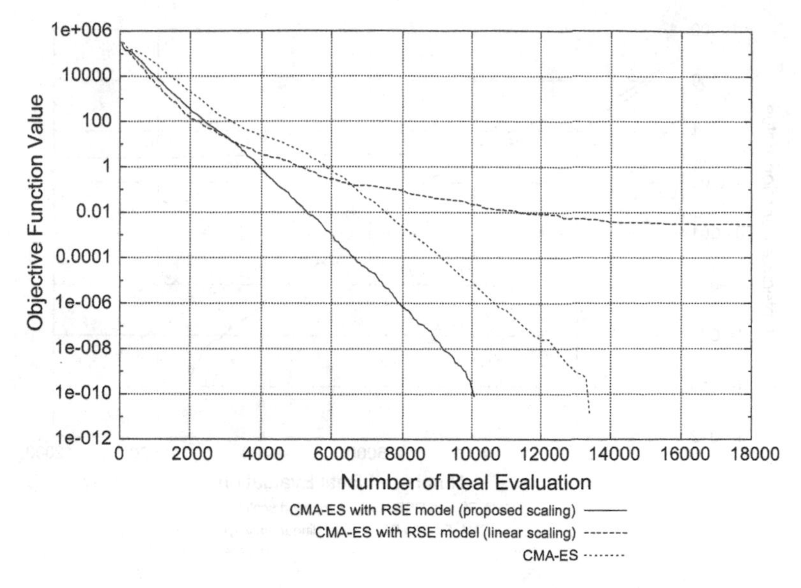

Fig. 4 Objective function value with fitness evaluations (for CMA-ES using RSE model with the k-tablet function)

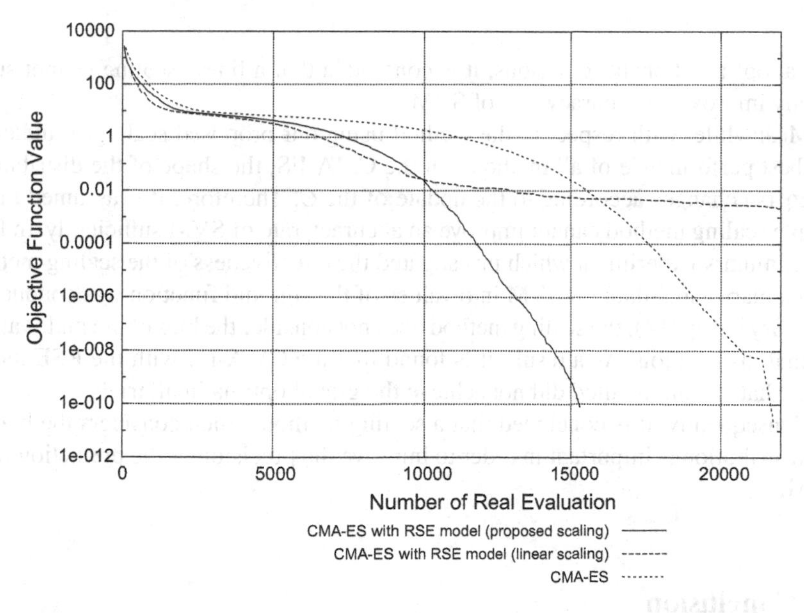

Fig. 5 Objective function value with fitness evaluations (for CMA-ES using RSE model with the Rosenbrock function)

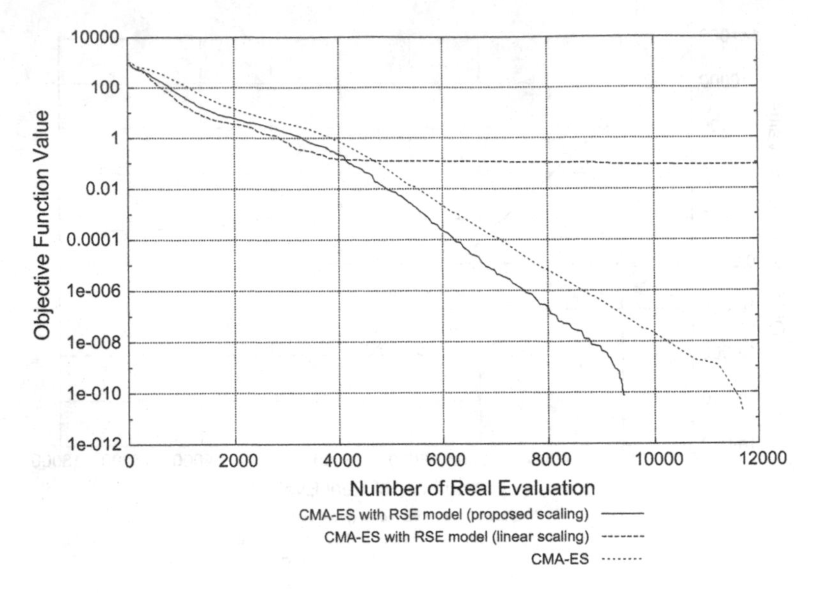

Fig. 6 Objective function value with fitness evaluations (for CMA-ES using RSE model with the Bohachevsky function)

global optima. For those reasons, it is concluded that a linear scaling cannot sufficiently improve an accuracy rate of SVM.

Meanwhile, with respect to the method using our proposed scaling, it indicates the best performance of all methods. In the CMA-ES, the shape of the distribution in Eq. (4) changes according to the update of the $\boldsymbol{C}$. Therefore, it is assumed that a simple scaling method cannot improve an accuracy rate of SVM sufficiently. In fact, a preliminary experiment which investigated the effectiveness of the scaling method which maps input data for SVM into values of the sigmoid function was conducted. Contrary to Eq. (14), the scaling method does not consider the bias of the multivariate normal distribution. As a result, it is found that the CMA-ES with the RSE model using that scaling method did not achieve the global optima in all trials.

Consequently, it is concluded that a scaling method which considers the bias of the distribution is important in order to improve the precision of the prediction from SVM.

6 Conclusion

This paper demonstrated the effectiveness of our proposed surrogate model:RSE model by means of CMA-ES with it. We also proposed the algorithm to introduce the RSE model to the CMA-ES and the scaling method which consider the bias in orientation of principle components of the distribution in Eq. (4) in order to improve the performance of machine learning part in FLLEC.

Moreover, the computer simulations were carried out so as to confirm the effectiveness of our proposed method. The results of these computer simulations show a satisfactory performance for FLLEC with CMA-ES.

Important future works are as follows.

- To confirm what a fitness landscape learned in training for SVM is.
- To devise a better training method for SVM in order to predict a fitness landscape correctly.
- To investigate the effectiveness of kernel functions of SVM for various optimization problems.
- To apply our method to optimization problems which involve constraints.

Acknowledgments A part of this work was supported by JSPS KAKENHI Grant, Grant-in-Aid for Scientific Research(C), 26330282, by JSPS KAKENHI Grant, Grant-in-Aid for JSPS Fellows, 16J10941, and by Program for Leading Graduate Schools of Ministry of Education, Culture, Sports, Science and Technology in Japan. We are grateful to Dr. Ilya Loshchilov for helpful discussions.

References

1. Y. Jin. A comprehensive survey of Ltness approximation in evolutionary computa- tion. Soft Comput., Vol. 9, No. 1, pp. 3–12, January 2005.
2. Taku Hasegawa, Kaname Matsumura, Kaiki Tsuchie, Naoki Mori, and Keinosuke Matsumoto. Novel virtual Ltness evaluation framework for Ltness landscape learning evolutionary computation. In Proceedings of the 2014 Conference Companion on Genetic and Evolutionary Computation Companion, GECCO Comp '14, pp. 107– 108, New York, NY, USA, 2014. ACM.
3. Kalyan Shankar Bhattacharjee and Tapabrata Ray. An Evolutionary Algorithm with ClassiLer Guided Constraint Evaluation Strategy for Computationally Expensive Optimization Problems, pp. 49–62. Springer International Publishing, Cham, 2015.
4. N. Hansen. The cma evolution strategy: A comparing review. Evolutionary Computation, pp. 75–102, 2006.
5. A. Ostermeier, A. Gawelczyk, and N. Hansen. Step-size adaption based on non-local use of selection information. In Proceedings of the International Conference on Evolutionary Computation. The Third Conference on Parallel Problem Solving from Nature: Parallel Problem Solving from Nature, PPSN III, pp. 189–198, London, UK, UK, 1994. Springer-Verlag.
6. N. Hansen, S. Muller, and P. Koumoutsakos. Reducing the time complexity of the derandomized evolution strategy with covariance matrix adaptation (cma-es). Evol. Comput., Vol. 11, No. 1, pp. 1–18, March 2003.
7. Y. Akimoto, J. Sakuma, I. Ono, and S. Kobayashi. Proposal and evaluation of functionally specialized cma-es. Vol. 24, pp. 58–68. The Japanese Society for ArtiLcial Intelligence, 2009.
8. Zongzhao Zhou, Yew S. Ong, Prasanth B. Nair, Andy J. Keane, and Kai Y. Lum. Combining global and local surrogate models to accelerate evolutionary optimization. Systems, Man, and Cybernetics, Part C: Applications and Reviews, IEEE Transactions on, Vol. 37, No. 1, pp. 66–76, 2007.
9. Yaochu Jin, Markus Olhofer, and Bernhard Sendhoff. On evolutionary optimization with approximate Ltness functions. In D. Whitley et al., editor, Proceedings of the Genetic and Evolutionary Computation Conference GECCO, pp. 786–793. Morgan Kaufmann, 2000.

An Evolutionary Simulating Annealing Algorithm for Google Machine Reassignment Problem

Ayad Turky, Nasser R. Sabar and Andy Song

Abstract Google Machine Reassignment Problem (GMRP) is a real world problem proposed at ROADEF/EURO challenge 2012 competition which must be solved within 5 min. GMRP consists in reassigning a set of services into a set of machines for which the aim is to improve the machine usage while satisfying numerous constraints. This paper proposes an evolutionary simulating annealing (ESA) algorithm for solving this problem. Simulating annealing (SA) is a single solution based heuristic, which has been successfully used in various optimisation problems. The proposed ESA uses a population of solutions instead of a single solution. Each solution has its own SA algorithm and all SAs work in parallel manner. Each SA starts with different initial solution which can lead to a different search path with distinct local optima. In addition, mutation operators are applied once the solution cannot be improved for a certain number of iterations. This will not only help the search avoid being trapped in a local optima, but also reduce computation time. Because new solutions are not generated from scratch but based on existing ones. This study shows that the proposed ESA method can outperform state of the art algorithms on GMRP.

Keywords Machine Reassignment Problem · Simulating annealing · Cloud computing · Evolutionary algorithm

A. Turky (✉) · A. Song
School of Computer Science and I.T., RMIT University, Melbourne, Australia
e-mail: ayad.turky@rmit.edu.au

A. Song
e-mail: andy.song@rmit.edu.au

N.R. Sabar
Queensland University of Technology, Brisbane, Australia
e-mail: nasser.sabar@qut.edu.au

G. Leu et al. (eds.), *Intelligent and Evolutionary Systems*,
Proceedings in Adaptation, Learning and Optimization 8,
DOI 10.1007/978-3-319-49049-6_31

1 Introduction

Cloud computing is a fast growing area which aims to provide online resources including storage, processing and network bandwidth to meet computational needs [1, 2]. Service providers like Google and Amazon need to find an efficient way to manage their large-scale data centers and ensure high quality services for large cohort of end users. Optimising resource allocation becomes more and more important in this industry [2]. A recent problem in the public domain is the Google Machine Reassignment Problem (GMRP) proposed at ROADEF/EURO challenge 2012 competition [3]. It is a combinatorial optimisation problem, aiming to improve resource utilization by reassigning a set of processes across a pool of servers while satisfying a set of constraints. A range of algorithms have been proposed for GMRP in the literature. These include simulated annealing [4], variable neighbourhood search [5], constraint programming-based large neighbourhood search [6], large neighbourhood search [7], multi-start iterated local search [8] and restricted iterated local search [9].

In this work, we propose an evolutionary simulating annealing (ESA) algorithm for GMRP. Simulating annealing (SA) is a single solution based heuristic, which has been successfully used in various optimisation problems. It is a probabilistic local search method that allows the acceptance of worse solutions in order to escape from the local optima. It starts with an initial solution and then generated a neighbourhood solution. The generated neighbourhood solution will be replaced with initial one if the solution quality is better or it satisfies the acceptance probability. In contrast the proposed ESA uses a population of solutions instead of a single solution. Hence, each solution has its own SA procedure. All SA procedures work in parallel. Each SA starts with different initial solution which can leads to a different search path with distinct local optima. Furthermore a range of mutation operators are applied once the solution cannot be improved for a certain number of iterations. This will not only help the search avoid being trap in a local optima, but also reduce the computational cost, because solutions are not built from scratch but based on existing solutions.

The performance of the proposed ESA algorithm is evaluated using 20 instances of the machine reassignment problem from ROADED/EURO 2012 challenge. These instances are very diverse in terms of size and data characteristics. To verify the effectiveness of our proposed algorithm, state of the art algorithms are included for comparison. Our aim is to have an algorithm outperform simulated annealing as well as algorithms designed for GMRP.

The rest of the paper is organised as follows: Sect. 2 presents the problem description. Section 3 describes the proposed algorithm in detail. Section 4 shows the experiment settings. The results are reported in Sect. 5. Finally, the conclusions are presented in Sect. 6.

2 Problem Description

GMRP is a recent problem proposed at ROADEF/EURO challenge 2012 [3]. It is a combinatorial optimisation problem. The main elements of this problem are a set of machines M and a set of processes P. The goal of this problem is to find the optimal way to assign process $p \in P$ to machines $m \in M$ in order to improve the usage of a given set of machines. One machine consists of a set of resources such as CPUs and RAM. One process can be moved from one machine to another in order to improve overall machine usage. The allocation of processes must not violate the following hard constraints:

- *Capacity constraints*: the sum of requirements of resource of all processes does not exceed the capacity of the allocated machine.
- *Conflict constraints*: processes of the same service must be allocated into different machines.
- *Transient usage constraints*: if a process is moved from one machine to another, it requires adequate amount of capacity on both machines.
- *Spread constraints*: the set of machines is partitioned into locations and processes of the same service should be allocated to machines in a number of distinct locations.
- *Dependency constraints*: the set of machines are partitioned into neighbourhoods. Then, if there is a service depends on another service, then the process of first one should be assigned to the neighbouring machine of second one or vice versa.

A feasible solution to GMRP is a process-machine assignment which satisfies all hard constraints and minimises the weighted cost function as much as possible. The cost can be calculated as follows:

$$\begin{aligned} f = &\sum_{r\in R} weight_{loadCost}(r) \times loadCost(r) \\ &+ \sum_{b\in B} weight_{balanceCost}(b) \times balanceCost(b) \\ &+ weight_{processMoveCost} \times processMoveCost \\ &+ weight_{serviceMoveCost} \times serviceMoveCost \\ &+ weight_{machineMoveCost} \times machineMoveCost \qquad (1) \end{aligned}$$

where R is a set of resources, *loadCost* represents the used capacity by resource r which exceeds the safety capacity, *balanceCost* represents the use of available machine, *processMoveCost* is the cost of moving a process from its current machine to a new one, *serviceMoveCost* represents the maximum number of moved processes over services and *machineMoveCost* represents the sum of all moves weighted by relevant machine cost. $weight_{loadCost}$, $weight_{balanceCost}$, $weight_{processMoveCost}$, $weight_{serviceMoveCost}$ and $weight_{machineMoveCost}$ define the importance of each individual cost.

For more details about the constraints, the costs and their weights can be found on the challenge documentation [3]. Note that the quality of a solution is evaluated by the given solution checker. Our proposed algorithm takes the returned value from the checker as the fitness measure of a solution. Another important aspect of this challenge is the time limit. It was stated that *"The maximum execution time will be fixed to 5 min by instance on a core2duo E8500 3.16Mhz with 4Go RAM on debian 64 or Win7 64 bits."* All methods have to finish within 5 min. This is to ensure the fairness of comparison.

3 Methodology

SA algorithm is a probabilistic local search method proposed by [10] that uses an initial solution as a starting basis and then generated a neighbourhood solution. The generated neighbourhood solution will be replaced with initial one if it better in term of the quality or satisfy the acceptance probability. Instead of using a single solution, we use a population of solutions and each solution has its own SA algorithm and all SAs work in parallel manner. Each SA starts with different initial solution which can leads to a different search path with distinct local optima. In addition, in order to help the search to avoid being trap in a local optima point and to reduce the computation time, a different mutation operators are applied once the solution cannot be improved for a certain number of iterations.

The overview of the proposed single SA algorithm is shown in Fig. 1. It first sets the parameter values (Sect. 3.1), generate the initial solution (Sect. 3.2) and calculate the fitness value (Sect. 3.2). For GMRP, Google provides an initial solution for every problem instance [3]. For each SA, the initial solution is generated by randomly modifying this solution. Next, the algorithm calls the SA procedure (Sect. 3.3) to generate better solution by iteratively modifying the generated initial solution. If the solution cannot be improved for a certain number of iterations, a mutation operator (Sect. 3.4) is applied. Otherwise, the algorithm checks the termination condition. If that is satisfied, the process stops and returns the best solution. Otherwise, the fitness value (Sect. 3.5) of the resulting solution is calculated. The solution update strategy is then applied to determine if it updates the current solution or not. This process will be repeated for a pre-defined number of iterations. Our proposed algorithm differs from the work in [11] in that we assign one SA for each solution whereas their idea is to parallelise and distribute the SA algorithm by dividing the main population into small subparts.

The following subsections discuss the main components of the proposed single SA algorithm.

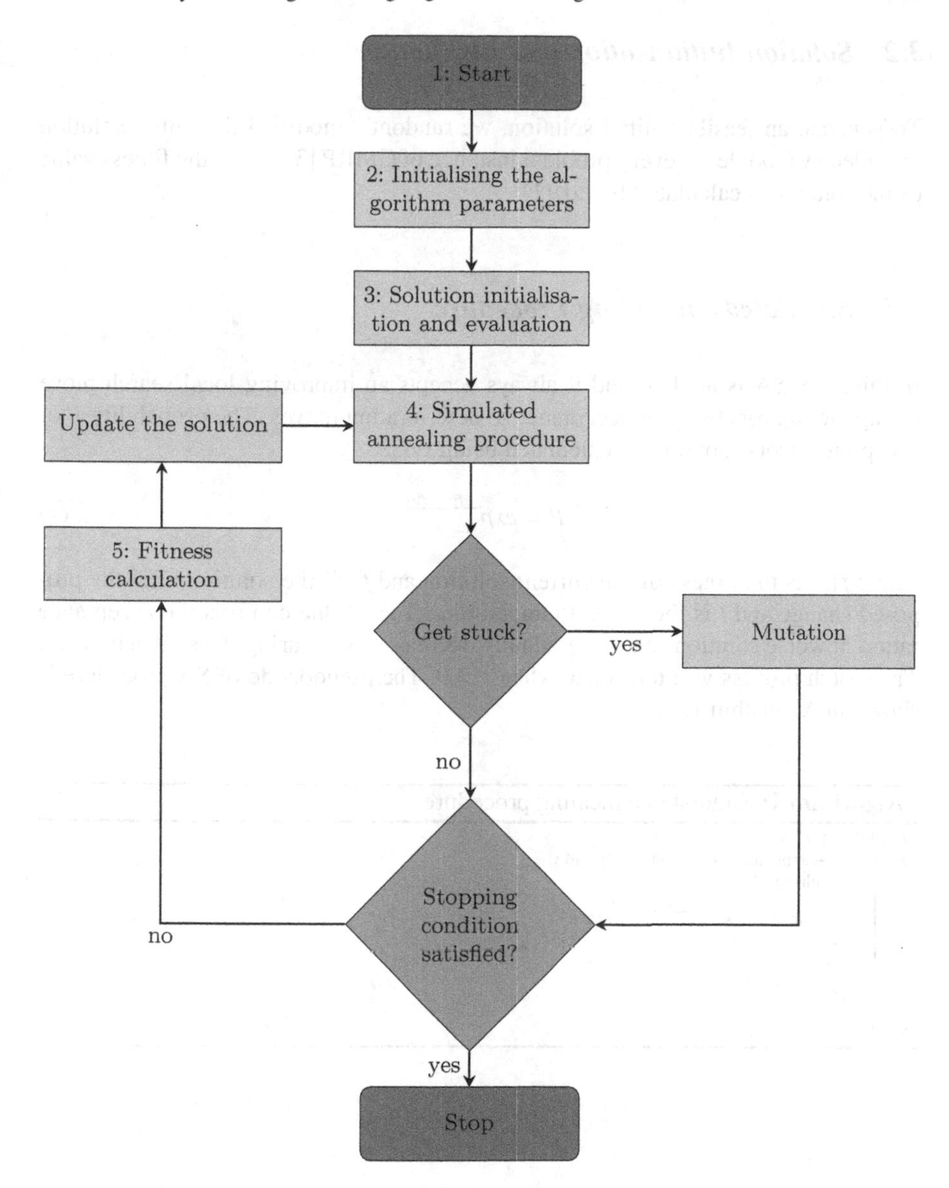

Fig. 1 The overall flowchart of the proposed single SA algorithm

3.1 Initialising the Algorithm Parameters

In this step, SA parameters are initialised which are initial temperature t, the cooling ratio α and final temperature t_f.

3.2 Solution Initialisation and Evaluation

To generate an feasible initial solution, we randomly modified the initial solution provided by Google for every problem instance of GMRP [3]. Next, the fitness value of the solution is calculated by Eq. (1).

3.3 Simulated Annealing Procedure

In this step, SA is applied and it always accepts an improving local search move but applies a probability of acceptance to deteriorating moves. The probability P of accepting a worse solution is calculated using Eq. 2

$$P = exp^{\frac{-(f(x')-f(x))}{t}} \tag{2}$$

where $f(x)$ is the fitness of the current solution and $f(x')$ the solution after the proposed change and t is the current temperature. The t value controls the acceptance ration of worse solutions and its gradually decreases by α during the search process. The search process will terminate when $t = 0$. The pseudocode of SA procedure is shown in Algorithm 1.

Algorithm 1: Simulated annealing procedure

```
1 while t_i > t_k do
2 |   x' ← apply neighbourhood operator to x;
3 |   calculate f(x');
4 |   if random[0, 1] < exp^(-(f(x')-f(x))/t) then
5 |   |   x' ← x;
6 |   end
7 |   t_{i+1} ← α * t_i;
8 |   i ← i + 1;
9 end
```

3.4 Mutation Operator

Mutation operator plays a big role on algorithm performance. Its main contribution is to help the search escape from a local optima point by providing a new starting point for the search. In this work, we use four different mutation operators. Each one is assigned to one SA algorithm. The reason of assigning different operators to each SA algorithm is because different problem instances and different stages of search react to mutation differently. These mutation operators are [8]:

- **Single swap**. It selects two processes from two different machines and interchanges them.
- **Double swap**. It selects four processes from two different machines and interchanges them.
- **Single move**. It selects a process from a machine and moves it to a different machine.
- **Double move**. It selects two processes from a machine and moves them to a different machine.

3.5 *Fitness Calculation*

The fitness value of a solution is calculated using the same equation shown in Eq. (1).

4 Experimental Settings

In this section, the GMRP instances used in our experiments are discussed. Parameter settings of the proposed ESA are also discussed.

4.1 *Problem Instances*

In total 20 different instances provided by Google for ROADEF/EURO 2012 challenge are used in the evaluation. They are divided into groups *a* and *b*. Theses instances have different characteristics in terms of number of machines, number of processes, neighbouring machines and so on. Table 1 shows the main characteristics of these instances. In the table, *R* is the number of processes; *TR* is the number of resources that need transient usage; *M* is the number of machines; *P* is the number of processes; *S* is the number of services; *L* is the number of locations; *N* is the number of neighbourhoods; *B* is number of triples and *SD* is the number of service dependencies.

4.2 *Parameters Settings*

The proposed ESA has three different parameters that need to be set by the user. These are: population size P ($P = 4$), the initial temperature t ($t = 10^8$) and the α ($\alpha = 0.7$). The t value controls the acceptance ration of worse solutions and its gradually decreases by α. The search process will terminate when $t = 0$. The suggested values of these parameters are obtained based on a preliminary test. We tested ESA with 31 independent runs using different parameter combination.

Table 1 The characteristics of the problem instances

Instance	R	TR	M	P	S	L	N	B	SD
a1_1	2	0	4	100	79	4	1	1	0
a1_2	4	1	100	1000	980	4	2	0	40
a1_3	3	1	100	1000	216	25	5	0	342
a1_4	3	1	50	1000	142	50	50	1	297
a1_5	4	1	12	1000	981	4	2	1	32
a2_1	3	0	100	1000	1000	1	1	0	0
a2_2	12	4	100	1000	170	25	5	0	0
a2_3	12	4	100	1000	129	25	5	0	577
a2_4	12	0	50	1000	180	25	5	1	397
a2_5	12	0	50	1000	153	25	5	0	506
b_1	12	4	100	5000	2512	10	5	0	4412
b_2	12	0	100	5000	2462	10	5	1	3617
b_3	6	2	100	20000	15025	10	5	0	16560
b_4	6	0	500	20000	1732	50	5	1	40485
b_5	6	2	100	40000	35082	10	5	0	14515
b_6	6	0	200	40000	14680	50	5	1	42081
b_7	6	0	4000	40000	15050	50	5	1	43873
b_8	3	1	100	50000	45030	10	5	0	15145
b_9	3	0	1000	50000	4609	100	5	1	43437
b_10	3	0	5000	50000	4896	100	5	1	47260

5 Results and Comparison

In this section, we evaluate the effectiveness of the proposed ESA by comparing it with SA and state of the art algorithms for GMRP. The approaches for comparison are:

1. **VNS**:Variable neighbourhood search [5].
2. **CLNS**: CP-based large neighbourhood search [6].
3. **LNS**: Large neighbourhood search [7].
4. **MILS**: Multi-start iterated local search [8].
5. **SA**: Simulated annealing [4].
6. **RILS**: Restricted iterated local search [9].

The computational results over 31 runs of the proposed ESA algorithm for all instances from both group *a* and group *b* are summarised in Table 2. The results are compared in term of the cost of the best solution and the search has to be finished within 5 min. In the table, the best result among all the algorithms is highlighted in bold. For GMRPs the lower the cost the better the solution. Table 2 reveals that,

Table 2 The results ESA compared to the state of the art algorithms

Instance	ESA	VNS	CLNS	LNS	MILS	SA	RILS
a1_1	**44, 306, 501**	**44, 306, 501**	**44, 306, 501**	44, 306, 575	**44, 306, 501**	44, 306, 935	–
a1_2	**777, 533, 310**	777, 536, 907	778, 654, 204	788, 074, 333	780, 499, 081	777, 533, 311	–
a1_3	**583, 005, 814**	583, 005, 818	583, 005, 829	583, 006, 204	583, 006, 015	583, 009, 439	–
a1_4	**251, 015, 178**	251, 524, 763	251, 189, 168	278, 114, 660	258, 024, 574	260, 693, 258	–
a1_5	727, 578, 311	**727, 578, 310**	727, 578, 311	727, 578, 362	727, 578, 412	727, 578, 311	–
a2_1	**167**	199	196	1, 869, 113	**167**	222	–
a2_2	**720, 671, 545**	720, 671, 548	803, 092, 387	858, 367, 123	970, 536, 821	877, 905, 951	–
a2_3	1, 194, 261, 501	**1, 190, 713, 414**	1, 302, 235, 463	1, 349, 029, 713	1, 452, 810, 819	1, 380, 612, 398	–
a2_4	**1, 680, 587, 592**	1, 680, 615, 425	1, 683, 530, 845	1, 689, 370, 535	1, 695, 897, 404	1, 680, 587, 608	–
a2_5	**310, 243, 641**	309, 714, 522	331, 901, 091	385, 272, 187	412, 613, 505	310, 243, 809	–
b_1	**3, 305, 899, 957**	3, 307, 124, 603	3, 337, 329, 571	3, 421, 883, 971	3, 516, 215, 073	3, 455, 971, 935	3, 511, 150, 815
b_2	**1, 015, 489, 174**	1, 015, 517, 386	1, 022, 043, 596	1, 031, 415, 191	1, 027, 393, 159	1, 015, 763, 028	1, 017, 134, 891
b_3	156, 978, 415	**156, 978, 411**	157, 273, 705	163, 547, 097	158, 027, 548	215, 060, 097	161, 557, 602
b_4	4, 677, 819, 354	4, 677, 961, 007	**4, 677, 817, 475**	4, 677, 869, 484	4, 677, 940, 074	4, 677, 985, 338	4, 677, 999, 380
b_5	**923, 299, 290**	923, 610, 156	923, 335, 604	940, 312, 257	923, 857, 499	923, 299, 310	923, 732, 659
b_6	**9, 525, 859, 941**	9, 525, 900, 218	9, 525, 867, 169	9, 525, 862, 018	9, 525, 913, 044	9, 525, 861, 951	9, 525, 937, 918
b_7	14, 835, 122, 181	**14, 835, 031, 813**	14, 838, 521, 000	14, 868, 550, 671	15, 244, 960, 848	14, 836, 763, 304	14, 835, 597, 627
b_8	**1, 214, 416, 703**	1, 214, 416, 705	1, 214, 524, 845	1, 219, 238, 781	1, 214, 930, 327	1, 214, 563, 084	1, 214, 900, 909
b_9	**15, 885, 545, 712**	15, 885, 548, 612	15, 885, 734, 072	15, 887, 269, 801	15, 885, 617, 841	15, 886, 083, 835	15, 885, 632, 605
b_10	18, 051, 241, 6401	**18, 048, 499, 616**	18, 049, 556, 324	18, 092, 883, 448	18, 093, 202, 104	18, 049, 089, 128	18, 052, 239, 907

the proposed ESA algorithm outperforms SA proposed by [4] on all 20 instances. In addition, ESA is very competitive compared to the reference algorithms. Specifically, the proposed ESA obtained new best results for 12 out of 20 tested instances and equal the best known result on two instances. Although Table 2 suggests that the proposed ESA could not find new best results for 8 instances, nevertheless, the difference is quite small.

A further analysis was carried out by conducting a statistical test to show the significant differences between our algorithm and the algorithms in the literature. As a statistical analysis, Friedman's test [12] was first applied, followed by Holm and Hochberg tests as post hoc methods (if significant differences are detected) to obtain the adjusted p-values for each comparison between the control algorithm (the best-performing one) and the rest. Table 3 summarises the ranking obtained by the Friedman's test. Note the averages are not available, hence the comparison is purely based on the best results.

Note that the RILS algorithm is not considered in the comparison due to the reason that this algorithm does not tested on set A instances.

Table 3 shows that ESA ranks first, followed by the VNS, CLNS, SA, MILS and LNS. The p-values computed by the Friedman's test is 0.000, which is below the significance interval of 95 % ($\alpha = 0.05$). This value shows that there is a significant difference among the observed results. Post hoc methods (Holm's and Hochberg's test) were also performed on the ESA algorithm. Table 4 shows the adjusted p-values (Friedman). We can see that Holm's and Hochberg's procedures reveal significant differences when using ESA algorithm as a control algorithm, where ESA algorithm

Table 3 Average ranking of Friedman test

#	Algorithm	Ranking
1	ESA	1.7
2	VNS	2.375
3	CLNS	3.175
4	SA	4
5	MILS	4.8
6	LNS	4.95

Table 4 Adjusted p-values (Friedman) of the compared methods

#	Algorithm	Unadjusted p	P Holm	P Hochberg
1	LNS	0	0	0
2	MILS	0	0.000001	0.000001
3	SA	0.000101	0.000304	0.000304
4	CLNS	0.01266	0.025319	0.025319
5	VNS	0.253887	0.253887	0.253887

is better than LNS, MILS, SA and CLNS algorithms, with $\alpha = 0.05$ (4/5 algorithms). Although the statistical tests show that the ESA is not better than VNS, the results in Table 2 shows that the ESA achieves better cost than VNS over 14 and 20 instances.

6 Conclusions

This paper introduced an evolutionary simulating annealing (ESA) algorithm for Machine Reassignment Problem from ROADEF/EURO 2012 challenge. ESA uses a population of solutions instead of a single one. Four SA algorithms search in parallel to solve the problem. Each SA starts with a different initial solution. Each leads to a different search path with distinct local optima. In addition, four different mutation operators are introduced once a solution cannot be improved for a certain number of iterations. This mutation mechanism not only helps the algorithm avoid getting stuck in a local optimum but also reduce computation time. In comparison with the state-of-the-art algorithms, ESA produces high-quality solutions and outperform other methods. We conclude that the ESA is an effective method for optimisation on Google machine reassignment problem.

References

1. Michael Armbrust, Armando Fox, Rean Griffith, Anthony D Joseph, Randy Katz, Andy Konwinski, Gunho Lee, David Patterson, Ariel Rabkin, Ion Stoica, et al. A view of cloud computing. *Communications of the ACM*, 53(4):50–58, 2010.
2. Rodrigo N Calheiros, Rajiv Ranjan, Anton Beloglazov, César AF De Rose, and Rajkumar Buyya. Cloudsim: a toolkit for modeling and simulation of cloud computing environments and evaluation of resource provisioning algorithms. *Software: Practice and Experience*, 41(1):23–50, 2011.
3. Roadef/euro challenge 2012: Machine reassignment. http://challenge.roadef.org/2012/en/.
4. Marcus Rolf Peter Ritt. *An Algorithmic Study of the Machine Reassignment Problem*. PhD thesis, UNIVERSIDADE FEDERAL DO RIO GRANDE DO SUL, 2012.
5. Haris Gavranović, Mirsad Buljubašić, and Emir Demirović. Variable neighborhood search for google machine reassignment problem. *Electronic Notes in Discrete Mathematics*, 39:209–216, 2012.
6. Deepak Mehta, Barry O'Sullivan, and Helmut Simonis. Comparing solution methods for the machine reassignment problem. In *Principles and practice of constraint programming*, pages 782–797. Springer, 2012.
7. Felix Brandt, Jochen Speck, and Markus Völker. Constraint-based large neighborhood search for machine reassignment. *Annals of Operations Research*, pages 1–29, 2012.
8. Renaud Masson, Thibaut Vidal, Julien Michallet, Puca Huachi Vaz Penna, Vinicius Petrucci, Anand Subramanian, and Hugues Dubedout. An iterated local search heuristic for multi-capacity bin packing and machine reassignment problems. *Expert Systems with Applications*, 40(13):5266–5275, 2013.
9. Ramon Lopes, Vinicius WC Morais, Thiago F Noronha, and Vitor AA Souza. Heuristics and matheuristics for a real-life machine reassignment problem. *International Transactions in Operational Research*, 22(1):77–95, 2015.

10. Scott Kirkpatrick, C Daniel Gelatt, Mario P Vecchi, et al. Optimization by simulated annealing. *Science*, 220(4598):671–680, 1983.
11. M Emin Aydin and Terence C Fogarty. A distributed evolutionary simulated annealing algorithm for combinatorial optimisation problems. *Journal of Heuristics*, 10(3):269–292, 2004.
12. Salvador García, Alberto Fernández, Julián Luengo, and Francisco Herrera. Advanced non-parametric tests for multiple comparisons in the design of experiments in computational intelligence and data mining: Experimental analysis of power. *Information Sciences*, 180(10):2044–2064, 2010.

Mobile Agent Based Obstacle Avoidance in Multi-robot Hunting

Shiyou Uehara, Munehiro Takimoto and Yasushi Kambayashi

Abstract Multi-robot hunting is a problem in which multiple robots cooperatively search for a target that emits signals in the all directions. Robots proceed in the guidance of the signals, and finally, some of them reach the target to capture. In order to solve the problem, an approach based on Particle Swarm Optimization (PSO), which is one of meta-heuristics, has been proposed. The PSO based approach is well known that it works well in fields with no obstacle. It is, however, not assumed to be used in practical situations with obstacles. In order to lift this restriction, we propose a new PSO based approach that enables particles search and capture the target while getting around the obstacles. In our approach, each robot records its moving trace in a fixed period. Once a robot is blocked by obstacles and cannot proceed, it creates a mobile software agent that migrates to other robots around it through Wi-Fi. The mobile software agent selectively migrates to some robots whose traces have some intersection points. Since a sequence of the traces gives a detour route through which a robot can go, the agent can just inform the detour route to its home robot. We have implemented a simulator based on our approach, and conducted experiments. The experimental results show that our approach is remarkably more effective than the original PSO based approach.

S. Uehara (✉) · M. Takimoto
Department of Information Sciences, Tokyo University of Science,
2641 Yamazaki, Noda, Tokyo 278-8510, Japan
e-mail: s-uehara@cs.is.noda.tus.ac.jp

M. Takimoto
e-mail: mune@cs.is.noda.tus.ac.jp

Y. Kambayashi
Department of Computer and Information Engineering, Nippon Institute of Technology,
4-1 Gakuendai, Miyashiro, Minami-Saitama 345-8501, Japan
e-mail: yasushi@nit.ac.jp

G. Leu et al. (eds.), *Intelligent and Evolutionary Systems*,
Proceedings in Adaptation, Learning and Optimization 8,
DOI 10.1007/978-3-319-49049-6_32

1 Introduction

Today, it is possible to address complicated problems by a sophisticated robot and solve the problems, while several efforts have been made to achieve the same problem-solving using multiple robots with relatively simple functionalities. Efforts have been made to control multiple robots based on Swarm Intelligence, which enables the multiple robots to achieve self-organized collective behaviors, while each robot behaves along the same simple algorithm. We call the multiple robots behaving based on swarm intelligence swarm robots.

One of the important collective behaviors of the swarm robots is hunting targets. In order to achieve target-hunting, some approaches have been proposed. Especially, Multi-robot Hunting based on Particle Swarm Optimization (PSO) has shown that no less than eight PSO based swarm robots can successfully find a target in an open space without collision [8]. Also, an extension of the PSO based approach has shown that it can save the total energy consumption and reduce the total communication costs between robots without sacrificing the success rate, using swarm robots controlled by mobile agents [4]. These PSO based huntings assume the work field has no obstacle so that the movements of robots are not blocked, and the view of the device such as a camera is always clear. The assumption would be too strong in terms of practical uses.

We propose a new multi-robot hunting approach that works in a field with obstacles. In our approach, robots behave based on the original PSO based approach. In the PSO, each particle, which corresponds to a robot, is given its own estimated value at the current location by a *fitness function*. The next location is determined by the estimation of its previous location and the current locations of the other particles. In this way, as time elapses, the swarm converges to the particle located at the most suitable position. On the other hand, robots have some volume unlike particles in PSO and hence, they would be strongly influenced by obstacles around them. The robots cannot easily avoid obstacles and have to make a detour based on just PSO. In order to solve this problem, we make mobile agents traverse robots with intersection points on their traces, which compose a detour route for the obstacles. The technique is based on the method proposed by Avilés et al. [1]. Each robot records its own trace as a log in a certain fixed period, so that each robot can always find some detour routes leading it to a positon beyond obstacles through its trace based on the migration manner of mobile agents. Therefore the swarm robots can find the target by simply following the PSO algorithm in a field with obstacles.

The rest of this paper is organized as follows. We describe the related works in the second section. We explain the fundamental concepts of PSO in the third section. We describe the details of the mobile agent system we use in the fourth section. We show the experimental results on a simulator we have implemented based on the proposed method and discuss about the results in the fifth section. Finally we conclude our discussion in the sixth section.

2 Related Works

Swarm Intelligence (SI) is the optimization method inspired by the behaviors of swarms in nature. Swarms that have been studied are ants, birds, insects, etc. and these swarms have been composed on computer as swarm models. SI based optimization is performed by the swarm of individuals located in search space. This swarm is generally composed of individuals of uniform agent and these individuals interact with environment as well as each other. Even though there is no central control system of the swarm, the chain of local interactions of neighborhood individuals triggers a large scale movement of the swarm as whole individuals. SI based optimization is the method to search optimal solution through the emergence of such a movement.

PSO is a typical algorithm of SI based optimization proposed by Kennedy and Eberhart [6]. In the PSO, particles are randomly located in the search space at the initialization phase and these particles compose a swarm. Each particle determines its next behavior by combining its own information and common information of the swarm. Thus all the particles slowly come near the most optimal position among the individuals. The fitness value of a position in search space is estimated by a function called "fitness function".

In the traditional PSO, a particle's movement is determined through calculating its velocity vector. Kennedy studied a method that removes the concept of velocity vector and then performs each particle's movement stochastically [5]. Kennedy proposed that each particle's next position is determined by creating a Gaussian random vector dimension-by-dimension since an observation of particles using standard vector strategy produced a bell-curved histogram of position being tested. This method is called Bare_Bones_Particle_Swarm (BBPS), and swarm following BBPS shows continuous improvement throughout the steps.

Bell et al. abstracted BBPS above and they called the method Particle_Field_Optimization (PFO) [2]. In PFO, each particle is always located at its own best position, and position updating at each step is not necessarily cover all the particles. A candidate point is generated as a Gaussian random vector from the Gaussian distribution of a particle probabilistically sampled from the swarm, and in this way a fixed number of candidate points at a step are generated. If there are candidate points generated from a particle and a candidate more suitable than the particle exists, then the particle move to the point. With suitable parameter setting, this method produces better solution than BBPS does.

There are quite a few multi-robot applications that take advantages of PSO. Zhu et al. propose an algorithm that guides mobile robots to search a target without precise global information [10]. They demonstrate that when some limited number of robots form a small group and explore cooperatively, the cooperation display remarkable performance. They also demonstrate that their multi-robot system is applicable to map-building through exploring unknown terrain [9]. For avoiding obstacles, Masehian and Sedighizadeh propose a variant of PSO that employs Probabilistic Roadmap Method [7]. They have reported that their method has merit on computational complexity. The most closely related research project to ours is the one conducted by Dadgar et al. [3]. They propose an extension of PSO called Adaptive Robotic PSO (A-RPSO) that provides control mechanism for multiple robots. Like our system, A-RPSO takes obstacle avoidance into consideration as well as has mechanism to escape from local optima.

3 PSO Based Hunting

PSO based hunting works by iteratively updating two vectors velocities and positions of particles that are initially scattered in a search field. These vectors are initialized to random values, and updated along the following equations.

$$\begin{aligned} V_i^{t+1} &= \omega * V_i^t \\ &+ C_1 * rand_1 * (P_i^t - X_i^t) \\ &+ C_2 * rand_2 * (P_g^t - X_i^t) \end{aligned} \tag{1}$$

$$X_i^{t+1} = X_i^t + V_i^{t+1} \tag{2}$$

In Eq. (1), t is the current iteration number, V_i^t and X_i^t are the velocity and the position vector of particle i in the current iteration, P_i^t is the most fitted location of the particle i, P_g^t is the location of the particle that is most closest to the target at current iteration. ω, C_1 and C_2 are constants, and $rand_1$ and $rand_2$ are two uniformly distributed random numbers in the range [0, 1].

We assume that robots corresponding to particles move in a square search field. We also assume that the target is located at the center of the field. The robots are located in the field at random manner.

The target emits signals in every direction. Robots are able to receive the signals and measure their strength, which is used as information for searching the target. The hunting successfully finishes once some robots capture the target, i.e. they reach the area within 30 units from the target.

Robots share the following information:

1. The position of the robot that is closest to the target
2. Strength of the signal that reaches the current location.

Each robot is equipped with a camera and a GPS device. It can detect other robots within 100 units from the current position through the camera if they are not hidden by any obstacle. On the other hand, it cannot directly recognize the target through any sensors including the camera. Also, each robot can know the precise information of its own position through the GPS device.

4 Agent Based Detour Detection

Once a robot collides with an obstacle, it stops moving, and then starts searching a detour to get around the obstacle. The robot periodically generates mobile agents and sends them to one of the neighbor robots at a time. In the process of searching a detour, we call this robot that generates agents the *start robot*. At this time, an agent can just migrate from the start robot to one of its neighbors that are reachable with Wi-Fi at the current position. Once the agent migrates to a neighbor robot, it checks a migration condition that is the GPS log of the robot has some intersection points with the GPS logs of the robots from which the agent has migrated(source robot), or the robot is within the view range of the source robot without being blocked by any obstacle. If the neighbor robot satisfies one of the two conditions, the agent records the position, velocity and GPS log of it, and try further migrations; otherwise it abandons the robot, and migrates another neighbor of the source robot to check the conditions. The agent repeats the migrations without revisiting the same robots until it reaches the target robot that satisfies the following target condition decided by parameters L and D that are arbitrarily or randomly defined in the experiments:

1. The total distance of a sequence of collected GPS logs, which include the straight lines between robots within the view range of the other, is more than L, and
2. The direction of the velocity of the target robot is within angle D around the direction of the velocity of the start robot.

The agent that has reached the target robot traverses the sequence of robots on which it has migrated, in reverse order to go back to the start robot. Once the agent is back to the start robot, it gives the start robot the collected sequence of GPS logs as a result (Fig. 1).

Immediately after the start robot receives the result of searching a detour searching, it stops generating agents and composes a detour route based on the GPS logs. Finally, the start robot move toward the position where the target robot is along the

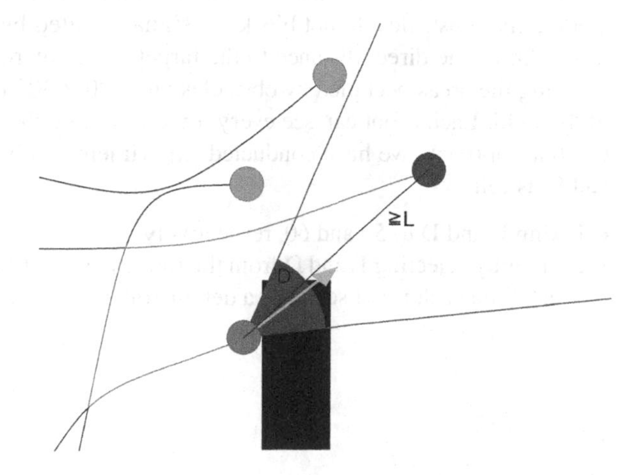

Fig. 1 Screenshot of robots' relative position. *Pink* node is the robot generating agents, *orange arrow* is the velocity vector that agent remember, the candidate goal area is composed of the *white* area surrounded by the *blue lines*. The deep *blue* node is the target robot

detour route, and then, restarts hunting based on PSO algorithm. Since the detour route is a chain of GPS logs with intersection points or the straight lines without any obstacle between robots that are within the view range of the other, the start robot can move along the detour route without being blocked by any obstacle.

The robot blocked by an obstacle also keeps calculating their own velocity in searching a detour routes. If the robot fortunately turns toward the direction of non-obstacle area before finding a detour route, the robot stops detour searching and resumes the searching behaviors following the PSO algorithm, immediately.

5 Numerical Experiments

We have implemented a simulator in order to demonstrate the effectiveness of our approach. Figure 2 is the screenshot of our simulator. In this figure, deep blue nodes represent robots that follow our PSO algorithm to move, light blue nodes are robots in search of detour route, pink nodes are robots taking a detour, and yellow lines are GPS logs of each robot, and red lines are detour routes that pink nodes take at current iteration. In this simulation, we measure the capability of our hunting approach with the different number of robots.

5.1 Simulator

This simulation field, which is depicted in Fig. 2, has 600×600 size and obstacles are located in centrosymmetric manner. Only one target is located at the center of this field, of which the coordinate is unknown to the searching robots. This target emits signal in every direction, which robots can receive, and measure the distance to the target based on its strength of the signal at the position. In the field, robots can move in a white area and cannot get into the black areas that represent obstacles. We assume that obstacles do not block the signal emitted by the target, through which robots know the direct distance to the target. Initially, robots are randomly located avoiding the areas occupied by obstacles and 300×300 rectangle area at the center of the field. Each robot can see every direction through an omnidirectional camera. For our approach, we have conducted experiments with two kinds of settings of L and D as follows:

- Fixing L and D to 50 and 60, respectively.
- Randomly selecting L and D from the ranges [30, 100] and [30, 180] respectively, every time each robot searches a detour route.

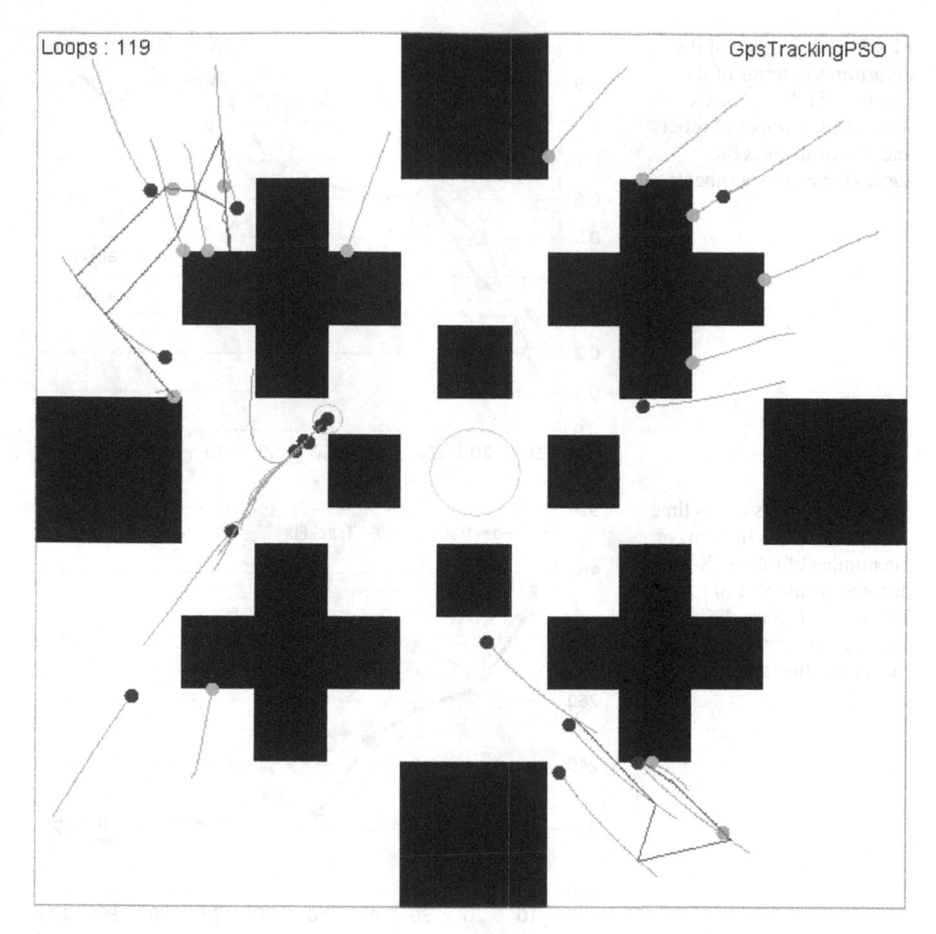

Fig. 2 Screenshot of the simulator

The settings of the other parameters were as follows:

- $\omega = 0.9$, $C_1 = 2.0$, $C_2 = 2.0$ on Eq. 1
- Robots searching a detour route periodically generate agents per 100 steps.
- The maximal number of steps in a simulation is 600.

5.2 *Results of the Experiments*

In the simulation, the number of robots is changed in the range of [10, 100] in units of 10, and the simulation is performed 50 times for each case.

Figure 3 shows the success rates in our approaches and traditional PSO based approach. Figure 4 shows the average number of time units taken until each simula-

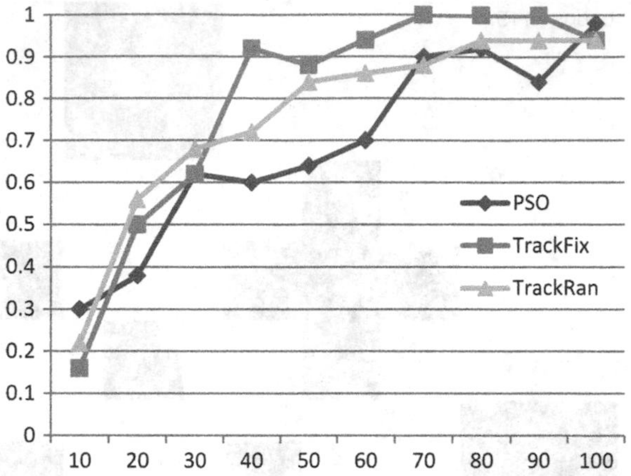

Fig. 3 Success rate of the algorithms in terms of the number of robots. X-axis denotes the number of robots and Y-axis denotes the success rate of experiments

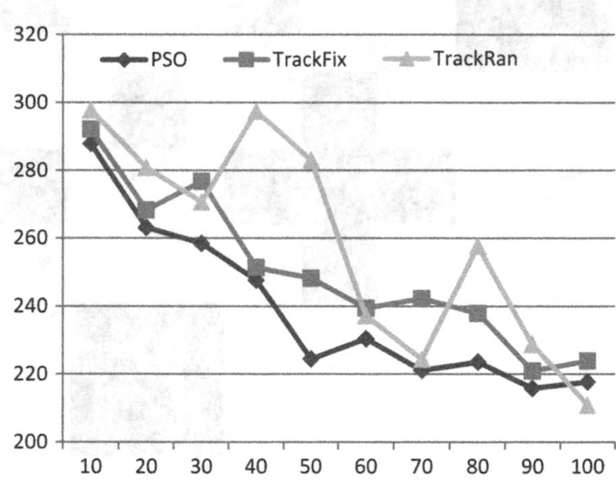

Fig. 4 Average success time of the algorithms in terms of the number of robots. X-axis denotes the number of robots and Y-axis denotes the average time spent until robots find the target

tion is successfully completed. We can observe that the traditional approach makes a robot stop moving without restarting once it collides with an obstacle.

In these figures, results of the traditional approach and our approaches with fixed parameters and randomly set parameters are shown by "PSO", "TrackFix", and "TrackRan", respectively. As shown in Fig. 3, TrackFix is the most successful i.e. 90 % success, when the number of robots is more than or equal to 40, while success rate of PSO is 60 % for 40 robots. On the other hand, as shown in Fig. 4, TrackFix roughly takes 20 more steps than traditional approach in order to reach the point improving success rate. TrackRan tends to be more successful with smaller number of robots but its success time is not stable.

Our approaches for both parameter settings give mostly better results than the traditional approach. In details, in the case where sufficiently many robots in search field, TrackFix tends to have an advantage, since each robot can steadily compose a detour route. Conversely, it is hardly able to find any target robot for composing a detour route for the small number of robots. In such a case, TrackRan has an advantage because the condition of parameters L and D may be weakened by random selection, giving some opportunities for searching a detour route in wider area.

It is worth investigating the roles that the detour routes play in our approach. The most effective role of them is to lead the robots that are blocked by obstacles. In the traditional approach, only a few robots can travel a long distance without colliding with obstacles because each robot does not take the obstacles into consideration. That is, taking advantage of the traditional approach in a search field with obstacles is difficult and we need to introduce new means to cope with this obstacle-present situation. In other words, the obstacles make robots as particles in PSO less efficient and less accurate. In contrast, since our approach rescues the robots that are blocked by obstacles and makes them re-active as particles in PSO, it makes robots more efficient and makes go further toward the target than those in the traditional PSO.

6 Conclusions

In this paper, we have proposed a novel multi-robot hunting approach based on PSO and GPS tracking. This approach leads the robots blocked by obstacles in front of them to detour routes. The detour routes are composed with GPS logs of adjacent robots that are collected by mobile software agents. In other words, our approach rescues the blocked robots and makes them re-activate as particles of PSO. We have implemented a simulator to demonstrate the effectiveness of our approach, and have conducted numerical experiments on it. As a result, we have shown that our approach remarkably increase success rate of searching a target, while it requires a little more iterations to complete the mission than the traditional PSO based approach.

As future directions, we are investigating other environmental parameters that are practically required, and the effects of the arrangements of obstacle map, or the topology of PSO.

Acknowledgments This work was supported by Japan Society for Promotion of Science (JSPS), with the basic research program (C) (No. 26350456), Grant-in-Aid for Scientific Research.

References

1. Avilés, A., Takimoto, M., Kambayashi, Y.: Distributed evacuation route planning using mobile agents. Transactions on Computational Collective Intelligence XVII pp. 128–144 (2014)
2. Bell, N., Oommen, B.J.: Particle field optimization: A new paradigm for swarm intelligence. In: Proceedings of the 14th International Conference on Autonomous Agents and Multiagent Systems. pp. 257–265 (2015)
3. Dadgar, M., Jafari, S., Hamzeh, A.: A pso-based multi-robot cooperation method for target searching in unknown environments. Neurocomputing 177, 62–74 (2016)
4. Ishiwatari, N., Sumikawa, Y., Takimoto, M., Kambayashi, Y.: Multi-robot hunting using mobile agents. In: Agent and Multi-Agent Systems: Technologies and Applications. pp. 223–232. Springer International Publishing (2014)

5. Kennedy, J.: Bare bones particle swarms. IEEE Swarm Intelligence Symposium pp. 80–87 (2003)
6. Kennedy, J., Eberhart, R.: Particle swarm optimization. Proceedings of IEEE International Conference on Neural Networks 4, 1942–1948 (1995)
7. Masehian, E., Sedighizadeh, D.: An improved particle swarm optimization method for motion planning of multiple robots. In: Distributed Autonomous Robotic Systems, pp. 175–188. Springer Tracts in Advanced Robotics vol. 83 (2013)
8. Nighot, M.K., Patil, V.H., Mani, G.S.: Multi-robot hunting based on swarm intelligence. Hybrid Intelligent Systems (HIS), 2012 12th International Conference pp. 203–206 (4-7 Dec 2012)
9. Zhou, Y., Xiao, K., Wang, Y., Liang, A., Hassanien, A.E.: A pso-inspired multi-robot map exploration algorithm using frontier-based strategy. International Journal of System Dynamics Applications 2, 1–13 (2013)
10. Zhu, Q., Liang, A., Guan., H.: A pso-inspired multi-robot search algorithm independent of global information. In: Proceedings of IEEE Symposium on Swarm Intelligence. pp. 1–7 (2011)

Communication-Less Cooperative Q-Learning Agents in Maze Problem

Fumito Uwano and Keiki Takadama

Abstract This paper introduces a reinforcement learning technique with an internal reward for a multi-agent cooperation task. The proposed method is an extension of Q-learning which changes the ordinary (external) reward to the internal reward for agent-cooperation under the condition of no communication. To increase the certainty of the proposed methods, we theoretically investigate what values should be set to select the goal for the cooperation among agents. In order to show the effectiveness of the proposed method, we conduct the intensive simulation on the maze problem for the agent-cooperation task, and confirm the following implications: (1) the proposed method successfully enable agents to acquire cooperative behaviors while a conventional method fails to always acquire such behaviors; (2) the cooperation among agents according to their internal rewards is achieved no communication; and (3) the condition for the cooperation among any number of agent is indicated.

Keywords Multi-agent system · Q-learning · Cooperation · Theory

1 Introduction

Multi-agent reinforcement learning (MARL) is a useful approach to tackle multi-agent cooperation tasks, such as multi-robot cooperation and traffic signal control [1–6]. However, a MARL has a difficulty of deriving good performance without communication among the agents because the agents do not know how they cooperate with each other [7, 8] due to the fact that the behavior of agents affect the behavior of other agents and vice versa. To address this issue, most of the conventional methods such as Tan's research [9] promote agents to cooperate with each

F. Uwano (✉) · K. Takadama
The University of Electro-Communications, 1-5-1 Chofugaoka Chofu,
Tokyo W6-309, Japan
e-mail: uwano@cas.hc.uec.ac.jp

K. Takadama
e-mail: keiki@inf.uec.ac.jp

G. Leu et al. (eds.), *Intelligent and Evolutionary Systems*,
Proceedings in Adaptation, Learning and Optimization 8,
DOI 10.1007/978-3-319-49049-6_33

other by using other agents' information through communication. Such information is very useful for the cooperation among the agents, but it goes without saying that the agents cannot guarantee to acquire all information required to cooperate with each other through communication. From this fact, it is important to explore a reinforcement learning method based on "no communication" for the cooperation among the agents.

For this issue, the conventional methods can be classified with the following two viewpoints: (1) the agents learn their behaviors empirically or theoretically; (2) information of other agents can be (partially) shared or not to cooperate with each other. As the empirically approach (not theoretical approach) employing information of other agents, Ono et al. proposed the modular Q-learning that combines the modular architecture with Q-learning and showed that the modular Q-learning agents succeeded to synthesize decision policies for cooperation among the agents by using information of other agents [10]. However, the cooperative behavior cannot be guaranteed because the decision policies are determined heuristically. As the empirically approach without sharing information of other agents, Iima et al. proposed the swarm reinforcement learning that improves the performance of the collective agents [3]. In this method, the agents do not need the information of other agents, but the optimal behavior of the agents cannot be guaranteed to be acquired. As the theoretical approach employing information of other agents, on the other hand, Elidrisi et al. proposed the fast adaptive learning in the stochastic game (FAL-SG) [5], which enables the agents to choose the optimal action by observing their actions each other. However, this method requires the complete information of all agents for the cooperation among them.

Compared with these methods, it is clear that there is no method based on the theoretical approach without sharing information of other agents, even though such methods are really needed as the number of agents increases. To tackle this issue, this paper aims to explore the theoretical approaches for cooperation among agents which do not require the communications during the learning. For this purpose, we propose the "internal reward" which is changed from the ordinary (external) reward. Specifically, the agents in our approach theoretically learn their behaviors according to the internal reward instead of the ordinary (external) reward. For this issue, we employ Q-learning agents in this paper because Q-learning is well studied and analyzed mathematically (i.e., the convergence of Q-value is proofed in the single agent environment [11]).

This paper is organized as follows. Section 2 describes the multi-agent cooperation task addressed in this paper. Our method is proposed in Sect. 3. Section 4 conducts the experiment and analyzes the obtained results. Finally, our conclusion is given in Sect. 5.

2 Multi-agent Cooperation Task

2.1 *Cooperation on Maze Problem*

This section introduces a multi-agent cooperation task using a 3×8 grid maze problem. Figure 1 shows an example of 3×8 grid maze. On the maze this paper uses, as shown in the figure, we define there are two possible start states (A, B) where agents will be initially placed before learning; and two goal states (S, L) where the agents attempts to reach. A difficulty of multi-agent cooperation task on the maze, is each agent attempts to selfishly learn the minimum step for him that does not find a cooperating behavior with other agents. For instance, on the maze problem shown by Fig. 1, when two agents start from states A and B respectively, they attempt to reach the same goal state S since their minimum steps to goal can be achieved by reaching it. In this case, the agent starting from the state B can potentially reach goal state S faster than another agent starting from the state A. This result suggests that the agent starting from the state A should reach the goal state L. This is the best (but selfish) solution for the agent starting from the state B while the worst solution for agent starting from the state A since he should take the longest step to reach goal. The cooperating behavior can be selected as; the agents starting from state A and B reach the goal state S and L respectively. This difficulty is often called as a dilemma problem.

2.2 *Definition of the Cooperation Task*

For the theory, we define the cooperation task of maze problem mathematically by using minimum steps from start states to goal states. Note that two Q-learning agents are employed and we assume that all minimum steps are different from each other. In this section, we define t_{AS}, t_{AL}, t_{BS} and t_{BL} as the minimum steps. The minimum step to Goal S for agent A indicates t_{AS}, and the other steps indicate at the same way. When the agents starting from states A and B select goal states S and L respectively after learning, if this situation is the dilemma problem, the minimum step spend by this selection is larger than another selection because either agent spends long

Fig. 1 Maze Problem

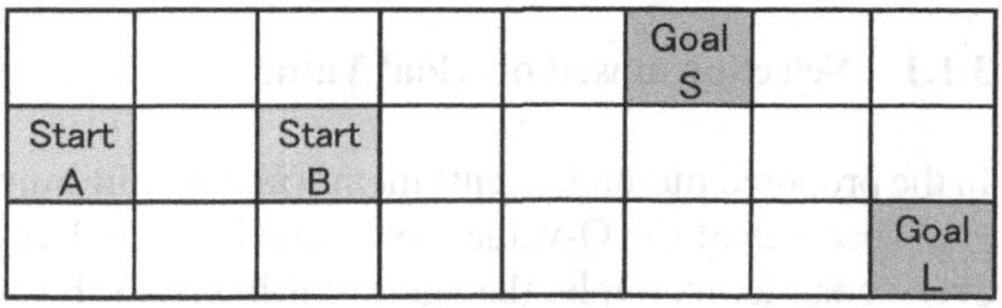

					Goal S		
Start A		Start B					
							Goal L

steps. Specifically, if the agent starting from the state A cannot reach the goal state L and reaches the goal state S because of another agent, the following equation is established.

$$t_{AL} < t_{AS} \tag{1}$$

$$t_{BL} < t_{BS} \tag{2}$$

$$t_{AL} > t_{BL} \tag{3}$$

$$t_{AS} > t_{BS} \tag{4}$$

the Eqs. 1 and 2 indicates the selfish results of both agents, and the Eq. 3 indicates that the agent starting from the state A cannot reach the goal state L because of another agent starting from the state B. Since t_{BL} is shortest from these equations, if this is dilemma problem, t_{AS} is bigger than t_{BS} (Eq. 4) because, by yielding of the agent starting from the state B, the agents starting from state A and B reach the goal state S and L respectively in minimum step shorter than that in this selection as another selection.

Therefore, if this situation is the dilemma problem, the Eqs. 1, 2, 3 and 4 are established. In addition, the equations same as these equations can be established by the same way in other situations.

3 Proposed Method

In the dilemma maze problem like Fig. 1, in order to achieve the cooperation of agents, we proposed the learning method as shown in Fig. 2. Agents repeats the learning cycle (observe states, select action, state transition, learn) on each square (state) of this maze. The agents obtain reward at the goal point and calculate Q-value from the reward during this cycle. The proposed method enables the agents to predict the optimal goals for cooperation, and get the agents to have own reward (internal reward) to reach the selected goals. The remain of this section explains the mechanism of the cooperation method (goal selection and internal reward design).

3.1 Goal Selection

3.1.1 Selection Based on Goal Value

In the proposed method, agents memorize the minimum steps from own start to every goal (because of the Q-value table), and have goal values which indicate priority of every goal, for example, the agent starting from the state A has this Table 1. In this table, bid_{AS} and bid_{AL} indicates the goal values for the goal state S and L respectively. The agent starting from the state B has the thing same as this Table 1, and bid_{BS} and

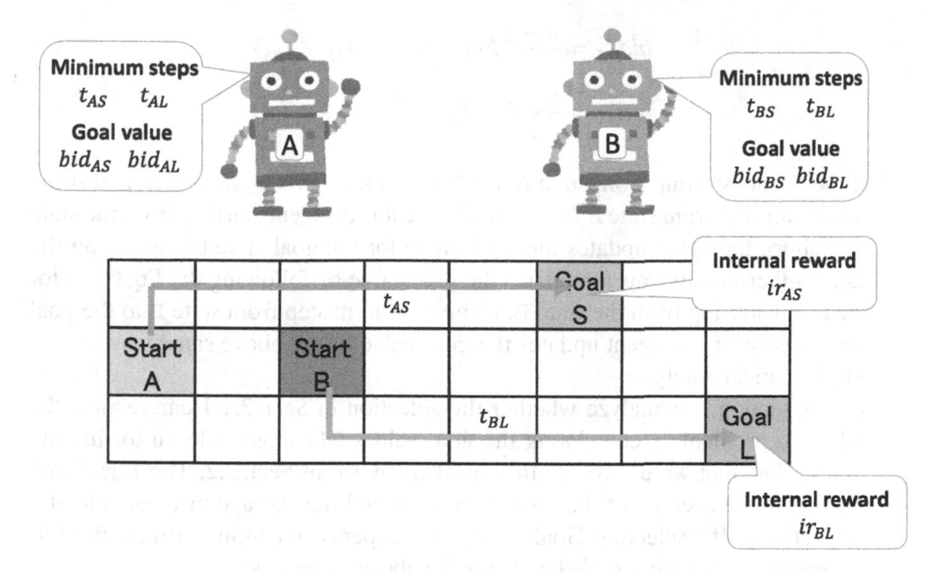

Fig. 2 Proposed method

Table 1 Table memorized by agent at state A

	Goal S	Goal L
Minimum step	t_{AS}	t_{AL}
Goal value	bid_{AS}	bid_{AL}

bid_{BL} indicates the goal values for the goal state S and L. The goal value is the real number and indicates whether the agent select the goal. In this process, the agents select the goal states by using the goal values and update the goal values from the result of this selection.

1. Selecting Goal
   First, the agents select the goals from the goal values. the agents select the goal which has the largest goal value, for example, if $bid_{AS} < bid_{AL}$ is established, the agent starting from the state A selects the goal state L. After that, in the `process 9`, the agent estimates internal reward to reach the goal selected in this process. Note that, in implementation, the agents can select the goals randomly in some probability to update the goal values evenly.
2. Update the Goal Value
   Second, the agents update the goal values from the result of above selection. The agents update by following the equations same as the Eqs. 5 and 6 for the agent starting from the state A.

$$bid_{AS} = \frac{n-1}{n} bid_{AS} + \frac{t_{AS}}{n} \quad (t_{AS} < t_{BS}) \tag{5}$$

$$bid_{AS} = \frac{n-1}{n} bid_{AS} + \frac{0}{n} \quad (t_{AS} > t_{BS}) \tag{6}$$

If the agents starting from state `A` and `B` select the same goal state `S`, and if the minimum step from state `A` to that goal state for the agent starting from the state `A` is short, the agent updates the goal value for the goal state by following the Eq. 5; otherwise, the agent updates the goal value by following the Eq. 6. As for the agent starting from the state `B`, if the minimum step from state `B` to the goal state `L` is short, the agent updates the goal value by the above same way.

3. Mathematical Analysis
In this section, we analyze whether the selection in Sect. 3.1.1 can resolve the dilemma problem by calculating the goal values when agents learn for infinite times. Note that we use the definition of the maze in Sect. 2.2. The Eqs. 5 and 6 indicate the average of the minimum steps and the average of those added 0 respectively. If "Selecting Goal" phase is happened for infinite times, the following equations are established from the above equations.

$$\lim_{n \to \infty} bid_{AS} = t_{AS} \quad (t_{AS} < t_{BS}) \tag{7}$$

$$\lim_{n \to \infty} bid_{AS} = 0 \quad (t_{AS} > t_{BS}) \tag{8}$$

If the agents starting from state `A` and `B` will reach the goal states `S` and `L`, respectively in dilemma problem, same way as the Sect. 2.2, the each goal value is put on the following Table 2. The goal value of the goal state `S` for the agent starting from the state `A` becomes 0 by the Eq. 8 because the Eq. 4 is established in this situation. While, though the Eq. 3 is established in this situation, the goal value of the goal state `L` for the agent starting from the state `A` becomes t_{AL} because the agent starting from the state `B` will only select the goal state `S`. After all, from the goal values of Table 2, the agent starting from the state `A` selects the goal state `L` and the agent starting from the state `B` selects the goal state `S` by following the Eq. 2. In addition, it is clear that this selection resolves the dilemma problem in this situation.

Table 2 Goal value table for two agents

	Goal S	Goal L
Agent A	0	t_{AL}
Agent B	t_{BS}	t_{BL}

3.2 Internal Reward Design

In this process, each agent designs the internal reward to reach the goal state selected in first step by using the method. Figure 2 shows a way for two agent to cooperate each other. t_{BS} indicates minimum step from start state B to goal state S and t_{BL} indicates minimum step count from start state A to goal state L. Red star represents the turning point to select whether the agent reaches the goal state S or L. Yellow directional arrows represent the agent's mainly action for each goal states, and arrows' thickness indicates Q-value. Note that each agent knows the goal with the process introduced in Sect. 3.1. The agent estimates Q-value to reach the optimal goal selected in Sect. 3.1 by an internal reward for Q-value described above.

In Fig. 2, the agent starting from the state A should reach the goal state S and the agent starting from the state B should reach the goal state L. Note, it is not necessary for the agent starting from the state A to set internal reward, since it reaches the goal state S normally (maximizing reward for the agent A). However, the agent starting from the state B should set the internal reward; under the standard RL the agent starting from the state B would reach to the goal state S, while the agent starting from the state B need to reach the goal state L for maximizing the reward for all agents. Then, in the proposed method, the internal reward is added to reform the reward design of the agent starting from the state B to reach the goal state L.

In the turning point on Fig. 2, the Q-value of the action to reach the goal state S eventually converges to a value r and the Q-value of the action to reach the goal state L is $r\gamma^2$. If the agent starting from the state B uses the internal reward ir_S, ir_L, the Q-value of the action to reach the goal state S is ir_S and the Q-value of the action to reach the goal state L is $ir_L\gamma^2$. Since $ir_L\gamma^2 > ir_S$ is satisfied, if ir_L is $\frac{r}{\gamma^2} + 1$ and ir_S is r, the agent starting from the state B will reach the goal state L finally and be able to cooperate. We explains the general way to design internal reward in the following.

4 Experiment

To validate the performance of the proposed method, we executed two experiments. We applied two agents to one maze problem for cooperation among agents in first experiment. We applied three agents to one maze problem in second experiment.

4.1 Experimental Setting

To investigate the effectiveness of our RL mechanism, this paper applies it into two maze problems as the multi-cooperation tasks. Specifically, in the first problem, we employ a 3×8 grid maze, as shown in Fig. 1, where two agents should cooperate with each other in order not to fall into dilemma (i.e., the local minimum). In the second

Fig. 3 Maze problem for three agents

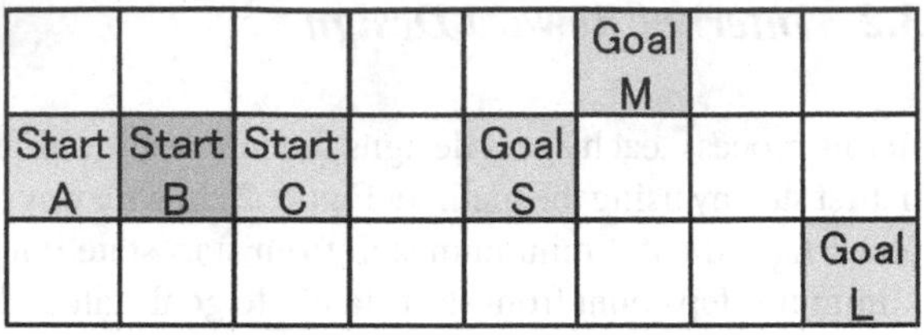

problem, we employ a 3×8 grid maze, as shown in Fig. 3, where three agents should cooperate with each other.

As the evaluation criterion, this paper evaluates the success rate for the cooperation among two or three agents. In addition to this criterion, this paper also investigates the effect of the random selection and the constant δ, respectively. Concretely, we conduct the simulation by employing the percentage for random action η (5) and constant δ (1). The parameters for Q-learning as summarized in Table 3. In detail, note that the parameter of both experiments is the same. In this table, the learning iterations and step counts are limited to 30000 and 100 as threshold showed in first and second lines of Table 3, respectively. If the iterations come over the threshold, the experiment is finished, and if the step counts come over, the present learning is stopped and next learning is started. We initialize Q-values at all states to 0 in third line of Table 3. The agents learn according to the parameter $\alpha = 0.1$ and $\gamma = 0.9$ in 4th and 5th lines of Table 3. In this experiment, agents select an action by $\epsilon - greedy$ of $\epsilon = 0.7$, and if the agent reaches goal states, the agent acquires the ordinary (external) reward 10 in 7th line of Table 3. Constant δ and random goal selection η in 8th and 9th lines of Table 3 indicate the difference of the internal reward for cooperation among the agents and the percentage that the agents select the goal states randomly, respectively.

Table 3 Parameters

	Q-learning
Iterations	30000
Maximum steps	100
Initial Q-value	0
Learning rate α	0.1
Discount rate γ	0.9
Action selection	ϵ-greedy, $\epsilon = 0.7$
Reward	10
Constant δ	1
Random goal selection η	5

4.2 *Result*

Figures 4 and 5 show the success rate of cooperation between two agents by the conventional method and the proposed methods. The vertical axis indicates the success rate, while the horizontal axis indicates the iteration of learning. As shown in Figs. 4 and 5, the success rate of the conventional Q-learning is low, while the proposed Q-learnings based on the internal reward increases as the iterations increases and finally becomes 1. From these results, it is clear that the proposed methods enable Q-learning agents to cooperate with each other.

4.2.1 Verification of the Goal Selection

Figures 6 and 7 show the estimated goal values by the proposed method and the theoretical ones, respectively. In these figures, the vertical axis indicates the averaged

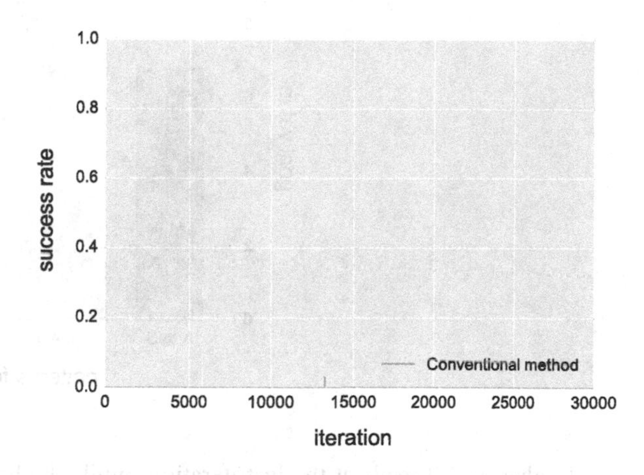

Fig. 4 Success rate of conventional Q-learning (two agents)

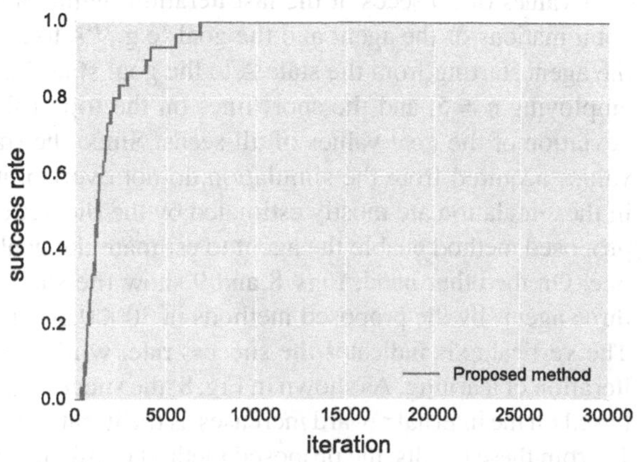

Fig. 5 Success rate of the proposed method (two agents)

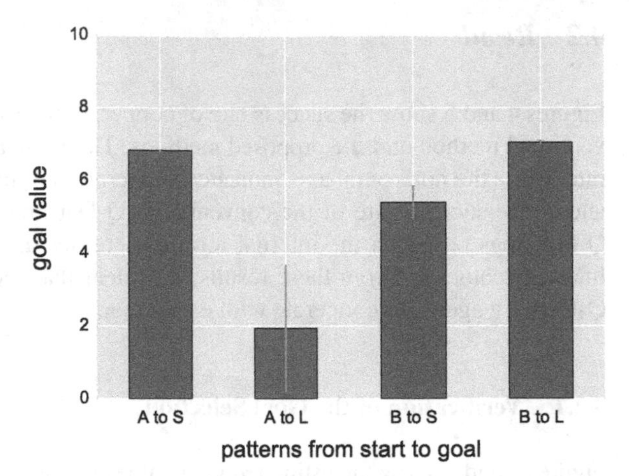

Fig. 6 The estimated goal value (two agents)

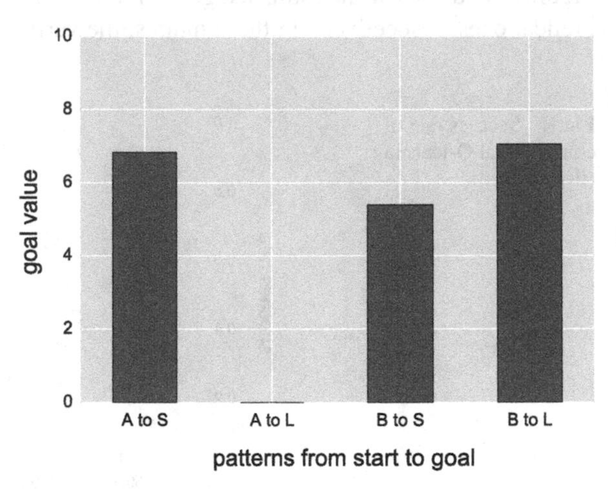

Fig. 7 The theoretic goal value (two agents)

goal values of 30 seeds at the last iteration, while the horizontal axis indicates the combinations of the agent and the goal (e.g., "A to S" means that the goal value of the agent starting from the state A to the goal state S). Bars indicates the goal value employing $\eta = 5$, and the short lines on the top of the bars indicates the standard deviation of the goal values of all seeds. Since the standard deviations of the goal values acquired from the simulation do not overlap, it is clear that the goal values in the simulation are mostly estimated by the theoretical ones. From the figures, the proposed method enable the agent to estimate the goal value close to the theoretical one. On the other hand, Figs. 8 and 9 show the success rate of cooperation among three agents by the proposed methods in 30000 and 150000 iterations, respectively. The vertical axis indicates the success rate, while the horizontal axis indicates the iteration of learning. As shown in Fig. 8, the success rate of the proposed Q-learnings based on the internal reward increases as the iterations increases, and finally becomes 1. From these results, the proposed method is effective in the situation of three agents

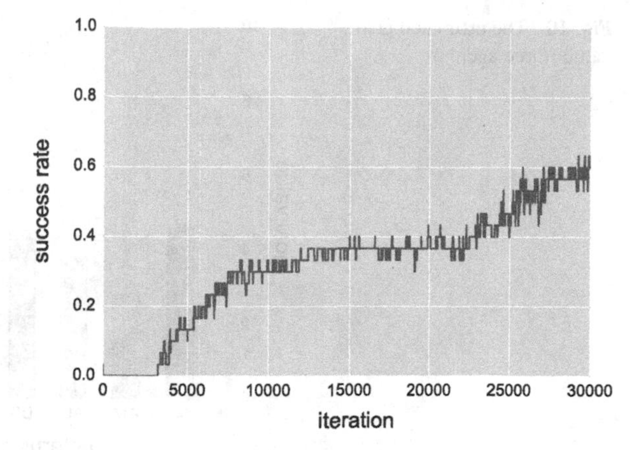

Fig. 8 Success rate of the proposed method (three agents)

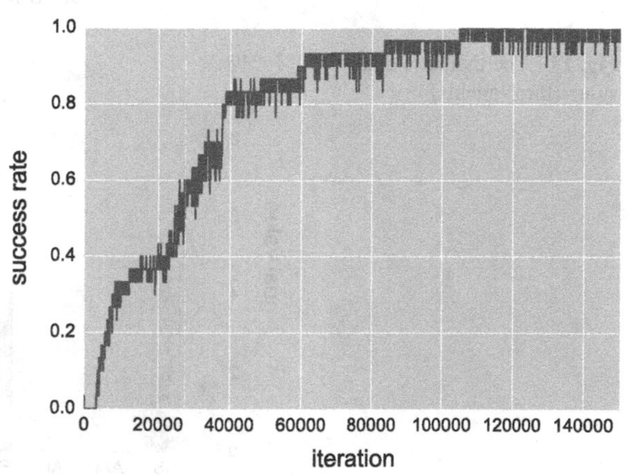

Fig. 9 Success rate of Fig. 8 in 150000 iterations (three agents)

perfectly. In addition, Figs. 10 and 11 show the estimated goal values by the proposed method and the theoretical ones, respectively. In these figures, the vertical axis indicates the averaged goal values of 30 seeds at the last iteration, while the horizontal axis indicates the combinations of the agent and the goal (e.g., "AS" means that the goal value of the agent starting from the state A to the goal state S). The agents starting from the state A, B and C set the largest goal value to the goal state S, M and L, respectively. these combinations are the same as the combinations determined by theoretic goal value. However, the standard deviations of the goal values acquired from the simulation overlap one another, and the estimated goal values of the combinations whose theoretic goal value is 0 have not become 0. These suggest that the three agents want more learning iterations.

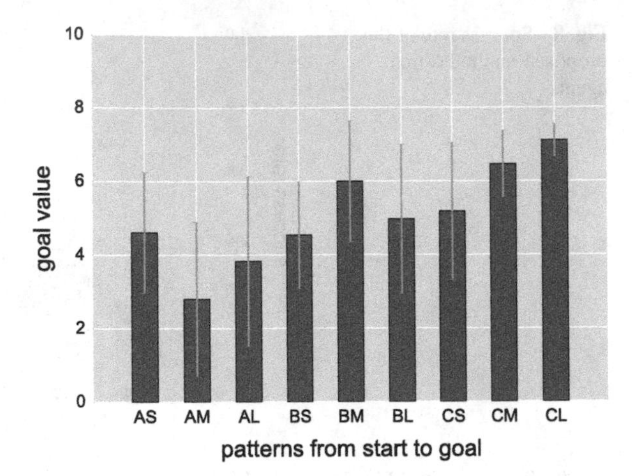

Fig. 10 The estimated goal value (three agents)

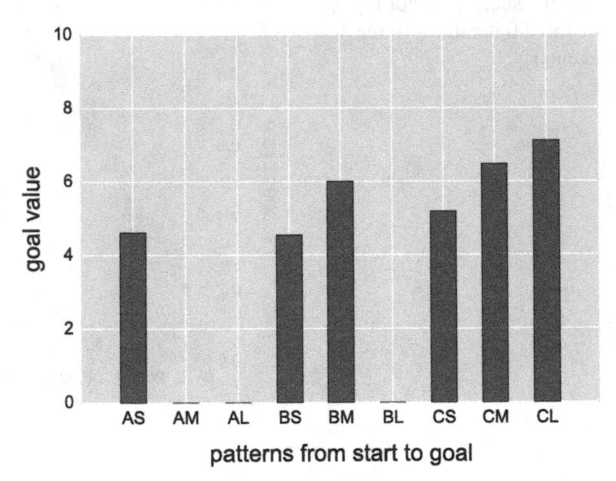

Fig. 11 The theoretic goal value (three agents)

4.3 Discussion

From the above results, it is clear that the proposed methods, i.e., the internal reward design based on the goal selection enable the agents to learn appropriate actions according to the theory, which needs to cooperate with each other. Since the agents should acquire the minimum steps to derive to the best performance, this subsection investigates the effectiveness of the proposed method.

4.3.1 Goal Selection

In the proposed method, the agents can select an appropriate goal for cooperation by estimating the goal value which converges to the number of the minimum steps towards the goal that the agent can reach. Since the agent far from goals, i.e., the agent starting from the state A, estimates the goal value of a far goal as 0 while estimates that of a near goal as the minimum steps in Fig. 7, the agent selects the near goal because the goal values of the near goal is largest in the situation of two agents (the agents can select appropriate goals by same process in the situation of three agents). This result suggests that it is important for the agents to select other (not optimal) goals. In other words, it is important that the agent near goals leads the agent far from goals to optimal goal. In the proposed method, η is related to this factor. If η is 5 (same as η in the experiment), the agents select goals randomly for five times per one iteration. If the number of the agent is two, since the agent starting from the state B can estimate all goal values accurately, the agent starting from the state A is led to the optimal goal by another agent. From the explanation, $\eta = 5$ is appropriate in the situation which there are two agents in one environment. This is shown in Figs. 4, 5, 6 and 7. In the situation which there are three agents in one environment, the result is worse than the result of the situation of two agents. The reason of this is that the goal value estimation among three agents is not easier than that among two agents, unlike the randomly goal selection. Since the goal value is converged from the near agent to goals, if the number of the agent is increased, to estimate the goal value is difficult. This is clearly a fact because if the learning iteration is increased, the success rate is improved in the experiment and converged to 1, such as Figs. 8 and 9.

4.3.2 The Number of Agent

From the result of the experiment of three agents, If the proposed method fulfills certain conditions, the proposed method can be effective in the situation which there are three agents in a same environment. This suggests that if the proposed method fulfills certain conditions, the proposed method can be effective in the situation which there are any agent in a same environment. Since the proposed method get the agents to estimate the goal value in order to reach far goal state, and the agents near the goal states lead the agents far from the goal states, the proposed method might mistake in some situation; the agents near the goal states select a wrong goal state. "far goal state" must be the goal state which is faraway from start points of all agents, in other words, each goal state has same magnitude correlation with one another against each agent (if a goal state is most faraway from start point of one agent, this goal state is most faraway from start points of all agents). Therefore, each goal state has same magnitude correlation with one another against each agent so that the proposed method is effective. Since the experiment of this paper meets this condition, the agents can cooperate with one another. This is clearly from the results of the experiments.

5 Conclusion

This paper focused on a multi-agent cooperation which is generally difficult to be achieved without communication among the agents, and proposed the Q-learning method for a multi-agent cooperation based on no communication. For this purpose, our proposed method, the extension of Q-learning, changes the ordinary (external) reward to the internal reward. To increase the certainty of the proposed methods, this paper theoretically investigate (1) how the internal rewards should be set to guarantee the cooperation among agents; and (2) how their values should be set to discover the goal for cooperation among agents. Through the intensive simulations on the cooperating task in the 3×8 grid maze where two or three agents are required to cooperate with one another, the following implications were revealed: (1) our proposed method successfully enables the agents to select their own appropriate cooperative behaviors which contribute to acquiring the minimum steps towards to their goals, while the conventional methods cannot always acquire such behaviors. In detail, the proposed method always enables agents to cooperate with each other without communication by prioritizing the goal to select the yielding action; and (2) the cooperation among agents according to their internal rewards is achieved even less communication or no communication. In detail, the proposed method enables the agents to cooperate with each other by designing the internal reward, and the agents can design the internal reward by the minimum steps acquired with no communication; and (3) the proposed method is effective in the situation which there are three agents in a same environment and we reveal the condition of the proposed method to be effective.

What should be noticed here is that the results have only been obtained from one simple gird maze problem with two agents. Therefore, further careful qualifications and justifications, such as an analysis of results with other tasks or an increase of the number of agents, are needed to generalize our results. Such important directions must be pursued in the near future in addition to the following future research: (1) an extension of the proposed method to any number of agent; and (2) an application of the proposed method into large scale environment.

References

1. Yong-Jae Kim Kui-Hong Park and Jong-Hwan Kim. Modular q-learning based multi-agent cooperation for robot soccer. *Robotics and Autonomous System*, pages 3026–3033, 2015.
2. Michael Camara, Oliver Bonham-Carter, and Janyl Jumadinova. A multi-agent system with reinforcement learning agents for biomedical text mining. In *Proceedings of the 6th ACM Conference on Bioinformatics, Computational Biology and Health Informatics*, BCB '15, pages 634–643, New York, NY, USA, 2015. ACM.
3. H. Iima and Y. Kuroe. Swarm reinforcement learning methods improving certainty of learning for a multi-robot formation problem. *CEC*, pages 3026–3033, May 2015.
4. Y. Ichikawa and K. Takadama. Designing internal reward of reinforcement learning agents in multi-step dilemma problem. *Journal of Computational Intelligence and Intelligent Informatics, JACIII*, 17(6):926–931, 2013.

5. M. Gini M. Elidrisi, N. Johnson and J. Crandall. Fast adaptive learning in repeated stochastic games by game abstraction. *AAMAS*, pages 1141–1148, May 2014.
6. Prabuchandran K. J., Hemanth Kumar A. N, and S. Bhatnagar. Multi-agent reinforcement learning for traffic signal control. In *Intelligent Transportation Systems (ITSC), 2014 IEEE 17th International Conference on*, pages 2529–2534, Oct 2014.
7. Katja Verbeeck Karl Tuyls and Tom Lenaerts. A selection-mutation model for q-learning in multi-agent systems. *Robotics and Autonomous System*, pages 3026–3033, May 2015.
8. Alessandro Lazaric Enrique Munoz de Cote and Marcello Restelli. Learning to cooperate in multi-agent social dilemmas. *AAMAS*, pages 783–785, May 2006.
9. Ming Tan. Multi-agent reinforcement learning: Independent vs. cooperative agents. In *In Proceedings of the Tenth International Conference on Machine Learning*, pages 330–337. Morgan Kaufmann, 1993.
10. N. Ono and K. Fukumoto, editors. *Multi-agent reinforcement learning: A modular approach*, 1996.
11. R.S. Sutton and A.G. Barto. *Reinforcement Learning*. Bradford Books/MIT Press, Cambridge, MA, 1998.

Autonomous Task Allocation for Swarm Robotic Systems Using Behavioral Decomposition

Yufei Wei, Toshiyuki Yasuda and Kazuhiro Ohkura

Abstract Swarm robotic systems (SRS) are a type of multi-robot systems, in which robots operate without any form of centralized control. In SRS, the generation of a complex swarm behavior resulting in robots being dynamically distributed over different sub-tasks requires an autonomous task allocation mechanism. It has been well recognized that evolutionary robotics with an evolving artificial neural network is a promising approach for generating collective swarm behavior. However, the artificial evolution often suffers from the bootstrap problem, especially when the underlying task is very complex. On the other hand, the behavioral decomposition, which is based on the divide-and-conquer thinking, has been reported to be effective for overcoming the bootstrap problem. In this paper, we describe how a behavioral decomposition based evolutionary robotics approach can be applied to synthesize a composite artificial neural network based controller for a complex task. The simulation results show the hierarchical strategy based evolutionary robotics approach is effective for generating autonomous task allocation behavior for a swarm robotic system.

Keywords Swarm robotic systems · Evolutionary robotics · Autonomous task allocation · Behavioral decomposition

Y. Wei · T. Yasuda · K. Ohkura (✉)
Graduate School of Engineering, Hiroshima University,
1-4-1, Kagamiyama, Higashi-Hiroshima, Hiroshima 739-8527, Japan
e-mail: ohkura@ohk.hiroshima-u.ac.jp

Y. Wei
e-mail: wei@ohk.hiroshima-u.ac.jp

T. Yasuda
e-mail: yasuda@ohk.hiroshima-u.ac.jp

G. Leu et al. (eds.), *Intelligent and Evolutionary Systems*,
Proceedings in Adaptation, Learning and Optimization 8,
DOI 10.1007/978-3-319-49049-6_34

1 Introduction

Swarm robotics (SR) [1] is a field in which many homogeneous robots coordinate behavior to accomplish a given task without any form of global control. The robots are relatively simple compared to the task they are dealing with, that their communication is usually local and sensory capabilities are limited. Therefore, the emergence of complex collective behaviors can be regarded as a result of local interactions between the robots, and between the robots and the environment. A swarm robotic system exhibits three advantages: (i) robustness: the system is inherently fault-tolerant, that even if some robots stop working during the task, the system is still functional, (ii) flexibility: the solution generated by SRS is flexible enough to deal with similar tasks, (iii) scalability: the system can operate with a wide range of group sizes.

SRS is mainly inspired by social insects. Social insects such as ants, bees and wasps exhibit collective behaviors to accomplish tasks beyond the capability of a single individual. This highly intelligent behavior implicitly shown in these social insects is sometimes called swarm intelligence [2]. Among those intelligent collective behaviors, task allocation is the one which results in individuals being engaged into different sub-tasks while solving a complex task, in which the allocation is not fixed but may change dynamically based on local observations of individuals. Since task allocation can be widely observed in almost all social livings, similar behavior can be beneficial for SRS in the same way.

The design methods for SRS can mainly be divided into two categories [3]. The first one is behavior-based design, in which individual-level behaviors are developed, tested and improved manually following a typical trial-and-error process until the required collective-level behaviors are obtained [4, 5]. However, in this approach, expertise knowledge of the undertaken task is required and the performance of the system is completely reliant on the human designer. The second approach is automatic design, wherein a certain automatic design method is adopted to reduce the effort of human developers. One promising automatic design approach is evolutionary robotics (ER) [6, 7], in which the controller of robots is developed in an iterative way utilizing the Darwinian principle. Typically, robot controllers in ER are represented by artificial neural networks (ANN) [8, 9], whose parameters are obtained through evolutionary algorithms (EA).

Currently, task allocation behaviors in SRS are mainly obtained through the use of behavior-based design methods. The difficulty that prevents ER from being applied for generating autonomous task allocation behaviors resides in the fact that for such complex tasks requiring proper task allocation, the design objective of the controller is too far beyond primitive capabilities of the controller. The gap results in all individuals in the first generations performing equally poorly, that the evolution process cannot start and no valid controller could be found [10]. This problem is also referred to as the bootstrap problem.

One methodology for overcoming the bootstrap problem is to assist the evolutionary process with human knowledge. In this context, three approaches have been widely adopted: incremental evolution, behavioral decomposition and human in-the-loop [10]. In incremental evolution [11], the robot controller is trained in a simplified or partial task first, then the difficulty of the task is gradually increased until the original goal is achieved. In behavioral decomposition [12, 13], the original task is decomposed into several relatively simple sub-tasks, and the robot controller is also divided into sub-controllers to deal with the corresponding sub-tasks, respectively. Differently, the key idea of human in-the-loop is to let developers indicate intermediate states directly in order to avoid local optima [14]. Considering the fact that complex tasks which require task allocation behavior are inherently composed of sub-tasks, we believe that the behavioral decomposition has more advantages.

In this study, we utilize a behavioral decomposition based evolutionary robotics approach to generate autonomous task allocation behavior for a robotic swarm in order to accomplish a complex task. The remainder of this study is organized as follows: related work are introduced in Sect. 2. Section 3 describes the behavioral decomposition. Experiment settings are explained in Sect. 4. Section 4.5 discusses the results. Finally, we conclude this study with Sect. 5.

2 Related Work

In most existing work, task allocation behaviors have been generated for scenarios where robots need to search for some objects in the environment and then operate on these objects (e.g. foraging).

In [5], the authors described a foraging scenario in which robots have to decide whether to perform a foraging task or rest in the nest according to the density of foods in the environment in order to maximize the net swarm energy.

Pini et al. [15] devised a task where robots need to transport objects from source to nest, which can be achieved by two ways: transport the objects directly through a long path, or take a shortcut using a cache, which allows only a few robots operating on it simultaneously. The robots therefore need to choose which way to use based on the cost involved.

Agassounon et al. [16] considered a scenario where robots gather randomly located objects in the field and then group them in a cluster. Since the probability for robots to find scattered objects reduces as they are clustered, redundant robots can rest in order to increase the efficiency of the swarm.

Conventionally, task allocation behavior is obtained through the use of either probabilistic or threshold methods.

Liu et al. [5] proposed a mathematical model to generate task allocation behavior for a foraging task. In their approach, each robot keeps two thresholds tracing the permissible maximum time of searching and resting. The two thresholds are dynamically updated over time in an adaptive way based on local interactions. In [17], Castello et al. utilized an adaptive response threshold to describe the sensitivity of a robot to the "need" for performing a task, and the relation between those "needs" and response is updated by a probabilistic method.

On the other hand, good results have been reported by several works which utilized behavior decomposition to address the bootstrap problem in ER. Lee [12] developed a hierarchical controller for a scenario in which the robot had to find a box and transport it to a light source. In their study, the original task was divided into three sub-tasks: "circle the box", "transport the box" and "explore for the box". In [13], Duarte et al. studied a complex task in which a robot had to rescue a teammate in a double T-maze. The rescue task is composed of three parts: exit the initial room, solve the double T-maze, and bring the teammate back to the initial room. A hierarchical strategy based controller was synthesized and achieved 92 % solve-rate.

3 Behavioral Decomposition

The key idea of behavioral decomposition is based on the divide-and-conquer thinking: if a task is too complex to be solved directly, one can decompose it into relatively simple sub-tasks and achieve them respectively. The implementation of behavioral decomposition proposed so far shares many similarities [12, 13]. Here, we describe a generalized version as follows:

Firstly, the original task is decomposed manually into relatively simple sub-tasks in a hierarchical and recursive way until all sub-tasks are solvable. The original task can be then represented by a tree-like graph: leaf nodes represent the simplest sub-tasks after decomposition, internal nodes are harder sub-tasks consisting of children sub-tasks, and then the root node is the original task to be solved. Importantly, the decomposition is based on the analysis of task specification rather than the robot individuals' abilities, which allows the designer to devise task structure by top-to-down schemes. In addition, if several sub-tasks have the same requirements for robots, they can be solved by the same sub-controller.

To accomplish those sub-tasks respectively, the robot controller is also organized in a hierarchical way similar to the task, in which sub-controllers have two types: behavior primitive and arbitrator [12]. A behavior primitive is used to solve the corresponding simplest sub-tasks sharing the same requirements, and an arbitrator combines several primitives or lower arbitrators together to achieve relatively harder sub-tasks. Primitives and arbitrators are usually represented by artificial neural networks receiving a part or all of the sensory inputs of the robot, respectively. The outputs of primitives have direct control of the robot, while the outputs of arbitrators are used to activate or restrain their primitives dynamically. An example of the robot controller architecture is as shown in Fig. 1.

The development of a robot controller is a hierarchical process following a bottom-to-top procedure, where lower sub-controllers are developed first, and then they are combined together through the development of its upper arbitrator. This process is repeated until all controllers are developed.

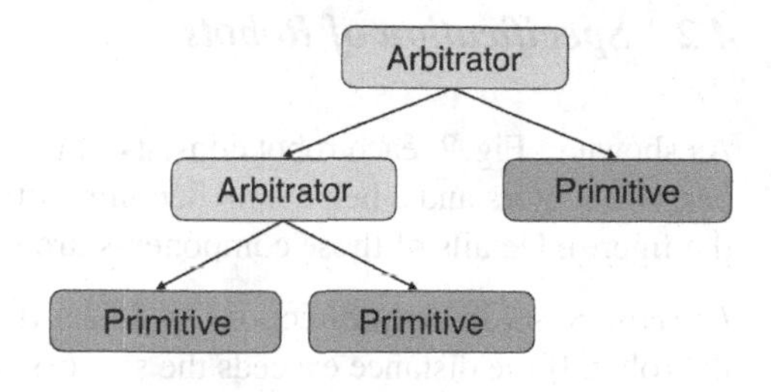

Fig. 1 Behavioral decomposition based controller

4 Experiments

4.1 A Complex Task

To demonstrate the effectiveness, we apply the proposed approach to a complex variation of classic foraging task as shown in Fig. 4a.

The experiment field is composed of three parts: nest (with barriers at the entrance, green area), central area (grey area) and split-up area (pink area). Thirty robots are located in the nest at the beginning of the experiment. There are ten small foods (S-foods, henceforth) and five large foods (L-foods, henceforth), each L-food is composed of seven S-foods in the central area. An S-food can be transported by a single robot, while an L-food is too heavy. Therefore, robots have to cooperate to move it. In addition, L-foods can be separated into S-foods automatically by moving them to the split-up area.

The goal task for robots is to transport all foods back to the nest. However, due to the existence of barriers at the entrance of the nest, L-foods must be moved to the split-up area first for decomposition. Furthermore, the sight of robots is limited, therefore they need to explore the field to find foods.

As discussed above, the goal task is actually composed of three sub-tasks, described as follows:

- Explore
  Since the sight of robots is limited, they have to explore the experiment field for foods.
- Transport L-foods to the split-up area
  L-foods can not be transported to the nest directly due to the existence of barriers. Instead, robots have to move L-foods to the split-up area to decompose them into S-foods.
- Transport S-foods to the nest
  S-foods should be transported to the nest in order to achieve the task.

4.2 Specification of Robots

As shown in Fig. 2, each robot consists of eight IR sensors, an omni-camera, a gripper, two motors and a behavioral decomposition based ANN controller (invisible in the figure). Details of those components are described as follows:

IR sensors Set in eight directions to detect distance between the nearest object and the robot. If the distance exceeds the sensors' maximum range, it will be recognized as the maximum range.

Omni-Camera Main sensorial component with limited range of sight, gathers information of:

- Number of robots, S-foods and L-foods in its sight, separately
- Angle of the nest, nearest S-food, nearest L-food, central area and split-up area (relative angle to the robot, represented by $\sin\theta$ and $\cos\theta$)

Gripper A pre-programmed component in the front of the robot, catches an S-food automatically when two conditions are met simultaneously: the robot is performing the "transport S-foods to the nest" sub-task and the gripper is touching an S-food, otherwise it will drop the caught S-food if there is any.

Motors Each accepts a target speed argument, and then accelerates until the target speed is achieved.

Behavioral decomposition based ANN controller Corresponding to the task decomposition, the controller is composed of three behavior primitives and one arbitrator, all represented by three-layered ANN whose hidden layer is fully linked (recurrent). The number of input nodes of these ANNs and the information they hold differ from each other according to specific sub-tasks, while they all have ten hidden nodes. Each primitive has two outputs that take control of the robots, dealing with three sub-tasks respectively. The arbitrator has three outputs determining which sub-task to perform at each moment. Figure 3 shows the architecture of this controller. Details of the arbitrator and three primitives are described as follows:

- Explore Primitive
  If this sub-controller is activated, the robot moves around the central area. Inputs consist of the angle of the central area and eight IR sensors.

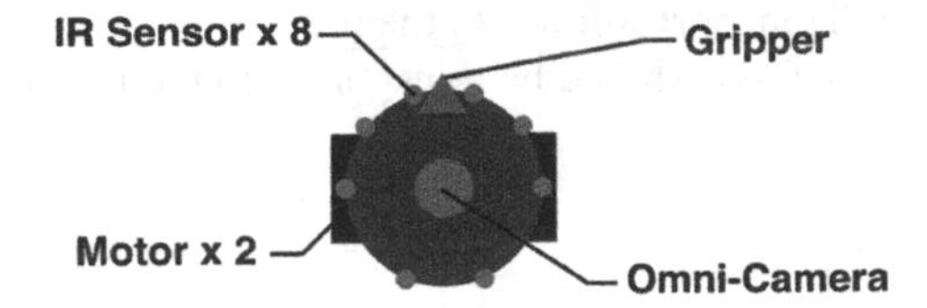

Fig. 2 Specification of Robots

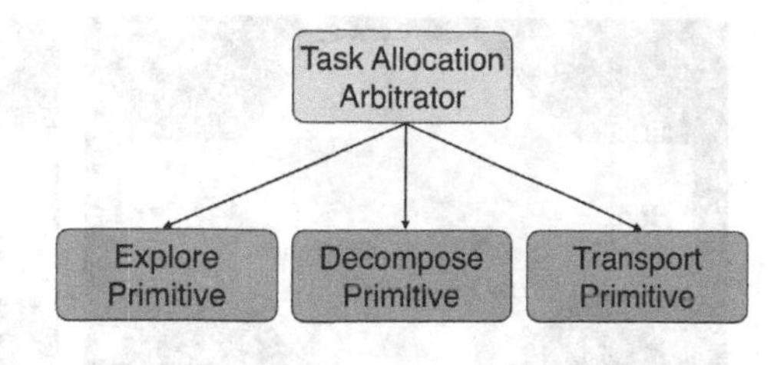

Fig. 3 Architecture of the Robot controller

- Decompose Primitive
  This sub-controller is used to perform the "transport L-foods to the split-up area" sub-task. It takes the angle of the split-up area and the nearest L-food as its inputs. The gripper will automatically catch an S-food in the front of the robot, if it is available. The robot will stop if there is no L-food in the robot's sight while this primitive is active,
- Transport Primitive
  When this sub-controller is active, the robot grabs and transport the nearest S-food to the nest. Inputs include status of the robot's gripper (catching an S-food or not), the angle of the nest and the nearest S-food and eight IR sensors. If the robot can not see any S-food in its sight, the robot stops.
- Task Allocation Arbitrator
  This arbitrator takes the number of robots, S-foods and L-foods in its sight, as well as its output in the last time step as its inputs. The corresponding primitive to the highest output will be activated.

4.3 Experimental Settings

As discussed in Sect. 3, to obtain the final controller, we first evolve three primitives in three independent experiments, and then evolve the task allocation arbitrator in the goal task to exploit those primitives. In addition, to increase the accuracy, all experiments are performed by simulation using Box2D.[1] Details of those experiments are described as follows:

4.3.1 Experiment for Explore Primitive

Since the sight of robots is limited, they have to explore the field for foods. To train this primitive, we devised the following experiment, as shown in Fig. 4b. There are five robots in the nest and twenty-five obstacles in the central area. Five foods are located in the central area and once robots detect a food, they get a bonus and the food will be marked detected meaning it can not be detected again. Foods will be refreshed

[1]Box2D is an open-source physics engine, which can be found on http://box2d.org.

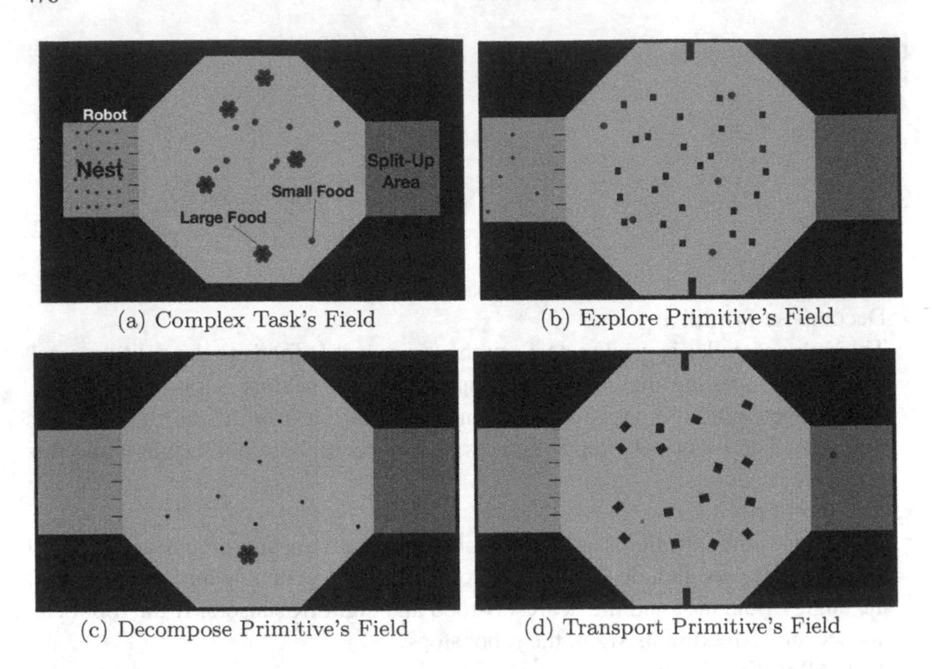

(a) Complex Task's Field (b) Explore Primitive's Field (c) Decompose Primitive's Field (d) Transport Primitive's Field

Fig. 4 Experiment fields for **a** complex task—the overall task, **b** explore primitive, **c** decompose primitive, and **d** transport primitive

Table 1 Fitness function for explore primitive

Action	Fitness
Bonus when detect a virtual food	+100
Robots touch obstacles	−180/s/robot

every 10 s in a new random position. The initial position of robots, obstacles and foods is also randomly selected. Each experiment lasts for 50 s.

During the artificial evolution, each individual candidate of this primitive controller is simulated 50 times, and the average fitness value is used as its final fitness value. Table 1 describes the fitness function.

4.3.2 Experiment for Decompose Primitive

The L-foods can not be moved to the nest directly due to the existence of barriers. Instead, robots need to transport them to the split-up area to be decomposed into S-foods. In this experiment, as shown in Fig. 4c, nine robots and one L-food are located in the central area. Their initial position is randomly selected. Additionally, the range of omni-camera in this experiment is temporally set to unlimited. Each experiment lasts for 50 s.

Table 2 Fitness function for decompose primitive

Action	Fitness
Robots touch the L-food	+6/robot/second
L-food moved toward the nest	+moved distance/original distance × 4500
L-food moved to the nest	+500 +remaining seconds × 600

Table 3 Fitness function for transport primitive

Action	Fitness
Robot gets close to the S-food	+final distance/original distance × 5000
Robot caught the S-food	+7000
S-food moved toward the nest	+moved distance/original distance × 11000
S-food moved to the nest	+1000 +remaining seconds × 600
Robot touches obstacles	−180/second

Each individual candidate of this primitive controller is simulated 50 times during the artificial evolution process, and the average fitness value is used as its final fitness value. The fitness function is shown in Table 2.

4.3.3 Experiment for Transport Primitive

The final objective for robots is to transport all S-foods to the nest. Therefore we perform this experiment to obtain the transport primitive. The experiment field is illustrated in Fig. 4d. There are sixteen obstacles, one robot and one S-food in the central area. The initial position is randomized at the beginning of each experiment, time limit of which is set to 50 s. Similarly, the range of omni-camera is set to unlimited.

The fitness function is shown in Table 3. Each controller candidate is evaluated 50 times, and the average fitness value is used as its final fitness value.

4.3.4 Experiment for Task Allocation Arbitrator

The task allocation arbitrator is evolved in the original task as shown in Fig. 4a. Details are described in the beginning of this section. The fitness function is shown in Table 4, some parts are derived from its primitives' fitness functions. This experiment lasts for 120 s for each candidate.

Table 4 Fitness function for task allocation arbitrator

Action	Fitness
L-foods moved toward split-up area	+moved distance/original distance × 4500
L-foods moved to split-up area	+500 +remaining seconds × 20
S-foods moved toward the nest	+moved distance/original distance × 11000
S-foods moved to the nest	+1000 +remaining seconds × 10
Robots stopped	−60/second/robot

4.4 Evolutionary Algorithm Settings

We adopted (μ, λ) evolution strategy for all experiments, sharing most parameters. The population size is set to 500, and the parent size to 75. The range of each ANN link is $\in [-1.0, 1.0]$ following the standard normal distribution mutation, where initial mutation step size is set to 0.05. The maximum generation for three primitive experiments is 500, and 1000 for the arbitrator. For all experiments, we run 10 trials.

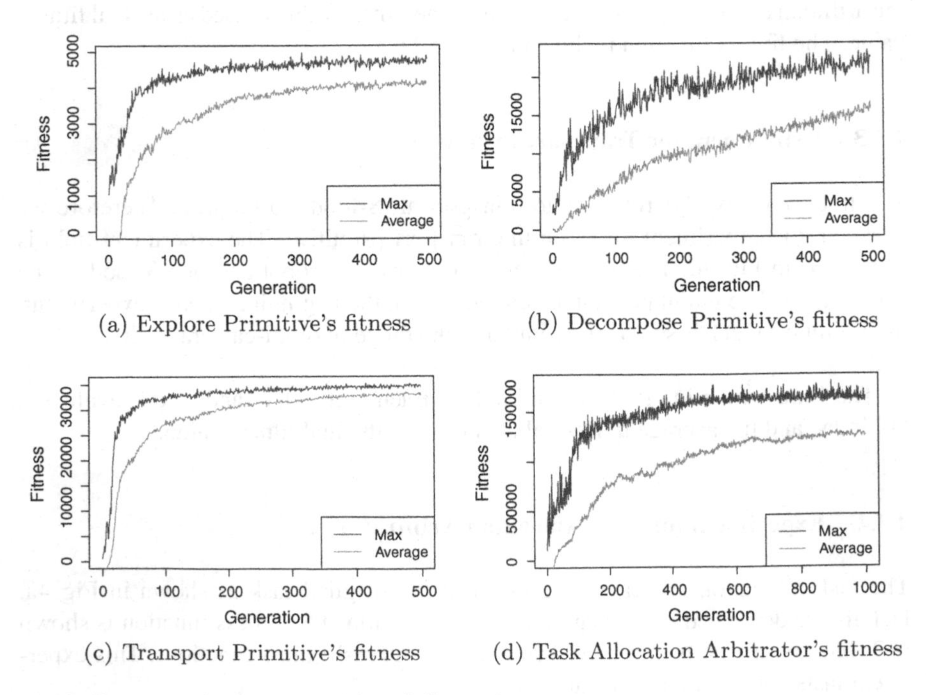

Fig. 5 Representative fitness transition of all experiments

4.5 Results

Figure 5 illustrates the most representative fitness transition of all experiments in 10 trials. The fitness of explore primitive, transport primitive, and the arbitrator grew smoothly and finally converged and reached the maximum fitness in the final stage. On the contrary, the fitness of decompose primitive was still growing in the final stage, which indicates that the maximum generation was not enough.

Snapshots of the generated autonomous task allocation behavior are shown in Fig. 6. In the first snapshot, robots start with performing the explore sub-task. After most robots moved to the central area, in snapshot 2, some robots changed to perform the "decompose L-foods" sub-task, and some robots performed the "transport

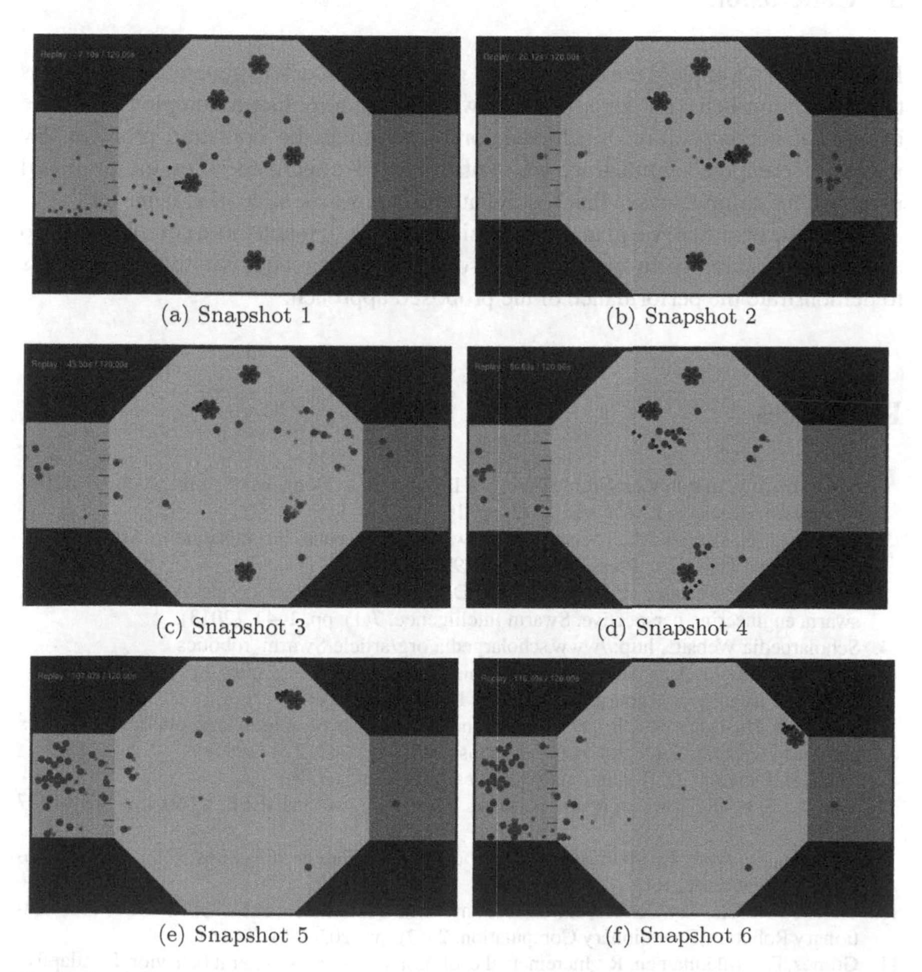

(a) Snapshot 1 (b) Snapshot 2

(c) Snapshot 3 (d) Snapshot 4

(e) Snapshot 5 (f) Snapshot 6

Fig. 6 Snapshots of a successful trial, showing that robots are able to change their roles dynamically to achieve the task

S-foods" sub-task. In snapshot 3, they moved an L-food to the split-up area and then started to transport the appeared S-foods after decomposition, but when they found another L-food, they changed soon to move that L-food (see the change between snapshot 3, 4). Snapshot 5 shows that after they transported S-foods to the nest, they turned back to perform the explore sub-task. In snapshot 6, most foods were transported to the nest and the experiment ended due to the time limit.

As shown in these snapshots, we obtained a behavioral decomposition based controller that successfully exhibited autonomous task allocation behavior to achieve this complex task by using the proposed approach.

5 Conclusion

In this paper, we applied the evolutionary robotics approach to generate autonomous task allocation behavior for a robotic swarm to accomplish a complex task. The behavioral decomposition is adopted for overcoming the bootstrap problem. By means of computer simulation, we confirmed the effectiveness of the proposed approach for complex tasks that require autonomous task allocation behavior.

As for the next step, we plan to apply this approach to much more complex tasks to examine its scalability. In addition, we also plan to conduct quantitative experiments to demonstrate the performance of the proposed approach.

References

1. Şahin, E.: Swarm Robotics: From Sources of Inspiration to Domains of Application. SAB2004 WS Swarm Robotics, LNCS, vol. 3342, pp. 10–20 (2005)
2. Bonabeau, E., Dorigo, M., Theraulaz, G.: Swarm Intelligence: From Nature to Artificial Systems. Oxford University Press, New York (1999)
3. Brambilla, M., Ferrante, E., Birattari, M., Dorigo, M.: Swarm robotics: A review from the swarm engineering perspective. Swarm Intelligence, 7(1), pp. 1–41, (2013)
4. Scholarpedia Website, http://www.scholarpedia.org/article/Swarm_robotics
5. Liu, W., Winfield, A.: Modelling and optimisation of adaptive foraging in swarm robotic systems. The International Journal of Robotics Research. (2010)
6. Harvey, I., Husbands, P., Cliff, D., Thompson, A., Jakobi, N.: Evolutionary robotics: the Sussex approach. Robotics and autonomous systems, 20(2), pp. 205–224 (1997)
7. Nolfi, S., Floreano, D.: Evolutionary robotics. MIT Press (2000)
8. Yao, X.: Evolving artificial neural networks. Proceedings of the IEEE, 87(9), pp. 1423–1447 (1999)
9. Floreano, D., Dürr, P., Mattiussi, C.: Neuroevolution: from architectures to learning. Evolutionary Intelligence, 1(1), pp. 47–62 (2008)
10. Silva, F., Duarte, M., Correia, L., Oliveriram S.M., Christensen, A.L.: Open Issues in Evolutionary Robotics. Evolutionary Computation, 24(2), pp. 205–236. (2016)
11. Gomez, F., Miikkulainen, R.: Incremental evolution of complex general behavior. In Adaptive Behavior, 5(3-4), pp. 317–342 (1997)

12. Lee, W. P., Hallam, J., Lund, H. H.: Learning complex robot behaviours by evolutionary computing with task decomposition. In Learning Robots, pp. 155–172. Springer Berlin Heidelberg (1997)
13. Duarte, M., Oliveira, S., Christensen, A.L.: Hierarchical evolution of robotic controllers for complex tasks. In IEEE International Conference on Development and Learning and Epigenetic Robotics (ICDL), pp. 1–6. IEEE Press, Piscataway (2012)
14. Celis, S., Hornby, G. S., Bongard, J.: Avoiding local optima with user demonstrations and low-level control. In IEEE Congress on Evolutionary Computation, pp. 3403–3410. IEEE. (2013)
15. Pini, G., Brutschy, A., Frison, M., Roli, A., Dorigo, M., Birattari, M.: Task partitioning in swarms of robots: An adaptive method for strategy selection. In Swarm Intelligence, 5(3–4), pp. 283–304. (2011)
16. Agassounon, W., Martinoli, A., Goodman, R.: A scalable, distributed algorithm for allocating workers in embedded systems. In IEEE International Conference on Systems, Man, and Cybernetics, Vol. 5, pp. 3367–3373. IEEE. (2001)
17. Castello, E., Yamamoto, T., Nakamura, Y., Ishiguro, H.: Task allocation for a robotic swarm based on an adaptive response threshold model. In Control, Automation and Systems (ICCAS), 2013 13th International Conference on (pp. 259–266). IEEE. (2013)

Acquisition of Cooperative Action by Rescue Agents with Distributed Roles

Mengchun Xie, Mitsutoshi Murata and Shoma Sato

Abstract Fully anticipating the overall effect on society is difficult due to the many as-yet-unrecognized factors at disaster sites. There is a need for autonomous disaster relief robots, which can learn from the conditions they encounter and then take independent actions. Reinforcement learning is one way that robots can acquire information about appropriate behavior in new environments. In the present study, we present the results of a disaster relief simulation that included multiple autonomous robots working as a multi-agent system. In order to assist in the use of reinforcement learning for the efficient acquisition of action rules, we divided the task into various sub-tasks. We propose an approach in which cooperative action is obtained by giving each agent a different reward; this encourages the agents to play different roles. We investigated how the various autonomous agents determined the appropriate action rules and examined the influence of providing separate rewards to different agents in the system. We also compared the values of various actions in different learning situations.

Keywords Multi-agent system • Reinforcement learning • Autonomous robot • Rescue agent • Distributed roles

M. Xie (✉) · M. Murata
Department of Electrical and Computer Engineering,
National Institute of Technology, Wakayama College, Gobo, Wakayama, Japan
e-mail: xie@wakayama-nct.ac.jp

M. Murata
e-mail: murata@wakayama-nct.ac.jp

S. Sato
Department of Computer Science and Engineering,
Toyohashi University of Technology, Toyohashi, Japan
e-mail: shoma.sato@gmail.com

G. Leu et al. (eds.), *Intelligent and Evolutionary Systems*,
Proceedings in Adaptation, Learning and Optimization 8,
DOI 10.1007/978-3-319-49049-6_35

1 Introduction

Research on disaster relief robots has been increasing since the Great East Japan Earthquake of March 11, 2011. Fully anticipating the overall effect on society is difficult due to the many as-yet-unrecognized factors at disaster sites. In addition, many different remote-controlled disaster relief robots have been developed. A further complication is that these robots must work in environments in which communication is not always secure. For these reasons, there is a need for autonomous disaster relief robots, which can learn from the conditions they encounter and then take independent actions.

Reinforcement learning is one way that robots can acquire information about appropriate behavior in new environments. This is a computational approach to understanding and automating goal-directed learning and decision-making. It is distinguished from other computational approaches by its emphasis on learning from direct interaction with the environment, without exemplary supervision or even complete models of that environment [1–3]. Thus, it enables the determination of the best practices, which can be difficult even for humans in an unknown environment. Recent research on disaster relief robots has included consideration of multi-agent systems, that is, systems that include two or more disaster relief robots.

These multi-agent systems include multiple interacting, intelligent agents that pursue a set of goals or perform a set of tasks [4]. In such systems, each agent must behave independently according to its current state and that of its environment, and, if necessary, it must cooperate with other agents in order to perform a given task. Multi-agent systems have greater robustness and flexibility than do conventional centralized management systems. However, it is difficult to predict in advance the actions of the various agents or to assign action rules to multi-agent systems because the interaction between agents is complex. A possible solution to this problem is the use of autonomous agents that are able to learn from the environment and implement independent actions [5–8].

In conventional multi-agent systems, such as the box-pushing problem and the pursuit problem, the task must be simple, and all agents share a common task [9–11]. However, disaster areas are complex, and thus it is difficult for agents to learn efficiently from their own actions.

In the present study, we present the results of a disaster relief simulation that included multiple autonomous robots working as a multi-agent system. In order to assist in the use of reinforcement learning for the efficient acquisition of action rules, we divided the task into various subtasks. We propose an approach in which cooperative action is obtained by giving each agent a different reward; this encourages the agents to play different roles, with some specializing in relief and others specializing in the removal of obstacles. We investigated how the various autonomous agents determined the appropriate action rules and examined the influence of providing separate rewards to different agents in the system. We also compared the values of various actions in different learning situations.

2 Multi-agent Systems and the Disaster Relief Problem

2.1 Multi-agent Systems

An *agent* is a computational mechanism that exhibits a high degree of autonomy and performs actions in its environment based on information (sensors, feedback) received from that environment [12]. A multi-agent system is one in which several interacting, autonomous agents pursue some set of goals or perform some set of tasks. A key pattern seen in the interactions of multi-agent systems is goal- and task-oriented coordination, in both cooperative and competitive situations. In a cooperative interaction, several agents combine their efforts to accomplish as a group what the individuals could not accomplish alone, and in a competition interaction, several agents combine their efforts to obtain something that only some of them can have.

In [13], the following major characteristics of multi-agent systems were identified:

- each agent has incomplete information and is restricted in its capabilities;
- the control of the system is distributed;
- the data are decentralized; and
- the computation is asynchronous.

2.2 Disaster Relief Problem

The present study considers a disaster relief problem, in which the injured are placed on a field of fixed size and agents must rescue the injured as quickly as possible. This can be considered to be a multi-agent system in which the task of the system is to efficiently rescue all of the injured, but in which each agent focuses on achieving its own target.

As shown in Fig. 1, the field is divided into an N × N lattice. Agents are indicated by solid black circles ●, their initial position is indicated by solid gray squares ■, the injured are indicated by solid red squares, and obstacles are indicated by solid blue triangles ▲. There are two types of obstacles ▲: those that can be removed and those that cannot; agents are able to recognize to which category a given obstacle belongs. The injured are to be conveyed to the upper-left corner of the field, and movable obstacles are to be conveyed to the upper-right corner of the field. A single step is defined such that each of the agents on the field completes a single action, and the field is re-initialized once all of the injured have been moved.

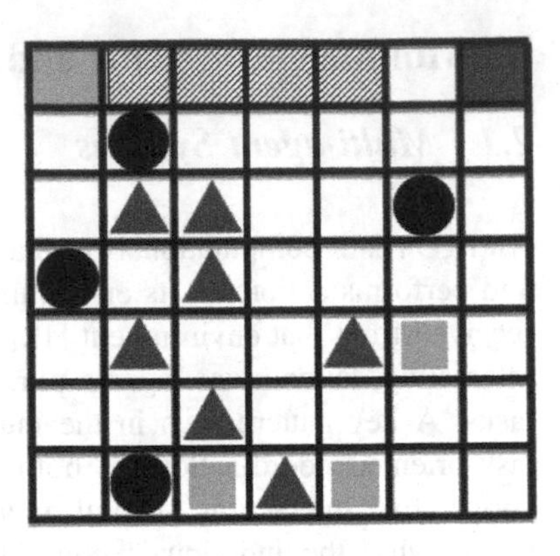

Fig. 1 Example of a disaster relief problem

2.3 *Actions of Available to Agents in Disaster Relief Problem*

The actions that an agent can take are as follows (Fig. 2):

- moving up, down, right, or left to an adjacent cell;
- remaining in the present cell.

No agent may move to a cell that is occupied by an obstacle that cannot be removed or by another agent, nor can an agent move outside the field; if there are no useful allowed moves, the agent remains in the current cell.

In addition, when an agent moves to a cell in which there is an injured person or a movable obstacle, that person or object will be taken for removal to the appropriate location (Fig. 3).

Once an agent has taken possession of an injured person and moved them to the appropriate destination, the subtask is completed (Fig. 4), and the agent can begin a new subtask. The overall task is completed when all of the injured on the field have been rescued.

(a) moving to a an adjacent cell

(b) remaining in the present cell

Fig. 2 Actions available to an agent

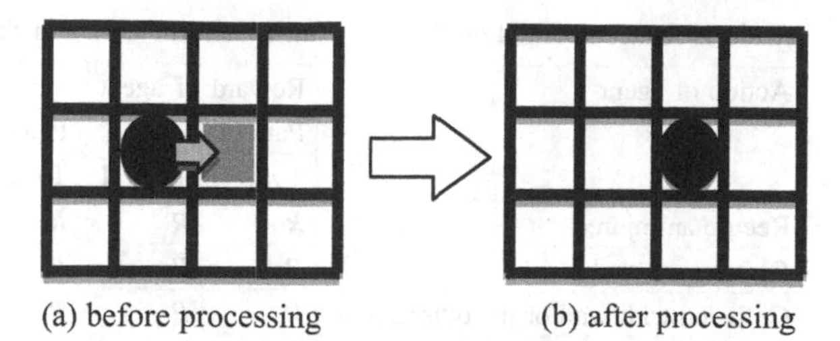

Fig. 3 Processing of injured person or movable obstacle

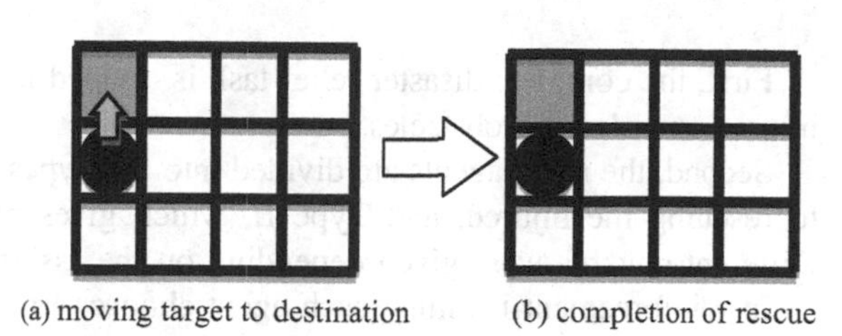

(a) moving target to destination
(b) completion of rescue

Fig. 4 Completing the rescue of an injured person

3 Cooperative Action Acquisition of Disaster Relief Multi-agents by Division of Roles

3.1 Cooperative Action of Agents

Multi-agent systems enable problems to be solved more efficiently. In addition, multi-agent systems can solve problems that may be impossible for an individual agent to solve, because multiple agents have a single common goal and can act cooperatively when adjusting to their environment and performing various actions. Occasionally, in order to proceed with a task in a changing environment, such agents must make precise judgements in order to make adaptive moves. In order to cooperate, it is necessary to have a rule that working as a group takes priority over the actions of any individual. Generally speaking, individual action is natural, whereas group action is acquired by learning to accomplish goals through cooperative actions [14–16].

In this study, in order to use reinforcement learning to efficiently acquire action rules, we divided the whole task into various subtasks. Cooperation was achieved by giving each agent a different reward, and assigning some agents to perform relief (*relief agents*) and others to remove obstacles (*removing agents*).

3.2 Cooperative Actions with the Role Division

For the disaster relief problem, the cooperative actions of the multiple agents are determined by using reinforcement learning and a division of roles.

Table 1 Rewards based on the task priority and the action of the agent

Action of agent	Reward of agent					
	Pattern 1		Pattern 2		Pattern 3	
	Type I	Type II	Type I	Type II	Type I	Type II
Recure an injured	R	R	R_h	R_i	R_h	0
Clear an obstacle	R	R	R_1	R_h	0	R_h
Collision with wall or the other agent	P	P	P	P	P	P

First, the complete disaster relief task is divided into two subtasks: rescuing the injured and clearing obstacles.

Second, the relief agents are divided into two types: Type I, which gives priority to rescuing the injured, and Type II, which gives priority to clearing obstacles. Different rewards were given depending on the assigned role.

In reinforcement learning, each agent chooses from among the various allowed actions, the one from which it will obtain the greatest reward. This means that by allocating a larger reward to a task, it is given higher priority; thus, it is possible to ensure that prioritize tasks when learning.

In this study, there were three patterns of rewards, which were assigned according to the priority of the task and the action of the agent. These are shown in Table 1, where R, R_h, and R_l are rewards, $R_h > R > R_l$, and P is a penalty ($P < 0$), which is given to an action that does not help achieve the task.

In Pattern 1, the same reward R is given to any agent who acts in a useful way; this is the same approach that is used in conventional methods. In this case, all agents learn that saving the injured and clearing obstacles have the same priority, and so the tasks are performed with the same frequency. In other words, all agents share a common two-factor task.

In Pattern 2, different rewards are assigned to agents depending on their type. Each agent chooses their actions based on the task that has been prioritized by being given a higher reward (R_h). If an agent chooses an action for a lower priority action, they receive a smaller reward (R_l). In this way, roles are assigned: Type I agents give priority to saving the injured, and Type II agents give priority to clearing obstacles.

In Pattern 3, only those agents who perform the highest priority action for their type receive a reward (R_h). In this way, there is complete role division between the different types of agents.

In all patterns, a penalty P is assigned to any agent that collides with a wall or another agent. In an actual disaster site, collision with other robots may lead to failure, and so during the learning phase, there are rewards for learning to avoid collisions.

4 Simulation and Results

To evaluate the effectiveness of the proposed method, we performed a simulation of a multi-agent disaster relief problem. We used the Java programming language.

We used a 15 × 15 grid for the field, and injured people and obstacles were arranged randomly. Each experiment was repeated 10,000 times, and each experiment began with a random initial state and continued until all agents had completed their tasks. The agents retain the information learned in all previous experiments.

Each agent has a constant but limited view of the field, and it can assess the surrounding environment. They are able to recognize other agents, obstacles, and the injured that appear within their range of vision. An example of their range of vision is shown in Fig. 5, where the grid squares in an individual agent's visual field are indicated by shading. The experimental conditions are listed in Table 2.

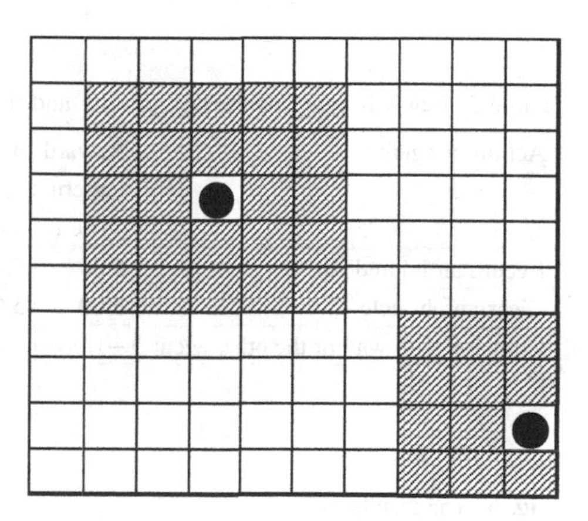

Fig. 5 Visual field of agents

Table 2 Experimental conditions

The setting of filed	
The field size	15 × 15
The number of rescue agents	2
The number of clearing agents	2
The number of injured individuals	5
The number of removal possible obstacles	15
The number of removal impossible obstacles	5
The setting of agent	
Range of vision of an agent	5 × 5
Learning rate α	0.1
Discount rate γ	0.9
Greedy policy ε	0.1

4.1 Results of Different Rewards

The three reward patterns are shown in Table 3. To compare their effectiveness, the total reward was the same for each pattern.

The learning results for three patterns are shown in Fig. 6. The horizontal axis indicates the number of experiments, and the vertical axis indicates the number of steps required to achieve the task and it is average of the previous 100 episodes of the step up to this episode. The average number of steps for each pattern in the final experiment is shown in Table 4.

Figure 6 and Table 4 show that Pattern 2 was more efficient than Pattern 1, which is the traditional method. With Pattern 1, there were delays in removing debris and saving the injured because no agents were taught to prioritize the clearing of obstacles. However, with Pattern 2, rescuing the injured, which it the goal, could be carried out effectively because agents were able to consider the whole task while still giving priority to their assigned task. Pattern 3 resulted in

Table 3 Rewards based on the task priority and the action of the agent

Action of agent	Reward of agent					
	Pattern 1		Pattern 2		Pattern 3	
	Type I	Type II	Type I	Type II	Type I	Type II
Recure an injured	5.00	5.00	6.67	3.33	10.00	0.00
Clear an obstacle	5.00	5.00	3.33	6.67	0.00	1.00
Collision with wall or the other agent	−1.00	−1.00	−1.00	−1.00	−1.00	−1.00

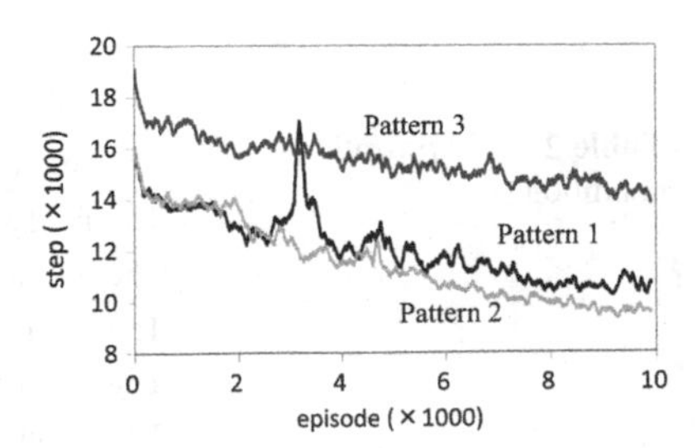

Fig. 6 The results of different rewards

Table 4 Average number of steps in the final experiment

	The average of steps
Pattern 1	10679.081
Pattern 2	9575.845
Pattern 3	13877.415

the greatest number of steps to complete the task because only those agents performing their assigned duty were rewarded; thus, for example, Type II agents did not attempt to save the injured.

Based on these experiment results, we find that Pattern 2 is effective for efficient learning. However, there is not much difference from the results for Pattern 1, which is the conventional method, and so there is room for improvement. The roles as currently assigned are not optimal for task completion, because, for example, Type II agents are rewarded for removing debris, even if this is unnecessary. This could be remedied by rewarding Type II agents only when they remove debris that is preventing the rescue of victims.

4.2 Q-Factors and Agent Cooperation

Almost all reinforcement learning are based on estimating value functions of state-action (i.e., Q-factor). In this study, to evaluate the effectiveness of the proposed method, we compared the Q-factors following learning.

Consider an agent that is in the state shown in Fig. 7; the Q-factor for each state, each pattern, and each type of agent is shown in Table 5. The reward is the value used to prioritize each behavior. In Fig. 7, a solid black circle ● indicates an agent, a solid red square ■ indicates an injured person, and a solid blue triangle ▲ indicates an obstacle.

In Table 5, we can see that for Pattern 2, the Q-factors were similar to the reward for both types of agents. However, for the state shown in Fig. 7a, for Pattern 1, the Q-factors were smaller than the reward value, which implies that there was not sufficient learning.

These results show that with the proposed method (Pattern 2), the agents performed their assigned roles and cooperated with each other. In addition, the learning time was sufficient.

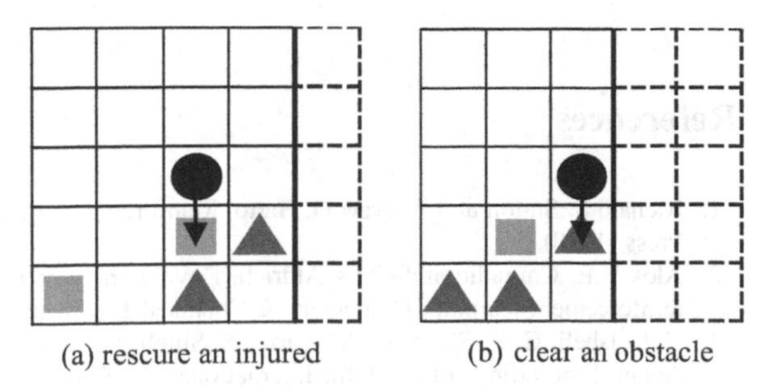

Fig. 7 State of an agent

Table 5 Q-factors of each rescue agent following learning

State	Method	Agent	Reward	Q-factor
Recure an injured	Pattern 1	Type I	5.000	1.720
		Type II	5.000	3.063
	Pattern 2	Type I	6.670	6.571
		Type II	3.330	3.035
Clear an obstacle	Pattern 1	Type I	5.000	5.000
		Type II	5.000	4.641
	Pattern 2	Type I	3.330	3.156
		Type II	6.670	6.079

5 Conclusion

We considered a disaster relief simulation in which multiple autonomous robots worked as a multi-agent system. We examined whether reinforcement learning would be an efficient method for the agents to learn their tasks and the environment. We proposed dividing the task into various subtasks and ensuring cooperation by giving agents different rewards for different tasks. In this way, tasks were divided between rescue agents and clearing agents.

The results of a simulation showed that the number of steps required to rescue an injured individual decreased as the number of learning iterations increased; thus, allowing the agents to learn had a positive effect. In addition, for the proposed method, the Q-factors were close to the reward for both types of agents, and the learning was efficient.

However, at an actual disaster site, the state of the injured and the amount of generated debris may change from moment to moment, and the situation will certainly be more complex. Therefore, in future research, we intend to further develop the simulation environment so that it can model more varied conditions.

Acknowledgments This research was supported by JSPS KAKENHI Grant Number JP16K01303.

References

1. Richard S. Sutton and Andrew G. Barto, Reinforcement Learning: An Introduction, The MIT Press (1998).
2. Alex V.E. Conradie and Chris Aldrich, Development of neurocontrollers with evolutionary reinforcement learning, Computers & Chemical Engineering, Vol. 30, pp. 1–17 (2005).
3. C. L. Isbell, C. R. Shelton, M. Kearns, S. Singh, P. Stone, A Social Reinforcement Learning Agent, Proceedings of the Fifth International Conference on Autonomous Agents, pp. 377–384 (2001).
4. Michael Wooldridge, Intelligent Agents, Multiagent Systems: A modern Approach to Distributed Artificial Intelligence, edited by Gerhard Weiss, The MIT Press (2000).

5. Peter Stone and Manuela Veloso. Using Machine learning in the Soccer Server, Proc. of IROS-96 Workshop on Robocup (1996).
6. Kumiko Tanaka-Ishii, Itsuki Noda and Ian Frank et.al, MIKE: An Automatic Commentary System for Soccer – System Design and Control-, Proceedings of International Conference on Multi-Agent Systems'98 (1998).
7. H. Matsubara, I. Frank and K. Tanaka et.al., Automatic Soccer Commentary and RoboCup, The 2nd Proceedings of RoboCap Workshop (1998).
8. Pereira, Renato de Pontes/ Engel, Paulo Martins, A Framework for Constrained and Adaptive Behavior-Based Agents, arXiv preprint arXiv:1506.02312v1, pp. 1–16 (2015).
9. K. Kovac, Simulation of multi-robot reinforcement learning for box-pushing Problem, Proceedings of the 12th IEEE Mediterranean, Vol. 2, pp. 603–606 (2004).
10. P. Legreneur, M. Laurin and V. Bels, Predator-prey interactions paradigm: a new tool for artificial intelligence, Adaptive Behavior, Vol. 20, pp. 3–9 (2012).
11. K. C. Jim and C. L. Giles, Talking Helps: Evolving Communicating Agents for the Predator-Prey Pursuit Problem, Artificial Life, No. 6, Vol.3, pp. 237–254 (2000).
12. L. Panait, and S. Luke, Cooperative Multi-Agent Learning: The state of the Art, Auto Agents Multi-Agent Syst (3), pp. 387–434 (2005).
13. N. Jennings, K. Sycara, and M. Wooldridge, A roadmap of agent research and development, Autonomous Agents And Multi-Agent Systems, Vol.1, pp. 7–38 (1998).
14. M. Xie and A. Tachibana, Cooperative Behavior Acquisition for Multi-agent Systems by Q-learning, Proceedings IEEE Symposium on Foundations of Computational Intelligence, 424–428 (2007).
15. F. Toriumi, H. Yamamoto and I. Okada, Effects of Controllable Facilitators on Social Media: Simulation Analysis Using Generalized Metanorms Games, Proceeding of Web Intelligence and Intelligent Agent Technologies, 2013 IEEE/WIC/ACM international Joint Conferences (2013).
16. L. Panait, and S. Luke, Cooperative Multi-Agent Learning: The state of the Art, Auto Agents Multi-Agent Syst (3), pp. 387–434 (2005).

An Evolutionary Framework for Bi-objective Dynamic Economic and Environmental Dispatch Problems

Forhad Zaman, Saber M. Elsayed, Tapabrata Ray and Ruhul A. Sarker

Abstract A dynamic economic and environmental dispatch (DEED) problem is a challenging bi-objective optimization problem that simultaneously minimizes both operating costs and gas emissions. To solve it, several evolutionary algorithms (EAs) have been used, each of which has pros and cons, with one performing better in an early stage of evolution and another later. In this paper, to solve such problems, an evolutionary framework is designed based on two EAs, a genetic algorithm (GA) and differential evolution (DE), dynamically configures the better of the two during the evolution. In it, two sub-populations are performed, one for each of GA and DE, and their sizes updated in each generation according to the respective algorithm's performance in previous generations. Moreover, a heuristic is employed to improve the performance of the proposed algorithm by repairing infeasible individuals towards feasible directions. To demonstrate its performance, two renewable-based DEED problems are solved using the proposed and state-of-the-art algorithms. An analysis of the simulation results reveals that the proposed algorithm is the best of those considered, with the heuristic enhancing its performances.

Keywords Renewable energy · Economic and emission dispatch · Multiobjective optimization · Differential evolution · Genetic algorithm

F. Zaman (✉) · S.M. Elsayed · T. Ray · R.A. Sarker
School of Engineering and Information Technology,
University of New South Wales Canberra, Canberra, Australia
e-mail: md.zaman@student.adfa.edu.au

S.M. Elsayed
e-mail: s.elsayed@unsw.edu.au

T. Ray
e-mail: t.ray@unsw.edu.au

R.A. Sarker
e-mail: r.sarker@unsw.edu.au

G. Leu et al. (eds.), *Intelligent and Evolutionary Systems*,
Proceedings in Adaptation, Learning and Optimization 8,
DOI 10.1007/978-3-319-49049-6_36

1 Introduction

Over the last decade, economic dispatch (ED) problems have been used to determine the allocation of electricity demand among fossil fuel-based thermal generating units to minimize operating costs subject to physical and technological constraints [13]. However, the excessive use of fossil fuels produces large amounts of atmospheric pollutants that are continuously released into the environment. Therefore, alternatives to thermal energy power generation have emerged, such as solar, wind and hydro energy, which are now widely used because of their lower production costs and environmentally friendly characteristics [14, 17]. Consequently, the economic and environmental dispatch (EED), a new bi-objective optimization problem, has been introduced to simultaneously minimize operating costs and air pollution [12].

A DEED problem is an extension of the conventional EED one that schedules generators for an operational cycle in a time horizon divided into multiple periods while taking into account the intrinsic links between two hours of the ramp limit of a thermal generator [15]. Although the DEED is a more realistic problem than the EED, its computational process is also more complex because of its large number of decision variables and chain of equality constraints [13]. Moreover, considering the valve point effect (VPE) of a thermal generator's cost function, it becomes a non-linear, non-smooth, non-convex and multi-modal bi-objective optimization problem which is difficult to solve using a classical optimization approach [13, 16]. Therefore, an efficient algorithm, such as an EA, is required because of its flexible, efficient and stochastic searching feature [13].

During the last decade, several meta-heuristic methods, such as GA [13], simulated annealing (SA) [8], particle swarm optimization (PSO) [18] and DE [13], PSO—sequential quadratic programming (PSO-SQP) [11] and modified hybrid evolutionary programming (EP)–SQP (MHEP-SQP) [11] have been effectively used to solve various single-objective DED problems. Also, several algorithms, such as a binary PSO [6], hybrid PSO, non-dominated sorting GA-II (NSGA-II) with a heuristic (H-NSGA-II) [12], infeasibility-driven EA (IDEA) with a heuristic (H-IDEA) [12] and gravitational search algorithm (GSA) with GA [5], have been used to solve bi-objective DEED problems. However, most solve the problems as single-objective optimization ones by aggregating two objectives to produce a single solution, not a Pareto frontier, with many runs required to generate a set of trade-off solutions [12].

In our previous research [13], it was found that, for solving different types of DED problems, one EA may perform well in an early stage of the optimization process but less well in later generations and vice versa. To efficiently solve a DED problem, multi-method EAs that integrate two or more optimization techniques in order to utilize their strengths and overcome their own and each other's weaknesses, have been developed. Similar ones, such as multiple operators of a GA in [10], a multi-operator evolutionary framework with various EAs in [4] and a general framework of two EAs (GA and DE) in [13] have been developed to solve various single-objective optimization problems. However, to the best of our knowledge, solving bi-objective DEED problems using a multi-EAs framework has not yet been explored.

In this paper, an evolutionary framework called GA-DE, in which two EAs (GA and DE) are run in parallel under two sub-populations, is designed to solve bi-objective DEED problems. Although the initial sub-population sizes are the same, they are dynamically varied in each generation based on the performance of each EA in previous generations. After a predefined number of generations (also called a cycle), only the better algorithm is allowed to run alone for a subsequent cycle. After that cycle is finished, both algorithms are run again for another cycle, with both using the same sub-population size. The process is continually repeated until a stopping criterion met. Moreover, rather than setting the control parameters of DE, self-adaptive mutation and crossover techniques that automatically configure the best ones in each generation are used. Also, a heuristic technique is employed to improve the convergence rate of each algorithm by rectifying infeasible individuals towards feasible directions. The results obtained by the proposed approach for solving two renewable-based bi-objective DEED problems, (i) hydro-thermal [1] and (ii) solar-thermal [16], are compared with those from recent state-of-the-art algorithms, with GA-DE shown to perform best.

The rest of this paper is organized as follows: Sect. 2 presents the problem formulation, Sect. 3 the proposed methodology, Sect. 4 the experimental results and analysis and Sect. 5 conclusion and future works.

2 Mathematical Formulations

The bi-objective hydro-thermal and solar-thermal DEED problems are formulated to determine the optimal level of power generation in each participating plant by minimizing both the fuel costs and greenhouse gas emissions while satisfying a number of constraints, as presented in this section.

2.1 Hydro-Thermal

In the hydro-thermal DEED problem, the objectives are to minimize both the operating costs and gas emissions subject to a number of equality and inequality constraints.

2.1.1 Objective Functions

Considering the VPE, the cost and emission functions of thermal generators are, respectively:

$$F_c\left(P_{T_{i,t}}\right) = \sum_{t=1}^{T}\sum_{i=1}^{N_T}\left(a_i + b_i P_{T_{i,t}} + c_i P_{T_{i,t}}^2 + \left|d_i \sin\left\{e_i\left(P_{T_{i,t}}^{\min} - P_{T_{i,t}}\right)\right\}\right|\right) \forall i,t \quad (1)$$

$$F_E(P_{T_{i,t}}) = \sum_{t=1}^{T}\sum_{i=1}^{N_T}\left(10^{-2}\left(\alpha_i + \beta_i P_{T_{i,t}} + \gamma_i P_{T_{i,t}}^2\right) + \eta_i e^{\lambda_i P_{T_{i,t}}}\right) \forall i,t \quad (2)$$

The first objective of Eq. (1) is to minimize the sum of all the fuel costs of the thermal power plants ($P_{T_{i,t}}$) under consideration (N_T) during an operational cycle (T), where, a_i, b_i, c_i, d_i and e_i are the cost coefficients. The second Eq. (2) is to minimize the gas emissions from the thermal plants, where, α_i, β_i, γ_i, λ_i and η_i are their emission coefficients.

2.1.2 Constraints

The hydro-thermal DEED problem includes the following constraints.

$$\sum_{i=1}^{N_T} P_{T_{i,t}} + \sum_{j=1}^{N_T} P_{H_{j,t}} = P_{D_t}\ t \in T \quad (3)$$

$$P_{H_{j,t}} = C_{1,j}V_{j,t}^2 + C_{2,j}X_{j,t}^2 + C_{3,j}V_{j,t}X_{j,t} + C_{4,j}V_{j,t} + C_{5,j}X_{j,t} + C_{6,i}\ j \in N_H,\ t \in T \quad (4)$$

$$V_{j,t+1} = V_{j,t} - X_{j,t} + I_{j,t} - S_{j,t} + \sum_{r=1}^{N_{up}}\left(X_{r,(t-t_{drj})} + S_{r,(t-t_{drj})}\right),\ j \in N_H \quad (5)$$

$$P_{H_j}^{\min} \le P_{H_{j,t}} \le P_{H_j}^{\max}\ j \in N_H,\ t \in T \quad (6)$$

$$P_{T_i}^{\min} \le P_{T_{i,t}} \le P_{T_i}^{\max}\ i \in N_T,\ t \in T \quad (7)$$

$$V_{H_j}^{\min} \le V_{H_{j,t}} \le V_{H_j}^{\max}\ j \in N_H,\ t \in T \quad (8)$$

$$X_{H_j}^{\min} \le X_{H_{j,t}} \le X_{H_j}^{\max}\ j \in N_H,\ t \in T \quad (9)$$

$$\left|V_{j,t}\right|^{t=0} = V_j^{ini},\ \text{and}\ \left|V_{j,t}\right|^{t=T} = V_j^{end}\ j \in N_H \quad (10)$$

Equation (3) is the power balance constraint, where P_{H_j}, X_j, and V_j are the hydro power generation, water storage rate and volume of jth hydro plant, respectively. N_H, $C_{k,j} \forall k$, I_j, S_j, N_{up}, and $t_{d_{r,j}}$ are the number of hydro power plants, the generation coefficient, natural water inflow rate, spillage water (assume zero, as in [1]), number of upstream plants and water transport delay from the rth to jth reservoir, respectively. The constraints in Eqs. (6)–(9) are the capacity limits of the hydro and thermal plants, water storage volume and water discharge rate, respectively, where, $P_{H_j}{}^{min}$ and $P_{H_j}{}^{max}$, V_j^{min} and V_j^{max}, and X_j^{min} and X_j^{max} are the minimum and maximum output power of the hydro power plant, water storage volumes and water discharge rates, respectively. The initial and final reservoir storage volumes, which must meet the requirements of all the reservoirs, are expressed in Eq. (10), where, V_j^{ini} and V_j^{end} are the initial and final water volumes of the jth reservoir, respectively.

2.2 Solar-Thermal

The solar-thermal DEED problem is considered a mixed-integer non-linear bi-objective optimization problem (MINP) [6] in which the solar and thermal units are represented as binary and continuous variables, respectively. Its objective functions and constraints are described below.

2.2.1 Objective functions

The objective functions of the solar-thermal DEED problem are to minimize both the operating costs and gas emissions, respectively, as:

$$\text{Min} : F_C\left(P_{T_{i,t}}, U_{s_{s,t}}\right) = \sum_{t=1}^{T}\left(\sum_{i=1}^{N_T}\left(F_{c_i}(P_{T_{i,t}})\right) + \sum_{s=1}^{N_S}\left(F_{S_S}(U_{S_{s,t}})\right)\right) \tag{11}$$

$$\text{where, } F_{c_i}(P_{T_{i,t}}) = a_i + b_i P_{T_{i,t}} + c_i P_{T_{i,t}}^2 + \left| d_i \sin\left\{e_i\left(P_{T_{i,t}}^{\min} - P_{T_{i,t}}\right)\right\}\right| \tag{12}$$

$$F_{S_s}(U_{s_{s,t}}) = PU_{\cos t_s} P_{S_{s,t}} U_{s_{s,t}},\ U_{S_{s,t}} \in \{0, 1\}\ s \in N_S\ t \in T \tag{13}$$

$$P_{S_{s,t}} = P_{r_s}\left\{1 + \Omega\left(T_{amb_{s,t}} - T_{ref_s}\right)\right\}\frac{Si_{s,t}}{1000} \tag{14}$$

$$\text{Min: } F_E\left(P_{T_{i,t}}\right) = \sum_{t=1}^{T}\sum_{i=1}^{N_T} h_i\left(F_{e_i}(P_{T_{i,t}})\right) = \sum_{t=1}^{T}\sum_{i=1}^{N_T}\left(\frac{F_{ci}\left(P_i^{\max}\right)}{F_{E_i}\left(P_i^{\max}\right)}\right)\left(F_{e_i}(P_{T_{i,t}})\right) \tag{15}$$

$$\text{where },F_{E_i}(P_{T_{i,t}}) = \alpha_i + \beta_i P_{T_{i,t}} + \gamma_i P^2_{T_{i,t}} + \eta_i e^{\lambda_i P_{T_{i,t}}} \; i \in N_T \, t \in T \tag{16}$$

The first objective function in Eq. (11) involves the operational costs of the solar and thermal generators, and the second in Eq. (15) the gas emissions from the thermal plants normalized to the cost function. Equation (13) indicates the operational costs of solar power generation ($P_{S_{s,t}}$), where, $U_{S_{s,t}}$ is a binary decision variable that determines whether a unit is turned on or off, and PU_{cost} the per unit cost of $P_{S_{s,t}}$, which expressed in Eq. (14), where P_{r_s} is the rated power, T_{ref_s} and $T_{amb_{s,t}}$ the reference and temperature, respectively, Ω the temperature coefficient and $Si_{s,t}$ the incident solar radiation of the sth plant at the tth time.

2.2.2 Constraints

The solar-thermal DEED problem has the following equality and inequality constraints.

$$\sum_{i=1}^{N_T} P_{T_{i,t}} + \sum_{s=1}^{N_S} P_{S_{s,t}} U_{S_{s,t}} = P_{D_t} + P_{loss_t} \; t \in T \tag{17}$$

$$P_{T_i}^{\min} \le P_{T_{i,t}} \le P_{T_i}^{\max} \; i \in N_T, \, t \in T \tag{18}$$

$$-DR_i \le P_{T_{i,t}} - P_{T_{i,t-1}} \le UR_i \; i \in N_T \, t \in T \tag{19}$$

$$\sum_{t=1}^{T}\sum_{s=1}^{N_S} P_{S_{s,t}} U_{S_{s,t}} \le 0.3 P_{D_t} \tag{20}$$

Equation (17) defines the power balance constraints, and Eqs. (18) and (19) the capacity and ramp constraints of the thermal generators, respectively, with UR and DR the upward and downward transition limits, respectively. The constraint in Eq. (20) is used to limit the solar share at any time based on a 30 % upper limit to avoid any uncertainty in terms of solar irradiance [6].

3 Proposed Bi-objective GA-DE Algorithm

In this research, an evolutionary framework is designed by configuring two optimization algorithms, namely GA and DE for solving the bi-objectives DEED problems. In the design, an initial population of size N_P is generated and then randomly divided into two subpopulations of equal size of N_{P1} and N_{P2} for GA and DE, respectively. In subsequent generations, the new individuals in GA and DE are generated from random individuals from either subpopulation (N_{P1} and N_{P2}) rather than only their own which results in information being exchanged between the two algorithms in each generation. Once the fitness functions of both the parents and children are evaluated, a non-dominated sorting approach [2] is applied to rank each individual, with the best N_P individuals selected for the next generation.

Based on the percentage of offspring surviving to the next generation, the success rate (*SR*) of each algorithm is calculated, and their subpopulation sizes subsequently updated considering their lower (N_{P1}^{min}) and upper (N_{P1}^{max}) bounds as in Eq. (24). This process is continued until a predefined number of generations (N_{gc}) is performed. Then, the best algorithm is determined, based on its average *SR* (*ASR*) during the last N_{gc}, and used to evolve all the N_P individuals from both subpopulations for the next N_{gc} generations. Once subsequent N_{gc} are completed, the final individuals are again equally and randomly assigned to both algorithms with two subpopulations (N_{P1} and N_{P2}) to evaluate next N_{gc} with the GA-DE algorithm terminating once the maximum number of generations, N_G is over. The pseudo code of the proposed GA-DE algorithm is shown in Algorithm 1.

3.1 *Initial Population*

The chromosomes or representations of the decision variables for both GA and DE are expressed as:

$$\vec{x}_p = \begin{cases} [P_{T_{i,t}}, X_{j,t}]_{1:N_x} & \text{for hydrothermal system} \\ [P_{T_{i,t}}, U_{S_{s,t}}]_{1:N_X} & \text{for solar-thermal system} \end{cases} \tag{21}$$

where, $i = 1, 2, \ldots, N_T, j = 1, 2, \ldots, N_H$, $s = 1, 2, \ldots, N_S$, $t = 1, 2, \ldots, T$, $U_{S_{s,t}} \in [0, 1]$, $p \in N_P$, with N_P the population size and N_x the number of decision variables as $T \times (N_T + N_H)$ for the hydrothermal system, and $T \times (N_T + N_S)$ for the solar-thermal one. Each individual is generated as:

$$\vec{x}_p = \vec{x}^{\min} + \left(\vec{x}^{\max} - \vec{x}^{\min}\right) \; lhs(N_x), \forall p = 1, 2, \ldots, N_P \tag{22}$$

where $\vec{x}^{\min}$ and $\vec{x}^{max}$ are the vectors of the lower and upper bound, respectively, and $\vec{x}_p$ the pth individual in the N_P population, with $lhs\left(N_x\right)$ random individuals generated using Latin hypercube sampling (LHS) rules [15].

Algorithm 1 GA-DE algorithm

Require: N_G, N_P, N_{P1}^{min} and N_{P1}^{max}
1: Set, $count_1 = count_2 = 0$
2: Randomly generate initial individuals using Eq. (21)
3: Evaluate the individuals after repairing the infeasible individuals using heuristic described in Sect. 3.3
4: Randomly distribute N_P individuals over two subpopulations with sizes of N_{P1} and N_{P2}, such that $N_{P1} = N_{P2}$
5: **for** $g = 1 : N_G$ **do**
6: Set, $count_1 = count_1 + 1$
7: **if** $count_1 \leq N_{gc}$ **then**
8: Generate N_{P1} and N_{P2} offspring from the all N_P parents using GA and DE operators, respectively
9: Repeat step 3 for both N_{P1} and N_{P2}
10: Determine best individuals from parents and offspring based on non-dominated selection approach described in Sect. 3.4
11: Calculate $SR_{1,g}$ and $SR_{2,g}$ based on numbers of offspring of GA and DE surviving to next generation, respectively,
12: Group selected individuals, $N_P \leftarrow N_{P1} + N_{P2}$
13: Update N_{P1} and N_{P2} according to Eqs. (24) and (25), respectively
14: **else**
15: set, $count_2 = count_2 + 1$
16: **if** $count_2 \leq N_{gc}$ **then**
17: Calculate average success rates of GA (ASR_1) and DE (ASR_2)
18: **if** $ASR_1 > ASR_2$ **then**
19: Perform GA, considering $N_{P1} \leftarrow N_{P1} + N_{P2}$
20: **else**
21: Perform DE, considering $N_{P2} \leftarrow N_{P1} + N_{P2}$
22: **end if**
23: **end if**
24: **if** $count_2 = N_{gc}$ **then**
25: Repeat step 4 and set again, $count_1 = count_2 = 0$
26: **end if**
27: **end if**
28: **end for**

3.2 GA-DE Search Operators

To update the individuals in GA-DE, we use either GA or DE search operators in various stages of an evolution, as previously discussed. Of the different operators available, simulated binary crossover (SBX) and non-uniform mutation (NUM) are used in GA and two self-adaptive mutation operators and one binomial crossover in DE because they showed superior performances for solving various DED problems in [3, 12, 13, 15]. Due to the limitation of this paper's number of pages, details of these operators are not provided but can be found in [15].

3.3 Heuristic for DEED Constraints

As previously mentioned, a DEED problem involves a number of equality and inequality constraints, all of which new solutions generated by an EA process may not satisfy, especially during the early stages of an evolution. To maintain feasibility throughout generations, in our previous research, we developed a heuristic for a single-objective DED problem [15]. In it, a DED problem with a 24-h load cycle is converted into 24 sub-problems, with the hourly infeasible individuals repaired in feasible directions based on a forward and backward slack-generation approach. In this paper, we employ this heuristic for a bi-objective DEED problem, with its detailed steps provided in [15].

3.4 Selection Process

To rank the chromosomes, firstly, the parents and offspring are grouped together and the best N_P individuals among them selected for the next generation. To do this, we use a popular constraint-handling approach with a non-dominated sorting technique [2] in which an additional objective is considered based on the amount of relative constraint violations (CVs). Then, a crowding sorting technique and non-dominated mechanism are used to preserve diversity and elitism among the population members. The advantages of having an additional objective for constrained optimization problems are explicitly demonstrated in [9], with that of each individual of each algorithm expressed as:

$$CV_p = \sum_{k=1}^{K} \max\left(0, G_k\left(\vec{x}_p\right)\right) + \sum_{e=1}^{E} \max\left(0, H_e\left(\vec{x}_p\right) - \varepsilon_g\right) \ \forall p \in N_P \tag{23}$$

where $\vec{x}_p$ represents the pth individual in a sub-population, G and H their inequality and equality constraints, respectively, K and E their numbers of inequality and equality constraints, respectively, for a DEED problem.

Based on the number of individuals selected from the offspring, the *SR* of each algorithm is calculated; for example, if 30 % of the offspring of GA survive to the next generation, SR_1 is 30 %. Then, the subpopulation sizes (N_{P1} and N_{P2}) are updated for the next generation according to their normalized SRs as:

$$N_{P1} = \max\left[N_{P1}^{\min}, \min\left\{N_P \frac{SR_{1,g}}{SR_{1,g} + SR_{1,g}}, N_{P_1}^{\max}\right\}\right] \tag{24}$$
$$SR_{1,g} \cup SR_{2,g} \neq 0,\ g \in N_G$$

$$N_{P2} = N_P - N_{P1} \tag{25}$$

Equations (24) and (25) are applied when at least one SR is nonzero, whereas, if both are zero, the values of N_{P1} and N_{P2} remain the same as in the immediate previous generation.

4 Experimental Results and Analysis

For the experimental study, two standard benchmarks, (i) a 7-unit hydro thermal power system from [1, 12]; and (ii) a 19-unit solar-thermal power system from [6, 13, 14], for a 24-h planning horizon in one-hour time period are solved using our proposed and state-of-the-art algorithms with and without considering the heuristic, as follows:

1. Non-dominated sorting GA-II (NSGA-II) without heuristic,
2. Multi-objective DE (MODE) without heuristic,
3. Proposed GA-DE without heuristic,
4. NSGA-II with heuristic (H-NSGA-II),
5. MODE with heuristic (H-MODE),
6. Proposed GA-DE with heuristic (H-GA-DE),

Based on [13], N_P, N_g are set to 200, 500, and 100, 1000 for the hydro-thermal and solar-thermal systems, respectively, and NP_1^{min}, NP_1^{max}, and N_{g_c} to 20, 80, and 50, respectively for both. It is also noted that each algorithm evaluates an equal number of fitness functions for a fairer comparisons. Also, each one runs 30 times using a desktop personal computer which has a 3.4 GHZ Intel Core i7 processor with 16 GB of RAM using the MATLAB (R2014a) environment.

4.1 Hydro-Thermal DEED

In this section, a 7-unit bi-objective hydro-thermal DEED problem comprising 3 thermal and 4 hydro units is solved using the proposed and state-of-the-art algorithms on the same platform. Once the 30 random runs of each algorithm are completed, their hyper-volume (HV) values are calculated based on their normalized fitness values as [7]:

$$f_{norm} = \frac{f - f_{ideal}}{f_{Nadir} - f_{ideal}} \tag{26}$$

where, f_{norm} and f are the normalized and actual function values, respectively, and f_{ideal}, and f_{Nadir} the ideal and nadir points [7] for this problem, respectively, which are found to be (7.17E+4,10.09) and (1.28E+5,142.95), respectively from all the runs of all the algorithms considered. The best, mean, median, worst, and standard deviation (STD) of the HV values obtained from algorithm with and without the heuristic

Table 1 Comparison of performances of algorithms for hydro-thermal DEED

Algorithm	HV (reference: [1,1])					Time (sec)	MR
	Best	Mean	Median	Worst	STD		
NSGA-II	0.59	0.53	0.54	0.44	0.05	56.81	1.60
MODE	0.49	0.43	0.44	0.35	0.05	**48.91**	1.70
GA-DE	0.71	0.67	0.68	0.63	0.03	53.75	2.70
H-NSGA-II	0.84	0.81	0.81	0.79	0.01	237.82	4.90
H-MODE	0.81	0.77	0.78	0.71	0.03	232.55	4.10
H-GA-DE	**0.91**	**0.89**	**0.89**	**0.87**	**0.01**	234.41	6.00

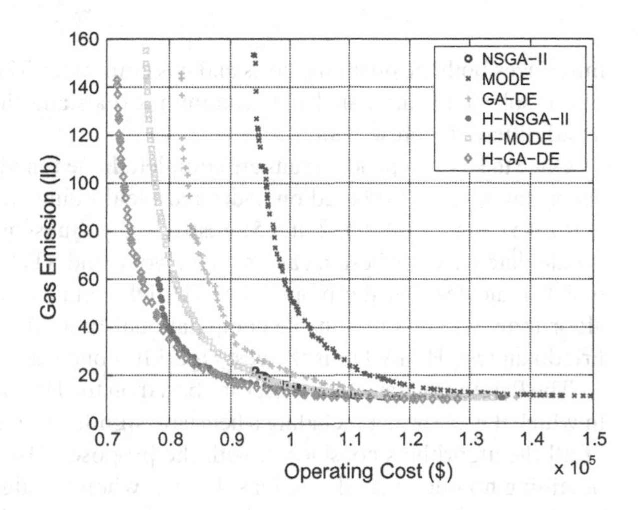

Fig. 1 Pareto-frontiers for hydro-thermal problem

are shown in Table 1. It is indicated that the proposed approach with the heuristic (H-GA-DE) obtains the best and most consistent results of all the algorithms in a reasonable computational time. The Pareto-frontiers of the best runs based on the HV values for all the algorithms are plotted in Fig. 1 which also shows the superiority of the proposed algorithm. In fact, the GA-DE approach obtains the best non-dominated solutions, both inclusive and exclusive of the heuristic, with H-GA-DE the best algorithm of all. Also, a Friedman test is performed considering the HV of each run of each algorithm with their mean ranks (MRs) are listed on Table 1 which proved that the H-GA-DE is the best algorithm.

4.2 *Solar-Thermal DEED*

To demonstrate the performances of the six algorithms, with and without the heuristic, on larger problems, in this section, we solve a 19-unit solar-thermal DEED problem formulated as a mixed-integer, non-linear, bi-objective optimization one that

Table 2 Comparison of performances of algorithms for solar-thermal DEED

Algorithm	HV (reference: [1,1])					Time (sec)	MR
	Best	Mean	Median	Worst	STD		
NSGA-II	0.18	0.17	0.17	0.15	0.01	64.47	1.00
MODE	0.21	0.20	0.21	0.18	0.01	**52.21**	2.80
GA-DE	0.20	0.19	0.20	0.18	0.01	75.38	2.20
H-NSGA-II	0.52	0.51	0.51	0.51	0.00	158.21	5.00
H-MODE	0.50	0.49	0.49	0.47	0.01	148.30	4.00
H-GA-DE	**0.56**	**0.55**	**0.55**	**0.54**	**0.00**	212.48	6.00

minimizes both the operating costs and gas emissions. The binary decision variables of the solar units are handled as continuous ones and then rounded off in order to avoid different representations.

Once the 30 independent runs are completed, the functions' values are normalized according to Eq. (26) based on nadir and ideal points, and found to be, (8.17E+5, 2.36E+5) and (3.08E+5, 2.0E+5), respectively. Subsequently, the HV of each run is calculated and the best, mean, median, worst and STD values presented in Table 2 which indicates that the proposed H-GA-DE obtains the best solutions of all the algorithms within a reasonable computational time. Also, based on the MR of the Friedman test, H-GA-DE is the best algorithm once again.

The Pareto frontiers of the best runs based on the HV values are presented in Fig. 2 in which it is clear that including a heuristic significantly improves the performances of all the algorithms considered, with the proposed H-GA-DE the best in terms of obtaining non-dominated solutions. In fact, when the algorithms do not include the heuristic, as their numbers of feasible solutions are very limited, the range of Pareto

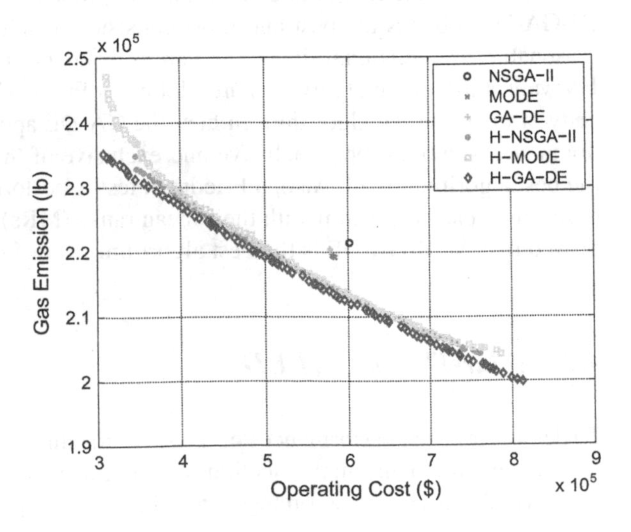

Fig. 2 Pareto-frontiers for solar-thermal problem

frontiers is very narrow. Conversely, when the heuristic is applied to rectify infeasible solutions towards a feasible direction, the algorithms quickly obtain non-dominated feasible solutions while simultaneously minimizing both objectives.

5 Conclusion and Future Work

In this paper, an evolutionary framework based on the automatic configuration of GA and DE was designed to solve bi-objective DEED problems. In it, random individuals from the initial population were evaluated in parallel through two different sub-populations, one using GA and the other DE. The sub-population sizes were dynamically updated during the evolutionary process based on their prior performances, with the better-performing algorithm receiving more individuals to evolve and vice versa. To enhance the performance of the proposed algorithm, a heuristic was employed to rectify infeasible individuals. The proposed framework was tested by solving two renewable-based bi-objective DEED problems using the proposed and state-of-the-art algorithms. A comparison indicated that the proposed GA-DE framework consistently performed better than all the other algorithms, with the heuristic greatly enhancing all their performances.

In future, bi-objective DEED problems could be solved using this configuration but with more algorithms and the uncertainty factors of renewable sources incorporated in the model.

References

1. Basu, M.: Economic environmental dispatch of hydrothermal power system. International Journal of Electrical Power & Energy Systems 32, 711–720 (2010)
2. Deb, K., Pratap, A., Agarwal, S., Meyarivan, T.: A fast and elitist multiobjective genetic algorithm: Nsga-ii. IEEE Transactions on Evolutionary Computation 6, 182–197 (2002)
3. Elsayed, S., Zaman, M.F., Sarker, R.: Automated differential evolution for solving dynamic economic dispatch problems. In: Intelligent and Evolutionary Systems, pp. 357–369. Springer (2016)
4. Elsayed, S.M., Sarker, R.A., Essam, D.L.: Multi-operator based evolutionary algorithms for solving constrained optimization problems. Computers & Operations Research 38, 1877–1896 (2011)
5. Gjorgiev, B., Acepin, M.: A multi-objective optimization based solution for the combined economic-environmental power dispatch problem. Engineering Applications of Artificial Intelligence 26, 417–429 (2013)
6. Khan, N.A., Awan, A.B., Mahmood, A., Razzaq, S., Zafar, A., Sidhu, G.A.S.: Combined emission economic dispatch of power system including solar photo voltaic generation. Energy Conversion and Management 92, 82–91 (2015)
7. Kim, I.Y., de Weck, O.L.: Adaptive weighted sum method for multiobjective optimization: a new method for pareto front generation. Structural and Multidisciplinary Optimization 31, 105–116 (2006)

8. Panigrahi, C.K., Chattopadhyay, P.K., Chakrabarti, R.N., Basu, M.: Simulated annealing technique for dynamic economic dispatch. Electric Power Components and Systems 34, 577–586 (2006)
9. Ray, T., Singh, H., Isaacs, A., Smith, W.: Infeasibility driven evolutionary algorithm for constrained optimization. In: Mezura-Montes, E. (ed.) Studies in Computational Intelligence, vol. 198, pp. 145–165. Springer Berlin Heidelberg (2009)
10. Spears, W.M.: Adapting crossover in evolutionary algorithms. In: Evolutionary programming. pp. 367–384 (1995)
11. Victoire, T.A.A., Jeyakumar, A.E.: A modified hybrid ep-sqp approach for dynamic dispatch with valve-point effect. International Journal of Electrical Power & Energy Systems 27, 594 (2005)
12. Zaman, F., Sarker, R.A., Ray, T.: Solving an economic and environmental dispatch problem using evolutionary algorithm. In: IEEE International Conference on Industrial Engineering and Engineering Management (IEEM). pp. 1367–1371 (2014)
13. Zaman, M., Elsayed, S., Ray, T., Sarker, R.: Configuring two-algorithm-based evolutionary approach for solving dynamic economic dispatch problems. Engineering Applications of Artificial Intelligence 53, 105–125 (2016)
14. Zaman, M.F., Elsayed, S., Ray, T., Sarker, R.: An evolutionary approach for scheduling solar-thermal power generation system. In: International Conference on Computers & Industrial Engineering (CIE45). vol. 45. Metz, France, (2015)
15. Zaman, M.F., Elsayed, S.M., Ray, T., Sarker, R.A.: Evolutionary algorithms for dynamic economic dispatch problems. IEEE Transactions on Power Systems, 31, 1486–1495 (2016)
16. Zaman, M.F., Elsayed, S.M., Ray, T., Sarker, R.A.: A double action genetic algorithm for scheduling the wind-thermal generators. In: Ray, T., Sarker, R., Li, X. (eds.) Artificial Life and Computational Intelligence: Second Australasian Conference, ACALCI 2016, Canberra, ACT, Australia, February 2–5, 2016, Proceedings, pp. 258–269. Springer International Publishing (2016)
17. Zaman, M., Elsayed, S.M., Ray, T., Sarker, R.A.: Evolutionary algorithms for power generation planning with uncertain renewable energy. Energy 112, pp. 408–419 (2016)
18. Zhang, Y., wei Gong, D., Geng, N., yan Sun, X.: Hybrid bare-bones pso for dynamic economic dispatch with valve-point effects. Applied Soft Computing 18, 248–260 (2014)

Author Index

G. Leu et al. (eds.), *Intelligent and Evolutionary Systems*,
Proceedings in Adaptation, Learning and Optimization 8,
DOI 10.1007/978-3-319-49049-6

Printed in the United States
By Bookmasters

Printed in the United States
By Bookmasters